KU-285-070

PROTEINS

BIOCHEMISTRY AND BIOTECHNOLOGY

Books are to be returned on

LIVERPOOL JMU LIBRARY

3 1111 01079 1935

LIVERPOOL
JOHN MOORES UNIVERSITY
AVRIL ROBARTS LRC
TITHEBARN STREET
LIVERPOOL L2 2ER
TEL. 0151 231 4022

PROTEINS

BIOCHEMISTRY AND BIOTECHNOLOGY

Gary Walsh

Industrial Biochemistry Programme,
CES Department,,
University of Limerick, Ireland

JOHN WILEY & SONS, LTD

Copyright © 2002 by John Wiley & Sons Ltd,
Baffins Lane, Chichester,
West Sussex PO19 1UD, England

National 01243 779777
International (+44) 1243 779777

e-mail (for orders and customer service enquiries): cs-books@wiley.co.uk

Visit our Home Page on http://www.wiley.co.uk
or
http://www.wiley.com

Reprinted August 2002

All rights reserved. No part of this publication may be reproduced, stored in a retrieval
system, or transmitted, in any form or by any means, electronic, mechanical, photocopying,
recording, scanning or otherwise, except under the terms of the Copyright, Designs and
Patents Act 1988 or under the terms of a licence issued by the Copyright Licensing Agency,
90 Tottenham Court Road, London, UK W1P 9HE, without the permission in writing of the
publisher.

Other Wiley Editorial Offices

John Wiley & Sons, Inc., 605 Third Avenue,
New York, NY 10158–0012, USA

Wiley-VCH Verlag GmbH, Pappellallee 3,
D-69469 Weinheim, Germany

John Wiley & Sons (Australia) Ltd, 33 Park Road, Milton,
Queensland 4064, Australia

John Wiley & Sons (Asia) Pte Ltd, 2 Clementi Loop #02-01,
Jin Xing Distripark, Singapore 0512

John Wiley & Sons (Canada) Ltd, 22 Worcester Road,
Rexdale, Ontario M9W 1L1, Canada

Library of Congress Cataloging-in-Publication Data

Walsh, Gary.
    Proteins : biochemistry and biotechnology / Gary Walsh.
        p. cm.
    includes bibliographical references and index.
    ISBN 0-471-89906-2 (hbk.) – ISBN 0-471-89907-0 (pbk.)
        1. Proteins–Biotechnology. 2. Proteins. I. Title.

TP248.65.P76 W353 2001
660.6'3–dc21   2001046998

British Library Cataloguing in Publication Data

A catalogue record for this book is available from the British Library
ISBN 0-471-899062 (Hardback)   0-471-899070 (Paperback)

Typeset in 10/12 pt Times by Kolam Information Services Pvt. Ltd, Pondicherry, India
Printed and bound in Great Britain by Antony Rowe Ltd, Chippenham, Wiltshire
This book is printed on acid-free paper responsibly manufactured from sustainable forestry,
in which at least two trees are planted for each one used for paper production.

Contents

10 Industrial enzymes: an introduction — **393**

11 Industrial enzymes: proteases and carbohydrases — **419**

Preface

This text has evolved from an earlier book, *Protein biotechnology*, first published by John Wiley & Sons in 1994. It aims to provide a comprehensive and up-to-date overview of proteins, both in terms of their biochemistry and applications. Chapters 1–3 are largely concerned with basic biochemical principles. In these chapters issues relating to protein sources, structure, folding, stability, quantification, purification and characterization are addressed. The remaining 10 chapters largely focus upon the production of proteins and their applications in medicine, analysis and industry.

The text caters mainly for students of biotechnology but it should also be of value to students pursuing degrees in biochemistry, microbiology, or any branch of the biomedical sciences. Its scope also renders it useful to those currently working in the biotechnology sector.

A sincere note of thanks is due to a number of people who have contributed to the successful completion of this project. To Sandy, who by now can read my mind, and who converted many of my original hand drawings (often bordering on abstract art) into clearly understandable figures. Also to Kate and Grainne, for typing up some tables when they probably should have being doing other things. Thank you also to Gerard Wall for some useful discussions regarding protein engineering, and to Nancy for biting her tongue when I explained I couldn't help with the housework because I needed to work on the book. I also wish to pay tribute my co-author of *Protein biotechnology*, Denis Headon, for being a good teacher and friend, and for his constant encouragement over the years. I am grateful too to John Wiley & Sons for their professionalism and efficiency, and for not suing me when I over-ran on the manuscript submission date, twice. Thank you also to all those publishers who granted me permission to reproduce certain copyrighted material.

Finally I dedicate this book to the most amazing and precious group of proteins I know, my beautiful baby daughter Eithne.

Gary Walsh
June, 2001

1

Protein structure

INTRODUCTION

The subsequent chapters of this book seek to overview issues relating to protein sources, purification, characterization and application. In this introductory chapter we focus upon protein structure. A comprehensive treatment of such a topic would easily constitute a book on its own, and many such publications are available. The aim of this chapter is to provide a basic overview of the subject. The interested reader is referred to the further reading section, which lists several excellent specialist publications in the field. Much additional information may also be sourced via the websites mentioned within this chapter.

OVERVIEW OF PROTEIN STRUCTURE

Proteins are macromolecules consisting of one or more polypeptides (Table 1.1). Each polypeptide consists of a chain of amino acids linked together by peptide (amide) bonds. The exact amino acid sequence is determined by the gene coding for that specific polypeptide. When synthesized, a polypeptide chain folds up, assuming a specific three-dimensional shape (i.e. a specific conformation), which is unique to it. The conformation adopted is dependent upon the polypeptides' amino acid sequence, and this conformation is largely stabilized by multiple, weak interactions. Any influence (e.g. certain chemicals and heat) that disrupts such weak interactions results in disruption of the polypeptide's native conformation, a process termed denaturation. Denaturation usually

Table 1.1 Selected examples of proteins. The number of polypeptide chains and amino acid residues constituting the protein are listed, along with its molecular mass and biological function

Protein	No. polypeptide chains	Total no. amino acids	Molecular mass (Da)	Biological function
Insulin (human)	2	51	5 800	Complex, includes regulation of blood glucose levels
Lysozyme (egg)	1	129	13 900	Enzyme capable of degrading peptidoglycan in bacterial cell walls
Interleukin-2 (human)	1	133	15 400	T-lymphocyte derived polypeptide that regulates many aspects of immunity
Erythropoietin (human)	1	166	36 000	Hormone which stimulates red blood cell production
Chymotrypsin (bovine)	3	241	21 600	Digestive proteolytic enzyme
Subtilisin (from *Bacillus amyloliquefaciens*)	1	274	27 500	Bacterial proteolytic enzyme
Tumour necrosis factor (human TNF-α)	3	471	52 000	Mediator of inflammation and immunity
Haemoglobin (human)	4	574	64 500	Gas transport
Hexokinase (yeast)	2	800	102 000	Enzyme capable of phosphorylating selected monosaccharides
Glutamate dehydrogenase (bovine)	~40	~8 300	~1 000 000	Enzyme-interconverts glutamate and α-ketoglutarate and NH_4^+

Table 1.2   The major fibrous proteins and their biological function(s)

α-Keratin	Major protein constituent of hair, wool, feathers, nails and horns
Collagen	Major protein found in connective tissue (e.g. cartilage and tendons)
Elastin	Constitutes the elastic tissue of ligaments and the elastic connective tissue layer found in the walls of arteries
β-Keratins	Constitutes the major structural component of spider's web and silk.

results in loss of functional activity, clearly demonstrating the dependence of protein function upon protein structure. A protein's structure currently cannot be predicted solely from its amino acid sequence. Its conformation can however be determined by techniques such as X-ray diffraction and nuclear magnetic resonance spectroscopy (NMR).

Polypeptides are often characterized as being either 'fibrous' or 'globular'. Fibrous proteins adopt simpler three-dimensional shapes and are usually quite elongated. Such proteins normally play structural/protective roles within biological systems (Table 1.2). Globular proteins adopt a more complex three-dimensional shape, in which the amino acid chain is generally tightly folded into an approximately spherical shape. The diameter of such globular polypeptides is generally of the order of 50–100 Å. Proteins are also sometimes classified as 'simple' or 'conjugated'. Simple proteins consist exclusively of polypeptide chain(s) with no additional chemical components being present or being required for biological activity. Conjugated proteins, in addition to their polypeptide components(s), contain one or more non-polypeptide constituents known as prosthetic group(s). The most common prosthetic groups found in association with proteins include carbohydrates (glycoproteins), phosphate groups (phosphoproteins), vitamin derivatives (e.g. flavoproteins) and metal ions (metalloproteins).

Primary structure

Polypeptides are linear, unbranched polymers, potentially containing up to 20 different monomers (i.e. amino acids) linked together in a precise predefined sequence. The primary structure of a polypeptide refers to its exact amino acid sequence, along with the exact positioning of any disulphide bonds present. The 20 commonly occurring amino acids are listed in Table 1.3, along with their abbreviated and one-letter designations. The structures of these amino acids are presented in Figure 1.1. Nineteen of these amino acids contain a central (α) carbon atom, to which is attached a hydrogen atom (H), an amino group (NH_2) a carboxyl group (COOH), and an additional side chain (R) group – which differs from amino acid to amino acid. The amino acid proline is unusual in that its R group forms a direct covalent bond with the nitrogen atom of what is the free amino group in other amino acids (Figure 1.1).

Table 1.3   The 20 commonly occurring amino acids

R group classification	Amino acid	Abbreviated names (three letter)	Abbreviated names (one letter)	Molecular mass (Da)	% occurrence in 'average' protein
Non-polar, aliphatic	Glycine	Gly	G	75	7.2
	Alanine	Ala	A	89	8.3
	Valine	Val	V	117	6.6
	Leucine	Leu	L	131	9.0
	Isoleucine	Ile	1	131	5.2
	Proline	Pro	P	115	5.1
Aromatic	Tyrosine	Tyr	Y	181	3.2
	Phenylalanine	Phe	F	165	3.9
	Tryptophan	Trp	W	204	1.3
Polar but uncharged	Cysteine	Cys	C	121	1.7
	Serine	Ser	S	105	6.0
	Methionine	Met	M	149	2.4
	Threonine	Thr	T	119	5.8
	Asparagine	Asn	N	132	4.4
	Glutamine	Glu	Q	146	4.0
Positively charged	Arginine	Arg	R	174	5.7
	Lysine	Lys	K	146	5.7
	Histidine	His	H	155	2.2
Negatively charged	Aspartic acid	Asp	D	133	5.3
	Glutamic acid	Glu	E	147	6.2

The amino acids may be sub-divided into five groups on the basis of side chain structure. Their three- and one-letter abbreviations are also listed (one-letter abbreviations are generally used only when compiling extended sequence data – mainly to minimize writing space and effort). In addition to their individual molecular masses the % occurrence of each amino acid in an 'average' protein is also presented. This data was generated from sequence analysis of over 1000 different proteins.

As will be evident from the next section, peptide bond formation between adjacent amino acid residues entails the establishment of covalent linkages between the amino and carboxyl groups attached to their central (α) carbon atoms. Hence the free functional (i.e. chemically reactive) groups in polypeptides are almost entirely present as part of the constituent amino acids' R groups. In addition to determining the chemical reactivity of a polypeptide, these R groups also very largely dictate the final conformation adopted by a polypeptide. Stabilizing/repulsive forces between different R groups (as well as between R groups and the surrounding aqueous media) largely dictate what final shape the polypeptide adopts, as will be described later.

The R groups of the non-polar, aliphatic amino acids (Gly, Ala, Val, Leu, Ile and Pro) are devoid of chemically reactive functional groups. These R groups are noteworthy in that, when present in a polypeptide's

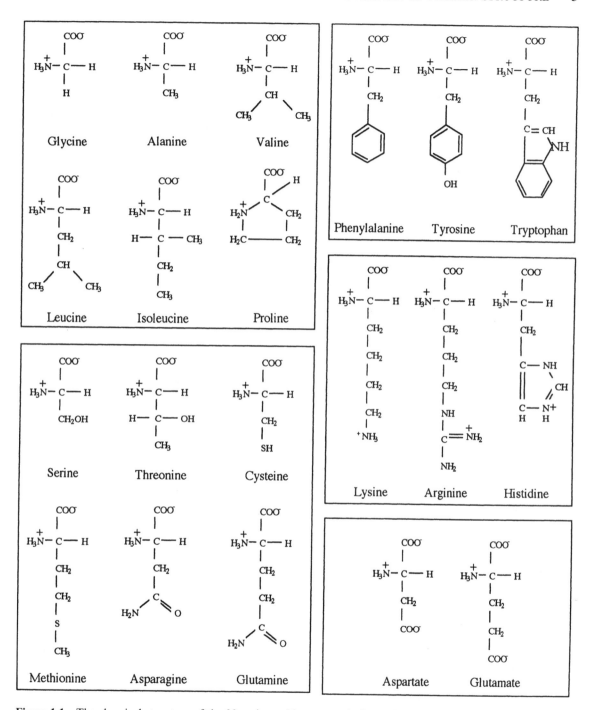

Figure 1.1   The chemical structure of the 20 amino acids commonly found in proteins

backbone, they tend to interact with each other non-covalently (via hydrophobic interactions). These interactions have a significant stabilizing influence on protein conformation.

Glycine is noteworthy in that its R group is a hydrogen atom. This means that the α-carbon of glycine is not asymmetric (i.e. is not a chiral centre). To be a chiral centre the carbon would have to have four different chemical groups attached to it, in this case two of its four attached groups are identical. As a consequence glycine does not occur in multiple stereoisomeric forms, unlike the remaining amino acids which occur as either D or L isomers. Only L-amino acids are naturally found in polypeptides.

The side chains of the aromatic amino acids (Phe, Tyr and Trp) are not particularly reactive chemically, but they all absorb ultraviolet (UV) light. Tyr and Trp in particular absorb strongly at 280 nm, allowing detection and quantification of proteins in solution by measuring the absorbance at this wavelength (Chapter 3).

Of the six polar but uncharged amino acids, two (cysteine and methionine) are unusual in that they contain a sulfur atom. The side chain of methionine is non-polar and relatively unreactive, although the sulfur atom is susceptible to oxidation. In contrast, the thiol ($-C-SH$) portion of cysteine's R group is the most reactive functional group of any amino acid side-chain. *In vivo* this group can form complexes with various metal ions and is readily oxidized, forming cystine residues in which two cysteine's are covalently linked (Figure 1.2). The formation of intrachain cystines (i.e. a disulfide linkage between two cysteine residues within the same polypeptide backbone) helps stabilize the three-dimensional structure of such polypeptides. Interchain disulfide linkages can also form, in which Cysteine's from two different polypeptides participate. This is a very effective way of covalently linking adjacent polypeptides.

Of the four remaining polar but uncharged amino acids, the R group of two (serine and threonine) contain hydroxyl (OH) groups while the R groups of asparagine and glutamine contain amide ($CONH_2$) groups. None are particularly reactive chemically, although, upon exposure to high temperatures or extremes of pH the latter two can deamidate, yielding aspartic acid and glutamic acid respectively.

Aspartic and glutamic acid are themselves negatively charged under physiological conditions. This allows them to chelate certain metal ions, and also to markelty influence the conformation adopted by polypeptide chains in which they are found.

Lysine, arginine and histidine are positively charged amino acids. The arginine R group consists of a hydrophobic chain of four CH_2 groups (Figure 1.1), capped with an amino (NH_2) group, which is ionized (NH_3^+) under most physiological conditions. However, within most polypeptides there is normally a fraction of unionized lysines, and these (unlike their ionized counterparts) are quite chemically reactive. Such lysine side chains can be chemically converted into various analogues. The arginine side chain is also quite bulky, consisting of three CH_2 groups, an amino group ($-NH_2$) and a ionized guanido group ($=NH_2^+$). The 'imidazole' side

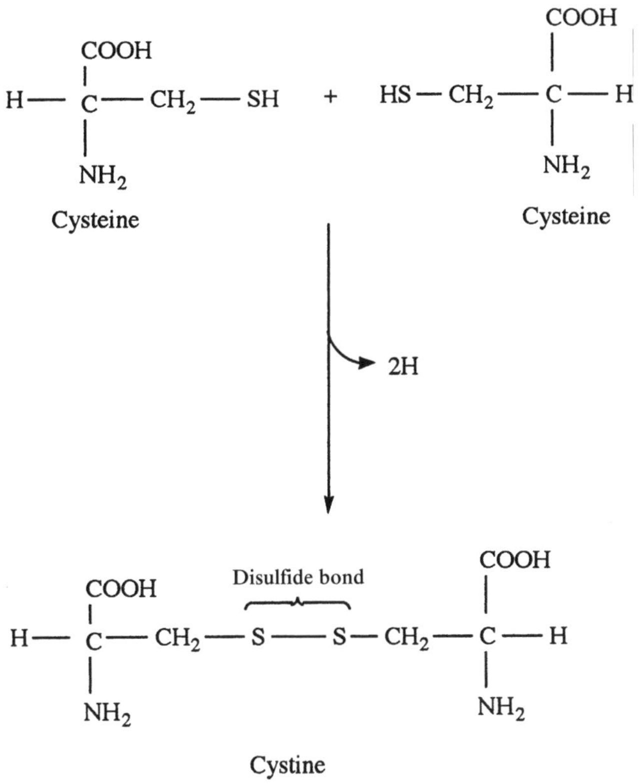

Figure 1.2 The formation of cystine via disulfide bond formation between two cysteines

chain of histidine can be described chemically as a tertiary amine (R_3—N), and thus it can act as a strong nucleophilic catalyst (the nitrogen atom houses a lone pair of electrons, making it a 'nucleus lover' or nucleophile; it can donate its electron pair to an 'electron lover' or electrophile). As such, the histidine side chain often constitute an essential part of some enzyme active sites (e.g. the protease 'subtilisin carlsberg', Chapter 11).

In addition to the 20 'common' amino acids, some modified amino acids are also found in several proteins. These amino acids are normally altered via a process of post-translational modification reactions (i.e. modified after protein synthesis is complete). Almost 200 such modified amino acids have been characterized to date. The more common such modifications are discussed separately later in this chapter. Other modified amino acids occur less commonly. These include 4–hydroxyproline and 5–hydroxylysine, found in some fibrous proteins, mainly collagen (Figure 1.3). These hydroxylation reactions occur in the endoplasmic recticulum and are catalysed by (vitamin C-dependent) hydroxylase enzymes. The modifications are necessary to ensure the proper folding and assembly of the mature fibrous proteins. γ-Carboxyglutamate (Gla) is an additional modified amino acid found in some proteins, most notably certain blood clotting factors. The enzyme responsible for the carboxylation of the glutamate side chains in such polypeptides is a

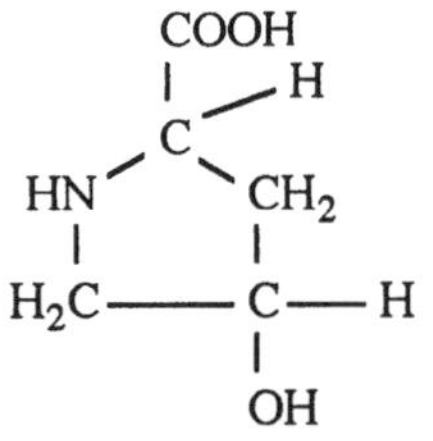

4-Hydroxyproline

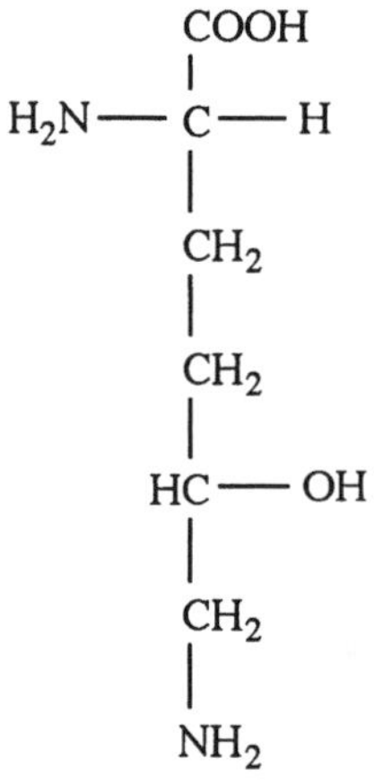

5-Hydroxylysine

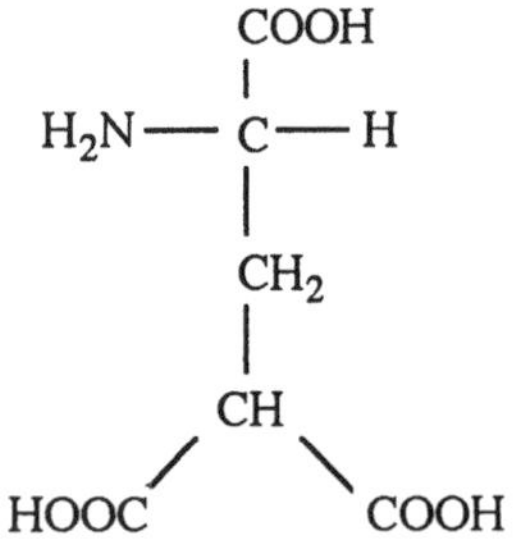

γ-Carboxyglutamate

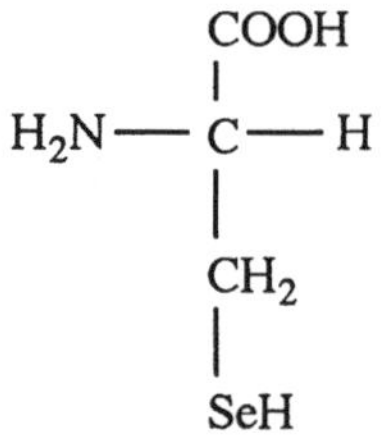

Selenocysteine

Figure 1.3 Structure of some modified amino acids

vitamin C-dependent carboxylase. Again, this enzyme is found in the endoplasmic reticulum. γ-Carboxyglutamate side chains are very effective in binding Ca^{2+} ions, a process necessary for the effective functioning of certain blood clotting factors (Chapter 5).

A final example of a modified amino acid is that of selenocysteine (Figure 1.3). Unlike the examples thus far described it is not generated via the post-translational modification route, but it exists as a preformed amino acid in its own right. (Selenocysteine is sometimes called the '21st amino acid'.) Selenium in this form is an essential component of a small number of enzymes. The nucleotide sequence of the genes coding for such enzymes contains a 'UGA' codon which codes for selenocysteine. In non-selenocysteine proteins UGA normally functions as a termination codon. The reading of UGA as a selenocysteine rather than the more usual stop codon is apparently dependent upon the nucleotide sequence on either side of the codon. As such, differential UGA recognition may be a function of mRNA conformation.

The peptide bond

Successive amino acids are joined together during protein synthesis via a 'peptide' (i.e. amide) bond (Figure 1.4). This is a condensation reaction, as a water molecule is eliminated during bond formation. Each amino acid in the resultant polypeptide is termed a 'residue', and the polypeptide

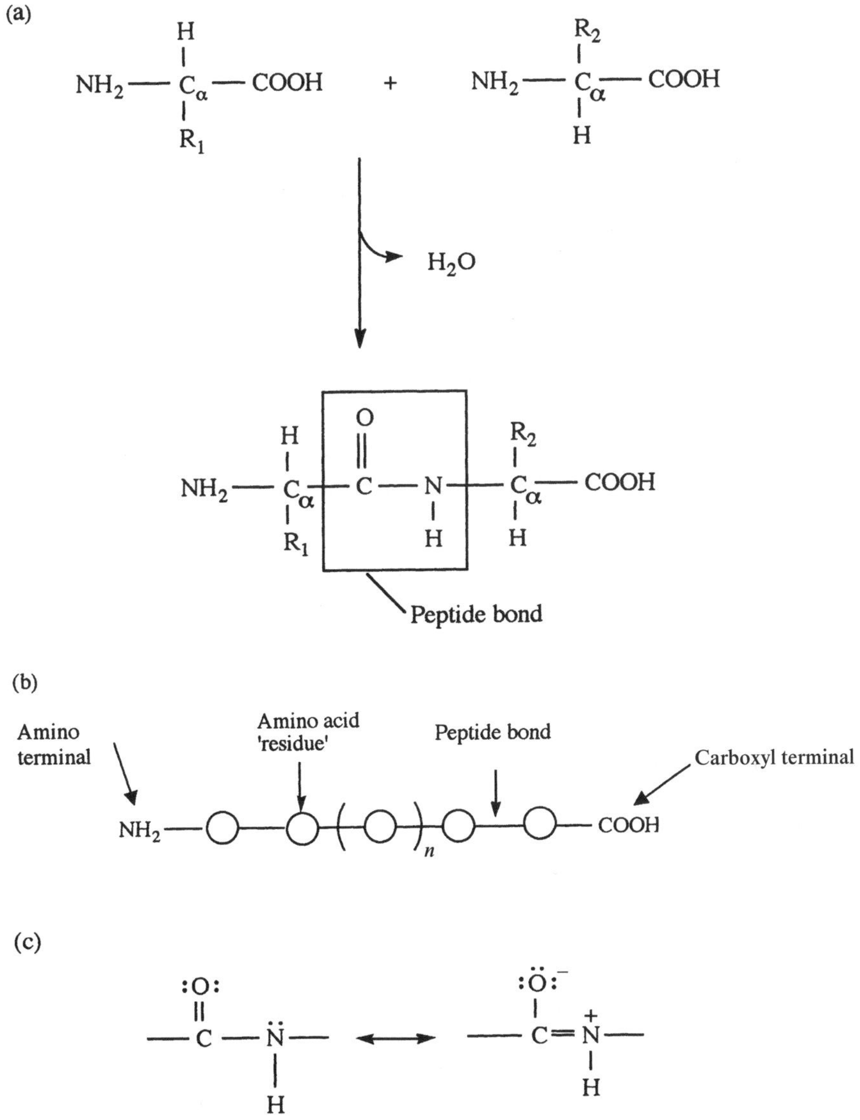

Figure 1.4 Peptide bond formation (a). Polypeptides consist of a linear chain of amino acids successively linked via peptide bonds (b). The peptide bond displays partial double bonded character (c)

chain will display a free amino (NH_2) group at one end and a free carboxyl (COOH) group at the other end. These are termed the amino and carboxyl termini, respectively.

The peptide bond has a rigid, planar structure and is in the region of 1.33 Å in length. Its rigid nature is a reflection of the fact that the amide nitrogen lone pair of electrons are delocalized across the bond (i.e. the bond structure is a halfway house between the two forms illustrated in Figure. 1.4c). In most instances, peptide groups assume a '*trans*' configuration (Figure. 1.4a). This minimizes steric interference between the R groups of successive amino acid residues.

While the peptide bond is rigid, the other two bond types found in the polypeptide backbone (i.e. the N—Cα bond and the Cα—C bond, Figure. 1.5) are free to rotate. The polypeptide backbone can thus be viewed as a series of planar 'plates' which can rotate relative to one another. The angle of rotation around the N—Cα bond is termed φ (phi) while that around the Cα—C bond is termed ψ (psi) (Figure 1.5). These angles are also known as rotation angles, dihedral angles or torsion angles. By convention, these angles are defined as being 180° when the polypeptide chain is in its fully extended, *trans* form. In principle, each bond can rotate to any value between −180° and +180°. However, the degrees of rotation actually observed are restricted because of the occurrence of steric hindrance between atoms of the polypeptide backbone and those of amino acid side chains.

For each amino acid residue in a polypeptide backbone, the actual φ and ψ angles that are physically possible can be calculated, and these

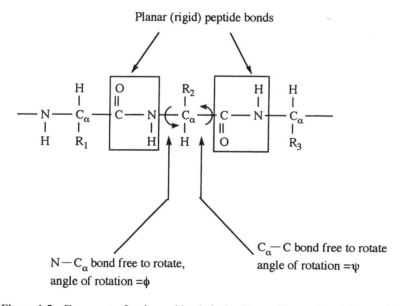

Figure 1.5   Fragment of polypeptide chain backbone illustrating rigid peptide bonds and the intervening N—Cα and Cα—C backbone linkages, which are free to rotate

angle pairs are often plotted against each other in a diagram termed a Ramachandran plot. Sterically allowable angles fall within relatively narrow bands in most instances. A greater than average degree of ϕ / ψ rotational freedom is observed around glycine residues, due to the latter's small R group – hence steric hindrance is minimized. On the other hand, bond angle freedom around proline residues is quite restricted due to this amino acid's unusual structure (Figure. 1.1). The ϕ and ψ angles allowable around each Cα in a polypeptide backbone obviously exerts a major influence upon the final three-dimensional shape assumed by the polypeptide.

Amino acid sequence determination

The amino acid sequence of a polypeptide may be determined directly via chemical sequencing, or by physical fragmentation and analysis. The sequence may be determined indirectly via determination of the nucleotide sequence of its coresponding structural gene. Direct chemical sequencing was the only method available until the 1970s. Insulin was the first protein to be sequenced by this approach (in 1953), requiring several years and several hundred grams of protein to complete. The method has been refined and automated over the years, such that today polypeptides containing 100 amino acids or more can be automatically sequenced within a few days, using μg–mg levels of protein.

The actual chemical sequencing procedure employed is termed the Edman degradation method. Briefly this entails sequential labelling, removal and identification of amino acid residues, beginning at the N-terminal end of the polypeptide (Figure 1.6). The polypeptide is first incubated with the reagent phenylisothiocyanate (PITC) which, at high pH values, reacts with the polypeptide N-terminal amino group yielding a phenylthiocarbamyl (PTC) derivative. Subsequent addition of anhydrous trifluoroacetic acid cleaves the PTC derivative, yielding a thiazolinone derivative (consisting of the PITC-N terminal amino acid derivative) and a polypeptide chain lacking its N-terminal amino acid. The thiazolinone derivative is then solvent-extracted and converted into the more stable phenylthiohydantoin (PTH) derivative, by acidification of the media. The PTH derivative is compared chromatographically to PTH-amino acid standards, in order to identify the amino acid moiety. The shortened polypeptide is then subjected to a second round of the cycle in order to identify amino acid number two of the original polypeptide. Further rounds of the cycle allows determination of the entire sequence.

Prior to commencement of Edman sequencing the individual polypeptides of multi-polypeptide proteins must be separated, and each independently sequenced. In addition, any disulfide linkages must be broken, usually by incubation of the polypeptide with a suitable reducing agent.

Theoretically, the Edman method should facilitate sequence analysis of proteins of any size. However, with each repeat of the sequencing cycle, a

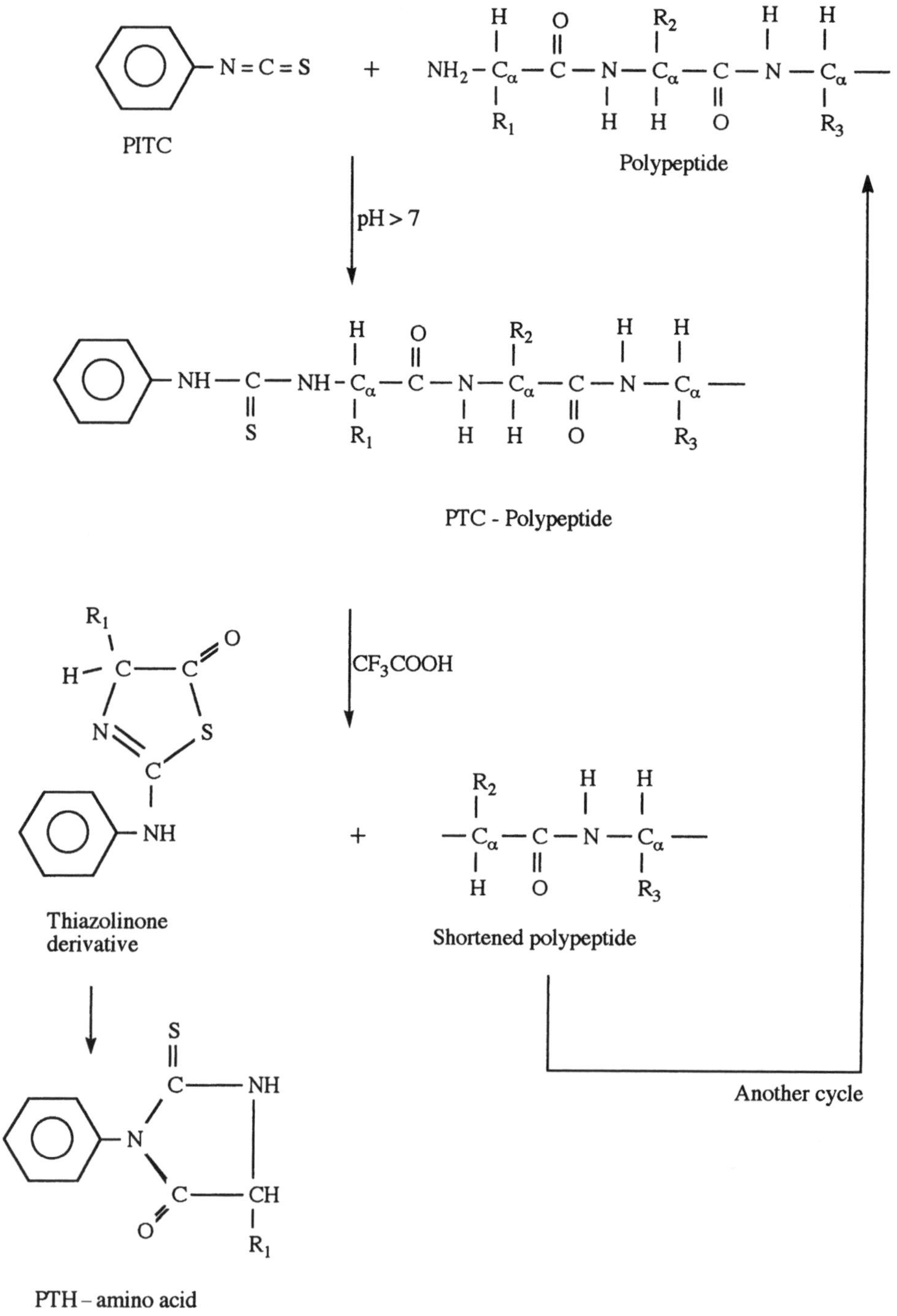

Figure 1.6   The Edman method of polypeptide sequence analysis. Refer to text for further details

small number of unreacted N-terminal amino acid residues will remain. Such errors are obviously cumulative, and the background 'noise' increases steadily until a point is reached where the amino acid residues being processed can no longer be assigned with absolute certainty. As previously mentioned, few systems can fully sequence polypeptides containing 100 or more amino acid residues. Larger polypeptides must first be fragmented into shorter peptides. The fragments are then separated and independently sequenced. Fragmentation may be achieved chemically or enzymatically and a range of chemical reagents and proteolytic enzymes capable of hydrolysing particular backbone peptide bonds are available (Table 1.4). The peptide fragments are then separated, either by one or two dimensional electrophoresis, or more commonly by high pressure liquid chromatography (HPLC).

In order to determine the correct order in which the (now sequenced) peptide fragments occurred in the intact polypeptide it is necessary to independently fragment a fresh sample of the polypeptide, using an alternative fragmentation reagent. The new set of fragments generated are then separated and sequenced. Identification of overlapping sequences between the original and new set of fragments allows determination of the full sequence data.

Sequencing can also be used to determine the location of disulfide linkages within primary sequence data. In this case, a fresh sample of protein is fragmented without prior reduction (i.e. breaking) of any disulfide linkages. The presence of a disulfide bond results in the generation of a peptide mix which will contain a large polypeptide 'fragment' actually consisting of two fragments linked by the disulfide bond. Upon its isolation the 'fragment' is reduced, and the two fragments released are independently sequenced. This pinpoints the cysteine molecules participating in the disulfide linkage.

Mass spectrometry provides another means by which a polypeptide's primary sequence may be elucidated. This application of mass spectrometry is discussed towards the end of Chapter 3.

An alternative approach to amino acid sequence determination is to sequence its structural gene/cDNA. The amino acid sequence can be inferred from the nucleotide sequence obtained. This approach has gained favour in recent years. Refinements to DNA sequencing methodologies and equipment has made such sequence analysis both rapid and relatively inexpensive. The ongoing genome projects continue to generate enormous amounts of sequence data. By the end of 2000, substantial/complete sequence data for some 273 organisms was available (Table 1.5). As a result, the putative amino acid sequence of an enormous number of proteins – most of unknown function and/or structure – had been determined.

Upon its generation, sequence information is normally submitted to various databases. The major databases in which protein primary sequence data are available are listed in Table 1.6. Included in this table also are the major nucleic acid sequence databases, as amino acid sequence information can potentially be derived from these.

Table 1.4 Various chemical and enzymatic reagents which may be used to fragment polypeptides by promoting hydrolysis of specific backbone peptide bonds. The peptide bond(s) specifically targeted by each cleavage reagent is also listed. Reprinted from; Creighton, T. (1993) *Proteins: Structures and Molecular Properties*, 2nd edn., by permission of Freeman, NY

Sequence cleaved[a]	Procedure or enzyme
Ala–Yaa	Elastase, bromelain
Arg–Yaa	Trypsin, endoproteinase Arg-C, clostripain
Asn–Gly	Hydroxylamine
Asp–Yaa	V-8 protease
Asp–Pro	Mild acid
Xaa–Asp	Asp-N protease
Xaa–Cys	Cyanylation
Glu–Yaa	V-8 protease
Gly–Yaa	Elastase
Leu–Yaa	Pepsin
Xaa–Leu	Thermolysin
Lys–Yaa	Trypsin, endoproteinase Lys-C, bromelain
Met–Yaa	CNBr
Phe–Yaa	Chymotrypsin, pepsin
Xaa–Phe	Thermolysin
Pro–Xaa	Prolylendopeptidase
Trp–Yaa	Iodosobenzoic acid, *N*-chlorosuccinimide, chymotrypsin
Tyr–Yaa	Chymotrypsin, bromelain

[a] The specific residues are indicated by bold type; Xaa and Yaa can be almost any amino acid, except for Pro in many instances. Cleavage is C-terminal to residue Xaa, N-terminal to residue Yaa.

The Swiss-Prot database is probably the most widely used protein database. It is maintained collaboratively by the European Bioinformatics Institute (EBI) and the Swiss Institute for Bioinformatics. It is relatively easy to access and search via the World Wide Web (Table 1.6). A sample

Table 1.5 Representative organisms whose genomes have been or will soon be completely/almost completely sequenced. Data largely taken from http://wit.integratedgenomics.com/GOLD/eucaryoticgenomes.html and http://www.tigr.org/tdb/mdb/mdcomplete.html Updated information is available on these sites

Organism	Classification	Genome size (Mb)	Organism	Classification	Genome size (Mb)
Aeropyrum pernix	Archaea	1.67	*Treponema pallidum*	Eubacteria	1.14
Archaeoglobus fulgidus	Archaea	2.18	*Vibrio cholerae*	Eubacteria	4.0
Pyrococcus horikoshii	Archaea	1.80	*Aspergillus nidulans*	Fungi	31.0
Pyrococcus furiosus	Archaea	2.10	*Candida albicans*	Fungi	15.0
Sulfolobus solfataricus	Archaea	2.99	*Neurospora crassa*	Fungi	47.0
Thermoplasma acidophilum	Archaea	1.56	*Schizosaccharomyces pombe*	Fungi	14.0
Aquifex aeolicus	Eubacteria	1.50	*Babesia bovis*	Protozoa	NL
Bacillus subtilis	Eubacteria	4.20	*Cryptosporidium parvum*	Protozoa	10.4
Bacillus anthracis	Eubacteria	4.50	*Leishmania major*	Protozoa	33.6
Bordetella pertussis	Eubacteria	3.88	*Arabidopsis thaliana* (thale cress)	Plant	70.0
Brucella suis	Eubacteria	3.30	*Hordeum vulgare* (barley)	Plant	5.0
Chlamydia pneumoniae	Eubacteria	1.23	*Gossypium hirsutum* (cotton)	Plant	NL
Clostridium tetani	Eubacteria	4.40	*Triticum aestivum* (wheat)	Plant	NL
Corynebacterium diphtheriae	Eubacteria	3.10	*Zea mays* (maize)	Plant	NL
E. coli	Eubacteria	5.23	*Danio rerio* (Zebra fish)	Fish	NL
Lactobacillus acidophilus	Eubacteria	1.90	*Gallus gallus* (chicken)	Bird	NL
Listeria monocytogenes	Eubacteria	2.94	*Bos taurus* (cow)	Mammal	NL
Mycobacterium leprae	Eubacteria	2.80	*Canis familiaris* (dog)	Mammal	NL
Mycobacterium tuberculosis	Eubacteria	4.40	*Rattus norvegicus* (rat)	Mammal	NL
Neisseria meningitidis	Eubacteria	2.18	*Ovis aries* (sheep)	Mammal	NL

continues overleaf

Table 1.5   (*continued*)

Organism	Classification	Genome size (Mb)	Organism	Classification	Genome size (Mb)
Pseudomonas aeruginosa	Eubacteria	6.3	*Sus scrofa* (pig)	Mammal	NL
Salmonella enterica	Eubacteria	NL	Ape	Primate	NL
Staphylococcus aureus	Eubacteria	2.80	*Homo sapiens*	Primate	NL
Streptococcus pneumoniae	Eubacteria	2.04			

NL = not listed in source publication.

Table 1.6   The major primary sequence (protein and nucleic acid) databases, and website addresses from which they may be accessed

Database	Web address
Protein	
PIR	http://www-nbrf.georgetown.edu/
Swiss-Prot	http://www.ebi.ac.uk/swissprot/
MIPS	http://www.mips.biochem.mpg.de/
NRL-3D	http://www-nbrf.georgetown.edu/pirwww/dbinfo/nrl3d.html
Tr EMBL	http://www.ebi.ac.uk/index.html
Owl	http://www.bis.med.jhmi.edu/Dan/proteins/owl.html
Nucleic acid	
EMBL	http://www.ebi.ac.uk/embl/index.html/
GenBank	http://www.ncbi.nlm.nih.gov
DDBJ	http://www.ddbj.nig.ac.jp/

entry for human insulin is provided in Figure 1.7. Additional information detailing such databases is available via the web address provided in Table 1.6 and in the Bioinformatics publications listed at the end of this Chapter.

A polypeptide's amino acid sequence can thus be determined by direct chemical (Edman) or physical (mass spectrometry) means, or indirectly via gene sequencing. In practice these methods are complementary to one another, and can be used to cross-check sequence accuracy. If the target gene/mRNA has been previously isolated DNA sequencing is usually most convenient. However, this approach reveals little information regarding any post-translational modifications present in the mature polypeptide. In contrast, direct sequencing approaches will identify such modifications. If the gene/mRNA coding for the target polypeptide has not been isolated, direct sequencing of the (purified) polypeptide is the only alternative available. The generation of even partial sequence information will then allow the design of oligonucleotide probes which can be used to isolate/clone the polypeptide's gene.

Figure 1.7 Sample entry for human insulin as present in the Swiss-Prot database. Refer to text for further details. This SWISS-PROT entry is copyright. It is produced through a collaboration between the Swiss Institute of Bioinformatics and the EMBL outstation – the European Bioinformatics Institute. There are no restrictions on its use by non-profit institutions as long as its content is in no way modified and this statement is not removed

General information about the entry

Entry name	**INS_HUMAN**
Primary accession number	**P01308**
Secondary accession number(s)	None
Entered in SWISS-PROT in	Release 01, July 1986
Sequence was last modified in	Release 01, July 1986
Annotations were last modified in	Release 39, May 2000

Name and origin of the protein

Protein name	INSULIN [Precursor]
Synonym(s)	None
Gene name(s)	INS
From	*Homo sapiens (Human)* [TaxID: 9606]
Taxonomy	Eukaryota; Metazoa; Chordata; Craniata; Vertebrata; Euteleostomi; Mammalia; Eutheria; Primates; Catarrhini; Hominidae; Homo.

Features

SIGNAL	1	24	
CHAIN	25	54	INSULIN B CHAIN.
PROPEP	57	87	C PEPTIDE.
CHAIN	90	110	INSULIN A CHAIN.
DISULFID	31	96	INTERCHAIN.
DISULFID	43	109	INTERCHAIN.
DISULFID	95	100	
VARIANT	34	34	H → D (IN PROVIDENCE; FAMILIAL HYPERPROINSULINEMIA). /FTID = VAR_003971.
VARIANT	48	48	F → S (IN LOS-ANGELES; TYPE-II DIABETES MELLITUS). /FTID = VAR_003972.
VARIANT	49	49	F → L (IN CHICAGO). /FTID = VAR_003973.
VARIANT	89	89	R → H (IN FAMILIAL HYPERPROINSULINEMIA; IMPAIRS POSTTRANSLATIONAL CLEAVAGE). /FTID = VAR_003974.
VARIANT	89	89	R → L (IN KYOTO; FAMILIAL HYPERPROINSULINEMIA). /FTID = VAR_003975.
VARIANT	92	92	V → L (IN WAKAYAMA). /FTID = VAR_003976.
TURN	32	32	
HELIX	33	46	
STRAND	48	50	
HELIX	91	95	
TURN	96	97	
HELIX	102	108	
STRAND	109	109	

Feature table viewer

Sequence information

Length: 110 AA [This is the length of the unprocessed precursor]	Molecular weight: **11981 Da** [This is the Mw of the unprocessed precursor]	CRC64: **C2C3B23B85E520E5** [This is a checksum on the sequence]

10	20	30	40	50	60	
MALWMRLLPL	LALLALWGPD	PAAAFVNQHL	CGSHLVEALY	LVCGERGFFY	TPKTRREAED	
70	80	90	100	110		
LQVGQVELGG	GPGAGSLQPL	ALEGSLQKRG	IVEQCCTSIC	SLYQLENYCN		P01308 in *FASTA format*

Polypeptide synthesis

Full scale polypeptide characterization usually requires modest/large (mg–g) amounts of the purified target polypeptide. Even larger quantities are then generally required if the polypeptide has a commercial application. In some cases a polypeptide can be obtained in sufficient quantities by direct extraction from its natural producer source. However, polypeptides may also be produced by direct chemical synthesis, as long as their amino acid sequence (and any post-translational modifications) has been elucidated. Synthesis can be undertaken by chemical means or via a biological route (recombinant DNA technology). Both of these approaches are discussed in Chapter 2.

HIGHER LEVEL STRUCTURE

Thus far we have concentrated on the primary structure (amino acid sequence) of a polypeptide. Higher level protein structure can be described at various levels, i.e. secondary, tertiary and quaternary:

- Secondary structure can be described as the local spatial conformation of a polypeptide's backbone, excluding the constituent amino acid's side chains. The major elements of secondary structure are the α-helix and β-strands, as described below.

- Tertiary structure refers to the three-dimensional arrangement of all the atoms which contribute to the polypeptide.

- Quaternary structure refers to the overall spatial arrangement of polypeptide subunits within a protein composed of two or more polypeptides.

Secondary structure

By studying the backbone of most proteins stretches of amino acids that adopt a regular, recurring shape usually become evident. The most commonly observed secondary structural elements are termed the α-helix and β-strands. Fibrous proteins often display only one type of secondary structure, stretching fully from one end of the polypeptide to the other. The entire backbone of the fibrous protein α-keratin, for example, adopts an α-helical shape, while the entire backbone of another fibrous protein, fibroin, assumes the β conformation. In contrast, the backbone of globular proteins generally exhibit several stretches of regular secondary structure (α-helix and/or β strands), separated by stretches largely devoid of regular, recurring conformation. The α-helix and β sheets are commonly formed because they maximize formation of stabilizing intramolecular hydrogen bonds and minimize steric repulsion between adjacent side chain groups, while also being compatible with the rigid planar nature of the peptide bonds.

The α-helix is a right-handed one, containing 3.6 amino acid residues in a full turn (Figure 1.8). This approximates to a length of 0.56 nm along the long axis of the helix. The participating amino acid side chains protrude outward from the helical backbone. Amino acids most conducive with α-helix formation include alanine, leucine, methionine and glutamate. Proline, as well as the occurrence in close proximity of multiple residues with either bulky side groups, or side groups of the same charge tend to disrupt α-helical formation. The helical structure is stabilized by hydrogen bonding, with every backbone C=O group forming a hydrogen bond with the N—H group four residues ahead of it in the helix. Stretches of α-helix found in globular polypeptides can vary in length from a single helical turn to greater than ten consecutive helical turns. The average length is about three turns.

In globular proteins, stretches of α-helix are most often positioned on the protein's surface, with one face of the helix facing the hydrophobic interior, and the other facing the surrounding aqueous media. The amino acid sequence of these helices is such that hydrophobic amino acid residues are positioned on one face of the helix, while hydrophobic amino acids line the other. The transmembrane section of polypeptides which span biological membranes often display one (or more) α-helical stretches. In such instances, almost all the residues found in the helix display hydrophobic side chains.

β-Strands represent the other major recurring structural element of proteins. β-strands usually are five to ten amino acid residues in length, with the residues adopting an almost fully extended zig-zag conformation.

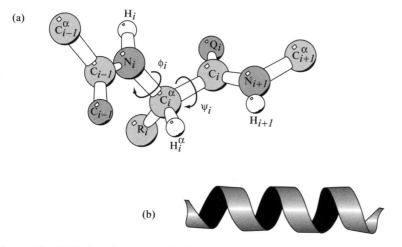

Figure 1.8 (a) Ball and stick and (b) ribbon representation of an α-helix. Reproduced from *Current protocols in Protein Science* by kind permission of the publisher, John Wiley & sons, Chichester

Single β-strands are rarely, if ever, found alone. Instead, two or more of these strands align themselves together to form a β-sheet. The β-sheet is a common structural element stabilized by maximum hydrogen bonding (Figure 1.9). The individual β-strands participating in β-sheet formation may all be present in the same polypeptide, or may be present in two polypeptides held in close juxtaposition. β-Sheets are described as being either parallel, antiparallel or mixed. A parallel sheet is formed when all

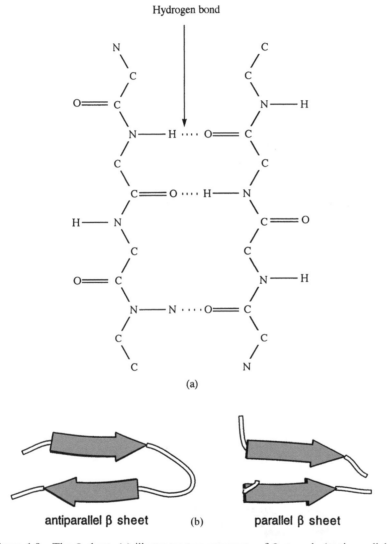

(a)

antiparallel β sheet   (b)   parallel β sheet

Figure 1.9   The β-sheet. (a) illustrates two segments of β strands (anti-parallel) forming a β-sheet via hydrogen bonding. The β strand is drawn schematically as a thick arrow. By convention the arrowhead points in the direction of the poly-peptide's C terminus. (b) Schematic illustration of a two-strand β-sheet in parallel and anti-parallel modes

the participating β stretches are running in the same direction (e.g. from the amino terminus to the carboxy terminus, Figure 1.9). An antiparallel sheet is formed when successive strands have alternating directions (N-terminus to C-terminus followed by C-terminus to N-terminus, etc.). A β-sheet containing both parallel and antiparallel strands is termed a mixed sheet.

In terms of secondary structure, most proteins consist of several segments of α-helix and/or β-strands separated from each other by various loop regions. These regions can vary in length and shape, and allow the overall polypeptide to fold into a compact tertiary structure. In general, loops are present on the surface of polypeptides. They are rich in polar/charged amino acid residues which, along with the N—H and C=O groups of their associated peptide bonds, hydrogen bond with the surrounding water molecules. Loop regions are usually quite flexible and longer loop regions can be susceptible to proteolytic cleavage. In addition to their obvious role in connecting stretches of regular secondary elements, loop regions themselves often participate or contribute directly to the polypeptide's biological function. The antigen binding region of antibodies, for example, are largely constructed from six loop regions. Such loops also often form the active site of enzymes.

One loop structure, termed a β-turn or β-bend, is unusual in that it is a characteristic feature of many polypeptides. The β-bend achieves a 180° alteration in backbone direction over the course of four amino acid residues, and is most often found between two stretches of antiparallel β strands (Figure 1.10). Glycine and proline residues are often found at β-turns. Glycine minimizes steric hindrance, due to its small side chain. Proline, by virtue of its unusual structure, naturally introduces a kink or bend in the polypeptide backbone. The loop is stabilized in part by the formation of a hydrogen bond between the C=O of the first residue and the NH of the fourth residue (Figure 1.10).

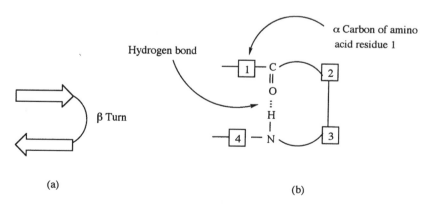

(a)   (b)

Figure 1.10   The β-bend or β-turn is often found between two stretches of antiparallel β strands, (a). It is stabilized in part by hydrogen bonding between the C=O bond and (b) the NH groups of the peptide bonds at the neck of the turn

Tertiary structure

As previously mentioned, a polypeptide's tertiary structure refers to its exact three-dimensional structure, relating the relative positioning in space of all the polypeptide's constituent atoms to each other. The tertiary structure of small polypeptides (approximately 200 amino acid residues or less) usually forms a single discrete structural unit. However, when the three-dimensional structure of many larger polypeptides is examined, the presence of two or more structural subunits within the polypeptide becomes apparent. These are termed domains. Domains, therefore, are (usually) tightly folded subregions of a single polypeptide, connected to each other by more flexible or extended regions (Figure 1.11). As well as being structurally distinct, domains often serve as independent units of function. For example, both domains of troponin C serve to bind calcium ions (Figure 1.11) and cell surface receptors usually contain one or more extracellular domains (some or all of which participate in ligand binding), a transmembrane domain (hydrophobic in nature and serving to stabilize the protein in the membrane) and one or more intracellular domains that play an effector function (e.g. generation of second messengers).

When closely studied, it becomes apparent that domains themselves are usually composed of 'building blocks' known as structural motifs. Structural motifs (sometimes called supersecondary structures) are composed of a few stretches of secondary structure (e.g. stretches of α-helix or β-strands), linked via loops and all arranged in a specific three-dimensional conformation. Any given single structural motif is often found in a wide variety of polypeptides which may have either related or unrelated biological functions. Some of the more commonly observed structural motifs include the helical bundle, the β-hairpin, the Greek key motif, the jelly roll, the β-sandwich and β-barrels (Figure 1.12).

The term helical bundles refers to the structural motif consisting of several stretches of α-helix separated by short bends or loops. The

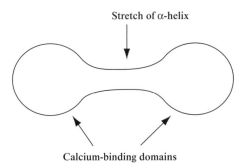

Figure 1.11 Schematic representation of the polypeptide troponin C, which functions as a calcium binding protein in muscle. It is composed of two tightly packed domains (each capable of binding calcium) separated by an extended stretch of α-helix

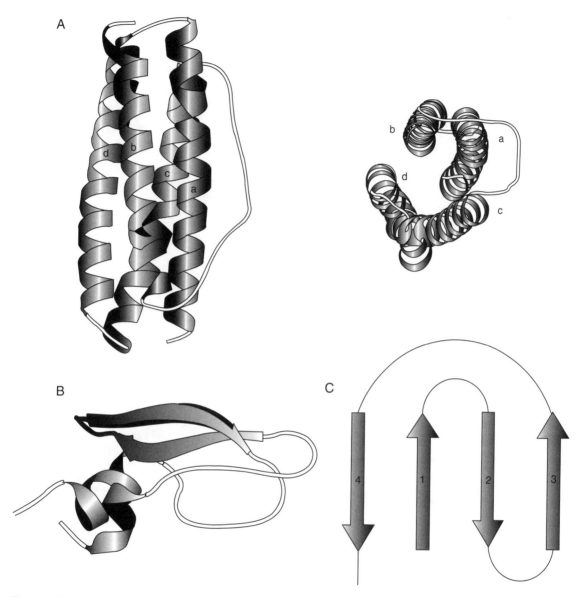

Figure 1.12 Some structural motifs commonly associated with (globular) polypeptides. (A) A four helix bundle; (B) a hairpin structure; (C) a β-sheet with a Greek key topology;

α-helical elements are usually (though not always) almost fully parallel or antiparallel to each other. The axis of the helical bundle is sometimes twisted, giving the entire bundle a twisted appearance. Helical bundles may also have different numbers of constituent helical stretches. Several haematopoietic cytokines, for example, display a four helical bundle, while cytochrome C oxidase displays a 22 helical bundle.

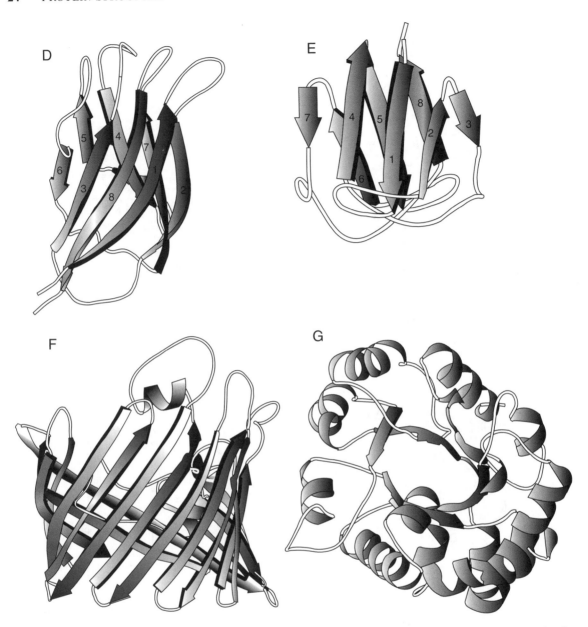

(D) a jelly roll motif; (E) a β sandwich; (F) a β barrel; and (G) an α/β barrel. Refer to text for further detail. Reproduced from *Current Protocols in Protein Science* by kind permission of the publisher, John Wiley & Sons, Chichester

The β-hairpin motif (also termed a β-ribbon or β−β unit) is a simple structural motif consisting of two stretches of β secondary structure connected by a loop. The axis of the resultant β-sheet is usually twisted (Figure 1.12). Hairpin motifs are found in a wide range of polypeptide types. Also common in many polypeptides is the Greek key motif, which is composed of four adjacent antiparallel β-strands folded into this characteristic structure (Figure 1.12). A jelly roll motif is actually composed of two closely associated Greek key motifs (Figure 1.12). In this instance, β-strands 1, 2, 7 and 8 form the first Greek key motif while strands 3, 4, 5 and 6 comprise the other. Overall, this forms a nearly fully closed barrel shape.

As the name suggests, a β-sandwich structural motif consists of two β-sheets packed face to face against each other. Variations of this sandwich structure are also found in some polypeptides. The αβα sandwich, for example, consists of a layer of β-sheet packed tightly between two α-helical stretches. This type of motif is often associated with nucleotide binding proteins.

β-Barrels are yet another motif type found in a wide variety of polypeptides. These are assemblages of stretches of β-strands, separated by loops and folded into a barrel-like structure. Depending upon the polypeptide, the 'barrel' can consist of anything from five to 16 individual β strands. A variation of this structure is the α / β barrel which, as the name suggests, is a barrel-like motif composed of alternating α and β stretches.

Polypeptides are most often characterized on the basis of their biological activity/function (e.g. enzymes, antibodies, etc.). An alternative categorization may be made on the basis of the polypeptide's domain structure. On this basis, polypeptide domains usually fall into one of three types: the α domain, the αβ domain and the β domain. In the case of α domains, the domain's core structure is built exclusively from stretches of α-helix. The single most common motif contributing to this is the four-helical bundle structure. α/β Domain structures consist of combinations of β−α−β motifs that form parallel β-sheets surrounded by stretches of α-helix. These are the most common domain types found in proteins. β-Domain structures display a core comprising of antiparallel β-sheets. There are usually two sheets present, packed against each other to form a distorted barrel-like structure.

Higher structure determination

There are three potential methods by which a protein's three-dimensional structure can be visualized: X-ray diffraction, NMR and electron microscopy. The latter method reveals structural information at low resolution, giving little or no atomic detail. It is used mainly to obtain the gross three-dimensional shape of very large (multi-polypeptide) proteins, or of protein aggregates such as the outer viral caspid. X-ray diffraction and NMR are the techniques most widely used to obtain high resolution protein structural information. Although recent advances in both the analytical equipment available and associated computing power renders these techniques more suited to polypeptide/protein structural analysis, the procedures and prerequirements necessary for their successful application means that the full three-dimensional structure of only a modest number of proteins has thus far been elucidated. A brief overview of these techniques is presented below. Appropriate further reading is listed at the end of this chapter.

Any structure can be visualized only if electromagnetic radiation of a wavelength comparable to its dimensions is used. In the case of proteins, the appropriate size is of the order of angstroms (10^{-10} m). The wavelength of X-rays approximate to this, hence the use of X-ray diffraction technology. X-ray diffraction entails bombarding a sample of the protein in crystalline form with a beam of X-rays (Figure 1.13). Most of these X-rays pass straight through the crystal but some are diffracted by the crystal. The resultant diffraction pattern, recorded on a detector, is a reflection of the three-dimensional structure of the protein molecules present in the crystal.

The initial pre-requisite to protein X-ray diffraction is the generation of protein crystals. Although many lower molecular mass substances crystallize relatively easily, this is not the case for the vast majority of globular proteins, which are extremely large and display irregular surfaces. Even if induced to crystalize, protein crystals will contain solvent-filled channels or pores between individual protein molecules. The solvent (usually water) generally occupies 30–80 per cent of crystal volume. Only a small proportion of the surface of individual proteins interact with each other and, as a result, the crystals are soft and easily destroyed.

A number of approaches may be adopted in order to grow protein crystals. Generally the protein must be very pure in order to crystallize successfully. A crystal forms when protein molecules are precipitated very slowly from supersaturated solutions, and this is usually achieved by vapour diffusion or by dialysis. Vapour diffusion, which is most commonly employed, entails slow concentration of protein molecules in the presence of a suitable precipitant such as polyethylene glycol (PEG; Fig. 1.14). Crystallization will also be affected by influences such as solution pH, the presence of specific buffer salts, metal ions, low molecular mass organic molecules, etc. The optimal crystallization conditions must usually be determined by direct experimentation. Crystals suitable for X-ray diffraction must have a minimum diameter of 50 μm, although even larger

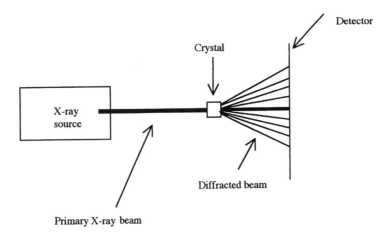

Figure 1.13 Overview of the principles of X-ray diffraction. Refer to text for details

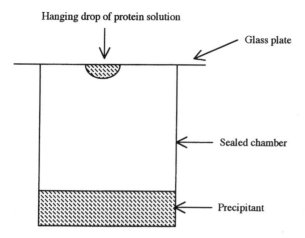

Hanging drop of protein solution

Glass plate

Sealed chamber

Precipitant

Figure 1.14 Growth of protein crystals by the vapour diffusion (hanging drop) method. A small (20 μl) drop of a concentrated purified protein solution containing a suitable precipitant (e.g. polyethylene glycol or ammonium sulfate) is placed on a glass surface. This is subsequently inverted and sealed (e.g. with vacuum grease) to the top of a chamber containing a reservoir of the precipitant. The apparatus is then incubated at a temperature of the order of 22 °C, resulting in slow evaporation of water from the protein-containing hanging drop. A supersaturated solution is slowly generated, which is conducive to crystal growth

crystals (500 μm) are required for some older X-ray crystallographic equipment. Such crystals may contain up to 10^{16} protein molecules.

X-rays are generated when a metal plate is bombarded by accelerating electrons. This is normally achieved in high voltage tubes, but more powerful X-ray beams may be generated in synchrotron storage rings, using electrons travelling close to the speed of light. An X-ray beam is allowed to escape from its source through a narrow window and it then passes through the protein crystal (Figure 1.13). The diffracted spots are recorded on an image plate, which is then scanned, and the diffraction pattern is stored on computer. Information from the diffraction pattern may then be analysed by a mathematical expression termed a Fourier transform. This yields information regarding the relative positioning of the atoms present in the protein. For various technical reasons, X-ray data from native protein crystals must be compared to X-ray data from crystals in which various atoms of the protein are complexed with heavy metals. This is usually achieved by first diffusing the metals into the pores or channels of pre-formed crystals. These can react with amino acid side chain groups such as cysteine's SH group.

A major bottleneck in the determination of protein structure by X-ray diffraction relates to difficulties encountered in inducing many proteins to crystallize. NMR may be used to determine the structure of proteins in free solution. However, the complexity of this technique, and particularly the data generated, limits the use of this approach to relatively small proteins (usually 25 kDa or less). However, technical advances in the field is likely to soon render feasible resolution of proteins up to 60–80 kDa.

NMR analysis entails applying a strong magnetic field to a sample of the protein of interest. Electromagnetic radiation in the radiofrequency (rf) range is then applied. NMR analysis is based on the fact that a number of atomic nuclei display a magnetic moment. These include ^{1}H, ^{13}C, ^{15}N and ^{31}P. This arises because nuclei behave as if they are spinning about an axis. As the above nuclei are positively charged such spinning nuclei act like tiny magnets, and will therefore interact with an applied magnetic field. If a protein is placed in a strong magnetic field the spin on such nuclei align along this field. This alignment can be converted to an excited state if radiofrequency energy of the appropriate frequency is applied. Subsequently the nuclei revert to their unexcited state, in the process emitting radiofrequency radiation, which may be detected and measured. The exact frequency emitted by any given nucleus is influenced by it's molecular environment, and these shifts in frequency emitted can be used to provide three-dimensional structural information regarding the protein.

Experimentally determined protein three-dimensional structural information is stored in various databases as sets of atomic coordinates. The principal such database is the Brookhaven protein database (PDB), which may be accessed via the Internet at http://www.pdb.bnl.gov. In addition, various Internet sites provide free software programs which can generate protein structural models from atomic coordinates. These structures can often be viewed interactively (e.g. rotated and zoomed). RasMol (http://www.umass.edu/microbio/rasmol/) is amongst the best known such programs. Another program, termed Mage is used to provide interactive protein structural models, called kinemages (http://www.faseb.org/protein/kinemages/kinpage.html).

PROTEIN POST-TRANSLATIONAL MODIFICATION

Many polypeptides undergo covalent modification, either during or after their ribosomal assembly. Amongst the most commonly observed modifications are proteolytic processing, glycosylation and phosphorylation, although a variety of other modifications can also occur (Table 1.7). Such modifications generally influence either the biological activity or the structural stability of the polypeptide. The fact that they also alter some physicochemical property or other of the polypeptide helps facilitate their identification.

Proteolytic processing

Proteolytic processing refers to limited and specific proteolytic cleavage of a polypeptide subsequent to its synthesis. The pre-cleaved ('pro') form

Table 1.7 Types of post-translational modifications that polypeptides may undergo. Refer to text for additional details

Modification	Example
Proteolytic processing	Various proteins become biologically active only upon their proteolytic cleavage (e.g. some blood factors)
Gylcosylation	For some proteins glycosylation can increase solubility, influence biological half life and/or biological activity
Phosphorylation	Influences/regulates biological activity of various polypeptide hormones
Acetylation	Function unclear
Acylation	May help some polypeptides interact with/anchor in biological membranes
Amidation	Influences biological activity/stability of some polypeptides
Sulfation	Influences biological activity of some neuropeptides and the proteolytic processing of some polypeptides
Hydroxylation	Important to the structural assembly of certain proteins
γ-Carboxyglutamate formation	Important in allowing some blood proteins bind calcium
ADP-ribosylation	Regulates biological activity of various proteins
Disulfide bond formation	Helps stabilize conformation of some proteins

of the polypeptide is generally inactive, with activation occurring upon proteolysis. Most proteins subject to such processing are destined for export into cellular organelles or for secretion from the cell. Examples include the mammalian digestive enzymes trypsin, chymotrypsin and pepsin. These are initially synthesized and stored in the pancreas as 'pro' or 'zymogen' precursors. Additional examples include a range of blood clotting factors (Chapter 5) and insulin (Chapter 7). Proteolytic activation is very specific and is generally irreversible.

Glycosylation

Glycosylation (the attachment of carbohydrates) is one of the most common forms of post-translational modifications associated with eukaryotic proteins, particularly eukaryotic extracellular and cell surface proteins. In the case of many glycoproteins removal of the sugar component has no detectable effect upon the biological properties (deglycosylated forms of the glycoprotein can be generated by including inhibitors of the glycosylation pathway (e.g. the antibiotic tunicamycin) in the cell growth medium, or by enzymatic degradation of the glycocomponent of preformed glycoproteins using glucosidase enzymes). However, in other cases, the sugar component plays a direct role in the biological activity of the glycoprotein. Native human chorionic gonadotrophin (hCG, Chapter 7), for example is a heavily glycosylated gonadotrophic hormone. Removal of its sugar components usually abolishes its ability to induce a biological response, although the hormone's binding affinity for its receptor remains unaltered – or is sometimes actually increased.

The sugar component of glycoproteins can also have several other potential functions, including one of targeting/recognition. Lysosomal enzymes, for example, are usually glycosylated, and their sugar components play a central role in directing the newly synthesized enzyme specifically to the lysosome. Several hormone–cytokine receptors are glycosylated and their glycocomponent appears to constitute an important part of receptor–ligand binding. Cell surface sugars may also play an important part in cell–cell adhesion and recognition. Such interactions would be critical to maintenance of tissue integrity in multicellular organisms, as well as processes such as fertilization, cellular differentiation and oncogenesis.

In yet other cases, the sugar component may help to stabilize the native glycoprotein conformation, render the protein more soluble or control its biological half life. The potential functions of the sugar components of glycoproteins are:

- Play a direct role in mediating the biological effects of some proteins (e.g. hCG and erythropoietin).

- Targeting: sugar component can help target proteins to specific cellular locations (e.g. lysosomal enzymes).

- Recognition: sugar component can often play a direct role in binding of a glycoprotein to its ligand (e.g. some cytokine receptors).

- Stabilization: sugar component may help stabilize glycoprotein conformation, as well as protecting it from proteolytic attack.

- Solubilization: removal of sugar component can render some proteins far less water soluble.

- Increasing biological half-life of protein (by decreasing its rate of clearance from the bloodstream).

Two types of glycosylation can occur: *N*-linked and *O*-linked. In the case of *N*-linked glycosylation, the sugar chain (the oligosaccharide) is attached to the protein via the nitrogen atom of an asparagine (Asn) residue, while in *O*-linked systems, the sugar chain is attached to the oxygen atom of hydroxyl groups, usually those of serine or theronine residues (Figure 1.15).

Monosaccharides most commonly found in the sugar side chain(s) include mannose, galactose, glucose, fucose, *N*-acetylgalactosamine, *N*-acetylglucosamine, xylose and sialic acid. These can be joined together in various sequences and by a variety of glycosidic linkages. The carbohydrate chemistry of glycoproteins is therefore quite complex. The structure of two such example oligosaccharide chains is presented in Figure 1.16.

N-Linked glycosylation is sequence specific, involving the transfer of a pre-synthesized oligosaccharide chain to an asn residue found in a characteristic sequence Asn–X–Ser, or Asn–X–Thr or Asn–X–Cys, where X represents any amino acid residue, with the exception of proline. An additional glycosylation determinant must also apply, as not all potential *N*-linked sites are glycosylated in some proteins. The determinants of *O*-linked glycosylation are even less well understood. Characteristic sequence recognition is not apparent in most cases and three-dimensional structural features may be more important in such instances.

Phosphorylation

Reversible phosphorylation represents yet another form of post-translational modification, and is undertaken primarily in eukaryotes but also in prokaryotes. The phosphate group donor is most often ATP and phosphorylation–dephosphorylation of the target protein is undertaken by substrate-specific protein kinase and protein phosphatase enzymes (Figure 1.17). The site of phosphorylation is usually the hydroxyl group of either serine, threonine or tyrosine residues (Figure 1.17), although the side chains of aspartate, lysine and histidine can also sometimes be phosphorylated. Again the exact site phosphorylated exhibits both a sequence-specific characteristic and a likely requirement for a characteristic three-dimensional shape.

In the vast majority of cases phosphorylation directly affects the biological activity of the target protein, with phosphorylation–dephosphorylation functioning as a reversible on–off switch. In some cases (e.g. the enzyme glycogen phosphorylase), phosphorylation results in activation, whereas in other cases (e.g. the enzyme glycogen synthase), phosphorylation results in inactivation. In a few instances, phosphorylation is not functionally important at all. For example, the phosphorylation of the milk protein casein (chapter 13) is of nutritional rather than functional importance. Phosphorylation of a protein alters its physicochemical properties, by increasing its number of negative charges. As a result, phosphorylation can often be detected by, for example, monitoring for a change in the rate of migration of a protein band in a (non-denaturing) electrophoretic gel.

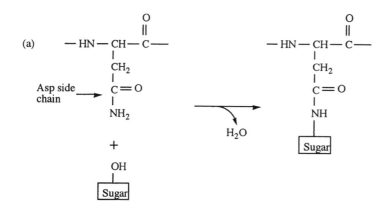

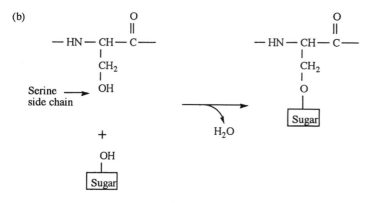

Figure 1.15   *N*-linked (a) versus *O*-linked (b) glycosylation. 'Sugar' represents an oligosaccharide chain, an example of which is provided in Figure 1.16

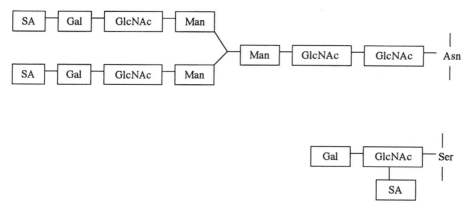

Figure 1.16   Structure of two sample oligosaccharide side chains (one *N*-linked the other *O*-linked) found in glycoproteins. Man, mannose; Gal, galactose; SA, sialic acid; GlcNAc, *N*-acetyl glucosamine; GalNAc, *N*-acetyl galactosamine; Ser, Serine; Asn, asparagine

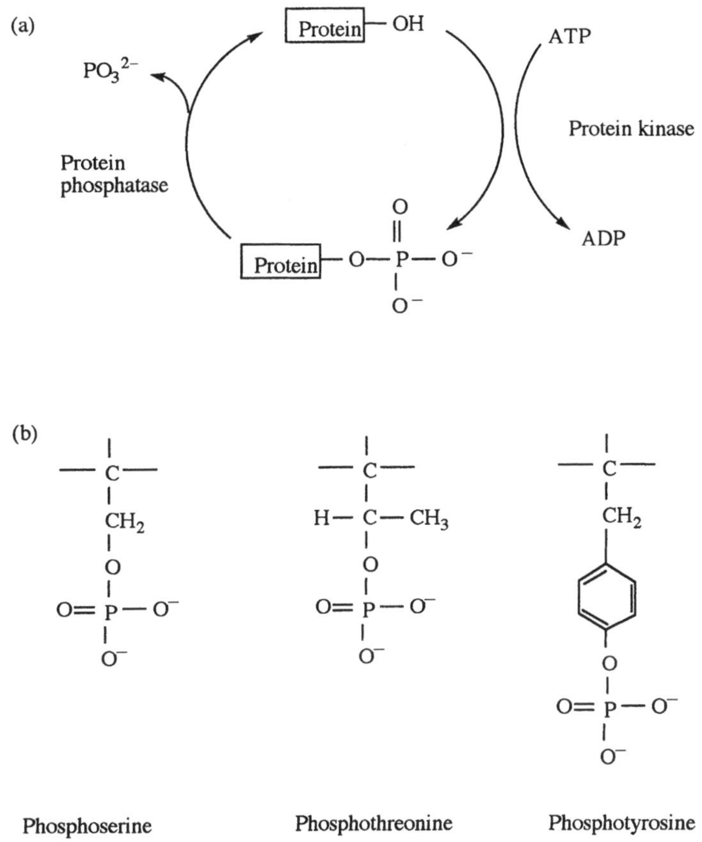

(a)

(b)

Phosphoserine   Phosphothreonine   Phosphotyrosine

Figure 1.17 Reversible phosphorylation of a protein substrate via a kinase – phosphatase mechanism. (b) phosphate groups are usually attached to the protein via the hydroxyl groups of serine, threonine or tyrosine residues

Acetylation, acylation and amidation

Well over 50 per cent of all eucaryotic polypeptides synthesized in the cytoplasm display an N-terminal acetyl group. The acetyl group donor is usually acetyl CoA, and the reaction is catalysed by an *N*-acetyl transferase enzyme (fig 1.18) which appears to be loosely associated with ribosomes. In some cases N-terminal acetylation appears to occur before the polypeptide is completely synthesized, while in other instances acetylation occurs post-translationally. The criteria which determine if a polypeptide is to be acetylated are not fully understood, nor is the exact function of this post-translational modification.

Protein acylation refers to the direct covalent attachment of fatty acids to a polypeptide backbone. The fatty acids most commonly found in association with acylated polypeptides are the 16-carbon saturated palmitic acid and the 14-carbon saturated myristic acid. Palmitic acid is usually

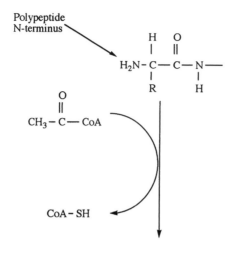

Figure 1.18   Acetylation of a polypeptide *N*-terminal amino group, as catalyzed by an *N*-acetyl transferase enzyme. The acetyl group donor is usually acetyl CoA

covalently linked via an ester or thioester bond to either a cysteine, serine or threonine residue, while myristic acid is invariably covalently attached to an N-terminal glycine residue via an amide bond (Figure 1.19). Acylated polypeptides appear to be ubiquitous in eukaryotes and are also found in many viruses.

Myristoylation (covalent attachment of myristic acid) appears to occur cotranslationally (i.e. before polypeptide synthesis is complete), and is promoted by the enzyme myristoyl CoA: protein *N*-myristoyl transferase. Enzyme-based attachment of palmitic acid is a more complex post-translational event, probably involving several enzymes. A wide range of protein types are acylated, including various cytoplasmic and membrane proteins, as well as viral structural proteins (Table 1.8). The exact function of the fatty acid component is not fully defined in most cases. Initially, it was believed to play a role in anchoring or allowing proteins to interact with biological membranes, but this is probably not its main function. The fatty acid component may play a more important role in promoting protein–protein interactions. Many such acylated polypeptides are components of multisubunit protein complexes, and in several cases removal of the fatty acid moiety negatively affects subunit interactions and/

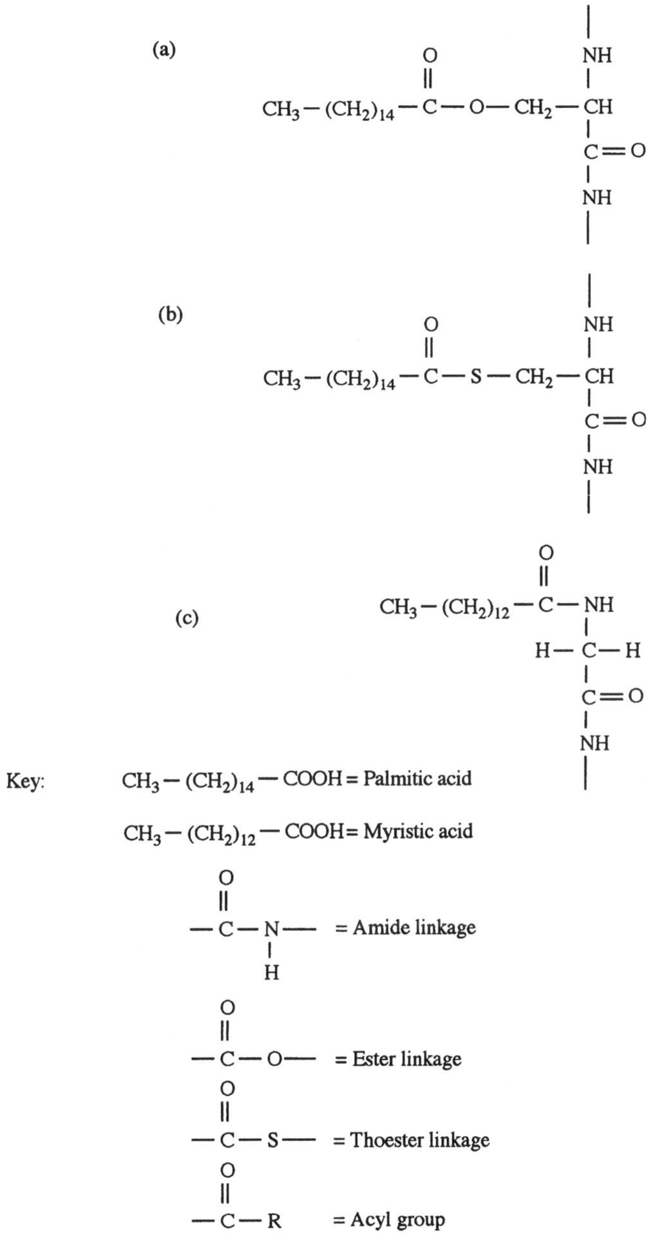

(a)

$$CH_3-(CH_2)_{14}-\overset{\overset{\textstyle O}{\|}}{C}-O-CH_2-\underset{\underset{\textstyle NH}{|}}{\overset{\overset{\textstyle NH}{|}}{CH}}$$

C=O

NH

(b)

$$CH_3-(CH_2)_{14}-\overset{\overset{\textstyle O}{\|}}{C}-S-CH_2-\underset{\underset{\textstyle NH}{|}}{\overset{\overset{\textstyle NH}{|}}{CH}}$$

C=O

NH

(c)

$$CH_3-(CH_2)_{12}-\overset{\overset{\textstyle O}{\|}}{C}-NH$$

H—C—H

C=O

NH

Key: $CH_3-(CH_2)_{14}-COOH$ = Palmitic acid

$CH_3-(CH_2)_{12}-COOH$ = Myristic acid

$$-\overset{\overset{\textstyle O}{\|}}{C}-\underset{\underset{\textstyle H}{|}}{N}-$$ = Amide linkage

$$-\overset{\overset{\textstyle O}{\|}}{C}-O-$$ = Ester linkage

$$-\overset{\overset{\textstyle O}{\|}}{C}-S-$$ = Thoester linkage

$$-\overset{\overset{\textstyle O}{\|}}{C}-R$$ = Acyl group

Figure 1.19 Linkage of (a) palmitic acid via an ester bond to a serine residue; (b) palmitic acid via a thioester bond to a cysteine residue; and (c) myristic acid to an N-terminal glycine reside via an amide bond

or biological activity of the multisubunit complexes. The fatty acid component of various acylated viral caspid proteins also appears to stabilize interactions between these structural proteins. Although palmitic and

Table 1.8 Representative examples of some acylated proteins. The fatty acid moiety attached as well as the locations of eucaryotic-derived acylated proteins is also listed

Protein	Cellular location	Fatty acid
cAMP-dependent protein kinase	Cytoplasm	Myristic acid
Cytochrome b_5 reductase	ER and mitochondria	Myristic acid
Gi and Go α-subunits	Plasma membrane	Myristic acid
Insulin receptor	Plasma membrane	Palmitic acid
Interleukin-1 receptor	Plasma membrane	Palmitic acid
Transferrin receptor	Plasma membrane	Palmitic acid
Rhodopsin	Disc membranes in retina	Palmitic acid
P55 & P28	HIV virus	Myristic acid
VP4	Picorna viruses	Myristic acid
P19 (gag)	HTLVI	Myristic acid
HA	Influenza virus	Palmitic acid
gE	Herpes simplex virus	Palmitic acid

HIV, human immunodeficiency virus; HTLVI, human T lymphotropic virus-1.

myristic acids are most commonly found in association with acylated proteins, additional lipid-based substances may also be used: these include stearic acid, phosphatidylinositol and farnesyl groups.

PROTEIN STABILITY AND FOLDING

Upon biosynthesis, a polypeptide folds into its native conformation, which is structurally stable and functionally active. The conformation adopted ultimately depends upon the polypeptide's amino acid sequence, explaining why different polypeptide types have different characteristic conformations. We have previously noted that stretches of secondary structure are stabilized by short-range interactions between adjacent amino acid residues. Tertiary structure on the other hand is stabilized by interactions between amino acid residues which may be far apart from each other in terms of amino acid sequence, but which are brought into close proximity by protein folding. The major stabilizing forces of a polypeptide's overall conformation are:

- hydrophobic interactions
- electrostatic attractions
- covalent linkages.

Hydrophobic interactions are the single most important stabilizing influence of protein native structure. The 'hydrophobic effect' refers to the tendency of non-polar substances to minimize contact with a polar solvent such as water. Non-polar amino acid residues constitute a significant proportion of the primary sequence of virtually all polypeptides. These polypeptides will fold in such a way as to maximize the number of such non-polar residue side chains buried in the polypeptide's interior – away from the surrounding aqueous environment. This situation is most energetically favourable.

Stabilizing electrostatic interactions include van der Waals forces (which are relatively weak), hydrogen bonds and ionic interactions. Although nowhere near as strong as covalent linkages (Table 1.9), the large number of such interactions existing within a polypeptide renders them collectively quite strong.

While polypeptides display extensive networks of intramolecular hydrogen bonds, such bonds do not contribute very significantly to overall conformational stability. This is because atoms hydrogen bonding with each other in a folded polypeptide can form energetically equivalent hydrogen bonds with water molecules if the polypeptide is in the unfolded state. Ionic attractions between (oppositely) charged amino acid side chains also contribute modestly to overall protein conformational stability. Such linkages are termed salt bridges and as one would expect, they are located primarily on the polypeptide surface.

Disulfide bonds represent the major covalent bond type which can help stabilize a polypeptide's native three-dimensional structure. Intracellular proteins, although generally harbouring multiple cysteine residues, rarely form disulfide linkages, due to the reducing environment which prevails within the cell. Extracellular proteins in contrast are usually exposed to a more oxidizing environment, conducive to disulfide bond

Table 1.9 Approximate bond energies associated with various (non-covalent) electrostatic interactions, as compared to a carbon-carbon single bond

Bond type	Bond strength (kJ/mol)
Van der Waals forces	10
Hydrogen bond	20
Ionic interactions	86
Carbon–Carbon bond	350

formation. In many cases the reduction (i.e. breaking) of disulfide linkages has little effect upon the polypeptides' native conformation. However, in other cases (particularly disulfide-rich proteins) disruption of this covalent linkage does render the protein less conformationally stable. In these cases the disulfide linkages likely serve to 'lock' functional and/or structurally important elements of domain–tertiary structure in place.

The description of protein structure as presented thus far may lead to the conclusion that proteins are static, rigid structures. This is not the case. A proteins' constituent atoms are constantly in motion and groups ranging from individual amino acid side chains to entire domains can be displaced via random motion by anything up to approximately 0.2 nm. A protein's conformation therefore displays a limited degree of flexibility and such movement is termed 'breathing'.

Breathing can sometimes be functionally significant by for example allowing small molecules to diffuse in/out of the protein's interior. In addition to breathing some proteins may undergo more marked (usually reversible) conformational changes. Such changes are usually functionally significant. Most often they are induced by biospecific ligand interactions (e.g. binding of a substrate to an enzyme or antigen binding to an antibody).

Currently there exists an enormous and growing deficit between the number of polypeptides whose amino acid sequence has been determined and the numbers of polypeptides whose three-dimensional structure has been resolved. Given the complexities of resolving three-dimensional structure experimentally it is not surprising that scientists are continually attempting to develop methods by which they could predict higher order structure from amino acid sequence data. While modestly successful secondary structure predictive approaches have been developed, no method by which tertiary structure may be predicted from primary data has thus far been developed.

Over 20 different methods of secondary structure prediction have been reported (Table 1.10). The approaches taken fall into two main categories: (1) empirical statistical methods, which are based upon data generated from studying proteins of known three-dimensional structure and

Table 1.10 Some secondary structure predictive methods currently used. Refer to text for further details

Method	Basis of prediction
Chou and Fasman	Empirical statistical method
Garnier, Osguthorpe and Robson (GOR) method	Empirical statistical method
EMBL profile neural network (PHD) method	Empirical statistical method
Protein sequence analysis (PSA) method	Empirical statistical method
Lim method	Physicochemical criteria

correlation of such proteins primary amino acid sequence with structural features; (2) methods based upon physicochemical criteria such as fold compactness (i.e. the generation of a folded form displaying a tightly packed hydrophobic core and a polar surface).

Most such predictive methods are at best 50–70 per cent accurate. The relatively large inaccuracy stems from the fact that the folded (tertiary) structure imposes constraints upon the nature and extent of secondary structure within some regions of the polypeptide chain. Any generalized 'rules' relating secondary structure to amino acid sequence data, by nature, will not take such issues into consideration.

The Chou and Fasman method remains one of the most popular predictive methods. These scientists studied structural details of a number of proteins whose three-dimentional structure had been experimentally determined by X-ray crystallography. This allowed them to construct 'conformational preference data' for each amino acid residue, which essentially relates the compatibility of the amino acid to α-helical stretches or β stretches (Table 1.11). By applying the preference data to the amino acid sequence of a polypeptide putative α-helical or β-strand regions can be assigned.

The analysis carried out by Chou and Fastman also allowed the following observations to be made:

- An α-helical stretch is usually initiated by a six-residue sequence containing at least four Hα or hα residues (Table 1.11).

- Proline residues, if present, are located at the amino terminus of the helix.

- Any group of four successive residues present in an α-helix will have an average Pα value greater than 1.0 (Table 1.11).

- A β-stretch is usually initiated by a five-residue sequence containing at least three Hβ or Hβ residues.

- Any group of four successive residues present in a β-stretch will have an average Pβ value greater than 1.0.

Accurate prediction of a polypeptide's three-dimensional structure from lower-order structural information remains to be achieved. Tertiary structure prediction directly from amino acid sequence data remains in the distant future, although a technique known as threading will likely support some progress towards this goal. Three-dimensional structural analysis has shown that only a limited number of stable protein folds exist and moreover, that many unrelated amino acid sequences can generate the same fold (a fold refers to a domain-like structure but which is common to many proteins). By analysing databases containing polypeptide tertiary structure information the various possible amino acid sequences that can give rise to any particular fold can potentially be determined.

Table 1.11   Conformational preferences and assignments of amino acid residues with regard to stretches of α-helix and β structure

α-helix			β strand		
Residue	P_α	Assignment	Residue	P_β	Assignment
Glu	1.44	Hα	Val	1.64	Hβ
Ala	1.39	Hα	Ile	1.57	Hβ
Met	1.32	Hα	Thr	1.33	hβ
Leu	1.30	Hα	Tyr	1.31	hβ
Lys	1.21	hα	Trp	1.24	hβ
His	1.12	hα	Phe	1.23	hβ
Gln	1.12	hα	Leu	1.17	hβ
Phe	1.11	hα	Cys	1.07	hβ
Asp	1.06	hα	Met	1.01	Iβ
Trp	1.03	Iα	Gln	1.00	Iβ
Arg	1.00	Iα	Ser	0.94	iβ
Ile	0.99	iα	Arg	0.94	iβ
Val	0.97	iα	Gly	0.87	iβ
Cys	0.95	iα	His	0.83	iβ
Thr	0.78	iα	Ala	0.79	iβ
Asn	0.78	iα	Lys	0.73	bβ
Tyr	0.73	bα	Asp	0.66	bβ
Ser	0.72	bα	Asn	0.66	bβ
Gly	0.63	Bα	Pro	0.62	Bβ
Pro	0.55	Bα	Glu	0.51	Bβ

Pα = propensity to form α helical regions; P_β = propensity to form β stretches; H_α = strong helix former; h_α = helix former; I_α = weak helix former; i_α = indifferent; b_α = helix breaker; B_α = strong helix breaker. Similar designations are used in the case of β formers, with 'β' replacing 'α'. Reproduced from *Current Protocols in Protein Science*, with kind permission of the publisher, John Wiley & Sons.

Threading essentially entails comparing the sequence of the polypeptide whose three-dimensional structure you wish to predict with the database sequences known to generate specific fold patterns. Computer programs can then be used to estimate the probability of the target sequence adopting each known folding structure.

Accurate tertiary structure prediction if elements of secondary structure have already been assigned is a more realistic – although still future – prospect. By studying known tertiary structures rules for packing arrangements of elements of secondary structure into higher order structures (e.g. specific motifs and domains) are slowly being elucidated.

The various forces which contribute to stabilization of a polypeptide in its native conformation have been discussed earlier. For a typical globular polypeptide the total sum of all the bonding energies associated with such stabilizing interaction (Table 1.10) add up to several thousand kJ/mol. Thermodynamic analysis, however, shows that the free energy difference between a typical 200 residue polypeptide in its folded versus denatured form is only of the order of 80–100 kJ/mol. This equates to just a few hydrogen bond equivalents. In the unfolded state intra-chain non-covalent (stabilizing) interactions are not maximized, but some new such interactions are formed which can 'stabilize' the denatured state. These include extensive hydrogen bonding between appropriate amino acid groups and surrounding water molecules. Moreover, the second law of thermodynamics states that it is more energetically favourable for a molecule to exist in a random order as opposed to a highly ordered state (the concept of entropy). The fact that proteins are only marginally more stable in their folded form is functionally significant. Structurally this renders the protein somewhat flexible, enabling them to more readily undergo various conformational changes central to their biological activity.

The factors influencing the intrinsic stability of native polypeptide conformation have largely been elucidated via the study of proteins which function under relatively mild environmental conditions. More recently the study of proteins derived from extremophiles (organisms living under extreme environmental conditions, Chapter 10) has further extended our understanding of conformational stability. The three-dimensional structure of a number of homologous proteins derived from various psycrophiles, mesophiles, thermophiles and hyperthermophiles (Chapter 10) have now been determined. This facilitates the identification of changes in structural features that help render the protein stable under its particular native physiological conditions. Thermodynamic analysis reveals that the principle of marginal stability between the native versus denatured state extends to proteins isolated from such extreme environments.

One might expect that proteins isolated from thermophiles and hyperthermophiles would exhibit an increased level of intramolecular stabilizing interactions, in order to compensate for the destabilizing influence of elevated temperature. Conversely, it could be predicted that, in order to remain at the appropriate degree of conformational flexibility, proteins from psycrophiles would display decreased levels of such stabilizing interactions. In broad terms these principles have been bourne out, although the methods by which this is achieved differs for different proteins. Increased thermal stability is generally related to one or more of the following structural adaptations:

- an increase in the number of intramolecular polypeptide hydrogen bonds

- an increase in the number of salt bridges

- increased polypeptide compactness (improved packing of the hydrophobic core).

- extended helical regions.

Conversely, enhanced stability/functional flexibility of proteins derived from psycrophiles, appear to be achieved by one or more of the following adaptations.

- fewer salt links

- reduced aromatic interactions within the hydrophobic core (reduction in hydrophobicity)

- increased hydrogen bonding between the protein surface and the surrounding solvent

- occurrence of extended surface loops.

Folding pathways

The above discussion focuses upon the factors stabilizing the fully folded protein structure. In this section we focus upon the folding event itself. Thermodynamically a protein folds from a higher energy unfolded state to a lower energy folded state. This process is a rapid one, typically lasting from under one second to several seconds. The speed of folding indicates that this process occurs via a directed pathway rather than going through random conformational searches until it stumbles upon the most stable structural arrangement. The folding pathway proceeds via the initial rapid formation of the more compact, partially folded, 'molten globule' followed by completion of the folding pathway at a slower pace (Figure 1.20).

The molten globule exhibits most secondary structural elements of the native protein, but only a limited degree of ultimate tertiary structure. Its formation, which requires only several milliseconds, is termed a hydrophobic collapse. As the name suggests, this process is primarily driven by the favourable energy change achieved by bringing hydrophobic amino acid residues into contact with one another and away from surrounding water molecules. Hydrophobic collapse in turn drives secondary structure formation; the internalization of hydrophobic residues also internalizes

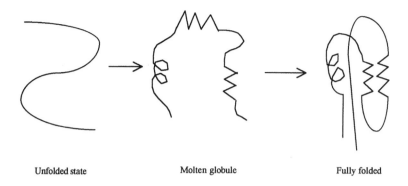

Unfolded state   Molten globule   Fully folded

Figure 1.20 Overview of the protein folding pathway. Refer to text for details

the polar −NH and C=O groups of their associated peptide bonds. This prevents the latter groups from forming stabilizing hydrogen bonds with surrounding water molecules. However, this energetically unfavourable situation can be counteracted if these groups are allowed hydrogen bond with each other in a situation maximized by the formation of α-helical or β-stretches.

In the subsequent and final phase of protein folding, elements of tertiary structure form. Initially these are most likely motifs or subdomains, which in turn lead to complete folding. While some proteins likely follow a single folding pathway others appear to have multiple pathways, which all ultimately lead to complete protein folding.

Many proteins can fold accurately and spontaneously *in vitro*, although some appear to fold more slowly or less accurately than they do *in vivo*. Although the primary sequence ultimately dictates tertiary structure, several obstacles to correct folding are known to exist. These include:

- aggregation of partially folded intermediates via intermolecular hydrophobic interactions

- isomerization of proline residues

- formation of disulphide linkages between incorrect pairs of cysteine residues.

Cells appear to have evolved various mechanisms by which such obstacles can be minimized or overcome.

Molecular chaperones are a class of protein which help polypeptides to fold into their correct native three-dimensional shape by preventing/correcting the occurrence of improper hydrophobic associations. They were first termed heat-shock proteins (Hsp) as their intracellular concentration increases at elevated temperatures. Two major classes of chaperones are now known to exist; Hsp 70 proteins and chaperonins. Both contain hydrophobic regions, which can interact with unfolded/misfolded polypeptides, helping them to fold correctly in an ATP-dependent process.

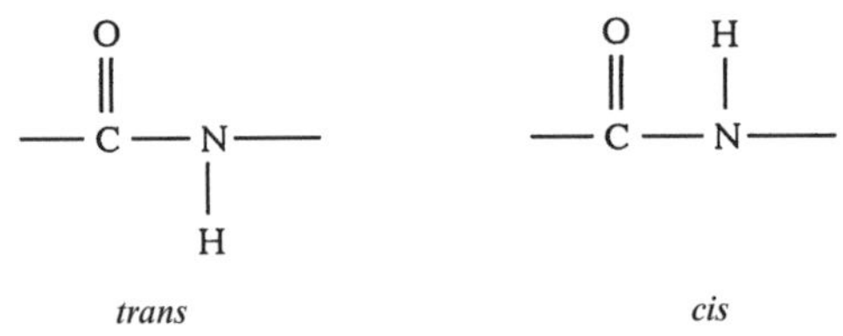

Figure **1.21** *Cis* versus *trans* forms of a peptide bond

The chaperonins of *Escherichia coli* have been studied in most detail. These consist of a large complex constructed from two protein types; the 60 kDa Hsp 60 protein (known as Gro EL) and the 10 kDa Hsp 10 protein (known as Gro ES). The intact structure consists of 14 subunits of Gro EL, which form a long hollow cylinder approximately 15 nm long and 14 nm in diameter. Seven subunits of the Gro ES interact and bind to one end of the cylinder, closing off the cavity at that end. The internal walls of the cavity display several regions rich in hydrophobic residues.

The intact chaperone will interact directly with unfolded or misfolded regions of polypeptide chains, but not with properly folded regions. The recognition mechanism is not fully understood but hydrophobic interactions between the presumably large exposed hydrophobic patches in unfolded or improperly folded regions and those in the internal surface of the chaperonin cavity likely play a central role. The exact molecular detail of how proper folding is subsequently assisted remains to be elucidated.

The enzymes peptidyl prolyl isomerases and protein disulfide isomerases also assist the efficient and proper folding of many proteins *in vivo*. The peptide bonds of polypeptide chains are generally in the trans conformation, as previously described. The *cis* conformation (Figure 1.21), although possible, is far less thermodynamically stable, and hence rarely occurs in native polypeptides. In some polypeptides, however, a peptide bond in which the NH group is derived from a proline residue does sometimes exist in the *cis* form. This often occurs at tight bends, and in such instances is required for structural flexibility. This *cis* form is stabilized by surrounding elements of three-dimensional structure. Prior to folding these stabilizing elements therefore do not exist.

In the unfolded state, an equilibrium between *cis* and *trans* forms of peptide bonds exist, but due to the latters stability the equilibrium greatly favours the trans form. Unaided *trans–cis* isomerization of target proline residues is thus a slow process and can be rate limiting to folding. Peptidyl propyl isomerases are enzymes found in most cell types, which can enhance the rate of isomerization of such proline based peptide bonds by rates of over 10^6. These enzymes therefore can eliminate this rate-limiting step to folding *in vivo*.

The folding of proteins which contain disulfide linkages can pose special problems for the cell; that of ensuring the correct pairing of cysteine residues in such linkages. This process is assisted by enzymes termed protein disulfide isomerases. These enzymes themselves have free thiol (SH) groups derived from constituent cysteine residues. These groups can form transient disulfide linkages with cysteine residues present in the target protein. As such they can assist in the intramolecular re-arrangements of incorrectly formed disulfide linkages.

Protein engineering

The advent of recombinant DNA technology renders straightforward the manipulation of a protein's amino acid sequence. This process, termed site-directed mutagenesis or protein engineering, entails the controlled alteration of the nucleotide sequence coding for the polypeptide of inter-est such that specific, predetermined changes in amino acid sequence are introduced (Box 1.1). Such changes can include insertions, deletions or substitutions. Protein engineering facilitates a greater understanding of the link between a polypeptides' amino acid sequence and its structure. It also provides a powerful method of studying the relationship between structure and its function. As such, this technique will help greatly in achieving the much pursued but still distant objective of *de-novo* protein design. Protein engineering is also used to tailor structural or functional attributes of commercially important proteins.

In recent times much effort has been directed towards engineering enzymes in order to confer on them enhanced catalytic activity, and/or temperature and pH stability. Studies using several model proteins have shown that replacing a specific lysine residue with arginine, or replacing glycine with alanine can result in increased thermal stability. For example replacement of lysine 253 with arganine in glucose isomerase resulted in enhanced thermal stability of the resultant engineered enzyme. Replace-ment of the lysine with arginine seemed to alter the enzymes conform-ation in such a way as to enhance the overall level of hydrogen bonding. This of course exerts a positive influence on protein stability. Glucose isomerase is utilized industrially in the production of high fructose syrup from glucose syrups (Chapter 11).

Introduction of cysteine residues which allow formation of disulphide bonds could also enhance protein stability. This has been illustrated in the case of lysozyme which has a cysteine residue at position 97. Conversion of an isoleucine residue at position 3 of the molecule to cysteine was shown to result in disulphide bond formation. This in turn enhances the stability of the enzyme. Protein engineering holds great promise for the future. The full potential of this technology however will not be attained until more is known about the forces which govern protein folding.

Human tissue plasminogen activator (t-PA) represents an important therapeutic agent which has been subject to modification by methods of protein engineering. tPA is a serine protease which plays an important role in promoting fibrinolysis, the degradation of blood clots (Chapter 5). Removal of a tripeptide sequence (Tyr-Phe-Ser) from a specific domain within tPA was shown to decrease significantly the plasma clearance rate

Box Figure 1.1

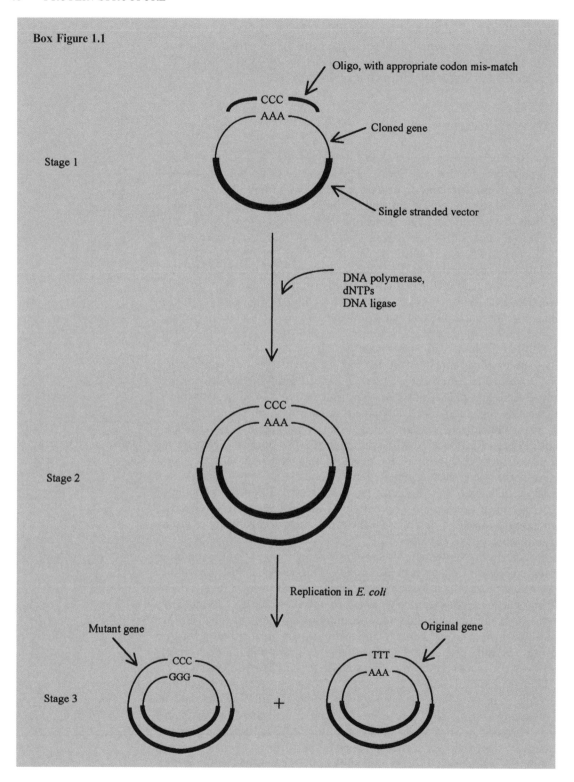

Box 1.1 Alteration of amino acid sequence by site-directed mutagenesis

This box provides an overview of the approach usually used to introduce a substitution, short insertion or short deletion into a protein's amino acid sequence by the technique of site-directed mutagenesis.

Firstly, the gene/cDNA coding for the native (unaltered) protein is cloned into a vector from which single stranded DNA can be prepared (e.g. various phage vectors). Next, an appropriate oligonucleotide (oligo) is designed and chemically synthesized. The two end segments of the oligo are complementary in sequence to either side of the section of the gene one wishes to alter. Its central portion, however, displays an altered nucleotide sequence corresponding to the alteration required. For example, in the case illustrated diagrammatically, a single amino acid substitution is desired; a change from lysine (corresponding to AAA in the gene sequence) to proline (corresponding to GGG on the coding strand). Upon annealing, the deliberately mismatched bases will 'loop out', but the sequences flanking each side will anneal fully via complementary base pairing.

The addition of DNA polymerase, dNTPs and DNA ligase facilitates DNA synthesis, forming a closed circular duplex (stage 2 in the associated diagram). This can then be introduced into *E. coli*, where both strands will be replicated (associated diagram, stage 3). The mutant gene can then be recovered and cloned into an appropriate expression vector, allowing the production of large amounts of the desired mutant protein by standard recombinant methods.

of this molecule. Native tPA exhibits a very short plasma half life. This necessitates its administration by prolonged i.v. infusion. An engineered protein exhibiting an extended plasma half life could be administered as a single i.v. injection. Further modifications have also been made to t-PA which not only increases its plasma half life but also renders it more potent in terms of promoting clot lysis.

Additional examples of engineered proteins which have found commercial application include chimaeric and humanized antibodies, long-or short-acting insulins and oxidation-resistant detergent enzymes. These and other examples will be discussed during the subsequent chapters of this book

FURTHER READING

Books

Angelitti, R. (1998) *Proteins*. Academic Press, London.

Attwood, T. (1999) *Introduction to Bioinformatics*. Longman, Harlow.

Baxevanis, A. (1998) *Bioinformatics*. John Wiley, Chichester.

Carey, P. (1996) *Protein Engineering and Design*. Academic Press, London.

Coligan, J. *et al.* (1998) *Current Protocols in Protein Science*. John Wiley, Chichester.

Dobson, C. (1996) *Protein Folding*. Cambridge University Press, Cambridge.

Fersht, A. (1999) Structure and Mechanism in Protein Science. Freeman, Basingstoke.

Higgins, D. (2000) Bioinformatics. *Oxford Unveristy Press*, Oxford.

Articles

Protein structure and folding

Al-Lazikani, B. *et al.* (2001) Protein structure predection. *Curr. Opin. Chem. Biol.* **5** (1), 51–56.

Arai, M. & Kuwajima, K. (2000) Role of the molten globule state in protein folding. *Adv. Protein Chem.* **53**, 209–282.

Bilsen, O & Matthews, C. (2000) Barriers in protein folding reactions. *Adv. Protein Chem.* **53**, 153–207.

Chasse, G. *et al.* (2001) Peptide and protein folding. *J Mol Struct. – Theochem.* **537**, 319–361.

Clarke, A. & Waltho, P. (1997) Protein folding pathways and intermediates. *Curr. Opini. Biotechnol.* **8**, 400–410.

Domingues, F. *et al.* (2000) The role of protein structure in genomics. *FEBS Lett.* **476**, (1–2), 98–102.

Osguthorpe, D. (2000) Ab initio protein folding. *Curr. Opin. Struct. Biol.* **10**, (2), 146–152.

Radford, S. (2000) Protein folding; progress made and promises ahead. *Trends Biochem. Sci.* **25**, (12), 611–618.

Robertson, A. & Murphy, K. (1997) Protein structure and the energetics of protein stability. *Chem. Rev.* **97** (5), 1251–1267.

Sadana, A. & Vo-Dinh, T. (2001) Biomedical implications of protein folding and mis-folding. *Biotechnol Appl. Biochem.* **33**, 7–16.

Protein structure determination

Beauchamp, J & Isaacs, N. (1999) Methods for X-ray diffraction analysis of macromolecular structures. *Curr. Opin. Chem. Biol.* **3**, (5), 525–529.

Clore, G. & Gronenborn, A. (1998) NMR structure determination of proteins and protein complexes larger than 20kda. *Curr. Opin. Chem. Biol.* **2**, 564–570.

Fischer, M. *et al.* (1998) Protein NMR relaxation; theory, applications and outlook. *Progr. Nucl. Magn. Res. Spectrosc.* **33**, (4), 207–272.

Kussmann, M & Roepstorff, P. (1998) Characterization of the covalent structure of proteins from biological material by MALDI mass spectrometry – possibilities and limitations. *Spectroscopy* **14**, (1), 1–27.

Smyth, M & Martin, J. (2000) X ray crystallography. *J Clin. Pathol. – Mol. Pathol.* **53**, (1), 8–14.

Uson, I & Sheldrick, G. (1999) Advances in direct methods for protein crystallography. *Curr. Opin. Struct. Biol.* **9** (5), 643–648.

Wiencek, J. (1999) New strategies for protein crystal growth. *Annu. Rev. Biomed. Eng.* **1**, 505–534.

Wilmouth, R. *et al.* (2000) Recent sucesses in time-resolved protein crystallography. *Natural Products Reports.* **17** (6), 527–533.

Protein stability and post-translational modifications

Bayle, J. & Crabtree, G. (1997) Protein acetylation; more than chromatin modification to regulate transcription. *Chem. Biol.* **4** (12), 885–888.

Elbein, A. (1991) The role of N-linked oligosaccharides in glycoprotein function. *Trends Biotechnol.* **9**, 346–352.

Imperiali, B. & O'Connor, S. (1999) Effect of N-linked glycosylation on glycopeptide and glycoprotein structure. *Curr. Opin. Chem. Biol.* **3** (6), 643–649.

Kobata, A. (1992) Structure and function of the sugar chains of glycoproteins. *Eur. J. Biochem.* **209**, 483–501.

Li, J. & Assmann, S. (2000) Protein phosphorylation and ion transport; a case study in guard cells. *Adv. Bot. Res.* **32**, 459–479.

McLihnney, R. (1990) The fats of life; the importance and function of protein acylation. *Trends Biochem. Sci* 387–391.

Merry, T. (1999) Current techniques in protein glycosylation analysis – a guide to their application. *Acta Biochem. Polon.* **46** (2), 303–314.

Parodi, A. (2000) Protein glycosylation and its role in protein folding. *Annu. Rev. Biochem.* **69**, 69–93.

Sietz, O. (2000) Glycopeptide synthesis and the effects of glycosylation on protein structure and activity. *Chembiochem* **1** (4), 215–246.

Toroser, D. & Huber, S. (2000) Carbon and nitrogen metabolism and reversable protein phosphorylation. *Adv. Bot. Res.* **32**, 435–458.

Ueda, K. (1985) ADP-ribosylation *Annu Rev. Biochem.* **54**, 73–100.

Yan, J. *et al.* (1998) Protein phosphorylation; technologies for the identification of phosphoamino acids. *J Chromatography A* **808** (1–2), 23–41.

Protein stability

Dahiyat, B. (1999) In silico design for protein stabilization. *Curr. Opin. Biotechnol.* **10** (4), 387–390.

Jaenicke, R. (1991) Protein stability and molecular adaptation to extreme conditions. *Eur. J. Biochem.* **202**, 715–728.

Lee, B. & Vasmatis, G. (1997) Stabilization of protein structures. *Curr. Opin. Biotechnol.* **8**, 423–428.

O'Fagain, C. (1995) Understanding and increasing protein stability. *Biochim. Biophys. Acta* **1252**, 1–14.

Protein engineering

Bryan, P. (2000) Protein engineering of subtilisin. *Biochim. Biophys. Acta – Prot. Struct. Mol. Enzymol.* **1543** (2), 203–222.

Harris, J. & Craik, C. (1998) Engineering enzyme specificity. *Curr. Opin. Chem. Biol.* **2**, 127–132.

Hartley, B. *et al.* (2000) Glucose isomerase: insights into protein engineering for increased thermostability. *Biochim. Biophys. Acta – Prot. Struct. Mol. Enzymol.* **1543** (2), 294–335.

Rubingh, D. (1997) Protein engineering from a bioindustry point of view. *Curr. Opin. Biotechnol.* **8** (4), 417–422.

Schulein, M. (2000) Protein engineering of cellulases. *Biochim. Biophys. Acta – Prot. Struct. Mol. Enzymol.* **1543** (2), 239–252.

Shanklin, J. (2000) Exploring the possibilites of protein engineering. *Curr. Opin. Plant Biol.* **3** (3), 243–248.

Steipe, B. (1999) Evolutionary approaches to protein engineering. *Combinat. Chem. Biol.* **243**, 55–86.

Svenden, A. (2000) Lipase protein engineering. *Biochim. Biophys. Acta – Prot. Struct. Mol. Enzymol.* **1543** (2), 223–238.

2
Protein sources

INTRODUCTION

A pre-resiquite to the isolation, characterization and/or utilization of any protein is the identification of a suitable protein source. In a few cases the desired protein may be unique to a specific species or be produced by a very restricted number of species (e.g. the gonadotrophic hormone pregnant mare serum gonadotrophin (PMSG) is found only in equids). Under such circumstances the choice of protein source is already made. In most cases however, the protein of interest will be produced by a range of species, thereby providing a choice of source.

The purpose for which the protein is required will also influence this choice. If it is to be studied for academic purposes the researcher may often choose from a wide range of sources. On the other hand some academic projects, by their very nature, will give the researcher no

freedom to choose the source. If the protein is to be used for an applied/
industrial purpose, choice of source must be made very carefully. Factors
impinging upon this choice include:

- *Technical characteristics of the desired protein*: which source provides a
  protein with physicochemical characteristics best suited to its intended
  application? For example various enzymes are added to animal feed in
  order to achieve specific nutritional goals (Chapter 12). As such
  enzymes must function in the animal's digestive tract they must display appreciable activity at 37 °C and at pH values characteristic of
  either the stomach or small intestine.

- *Source availability*: the protein source should be readily available or
  easily cultured.

- *Level of production*: a source which produces the desired protein at
  high levels is obviously most desirable.

- *Safety*: proteins must usually be obtained from organisms that do not
  produce any additional substances that might endanger health.

- *Regulatory constraints*: most protein products must be registered by
  an appropriate regulatory authority before they can be marketed. The
  regulatory authority must be satisfied that the product is safe and is
  capable of achieving its claimed effect before it will register the product. As part of its deliberations, the authority will consider the protein
  source (particularly with regard to safety issues, as eluded to in the
  previous point).

- *Patenting issues*: if the protein is to be patented it is essential to ensure
  that the patent application – which includes details of product production – does not infringe pre-existing patents.

- *Consumer perception*: in some cases consumer or public perception
  relating to the product source can influence its commercial success
  (e.g. recombinant versus non-recombinant production of food proteins).

From the above points it becomes obvious that both technical and non-
technical issues influence the choice of source of a protein produced for
commercial gain. Non technical factors such as patenting, regulatory
issues and public perception are less important/irrelevant if the protein
is required for purely academic purposes.

Recombinant versus non-recombinant production

Low natural expression levels has rendered difficult the isolation, study
and application of a range of proteins from native sources. This technical
hurdle has been overcome with the advent of recombinant DNA technol-

ogy. Today, in principle, the gene or cDNA coding for any protein (of known or unknown function) can be isolated and inserted into an appropriate expression system. A very large number of proteins are now produced by recombinant means, for both applied and academic purposes. Recombinant DNA technology has had a threefold impact upon protein production:

- *It overcomes problems of source availability*. Some proteins (e.g. most cytokines) are produced naturally at exceedingly low concentrations, rendering difficult or impossible their production in large quantities. The use of appropriate recombinant expression systems overcomes this.

- *It overcomes problems of source safety*. Many proteins are produced naturally by dangerous or pathogenic species (e.g. snakes, various microbial pathogens, etc.). This renders collection of raw material hazardous and opens the possibility of the presence of toxic contaminants in the final product. Recombinant production in a non-pathogenic, non-toxic host organism circumvents such potential difficulties.

- *It facilitates targeted modification of the protein's amino acid sequence*. As described in Chapters 1 and 11, protein engineering can be put to a variety of pure and applied uses.

Chapters 5–13 provide numerous specific examples of proteins now produced commercially by recombinant DNA technology. Selected examples are listed in Table 2.1.

MICROORGANISMS AS SOURCES OF PROTEINS

Many proteins of industrial interest are obtained from (non-recombinant) microbial sources (Table 2.2). The majority are synthesized by a limited number of microorganisms which are classified as GRAS (generally recognized as safe). GRAS microorganisms include bacteria such as *Bacillus subtilis*, *Bacillus amyloliquefaciens*, in addition to various other bacilli, lactobacilli and streptomyces species. GRAS-listed fungi include members of the species *Aspergillus*, *Penicillium*, *Mucor* and *Rhizopus*. Yeasts such as *Saccharomyces cerevisae* are also generally recognized as safe. GRAS-listed microbes are non-pathogenic, non-toxic and generally should not produce antibiotics. Microorganisms represent an attractive source of proteins as they can be cultured in large quantities over a relatively short time period by established methods of fermentation (Figure 2.1). Therefore, they can produce an abundant, regular supply of desired protein product. Microbial proteins are often more stable than analogous proteins obtained from plant or animal sources. Furthermore, microbes can be subjected to genetic manipulation more readily than animals or plants.

Table 2.1 Examples of proteins produced commercially by recombinant DNA technology

Protein	Application	Chapter
Blood proteins (e.g. factors VIII and IX)	Treatment of haemophilia and other blood disorders	5
Thrombolytic agents (e.g. tissue plasminogen activator, tPA)	Treatment of heart attacks	5
Recombinant 'subunit' vaccines (e.g. hepatitis B surface antigen)	Vaccination against specific diseases	5
Engineered antibodies (chimaeric and humanized antibodies)	Mainly used for the *in vivo* detection/treatment of tumours	6
Native and engineered insulins	Treatment of diabetes mellitus (type 1 diabetes)	7
Human growth hormone	Treatment of short stature	7
Interferons	Treatment of cancer, viral diseases	8
Cholesterol esterase	Used to determine blood cholesterol levels	9
Chymosin	Cheese production	11
Various proteases, lipases and amylases	Added to detergents to degrade 'biological' stains	11 and 12
Phytase	Added to animal feed to degrade dietary phytic acid	12

In some instances these commercial products are produced exclusively via recombinant technology. In other cases (e.g. insulin) both non-recombinant and recombinant versions of the product are commercially available. The chapters in which the various proteins listed are discussed are also indicated.

Many industrially significant proteins obtained by methods of fermentation are secreted by the producing microorganisms directly into the culture medium. Such extracellular protein production greatly simplifies subsequent downstream processing as there is no requirement to disrupt the microbial cells in order to effect protein release. Thus, there are fewer proteins from which to separate the product of interest. Whole cells may be removed from protein containing extracellular media by methods such as centrifugation or filtration. In many such cases few subsequent purification steps are required to yield a final product of the required purity. Specific examples of industrially important proteins secreted into the extracellular medium during fermentation include various amylolytic

Table 2.2 Some proteins obtained commercially from (non-genetically engineered) microorganisms. Sources include a range of bacteria, fungi and yeast, and the proteins have found medical, analytical and industrial uses. The chapters in which these proteins are discussed are also listed

Protein	Source	Application	Chapter
Streptokinase	Various haemolytic streptococci	Thrombolytic agent (degrades blood clots)	5
Staphylokinase	*Staphylococcus aureus*	Thrombolytic agent	5
Tetanus toxoid	Formaldhyde-treated toxin obtained from *Clostridium tetani*	Tetanus vaccine	5
Asparaginase	*Erwinia chrysanthemi* or *E. coli*	Cancer (leukaemia) treatment	6
Glucose oxidase	*Aspergillus niger*	Determination of blood glucose levels	9
Alcohol dehydrogenase	*Saccaromyceces cereviseae*	Determination of blood alcohol levels	9
Various Amylases	Various bacilli, *A. oryzae*	Degradation of starch	11
Various proteases	Various bacilli & aspergilli	Degradation of proteins for food, detergent and other applications	11
Cellulases	*Trichoderma* species *A. niger*, various actinomyces	Degradation of cellulose	11

and proteolytic enzymes produced by bacilli in addition to cellulases and other activities produced by fungi such as *Trichoderma veridiae*.

In some instances the protein of interest may be intracellular. In such cases it becomes necessary to disrupt the cells upon completion of fermentation and cell harvesting. Such an approach releases not only the protein of interest, but also the entire intracellular content of the cell. This, in turn, renders more complicated the subsequent purification procedures required to obtain the final product. Specific examples of intracellular proteins of industrial significance include asparaginase (Chapter 6), penicillin acylase (Chapter 12) and glucose isomerase (Chapter 11).

Traditionally, identification of the most suitable microbial protein source involved screening a wide range of candidate microorganisms. Obviously, the existence of a simple, rapid and sensitive assay to identify the protein of interest greatly facilitates such screening activities. Phytase, for example, is an enzyme which catalyses degradation of phytic acid (*myo*-inositol hexaphosphate, Chapter 12). The incorporation of phytic acid salts (e.g. sodium or calcium phytate) into nutrient agar renders the

Figure 2.1   Large-scale bioreactor used in the production of biotechnological products (photograph courtesy of Boehringer Mannheim, UK)

agar opaque. Rapid large-scale screening of microbes for phytase production can be undertaken by culturing on the phytate-containing plates. Phytase-producing colonies will generate a clearance zone in the agar, due to phytate degradation (Figure 2.2).

Initial screens serve to identify microbial species expressing the protein of interest, with further screens pin-pointing microbial species producing the largest quantities of the protein. Frequently, organisms found to produce elevated levels of the protein of interest are then subjected to mutational studies using chemical mutagens or UV light, in an effort to isolate over producing strains. Advantageous mutations can result in product enhancement in a number of ways. A mutation in the regulatory sequence of the gene encoding the desired protein can result in increased levels of expression of the gene product. A mutational event occurring in the actual gene itself can result in an altered amino acid sequence, which in some cases, may render the protein more functionally efficient or enhances its stability.

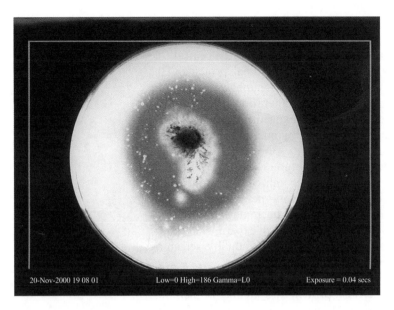

20-Nov-2000 19 08 01   Low=0 High=186 Gamma=L0   Exposure = 0.04 secs

Figure 2.2 Screening of a fungus for its ability to produce phytase by culture on agar media containing phytic acid. In this case phytase is produced as evidenced by a zone of substrate hydrolysis surrounding the fungal mycelium (refer to text for details). Photograph courtesy of Anne Casey, University of Limerick, Ireland

Microorganisms collected from many different environments are routinely screened in the hope of identifying more attractive sources of various industrially useful proteins. Few environments have as great a variety of microbial populations as fertile soil. Soil bacteria are amongst the most common group of organisms subject to routine screening. Soil bacilli (apart from the *Bacillus cereus* group) are suitable enzyme producers as they generally conform to GRAS requirements. They are also easily cultured in relatively simple media and produce a variety of industrially important enzymes extracellulary. Hundreds of different species of fungi also inhabit the soil, especially near the soil surface where aerobic conditions prevail. Such fungi are active in degrading a wide variety of biological materials present in the soil. They thrive on such material largely by secreting extracellular enzymes capable of degrading large polymeric plant molecules such as cellulose, hemicellulose and pectin, with subsequent assimilation of the liberated nutrients. Many such soil fungi thus represent attractive sources of cellulases and related hydrolytic enzymes.

Microbial sources of various industrially important proteins were identified and isolated largely by screening strategies such as those outlined above. Advances in the area of recombinant DNA technology have however, facilitated the development of an entirely new strategy with regard to the production of protein from microbial (or other) sources.

Protein production in genetically engineered microorganisms

Genetic manipulation by mutation and selection has played a central role in increasing expression levels of a myriad of microbial proteins. This approach, however, could be at best described as haphazard, with the researcher having little control over the genetic alterations achieved. The advent of recombinant DNA technology has changed this situation. This technology can be utilized in a highly directed manner to achieve specific genetic alterations. Rational improvements in source productivity can thus be achieved – a goal which mutational studies of the past realized only by chance. This is not to suggest that recombinant techniques will replace traditional mutational studies as the only method of strain improvement. Strain improvements by 'trial and error' mutational methods will continue to play a role, as large numbers of such experiments can be carried out conveniently and relatively inexpensively if suitable screening procedures are in place to detect the desired product.

Recombinant DNA technology can be used to increase the level of production of an endogenous microbial protein by a number of methods. These include: (a) introduction of additional copies of the relevant gene into the microorganism; or (b) introduction of a copy or copies of the relevant gene into the organism where control of expression has been placed under a more powerful promoter. Such strategies can result in a several-fold increase in production of the protein of interest. It must be noted, however, that failure will result if attempts to increase the levels of expression of the desired protein in any way compromise normal cellular function. Such possibilities can be minimized if the promoter used is inducible, as timing of gene expression can then be regulated.

The expression of recombinant proteins in cells in which they do not naturally occur is termed heterologous protein production. The vast majority of proteins produced by recombinant means in microorganisms are heterologous proteins. It is also possible to remove or isolate a gene/cDNA sequence derived from a particular microbial species and subsequently produce its gene product by recombinant means in that same species. This approach is termed homologous protein production. In this case the host cell is producing a recombinant protein it is naturally capable of producing anyway. However, by for example inserting multiple copies of the gene or altering the promoter, the gene product can be produced in greater quantities, or problems such as gene repression can be overcome. Specific examples of commercialized recombinant proteins produced by homologous means include the detergent lipase 'Lipomax', and the phytase enzyme 'Natuphos' (Chapter 12).

Heterologous protein production in *E. coli*

The bacterial species most commonly employed in the production of heterologous proteins is *E. coli*. This stems almost entirely from the fact

Table 2.3 Some promoters which have been used to control expression of recombinant proteins engineered in *E. coli* cells

Promoter	Method of induction	Comment
ara	Addition of arabinose to culture medium	Rapid induction and tight regulation, but repressed by glucose
cad	Adjustment of media to an acidic pH value	High expression levels achieved, cheap inducer
lac	Addition of isopropyl-β-D-thiogalactoside (IPTG) to media; High temperature	Low level expression, leaky expressin under non inducing conditions
tac	Addition of IPTG to media; High temperature	Well characterized, high-level but leaky expression
trp	Addition of indoleacrylic acid to media	Well characterized, high-level but leaky expression
T7	Addition of IPTG to media; High temperature	High-level expression but leaky expression under non-inducing conditions

that, traditionally, the study of prokaryotic genetics focused on *E. coli* as a model system. Hence, more was known about the genetic characteristics of *E. coli* than any other microorganism. Suitable plasmids were available, as were a variety of appropriate powerful and inducible promoters (Table 2.3)

In some instances, the foreign gene or cDNA coding for the protein of interest may be attached directly to a complete or partial *E. coli* gene, resulting in the production of a fusion protein. In other cases, bacterial regulatory elements, promoters and terminators are used to regulate transcription of the foreign gene. This approach is termed direct expression. Direct expression yields a protein whose amino acid sequence is identical to that of the protein obtained from its native source. Expression of a heterologous protein fused to a bacterial protein generally necessitates subsequent chemical or enzymatic cleavage in order to release the native protein.

Quite a number of therapeutically important proteins are now produced as heterologous proteins in *E. coli*. Many are in clinical use while others are undergoing clinical evaluation. The first heterologous protein to be employed clinically was human insulin produced in *E. coli*. This product was first approved for use in 1982 by the regulatory authorities in the UK, West Germany, the Netherlands and the USA. Additional examples of heterologous proteins produced in *E. coli* are listed in Table 2.4. Antibody fragments capable of binding antigen with the

Table 2.4   Some heterologous proteins produced in *E. coli*, and the levels of expression achieved

Protein	Level of expression achieved (% of total cellular protein)
Somatostatin	<0.05
Insulin A chain	20
Insulin B chain	20
Bovine growth hormone	5
Human growth hormone	5
α_1-Antitrypsin	15
Interleukin-2	10
Tumour necrosis factor	15
Interferon-β	15
Interferon-γ	25
Calf prochymosin	8

same affinity as the complete antibodies are now also routinely produced in *E. coli* expression systems. More examples are provided in various subsequent chapters of this book.

The majority of proteins synthesized naturally by *E. coli* are intracellular. Few of the cell's native proteins are exported to the periplasmic space and fewer still are exported into the extracellular medium. Thus the majority of heterologous proteins expressed in *E. coli* accumulate in the cell cytoplasm, where they can represent 25 per cent or more of total cellular protein (Table 2.4). Attempts have been made to construct recombinant protein production systems in which the foreign protein is secreted into the periplasmic space or extracellular medium. Such attempts have mainly focused on the fusion of the foreign gene with a signal peptide sequence from an endogenous *E. coli* gene, the protein product of which is normally secreted. So far such attempts have met with limited success. Extracellular secretion of recombinant protein products is considered desirable as it simplifies subsequent downstream processing.

Inclusion body formation   Extremely high levels of diverse heterologous proteins have been produced in *E. coli*. In most such cases, the resultant protein accumulates in the cell cytoplasm in the form of insoluble aggregates termed inclusion bodies (IB) or refractile bodies. It is believed that these aggregates are not derived from either native or fully unfolded forms of the protein but are composed of partially folded intermediates. Inclusion bodies may be readily viewed by dark field microscopy (typically their diameter is in the order of 1 μm and are composed predominantly of the expressed heterologous protein (contaminating proteins usually represent less than 15 per cent of inclusion body mass; major contaminants typically include RNA polymerase), *E. coli* outer membrane proteins and rRNA). It is not fully understood why recombinant proteins form inclusion bodies in *E. coli*. The most likely contributory factors include:

- very high local concentrations of the recombinant protein in the cytoplasm may lead to non-specific precipitation;

- presence of insufficient chaperones/folding enzymes with subsequent aggregation of partially folded intermediates;

- prevention of formation of disulfide linkages due to the reducing environment of the cytoplasm;

- lack of post-translational modifying enzymes may yield less stable (eukaryotic) heterologous protein products.

The formation of inclusion bodies should not be regarded as a wholly negative phenomenon. As inclusion bodies are very dense they tend to sediment readily under the influence of a low centrifugal force. This can be exploited to quickly and effectively partially purify the aggregated protein. Typically, this is achieved by a low speed centrifugation step, carried out immediately after cell homogenization. Inclusion bodies sediment more rapidly than cellular debris under the influence of a centrifugal force of the order of 500–1000 *g*. The inclusion body containing pellet can subsequently be resuspended and washed several times to reduce co-sedimenting cellular material. The carry-over of cellular debris can be problematic for several reasons:

- the presence of cell wall associated proteases/peptidases can potentially degrade the product;

- cell debris can clog (foul) chromatography columns used to subsequently purify the recombinant protein;

- the presence of lipopolysaccharides (undesirable for therapeutic proteins, Chapter 4) in the cell envelope.

After isolation of the inclusion body, denaturants are used to solubilize the aggregated polypeptides. Denaturants employed include urea, guanidinium chloride, various detergents and organic solvents, in addition to incubation under conditions of alkaline pH. Once solubilization of the inclusion body has been achieved, the denaturant is slowly removed by techniques such as dialysis, dilution or diafiltration. In this way, suitable conditions are induced which promote refolding of the protein into its native, biologically active conformation. Suitable conditions required for such renaturation can vary from protein to protein and recovery of high yields of activity is not always guaranteed. This is particularly true in the case of large proteins for which correct refolding into its active conformation is a complicated process. For this reason, many high molecular mass proteins such as antibodies and factor VIII may be produced more successfully in mammalian cell culture systems.

Several approaches may be undertaken in an effort to encourage production of soluble heterologous protein in *E. coli*. Some approaches are applicable to all protein types whereas others (e.g. inclusion of cofactors; see below) are appropriate only in certain specific instances. In general such approaches meet with only limited success. They include:

- Empirical determination of exact optimal expression system: the exact host strain, plasmid and plasmid copy number, as well as the promoter sequence used will influence the level of product expression and the propensity of product to accumulate as inclusion bodies.

- Expression of the desired product as a fusion protein: fusion to a (highly soluble) native host cytoplasmic protein can prevent appreciable inclusion body formation.

- Co-expression of chaperones can improve soluble yield of heterologous protein (see Chapter 1).

- Enhancing the endogenous production of a cofactor (or exogenously providing the cofactor) can increase soluble yields of a recombinant protein which requires the cofactor concerned.

- Growth of recombinant cells at sub optimal temperatures: a reduction of the *E. coli* growth temperature from 37 °C to 30 °C often discourages inclusion body formation.

Other characteristics of heterologous protein production in E. coli

E. coli, in theory, has the potential to become a recombinant source of virtually any protein of industrial interest. However, there still remain a number of disadvantages with regard to utilizing genetically modified *E. coli* as a source of such proteins (Table 2.5). As already discussed, one such disadvantage is inclusion body formation.

Another disadvantage is the inability of *E. coli* to perform post-translational modifications of recombinant eukaryotic proteins. *E. coli* do not have the ability to e.g. glycosylate, amidate or acetylate proteins (Chapter 1). Eukaryotic proteins normally subjected to such post-translational modifications when produced in their natural source are therefore devoid of such modifications when produced in *E. coli*. Lack of post-translational modification may or may not have a significant effect upon the biological activity of the heterologous protein. The normally glycosylated glycoproteins β-interferon and interleukin-2, when produced in *E. coli*,

Table 2.5 Some advantages and disadvantages of heterologous protein production in *E. coli*

Advantages	Disadvantages
Genetic characteristics of *E. coli* well characterized	Intracellular accumulation of recombinant protein as inclusion bodies
Suitable fermentation technology well established	Unable to undertake post-translational modification of proteins
Can generate potentially unlimited supplies of the recombinant protein	Adverse public perception of recombinant products
Economically attractive	

are biologically active despite the fact that they are not glycosylated. However, it should be stressed that glycosylation is a common feature of many eukaryotic proteins. Absence of glycosylation may, in some cases, alter the biological characteristics of the recombinant protein. This is particularly noteworthy with regard to proteins used for therapeutic purposes. Absence of a carbohydrate moiety, or other such modifications, can potentially result in modification of the biological activity of the molecule, alteration in its circulatory half-life value and/or altered immunogenicity. It may render the protein less soluble in aqueous media and/or render it more susceptible to attack by proteases.

Recombinant proteins produced in *E. coli* may also contain an extra amino acid residue (methionine) at their N-terminal end. This is due to the fact that translation in *E. coli* is always initiated by an *N*-formyl-methionine residue. Although *E. coli* have the enzymatic capability to deformylate and subsequently remove this additional amino acid, removal is less efficient with regard to processing of heterologous proteins. The presence of an extra N-terminal methionine in such recombinant proteins may alter their biological characteristics.

***Heterologous production in bacteria other than* E. coli**   A wide range of bacteria can serve as alternative expression systems to *E. coli*. In most instances they have been employed more slowly simply because historically they were genetically less well characterized than the latter. Bacteria which are listed as GRAS serve as obvious attractive alternative production systems. Various proteins have been produced (at a research level at least) in food bacteria such as *Lactococcus lactis* and *Corynebacterium glutamicum*. Recombinant protein production in various bacilli has also been undertaken. Bacilli naturally secrete various proteins in large quantities into their extracellular environment (many of which are of industrial significance; Chapters 11 and 12). Considerable experience has also accumulated with regard to their industrial scale culture. 'Lumafast' is the trade name given to a detergent lipase produced commercially as a recombinant product in a bacillus strain (Chapter 12). The main disadvantage of utilizing bacilli is their tendency to synthesize high levels of endogenous proteases which can potentially degrade the heterologous protein. The research literature also reports heterologous production of various proteins in species of *Streptomyces* (Gram-positive soil microbes). Again these represent non-pathogenic, relatively well-characterized bacteria with the capacity to secrete large quantities of protein extracellularly. However, in most cases reported recombinant expression levels were disappointingly low (of the order of mg/l fermentation media).

Heterologous protein production in yeast

Yeast cells have also become important host organisms for production of heterologous proteins. Yeasts are attractive hosts for the following reasons:

- they retain many of the advantages as outlined for *E. coli* expression systems;

- yeasts, such as *Saccharomyces cerevisiae*, are GRAS listed;

- many have played a central role in a range of traditional biotechnological processes such as brewing and baking; as a result a wealth of technical data has accumulated with regard to their fermentation and manipulation;

- the molecular biology of yeasts has been a focus of scientific research over a number of years;

- yeast cells, unlike bacteria such as *E. coli*, possess subcellular organelles and are thus capable of carrying out post-translational modifications of proteins.

Initially, the majority of heterologous proteins produced in yeast were produced in *Saccharomyces cerevisiae*. Because of its traditional industrial importance, this yeast is among the best studied of all organisms. Indeed, the first vaccine produced by recombinant DNA methods to be administered to humans (hepatitis B surface antigen; Chapter 5) was produced in this host.

Although many heterologous proteins have been successfully produced in *Saccharomyces cerevisiae* (Table 2.6) this system is also subject to a number of drawbacks (Table 2.7).

- Expression levels of heterologous proteins are often low, typically representing less than 5 per cent of total cellular protein. Such values compare unfavourably to heterologous protein production in *E. coli*.

Table 2.6 Selected proteins of therapeutic importance which have been produced (at a research level at least) by recombinant means in yeast

Therapeutic protein	Yeast-based expression system
Hepatitis B surface antigen	*Saccharomyces cerevisiae, Pichia pastoris, Hansenula polymorpha*
Influenza viral haemagglutinin	*S. cerevisiae*
Polio viral protein, VP2	*S. cerevisiae*
Insulin	*S. cerevisiae*
Human growth hormone	*S. cerevisiae*
Antibodies/antibody fragments	*S. cerevisiae*
Human nerve growth factor	*S. cerevisiae*
Human epidermal growth factor	*S. cerevisiae*
Interferon-α	*S. cerevisiae*
Interleukin-2	*S. cerevisiae*
Human factor XIII	*S. cerevisiae, Schizosaccharomyces pombe*
Human serum albumin	*S. cerevisiae, H. polymorpha, Kluyveromyces lactis*
Hirudin	*S. cerevisiae, H. polymorpha*
α$_1$-Antitrypsin	*S. cerevisiae*
Tissue plasminogen activator	*S. cerevisiae, K. lactis*
Streptokinase	*P. pastoris*
Tumour necrosis factor	*P. pastoris*
Tetanus toxin fragment C	*P. pastoris*

- Many heterologous proteins produced and secreted by *Saccharomyces cerevisiae* are not released into the culture medium, but are retained in the periplasmic space. This is especially true in the case of heterologous proteins of high molecular mass. In such cases downstream processing is rendered more complicated

- Although yeast systems have the ability to carry out post-translational modifications such as glycosylation, it has been noted that the level and type of modifications do not entirely resemble the modifications observed when animal proteins are produced by their natural source.

Table 2.7   Some advantages and disadvantages of heterologous protein production in yeast

Advantages	Disadvantages
Most are GRAS listed	Recombinant proteins usually expressed at low levels
Proven history of use in many biotechnological processes	Retention of many exported proteins in the periplasmic space
Fermentation technology is well established	Adverse public perception of products manufactured via recombinant technology
Ability to carry out post-translational modifications of recombinant proteins	Some post-translational modifications differ significantly from those achieved by animal cells

Such variations in post-translational modifications may or may not have a significant effect on the biological characteristics of the heterologous protein produced.

While *Saccharomyces cerevisiae* remains the most popular yeast host, several other yeasts are also utilized in the production of heterologous proteins of industrial interest. Popular alternatives to *Saccharomyces cerevisiae* include *Hansenula polymorpha*, *Kluyveromyces lactis*, *Pichia pastoris*, *Schizosaccharomyces pombe*, *Schwanniomyces occidentalis* and *Yarrowia lipolitica*. In some specific cases, the use of such alternative expression systems to *Saccharomyces cerevisiae* has resulted in improved glycosylation characteristics. This is, however, by no means true in all cases and the question of which yeast expression system is most appropriate to a given situation remains an open one.

Heterologous protein production in fungi

Filamentous fungi represent attractive hosts for heterologous gene expression for a number of reasons:

- they possess the enzymatic capability to carry out post-translational modifications;

- many are GRAS-listed;

- they have been extensively employed on an industrial scale for many decades in the production of a variety of enzymes as well as other primary and secondary products of metabolism (e.g. vitamins, various organic acids, antibiotics and alkaloids);

- they are capable of synthesizing and secreting large quantities of certain proteins into the extracellular medium, in marked contrast to *E. coli* or species of *Saccharomyces*;

- extracellular production of heterologous products is desirable as it simplifies subsequent product purification;

- some industrial strains of *Aspergillus niger* are known to naturally produce up to 20 g of the enzyme glucoamylase per litre of fermentation media;

- due to their industrial significance, large-scale fermentation systems for these fungi have been developed and optimized over a long time.

A wide variety of industrially and medically important proteins have now been produced as heterologous products in a number of fungal species, most notably in members of the *Aspergillus* species (Table 2.8). In many instances, the genetic information of interest has been coupled to a strong fungal promoter, for example the fungal α-amylase promoter, in an attempt to maximize protein production. While the resultant protein products with suitable signal sequences have invariably been secreted into the medium in biologically active form, the levels of expression have been somewhat disappointing. Levels ranging from a few hundred milligrams to in excess of one gram of heterologous protein per litre have been reported.

Low expression levels of biologically active recombinant (heterologous or homologous) proteins appears to be primarily a consequence of high-level natural protease expression. Most filamentous fungi produce a wide range of both intracellular and extracellular proteolytic activities. The spectrum of proteases produced by various fungi differs significantly. Both classic mutagenesis and targeted gene disruption techniques have been used to construct low-protease fungal strains. Many display protease levels of less than 5 per cent of normal values. Such strains may well prove more suitable hosts for expression of protease-sensitive recombinant proteins.

Table 2.8 Some proteins of industrial significance which have been expressed in recombinant fungal systems

Protein	Organism
Human interferons	*Aspergillus niger, A. nidulans*
Bovine chymosin	*A. niger, A. nidulans*
Aspartic proteinase (from *Rhizomucor michei*)	*A. oryzae*
Triglyceride lipase (from *Rhizomucor michei*)	*A. oryzae*
Lactoferrin	*A. oryzae, A. niger*

Limitations on the production of heterologous proteins in fungal (and other microbial) species can also potentially be due to codon usage, where specific codons for amino acids are used differentially in one species compared to the species from which the genetic information was obtained.

Human lactoferrin (hLF) is a protein of potential industrial significance which has been successfully expressed in *Aspergillus oryzae* and *A. niger*. hLF is an iron-binding glycoprotein of molecular mass, 78 kDa. Originally discovered in milk, it is also present in a variety of other external fluids such as tears and saliva. The exact biological role of hLF is subject to debate. Proposed functions include promotion of increased intestinal iron absorption and protection against microbial infection, as removal of iron would inhibit growth of iron-requiring microorganisms. Lactoferrin levels of the order of up to 1 g/l are normally associated with human milk. Bovine milk contains much lower levels of this protein. This may be one reason why breast-fed infants often perform better than their counterparts fed bovine milk-based diets. Inclusion of recombinant lactoferrin in the latter's diet may readdress this nutritional situation.

PROTEINS FROM PLANTS

Plants represent a traditional source of a wide range of biologically active molecules. Narcotics such as opium are perhaps the best known of such products. Crude opium consists of the dried milky exudate obtained from unripe capsules of certain species of plants mainly found in parts of Asia, including India and China. The most important medical constituents of opium are the alkaloids – the best known of which is morphine. Morphine is extracted and purified from crude opium preparations, generally by ion exchange chromatography.

For a number of reasons higher plants are not regarded as prolific producers of many commercially important proteins:

- Many industrially important proteins synthesized in plants are also found in other biological sources. In most cases, the alternative source becomes the source of choice for both technical and economic reasons.

- Plant growth is seasonal in nature and hence a constant source of material is not always obtainable.

- Higher plants also tend to accumulate waste substances in structures called vacuoles. Upon cell disruption these wastes, which include a number of powerful precipitating and denaturing agents, often irreversibly inactivate many plant proteins.

Despite such drawbacks, a number of industrially important proteins are obtained from plants. Two plant proteins – monellin and thaumatin –

are recognized as the sweetest-known naturally occurring substances. Such proteins have actual or potential uses in the food industry and will be reviewed later (Chapter 13). β-Amylases are also produced by many higher plants; the most widely used are obtained from barley. These enzymes play an important role in the starch-processing industry (Chapter 11). However, perhaps the best-known plant-derived protein produced on an industrial scale is still the proteolytic enzyme, papain.

Papain, also known as vegetable pepsin, is obtained from the latex of the green fruit and leaves of *Carica papaya*. It was first isolated and characterized in 1937 in the USA by A.K. Balls and colleagues. Papain is a cysteine protease (Chapter 11). Its active site contains an essential cysteine residue, which must remain in the reduced state if proteolytic activity is to be maintained. The purified enzyme exhibits broad proteolytic activity. It consists of a single polypeptide chain containing 212 amino acid residues, with a molecular mass of 23 kDa. The term papain is applied not only to the purified enzyme but also to the crude dried latex. Papain has a variety of industrial applications, the best known of which is its use as a meat tenderizing agent. During the tenderization process, the proteolytic activity of papain is directed at collagen, the major structural protein in animals, representing up to one-third of all vertebrate protein. It is the collagen present in connective tissue and blood vessels which renders meat tough (Chapter 11). Papain has a relatively high optimum temperature (65 °C) and retains activity at temperatures up to 90 °C. Because of its thermal stability, papain maintains its proteolytic activity even during the initial stages of cooking. This enzyme has also been used in other industrially important processes:

- meat tenderization;
- bating of animal skins;
- clarification of beverages;
- digestive aid;
- debriding agent (cleaning of wounds).

Ficin is another commercially available protease derived naturally from plant sources. It is generally extracted from the latex of certain tropical trees and, like papain, is a cysteine protease. It exhibits considerably higher proteolytic activity than papain and has similar industrial applications. Purified ficin has a molecular mass of approximately 25 kDa, though the term ficin is applied not only to the purified enzyme but also to the crude latex extract. Most large-scale industrial applications of papain and ficin do not require highly purified enzyme preparations. Plant enzymes, in particular those destined for application in the food processing industry, must be obtained only from non-toxic, edible plant species.

Production of heterologous proteins in plants

Advances in recombinant DNA technology facilitate genetic manipulation not only of microorganisms but also of eukaryotic cells. A number of different heterologous proteins and peptides are now produced in a variety of plants, at a research level at least. Genetic manipulation of plant systems may be undertaken for a number of reasons. Introduction of foreign genes or cDNAs may be undertaken in order to confer a novel function or ability on the resultant species. Novel DNA sequences may be introduced into plant cells by several means. These include use of *Agrobacterium* as a carrier or by direct injection of the DNA into certain plant cells. For example, using such techniques plants can be engineered to produce insecticides, which when expressed, may play a protective role. Plants may also be used to produce heterologous proteins of applied interest (Table 2.9).

Recombinant proteins produced by this strategy could be utilized in one of two ways; (a) the protein could be extracted from the plant tissue,

Table 2.9   Some proteins of industrial or medical interest which have been produced by recombinant means in plants

Protein	Original source	Expressed in	Production level achieved
α-Amylase	*Bacillus licheniformis*	Tobacco	0.3% of soluble leaf protein
Chymosin	Calf	Tobacco	0.5% of soluble protein
Cyclodextrin glycosyltransferase	*Klebsiella pneumoniae*	Potato	0.01% of soluble tuber protein
Erythropoietin	Human	Tobacco	0.003% of soluble protein
Glucoamylase	*Aspergillus niger*	Potato	Not reported
Growth hormone	Trout	Tobacco	0.1% of soluble leaf protein
Hepatitis B surface antigen	Hepatitis B virus	Tobacco	0.007% of soluble leaf protein
Hirudin	*Hirudo medicinalis* (a leech)	Canola	1.0% of seed weight
Interferon-β	Human	Tobacco	0.00002% of fresh weight
Lysozyme	Chicken	Tobacco	0.003% of leaf tissue
Phytase	*Aspergillus niger*	Tobacco	14.4% of soluble leaf protein
Serum albumin	Human	Potato	0.02% of soluble leaf protein
Xylanase	*Clostridium thermocellum*	Tobacco	4.1% of soluble leaf protein

In most cases this method of production has not been commercialized as yet. Reproduced with modifications from Kusandi, A. *et al.* (1997) Production of recombinant proteins in transgenic plants; practical considerations. *Biotechnol. Bioeng.* **56** (5), 473–483.

purified (if necessary) and then used for its applied purpose; (b) the recombinant plant tissue could be used directly as the protein source (e.g. enzymes added to animal feed which are currently produced by microbial fermentation (chapter 12) could be expressed in plant seeds, which could be fed directly to the animals).

Again recombinant protein production in plants displays both disadvantages and advantages (Table 2.10). Some of the disadvantages mirror those listed earlier in relation to direct extraction of native plant proteins for applied uses. Additional disadvantages include low expression levels, issues relating to post translational modification of proteins and lack of industrial experience with regard to extraction and purification of recombinant proteins from plants.

In most cases thus far reported recombinant protein expression levels achieved have been disappointingly low (Table 2.9). In order to approach commercial viability the recombinant protein should accumulate in the plant tissue at levels representing at least a few percent of total soluble proteins (TSP). To date in most cases expression levels achieved have fallen well below 1 per cent TSP. However, production levels achieved in the case of phytase (14.4 per cent TSP) and xylanase (4 per cent) illustrate that high expression values are attainable.

Plant-based expression systems achieve glycosylation patterns that differ (in extent and composition) to those achieved by animal cells.

Table 2.10 Summary of major advantages and disadvantages of recombinant protein production in transgenic plants

Advantages	Disadvantages
Likely economically attractive production costs	Low expression levels often reported
Ease of scale up	Glycosylation pattern achieved usually different from that observed on animal glycoproteins
Availability of established practices/ equipment for plant harvesting/storage	Lack of industrial experience or data on large-scale downstream processing of plant tissue
Elimination of downstream processing requirements if the plant material containing the recombinant protein can be used directly as the protein source	Seasonal or geographical nature of plant growth
Ability to target protein production/ accumulation to specific plant tissue	Presence of toxic substances in plant cell vacuoles
Ability to carry out post-translational modifications	Availability of established, alternative production systems

This point is important if an altered glycosylation pattern in any way negatively influences the recombinant protein product. This is especially important in the context of therapeutically important glycoproteins, where an altered glycosylation pattern could influence product safety and/or efficacy. Certain oligosaccharide epitopes commonly found on plant glycoproteins are highly immunogenic in mammals. This suggests that some mammalian glycoproteins intended for therapeutic application, if expressed in plant cells, could potentially be immunogenic.

Lack of industrial history or experience of large-scale recombinant protein production in plants is also a disadvantage from a practical standpoint. Also most companies producing proteins commercially do so using microbial or animal cell culture (or extraction from whole animal tissue). These companies have invested heavily in dedicated production equipment which would be redundant in the context of using plant based expression systems. Furthermore in the context of therapeutic proteins, regulatory authorities would have to be satisfied that plant expression systems yield safe and therapeutically effective products.

Despite such potential disadvantages a number of powerful arguments can be made in favour of plant-based recombinant expression systems. Once the transgenic plants have been generated (and are expressing the recombinant product at satisfactory levels), upstream processing costs (i.e. growth of the plants) would be low. Also scale up of production would simply involve increasing the acreage of the crop sown. Protein production would be independent of expensive facilities or production equipment and harvesting would be straightforward using pre-existing harvesting equipment. Depending upon the expression levels achieved and the variety of plant used, it has been estimated that upstream processing (i.e. growing) costs for recombinant protein synthesis in plant based systems would be 10–50 times cheaper that costs if recombinant *E. coli* was used. This would be economically significant, particularly if the protein is price sensitive and if the subsequent downstream processing required is minimal (e.g. if the harvested plant could be used directly as the recombinant protein source, without any purification). For proteins which require significant downstream processing (e.g. therapeutic proteins, which need to be purified to homogeneity), the overall savings may be insignificant. In such cases downstream processing costs would represent 80 per cent or more of total production costs. Production of the recombinant protein in plants would provide no downstream processing cost benefits over other expression systems, and in fact may be more costly.

The ability to target expression of recombinant proteins to a specific plant tissue can also be advantageous. It could reduce the potential toxicity of the protein (for the plant) and reduce environmental and regulatory concerns (for example, over release of the protein into the general environment via plant pollen). Targeted accumulation of the protein in plant seed is particularly attractive. The seeds of higher plants naturally contain high levels of storage protein. Seeds can be stored for extended periods after harvest, inexpensively and without causing protein degradation. In contrast, plant green tissue generally deteriorates rapidly after harvest. Recombinant production in green tissue would thus require immediate protein extraction after harvest, or storage or harvest under (expensive) refrigeration or frozen conditions).

Leu-enkephalin is an early example of transgenic peptide production in plants. This was achieved by inserting its DNA coding sequence into the gene coding for a seed storage protein termed 2S albumin. The family of 2S albumins are among the smallest seed storage proteins known, having a molecular mass in the order of 12 kDa. This family of proteins are derived from a group of structurally related genes – all of which exhibit both conserved and variable sequences. The variable regions vary not only in sequence but also in length. The strategy employed to produce leu-enkephalin involved substituting part of this variable sequence with a DNA sequence coding for the five amino acid neurohormone. The DNA construct was flanked on both sides by codons coding for tryptic cleavage sites. Expression of the altered 2S albumin gene resulted in production of a hybrid storage protein containing the leu-enkephalin sequence. The enkephalin was subsequently released from the altered protein by tryptic cleavage and purified by HPLC. Because of the incorporation of the tryptic cleavage sites, the purified product contained an extra lysine residue which was subsequently removed by treatment with carboxypeptidase C, a proteolytic enzyme which hydrolyses only the peptide bond at the carboxyl terminus of a peptide/polypeptide. Since then a number of larger polypeptides have also been expressed in the seeds of various plant varieties.

ANIMAL TISSUE AS A PROTEIN SOURCE

A wide variety of commercially available proteins are obtained from animal sources. This is particularly true with regard to numerous therapeutic proteins such as insulin and blood factors. Many such examples will be discussed in later chapters and thus are only briefly reviewed here. The existence of slaughterhouse facilities in which large numbers of animals are regularly processed greatly facilitates the collection of significant quantities of the particular tissue required as protein source.

Perhaps the best-known protein obtained from animal sources is insulin. This polypeptide hormone is produced in the pancreas by the beta cells of the islets of Langerhans. It is employed therapeutically in the treatment of insulin-dependent diabetes (type 1 diabetes, diabetes mellitus Chapter 7). Until the early 1980s insulin was obtained exclusively from pancreatic tissue derived from slaughterhouse cattle and pigs. The amount of insulin obtained from the pancreatic tissue of three pigs satisfies the requirements of one diabetic patient for approximately 10 days. The increasing worldwide incidence of diabetes raised fears that one day demand for insulin supplies could exceed supply from slaughterhouse sources. This is no longer of concern however, as potentially unlimited supplies of insulin can now be produced by recombinant means. Some additional examples of commercially available hormones obtained by direct extraction from animal sources are listed in Table 2.11.

Most industrially significant proteins obtained from human and other animal sources are destined for therapeutic use. One disadvantage with regard to the therapeutic utilization of such proteins from animal sources (especially human sources) relates to the potential presence of pathogens in the raw material. The large numbers of haemophiliacs who contracted acquired immune deficiency syndrome (AIDS) from human immuno deficiency virus (HIV)-infected blood transfusions stand as testament to this fact. Outbreaks of bovine spongiform encephalitis (BSE or mad cow disease) in cattle herds from various countries serve as another example. A number of precautions must thus be taken when animal tissue is used as a protein source. The most obvious precaution involves the use of tissue obtained only from disease-free animals. Downstream processing procedures employed in purifying the protein of interest must also be validated, thereby showing that the purification steps employed can eliminate pathogens which may be present in the starting material (Chapter 4).

Many pathogens, in particular viral pathogens, exhibit marked species specificity. Thus therapeutically used proteins obtained from a particular animal species are not usually administered to other animals of that same

Table 2.11 Some proteins of industrial and medical significance which have been traditionally obtained from animal sources

Protein	Source	Application
Insulin	Porcine/bovine pancreatic tissue	Treatment of type 1 diabetes
Glucagon	Porcine/bovine pancreatic tissue	Reversal of insulin induced hypoglycaemia
Follicle-stimulating hormone (FSH)	Porcine pituitary glands or	Induction of superovulation in animals
	Urine of post-menopausal women	Treatment of (human) reproductive dysfunction
Human chorionic gonadotrophin	Urine of pregnant women	Treatment of reproductive dysfunction
Erythropoietin	Urine	Treatment of anaemia
Blood factors	Human plasma	Treatment of haemophilia
Polyclonal antibodies	Human or animal blood	Various diagnostic and therapeutic applications
Chymosin (rennin)	Stomach of (unweaned) calves	Cheese manufacture

Many of the listed examples may now also be obtained via recombinant production.

species. For example, purified follicle stimulating hormone used to super-ovulate cattle is usually sourced from porcine and not bovine pituitary glands.

Heterologous protein production in transgenic animals

Over the past number of years exciting advances have been recorded in the field of transgenic animals. Initial experiments in this area concentrated on attempts to improve various animal characteristics. It was hoped that, for example, animal growth rates could be dramatically improved by genomic integration of extra copies of the growth hormone gene. Generation of such transgenic animals is normally achieved by direct microinjection of DNA into ova, although the success rate of this method is still somewhat low. One goal of such 'molecular farming' therefore is the introduction of specific functional genes into animals, thereby conferring on them desirable characteristics such as more efficient feed utilization, improved growth characteristics or generation of leaner meat.

Another goal of such transgenic technology is to confer on the transgenic animal the ability to produce large quantities of industrially important proteins. This has been achieved with varying degrees of success in mice, goats, sheep and cattle. Initial successes in expressing high levels of growth hormone in transgenic animals highlighted some potential problems associated with this technique. In many cases it was found that chronically high circulatory levels of growth hormone (significantly outside the normal physiological range) resulted in many adverse physiological effects. Elevated circulatory levels of many proteins of potential therapeutic value would also almost certainly promote similar adverse effects on normal transgenic animal metabolism. Specific animal tissues were also targeted as heterologous protein expression sites. One such tissue target is the mammary gland. By targeting expression of the foreign gene to the mammary gland the heterologous protein may be secreted directly into the milk. As such it is physically removed from the animal's circulatory system. Expression of the genetic information of interest can be targeted to this tissue by fusing the gene to the signal sequence of a milk protein. If this foreign DNA construct is successfully microinjected into an egg, the egg subsequently fertilized and implanted into a surrogate mother, and if the embryo is then brought to term and if the foreign DNA has been successfully incorporated into the newborn's chromosomal DNA and is expressed, the resultant transgenic animal is capable of producing the protein of interest and secreting that protein selectively into its milk.

The earliest success involving the production of a heterologous protein of considerable therapeutic potential was recorded in the mid-1980s, when the production of human tissue plasminogen activator (t-PA, Chapter 5) in the milk of transgenic mice was reported. This was achieved by injecting a DNA construct consisting of the promoter and upstream regulatory sequence from the mouse whey acidic protein gene to the gene coding for human tPA into mice embryos. Whey acidic protein is the most abundant whey protein found in mice milk. Biologically active tPA was recovered from the milk of the resultant transgenic mice.

High production levels of active human α_1-antitrypsin in the milk of transgenic sheep has also been achieved (Figure 2.3 and Chapter 5)

Production of pharmaceutically important proteins in the milk of suitable transgenic animals holds considerable future potential (Table 2.12). Apart from tissue plasminogen activator and α_1-antitrypsin, various other heterologous proteins have been successfully produced in milk. Such production systems have a number of potential advantages over alternative production methods such as animal cell culture (next section). Desirable features include the following.

- *High production capacities*. During a typical 5-month lactation period, one sheep can produce 2–3 l of milk per day. If the recombinant protein is expressed at a level of 1 g/l, a single sheep could produce in excess of 20 g product per week.

- *Ease of collection of source material*. This only requires the animal to be milked. Commercial automated milking systems are readily available. Such systems require only moderate design alterations as they are already designed to maximize hygienic standards during the milking process.

Figure 2.3 Photograph of 'Tracy', a transgenic sheep, and two of her lambs. Tracy expresses high levels of active human α_1-antitrypsin in her milk (photograph courtesy of Pharmaceutical Proteins Ltd, UK)

Table 2.12 Some proteins of therapeutic use which have been expressed in the milk of transgenic animals

Protein	Produced in	Use	Chapter
α_1-Antitrypsin	Sheep	Emphysema (some forms)	5
Factor IX	Sheep	Haemophilia B	5
Fibrinogen	Sheep	Blood disorders	5
Antithrombin III	Goats	Blood disorders	5
Various monoclonal antibodies	Goats	Various, including *in vivo* tumour detection	6
Human serum albumin	Cows	Plasma volume expander	5

Several proteins produced by this means are currently in clinical trials. The biochemistry and biotechnology of these proteins is considered in the chapters indicated.

- *Low capital investment requirements and low operational costs.* Traditional production methods yielding recombinant proteins require considerable expenditure on fermentation equipment. Using this technology such costs are reduced to raising and maintaining the transgenic herds.

- *Ease of production scale up.* Producer animal numbers can be expanded by breeding programmes.

A number of technical details relating to the production of therapeutically important proteins in the milk of transgenic animals remain to be optimized. Yields of heterologous proteins are often found to be extremely variable. In some cases, values of less than 1 mg/l have been recorded although much higher values are generally obtained. Mammary tissue is capable of carrying out a broad range of post-translational modifications of the proteins synthesized. Detailed characterization of the nature of such modifications, in particular in relation to glycosylation patterns, remains to be carried out.

Heterologous protein production using animal cell culture

Animal cell culture also represents an important source of several medically important proteins, virtually all of which are destined for therapeutic or diagnostic application. Monoclonal antibodies as well as various vaccines and interferons are among the best-known examples of proteins produced by such methods (Table 2.13). Increased interest in the production of proteins by cell culture has prompted an upsurge in research and development into animal cell manipulation and fermentation methods over the past two decades.

Table 2.13 Some recombinant pharmaceutical proteins approved for general medical use that are produced commercially via animal cell culture

Protein	Produced in	Medical application
Factor VIII	CHO cells, BHK cells	Haemophilia A
Factor IX	CHO cells	Haemophilia B
tPA	CHO cells	Heart attacks
FSH	CHO cells	Infertility
Erythropoietin	CHO cells	Anaemia
Interferon-β	CHO cells	Multiple sclerosis
Several monoclonal antibodies	Hybridoma cells	Various, including prevention of kidney transplant rejection and localization of tumours *in vivo*

CHO, Chinese hamster ovary; BHK, baby hamster kidney; tPA, tissue plasminogen activator; FSH, follicle-stimulating hormone.

Animal and microbial cells exhibit many basic differences in their cellular physiology and structure. Microbial cell fermentation technology has been adapted in order to promote successful culture of animal cells. Animal cells do not possess a cell wall, and thus are more susceptible to physical damage when compared to their microbial counterparts. Fermentation tanks in which animal cells are cultured usually contain agitation blades of modified design in order to reduce the damaging shear forces generated by such rotating blades. The potential physical damage caused to animal cells can be reduced still further if fermentation is conducted in air-lift reactors, in which liquid culture motion is promoted within the vessel by sparging an air–CO_2 mixture into the reactor at its base (Figure 2.4). Several other differences between animal and microbial cells also influence animal cell culture and fermentation design;

- the nutritional requirements of animal cells are more complex than those of microbial cells;

- animal cells tend to have lower oxygen requirements and grow more slowly than their microbial counterparts;

- far greater numbers of animal cells are required to effectively seed the fermentation tank.

Animal cells can be differentiated into anchorage-dependent and anchorage-independent cell types, based upon their mode of growth. Anchorage-dependent cells will grow only on a solid substratum. Such cells grow in monolayer fashion and exhibit contact inhibition. Anchorage-

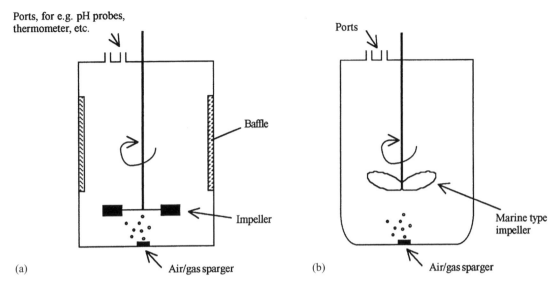

Figure 2.4 Design of a generalized microbial cell fermentation vessel (a), and an animal cell bioreactor (b). Animal cell bioreactors display several structural differences as compared to microbial fermentation vessels. Note in particular (i) the use of a marine type impeller (some animal cell bioreactors – air-lift fermenters – are devoid of impellers, and use sparging of air–gas as the only means of media agitation); (ii) the absence of baffles and; (iii) curved internal surfaces at the bioreactor base. These modifications aim to minimize damage to the fragile animal cells during culture. Note that various additional bioreactor configurations are also commercially available

independent cells, on the other hand, do not require a solid support for growth. Most established cell lines, in addition to transformed cell lines, fall into this latter category. Although both cell types may be grown in submerged cell culture fermenters, anchorage-dependent cells must be attached to suitable carrier beads.

Monoclonal antibodies are among the most notable protein type produced by animal cell culture. Because of their biological attributes these antibodies have numerous diagnostic, therapeutic and preparative uses. Their large-scale production is now a well established industrial process. Culture of hybridomas is still the most common method of monoclonal antibody production. However, many other mammalian cell lines also have the ability to successfully assemble and secrete antibodies, and recombinant expression of antibody genes in non-lymphoid mammalian cell lines, such as Chinese hamster ovary (CHO) cells, has been successfully achieved. Highly efficient mammalian cell expression systems can thus be used to produce numerous heterologous proteins including highly complex antibodies.

Various interferons have also been produced on an industrial scale by cell culture methods (Chapter 8). The potential therapeutic application of interferons as antiviral and anticancer agents were appreciated shortly after their initial discovery in the late 1950s. Their clinical application was limited for many years as a result of inadequate supplies. Until the early

1980s all of the interferon employed clinically was obtained directly from human leukocytes. Such supplies contained approximately 1 per cent interferon and were extremely limited. By the early 1980s, developments in cell culture techniques facilitated the production of large quantities of essentially pure interferon. Non-recombinant interferon-α, for example, is produced industrially by cell culture techniques in 8000 l bioreactors using a human lymphoblastoid cell line termed Namalwa. Recombinant interferon is also produced by fermentation of bacterial and mammalian cells.

As mentioned earlier, many animal cell lines exhibit relatively complex nutritional requirements. Basal tissue culture media typically contain a carbohydrate source, as well as various vitamins, mineral salts and amino acids. Traditionally, various other supplements have also been added to the culture media. Such supplements include antibiotics (to prevent growth of any contaminant microorganisms) and serum. Serum is added as a source of non-defined essential nutrients. Fetal calf serum was often found to be the most efficient serum type. It is usually added to the culture medium to a final concentration range of 0.5–25 per cent. The use of fetal calf serum in large-scale animal cell culture systems would be prohibitively expensive. It was often replaced in the past by serum obtained from new-born calves. More recent trends however, involve the substitution of the serum by more defined substances such as a mixture of bovine serum albumin, insulin and transferrin.

Although sometimes necessary, the addition of serum or substances derived from serum to animal cell culture media is undesirable for a number of reasons. The addition of such a complex mixture of proteins and other biomolecules can only render subsequent purification more complex. Addition of serum also risks contamination of the culture medium with blood borne pathogens.

Insect cell culture systems

A number of recombinant proteins have been produced at a research level by insect cell lines in culture. Expression of heterologous proteins in such cell lines often involves the use of a baculovirus expression system. This virus has the ability to infect and replicate in a number of established insect cell lines with the resultant production of high levels of a virally encoded structural protein termed polyhedrin. Polyhedrin synthesis can constitute up to 50 per cent of total proteins produced by the infected host insect cell. The strategy most often employed in producing heterologous proteins using this system involves placing the gene coding for the protein of interest under the influence of the polyhedrin promoter. Thus far, expression levels of heterologous proteins achieved using this system have been highly variable, generally falling within the range of 1–600 mg protein per litre of culture medium. Infection by baculovirus seriously compromises insect cell secretory pathways. As such most heterologous proteins expressed at high levels are intracellular. Attempts to overcome this problem has resulted in the development of plasmid-based expression systems. The use of efficient promoters and the ability to generate high copy numbers of the introduced plasmid allows high production levels of expressed proteins (100s of mg/l). Insect cells are relatively cheap and easy to culture and maintain (in comparison to animal cells). They are also capable of carrying out post-translational processing. As such, insect-cell-based systems will be of continued interest, although they are not used to any appreciable extent in industry as yet.

DIRECT CHEMICAL SYNTHESIS

All proteins used to date for medical or industrial purposes are synthe-sized in biological systems, be they recombinant or non-recombinant. The desired protein is then extracted from the biological source material, and purified if necessary. A range of peptides also find (mainly medical) use (Table 2.14). Peptides are usually produced naturally at extremely low levels, which renders difficult or impossible their large-scale preparation for applied purposes. For example, during the initial characterization of the peptide hypothalmic factor, tyrotrophin-releasing hormone, almost 4000 kg of brain tissue yielded only 1 mg of pure product. Peptides can however be manufactured economically and in bulk by direct chemical synthesis. The most commonly used method of synthesis, (Merrifield solid phase synthesis) is outlined in Box 2.1.

Table 2.14 Some peptides of potential and actual therapeutic use

Peptide	No. amino acids	Function/biological activity
Luteinizing hormone releasing hormone (LHRH)	10	Regulates synthesis/release of gonadotrophic hormones, thereby regulating fertility
Oxytocin	9	Promotes contraction of unterine muscle (important in childbirth/inducing labour)
Vasopressin	9	Promotes water reabsorption from the kidney
Bradykinin	9	Inhibits inflammation
Enkephalins	5	Pain relief
Angiotensin II	8	Regulates blood pressure
Peptide antibiotics	Various	Anti-microbial activity

Box 2.1   The chemical synthesis of peptides

A number of approaches may be adopted to achieve chemical synthesis of a peptide. The Merrifield solid phase synthesis method is perhaps the most widely used. This entails sequential addition of amino acids to a growing peptide chain anchored to the surface of modified polystyrene beads. The modified beads contain reactive chloromethyl ($-CH_2Cl$) groups.

The individual amino acid building blocks are first incubated with di-*tert*-butyl dicarbonate, thus forming a *tert*-butoxycarbonyl amide (BOC) amino acid derivative.

Amino acid   Di-*tert*-butyl dicarbonate

BOC – amino acid

The BOC group protects the amino acid amino group, thus ensuring that addition of each new amino acid to the growing peptide chain occurs via the amino acids carboxyl group (Step 1). Coupling of the first amino acid to the bead is achieved under alkaline conditions. Subsequent treatment with trifluoroacetic acid removes the BOC group (Step 2). Next, a second BOC-protected amino acid is added, along with a coupling reagent (DCC, Step 3). This promotes peptide bond formation. Additional amino acids are coupled by repeating this cycle of events. Note that the peptide is grown from its carboxy terminal end. After the required peptide is synthesized (a dipeptide in the example below), it is released from the bead by treatment with anhydrous hydrogen fluoride. Automated computer-controlled peptide synthesizers are available commercially, rendering routine peptide manufacture for pharmaceutical or other purposes.

Box Figure 2.1

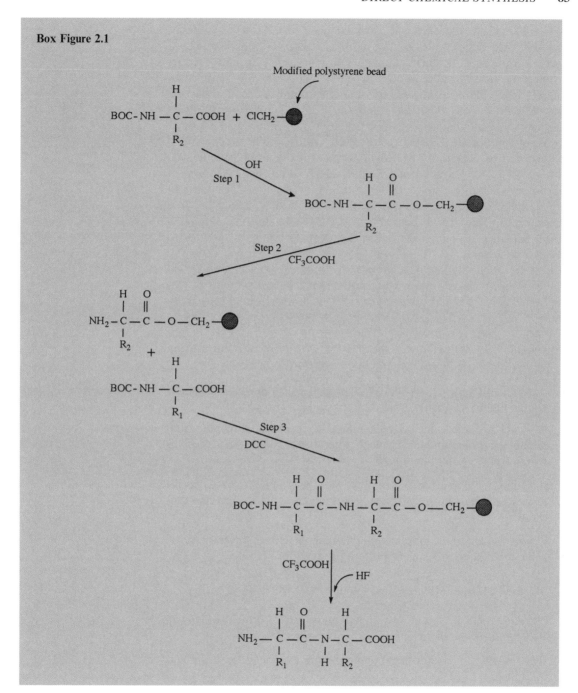

The Merrifield approach can now be used to routinely synthesize peptides/small polypeptides containing up to about 60 amino acid residues. Technical difficulties/economic considerations currently render problematic direct synthesis of polypeptides considerably larger than this. However, more recently an approach termed native chemical ligation has been developed which facilitates the direct chemical synthesis of polypeptides containing up to approximately 120 amino acid residues. By use of appropriate reagents, two peptides synthesized by the Merrifield method can be chemically ligated under aqueous conditions at neutral pH, yielding a single polypeptide chain. This method has been used at laboratory scale to synthesize up to 200 different proteins. X-ray diffraction and NMR analysis of several such proteins illustrated that they fold spontaneously into the expected (native) three-dimensional conformation, and such proteins usually display identical biological activity to the native molecule.

It is not clear as yet if economic or technical considerations will render attractive large scale synthesis of commercialized proteins by this method. Some potential advantages associated with this approach include:

- it would essentially eliminate the possibility of accidental transmission of disease via pathogen contaminated protein products;

- it would provide an alternative means of synthesizing proteins of altered amino acid sequence;

- it would provide a means of synthesizing proteins with unnatural amino acids/D-amino acids; such alterations could yield proteins with novel or useful functions;

- it would avoid the use of complex biological systems or extensive and expensive upstream processing equipment to achieve protein synthesis;

- it would likely prove more acceptable than genetic engineering to the general public as a means of protein production.

However, the approach does suffer from some drawbacks:

- currently there are limitations on maximum size of protein which can be manufactured via this approach;

- there is little or no current industrial-scale experience in using this approach;

- the economics of large-scale production are unclear;

- the inability of the approach to incorporate post-translational modifications associated with many native proteins.

CONCLUSION

Proteins of industrial interest have been traditionally obtained from a wide range of sources. The advent of recombinant DNA technology has rendered it technically feasible to produce virtually any protein of interest in a number of different organisms. However, it must be borne in mind that technical feasibility does not guarantee economic success.

Many microbial strains have been identified that are quite efficient at naturally producing a variety of industrially important proteins. These strains have been identified by employing standard screening selection and mutational techniques This is particularly true with regard to organisms producing enzymes such as amylases and proteases in large quantities.

Increased understanding of the factors governing gene expression and protein synthesis confers upon the scientific community the ability to render even more efficient protein production from most sources. Perhaps the greatest industrial impact of such genetic manipulation relates to the production of high value proteins for therapeutic and diagnostic purposes in novel hosts. Recombinant DNA technology now allows large quantities of such proteins to be produced, most notably in microbial production systems.

FURTHER READING

Books

Alcamo, E. (1999) *DNA Technology*. Academic Press Inc., London.

Bernd, G. (1995) *Peptides*. Academic Press, London.

Bishop, J. (1999) *Transgenic Mammals*. Longman, Harlow.

Butler, M. (1996) *Animal Cell Culture and Technology*. IRL Press, Oxford.

Dixon, R. (1995). *Plant Cell Culture*. IRL Press, Oxford.

Epton, R. (1992) *Innovation and Perspectives in Solid Phase Synthesis and Related Technology*. Intercept, Andover.

Freshney, R. (2000) *Animal Cell Culture*. Oxford University Press, Oxford.

Kung, S. (1992) *Transgenic Plants*. Academic Press Inc., London.

Mather, J. (1998) *Animal Cell Culture Methods*. Academic Press Inc., London.

Morgan, S. (1997) *Genetic Engineering*. Evans Brothers, London.

Russo, V. (1998) *Genetic Engineering*. Oxford University Press, Oxford.

Spier, R. (1994) *Animal Cell Biotechnology*. Academic Press Inc., London.

Stanburg, P. (1995) *Principles of Fermentation Technology*. Pergamon Press, Oxford.

Articles

Bacterial protein sources

Baneyx, F. (1999) Recombinant protein expression in *E. coli*. *Curr. Opin. Biotechnol.* **10** (5), 411–421.

Billman-Jacobe, H. (1996) Expression in bacteria other than *E. coli. Curr. Opin. Biotechnol.* **7**, 500–504.

LaVallie, E. *et al.* (1993). A thioredoxin gene fusion expression system that circumvents inclusion body formation in the *E. coli* cytoplasm. *Bio/Technology* **11**, 187–193.

Middelberg, A. (1996) Large-scale recovery of recombinant protein inclusion bodies expressed in *E. coli. J. Microbiol. Biotechnol.* **6** (4), 225–231.

Schein, C.H. (1989). Production of soluble recombinant proteins in bacteria. *Bio/Technology* **7** (11), 1141–1147.

Weickert, M. *et al.* (1996) Optimization of heterologous protein production in *E. coli. Curr. Opin. Biotechnol.* **7**, 494–499.

Yeast/fungal protein sources

Binnie, C. *et al.* (1997) Heterologous biopharmaceutical expression in *Streptomyces. Trends Biotechnol* **15**, 315–319.

Buckholz, R. & Gleeson, M. (1991) Yeast systems for the commercial production of heterologous proteins. *Bio/Technology* **9**, 1067–1071.

Cregg, J. *et al.* (2000) Recombinant protein expression in *Pichia pastoris. Mol. Biotechnol.* **16** (1), 23–52.

Cullen, D. *et al.* (1987). Controlled expressed and secretion of bovine chymosin in *Aspergillus nidulans. Bio/Technology* **5**, 369–375.

Kingsman, S. *et al.* (1985) Heterologous gene expression in *Saccharomyces cerevisiae. In Biotechnology and Genetic Engineering Reviews*, vol 3, pp. 377–416, G.E. Russell, ed. Intercept, Andover.

Rallabhandi, P. & Pak-Lam, Y (1996) Production of therapeutic proteins in yeasts, a review. *Aus Biotechnol.* **6** (4), 230–237.

Smith, R.A. *et al.* (1985) Heterologous protein secretion from yeast. *Science* **229**, 1219–1224.

Sudbery, P. (1996) The expression of recombinant proteins in yeasts. *Curr. Opin. Biotechnol.* **7**, 517–524.

Van Brunt, J. (1986) Fungi: the perfect hosts? *Bio/Technology*, **4**, 1057–1062.

Van den Hombergh, J. *et al.* (1997) Aspergillus as a host for heterologous protein production: the problem of proteases. *Trends Biotechnol.* **15**, 256–263.

Ward, P. *et al.* (1992) Production of biologically active recombinant human lactoferrin in *Aspergillus oryzae. Bio/Technology* **10**, 784–789.

Plant protein sources

Balls, A.K. *et al.* (1937) Crystalline papain. *Science* **86**, 379.

Fischer, R. *et al.* (1999a) Molecular farming of recombinant antibodies in plants. *Biol. Chem.* **380**, 825–839.

Fischer, R. *et al.* (1999b). Towards molecular farming in the future: using plant cell suspension cultures as bioreactors. *Biotechnol. App. Biochem.* **30**, 109–112.

Krebbers, E. and Vandekerckhove, J. (1990) Production of peptides in plant seeds. *Trends Biotechnol.* **8**, 1–3.

Kusnadi, A. *et al.* (1997) Production of recombinant proteins in transgenic plants: practical considerations. *Biotechnol. Bioeng.* **56**, 473–484.

Moffat, A., (1992) High-tech plants promise a bumper crop of new products. *Science* **256**, 770–771.

Pen, J. *et al.* (1993) Phytase-containing transgenic seeds as a novel feed additive for improved phosphorus utilization. *Bio/Technology* **11**, 811–814.

Animal/insect protein sources

Datar, R. *et al.* (1993) Process economics of animal cell and bacterial fermentations: a case study analysis of tissue plasminogen activator. *Bio/Technology* **11**, 340–357.

Finter, N.B. *et al.* (1984) Mass human cell culture as a source of interferons. *Lab. Technol.* **March–April** 57.

Fussenegger, M. *et al.* (1999) Genetic optimization of recombinant glycoprotein production by mamallian cells. *Trends Biotechnol.* **17** (1), 35–42.

Gordon, K. *et al.* (1987) Production of human tissue plasminogen activator in transgenic mouse milk. *Bio/Technology* **5** (11), 1183–1187.

Houdebine, L. (2000) Transgenic animal bioreactors. *Transgen. Res.* **9**, 305–320.

Hu, W.-S. & Aunins, J. (1997) Large scale mammalian cell culture, *Curr. Opin. Biotechnol.* **8**, 148–153.

Hu, W.S. Peshwa, M.V., (1993) Mammalian cells for pharmaceutical manufacturing. *Am Soc. Microbiol. News* **59**, 65–68.

Jones, I. & Morikawa, Y. (1996) Baculovirus vectors for expression in insect cells. *Curr. Opin. Biotechnol.* **7**, 512–516.

Mc Carroll, L. & King, L. (1997) Stable insect cell cultures for recombinant protein production. *Curr. Opin. Biotechnol.* **8**, 590–594.

Pfeifer, T. (1998) Expression of heterologous proteins in stable insect cell culture. *Curr. Opin. Biotechnol.* **9**, 518–521.

Rhodes, M. & Birch, J. (1988) Large-scale production of proteins from mammalian cells. *Bio/Technology* **6**, 518–523.

Van Brunt, J. (1988) Molecular farming: transgenic animals as bioreactors. *Bio/Technology* **6** (10), 1149–1154.

Weaver, J.F. *et al.* (1988) Production of recombinant human CSF-1 in an indusible mammalian expression system. *Bio/Technology* **6** (3), 287–290.

Wright, G. *et al.* (1991) High level expression of active human alpha-1-antitrypsin in the milk of transgenic sheep. *Bio/Technology* **9** (9), 30–34.

Chemical synthesis

Wilken, J. & Kent, S. (1998) Chemical protein synthesis. *Curr. Opin. Biotechnol.* **9**, 412–426

3

Protein purification and characterization

INTRODUCTION

As outlined in the previous chapter, proteins can be obtained from a wide variety of sources. Despite such diversity of origin, proteins (both native and recombinant) derived from any such source are usually purified using a similar overall approach and similar techniques (Figure 3.1). The exact details of the purification scheme for any given protein will depend upon a number of factors, including:

- exact source material chosen and location of the target protein (extracellular or intracellular);

- level of expression of target protein;

- physicochemical characteristics of the protein;

- purpose of purification.

The source material chosen will dictate the range and type of contaminants present in the starting material. If the protein is an intracellular one the exact method of cell disruption chosen will depend upon the nature of the source material (described later). Extracellular versus intracellular production will also influence the initial stages of the purification strategy applied. The level of expression of the target protein will also bear upon the purification protocol. Low-level expression for example, may necessitate extraction of very large quantities of source material, which subsequently requires significant concentration before chromatographic purification. On the other hand, high-level expression may render an initial concentration step redundant. The physicochemical properties of the protein will have an obvious influence upon the purification protocol. Proteins are fractionated from each other on the basis of differences in physicochemical

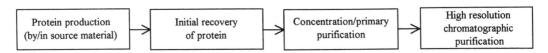

Figure 3.1 Generalized approach used to extensively purify a protein. Variations to this approach may also be pursued

characteristics such as solubility, size, surface hydrophobicity and charge. In the case of some recombinant proteins tags are attached, via means of protein engineering, in order to confer some very pronounced physicochemical characteristic on the protein. This renders its separation from contaminants more straightforward, as discussed later in this chapter.

An additional factor which significantly influences the purification protocol is the purpose for which the protein is being purified. If this is purely academic, purification to homogeneity is usually the most important goal, with issues such as the number of steps required, duration of procedure, cost and percentage yield of final product being of secondary importance. If the protein is being purified for a commercial application economic as well as technical factors will be of concern, and the protein will generally be purified only to the minimal level required (Chapter 4). The quantity of protein or level of purification required will also influence the protocol developed. Again this relates directly to the purpose of purification. Academic purposes usually require at most several hundred milligrams to 1 g of purified protein, in order to support functional and structural studies. Proteins destined for applied purposes on the other hand are usually required in much greater quantities (Table 3.1), and the level of purity demanded will vary according to application.

Table 3.1 Typical quantities of any given protein required to meet the indicated academic or applied purposes

Purpose	Quantity of protein required
Mass spectrometry studies	ng–µg levels
Protein sequencing/electrophoretic analysis	µg levels
General functional/physicochemical studies	mg–g levels
Specialized commercial proteins used for research-related applications	100s mg–100 g (moderately purified)
Commercial enzymes/antibodies used for *in vitro* diagnostic purposes	g–kg levels (moderately/highly purified)
Proteins used for *in vivo* therapeutic purposes; treatment of low-incident diseases (e.g. Gauchers disease)	100 g–10 kg levels (highly purified)
Proteins used for *in vivo* therapeutic purposes; treatment of high-incident diseases (e.g. diabetes, haemophilia)	10 kg–100s kg levels (highly purified)
Bulk industrial enzymes	> 1000 kg levels (usually un-purified)
Milk-derived proteins produced by/for bulk food applications	> 100 000 kg levels. (purity varies with production method)

Protein detection and quantification

The ability to detect and quantify (a) total protein concentration and (b) target protein levels is an essential preresiquite to the purification and characterization of any protein. Various methods may be used to determine protein concentration in a biological sample. The most commonly employed such methods, along with their mode of action, advantages and disadvantages are summarized in Table 3.2. Choice of assay method is made on the basis of assay sensitivity required, presence of interfering agents in the sample and, to a certain extent, personal preference. Although it is difficult to generalize, absorbance at 280 nm and the Bradford method are probably the most commonly employed.

Detection and quantification specifically of the protein of interest may be undertaken by one of two general methods: bioassay or immunoassay. As the name implies, bioassay involves monitoring the biological activity of the target analyte (protein). It is defined as the quantitative measurement of a response following the application of a stimulus (i.e. the target protein) to a biological system. The 'biological system' is most commonly microbial, animal or plant cells or tissue, or whole animals or plants, as appropriate. Some examples of specific bioassays are provided in Table 3.3.

Bioassays are generally comparative; in addition to the test sample, a 'standard' (i.e. a sample containing a known quantity of target analyte) is required. The standard will have been assigned a specific number of units of biological activity per mg, and such standards are often commercially available. Direct comparison of the magnitude of the biological response induced by the test sample to that induced by the standard allows assignment of activity levels present in the former.

Bioassays directly detect and quantify the true biological activity of the target protein. However, assay procedures can be costly, and of prolonged duration (hours – days). Furthermore, as the response measured is generated by a biological system, high inherent imprecision in the results is often a feature. The bioassay should of course also measure a biological response, which is triggered uniquely by the target protein.

Enzymes represent a key family of proteins, which continue to be a focus of scientific interest. Assay of enzyme activity is usually achieved by incubating the catalytic protein with its substrate for a specific period of time under defined environmental conditions (of temperature and pH). Catalytic activity is calculated by monitoring either the rate of substrate consumption or product generation. Two (related) units of enzyme activity are most commonly used;

- One unit (U) of enzyme activity is the amount of that enzyme which will catalyse the transformation of 1 μmol of substrate per minute (or where more than one bond of each substrate molecule is attached, one

Table 3.2 The various methods most commonly used to detect and quantify protein levels in a biological sample

Method	Sensitivity	Mode of action	Advantages	Disadvantages
Absorbance at 280 nm (UV method)	0.02–3.0 mg/ml	R-groups of aromatic amino acids (tyrosine, tryptophan and, to a lesser extent, phenylalanine) absorb at 280 nm	Simple and fast, non-destructive to sample	Low sensitivity, identical concentrations of different proteins will yield different absorbance values (reflection of different aromatic amino acid content)
Absorbance at 205 nm (far UV method)	1.0–100 µg/ml	Peptide bond absorbs at this wavelength	Simple, fast, sensitive, non-destructive to sample	Not all UV–Vis spectrophotometers are capable of measuring absorbance at 205 nm. Most commonly used buffers also absorbs at this wavelength, although if used at low strength (>10 mM) this may not cause a problem.
Acid digestion (ninhydrin method)	20–50 µg	Protein hydrolysed to constituent amino acids by incubation with H_2SO_4 at 100°C. Amino acid content quantified by subsequent reaction with ninhydrin, forming a derivitive which absorbs at 570 nm	Sensitive. Phenotic/aromatic compounds do not interfere with assay	Uses hazardous chemicals (H_2SO_4 at 100°C), extended total assay times (up to 20 h)
Bicinchonic acid method	20–100 µg/ml	Copper containing reagent which, when reduced by protein, reacts with bicinchonic acid yielding a derivitive which absorbs maximally at 562 nm	Sensitive and convenient, less susceptible to assay interference than some other methods	Cost: expensive reagent. Departure from exact assay protocol (e.g. duration, temperature, reagent concentration) can result in variable results
Biuret method	1–10 mg/ml	Copper containing reagent, which reacts quantitively with protein, yielding a product which absorbs maximally at 550 nm	Reagents are cheap and easy to prepare. Assay less susceptible to interference than some other methods	Poor sensitivity

continues overleaf

Table 3.2 (continued)

Method	Sensitivity	Mode of action	Advantages	Disadvantages
Bradford method	150–750 µg/ml	Based upon reversible, pH dependant binding of coomassie brilliant blue G-250 dye to protein. Absorbance max. at 595 nm	Moderately sensitive, easy and fast to run	Alkaline pH/buffers will interfere with assay. Subject to interference by detergents (e.g. SDS)
Dry weight method	Minimum of 2–4 mg/sample	Heat protein-containing sample to 106° for 4–6 h (min.), weigh dry content	Straightforward method. No chemical reagents required	Poor sensitivity. Sample must contain no non-volatile, non-protein substances
Fluorescence emission method	5–50 µg/ml	Based on fluorescence properties of aromatic amino acid residues in the protein	Sensitive, simple, fast, non destructive to sample	Identical concentrations of different proteins can yield different absorbance readings, due to varying aromatic amino acid content.
Hartree – Lowry method	30–150 µg/ml	Combination of copper and phosphomolybdic/phophotungstic acid reacts quantitatively with proteins, displaying an absorbency max. at 750 nm	Good sensitivity	Assay somewhat laborious. Subject to interference by detergents and chelating agents.
Silver binding procedure	150ng–20 µg/ml	Based on binding of silver ions to proteins	Extremely sensitive	Chelating agents, detergents and reducing agents interferes with assay.
Trichloroacetic acid precipitation method	Varies, depending upon detection method used	Entails addition of TCA to protein sample in order to precipitate it, followed by resuspension of the protein and assay by one of the above methods.	Used to remove (Non TCA precipitable) interfering agents from a protein sample before quantification	Laborious.

micro equilivant of the group concerned per minute) under the chosen assay conditions. A variation of this definition refers to one unit of activity as being the amount of enzyme required to produce 1 μmol of product / minute under the assay conditions.

- The Katal (kat) represents the SI unit of enzyme activity. One Katal is defined as that catalytic activity which will raise the rate of reaction by one mole per second in a specified assay system.

Enzyme assays directly detect and quantify the biological activity of these catalytic proteins. However, such assays are not considered to be true bioassays, as an enzyme's substrate is not considered a 'biological system'.

Immunoassays (Chapter 9) may also be used to detect and quantify a protein in solution. In this instance antibodies raised against the protein of interest must be available. Immunoassays display a number of positive characteristics, including a high degree of specificity and sensitivity, short assay duration and conduciveness to automation. Their major disadvantage is that they do not measure the protein's actual biological activity. Modifications to the protein (e.g. partial proteolysis or partial denaturation) which may decrease or abolish its biological activity, will not effect the immunoassay result if the modification does not destroy the three-dimensional structure of the specific part of the protein to which the antibodies bind. In most instances therefore it is preferential to detect and quantify a protein by direct measurement of its biological activity. If technical or economic considerations render this approach unattractive for routine use, progression of protein purification could be followed by immunoassay. A bioassay could then be carried out on at least the final purified product, and if possible also after particularly crucial purification steps.

Table 3.3 Example bioassays designed to detect and quantify the indicated proteins

Protein	Bioassay description
Interleukin-2	Ability to promote the proliferation of activated T-lymphocytes
Interferon-α	Ability to inhibit the cytopathic effect of certain viruses (e.g. vesicular stomatitis virus on certain human cell lines
Granulocyte colony-stimulating factor	Ability to promote proliferation of certain animal cell lines
Tumour necrosis factor	Ability to induce cytotoxic effect upon certain animal cell lines (e.g. murine fibroblast cell ines)
Erythropoietin	Ability to stimulate the proliferation of the TF-1 animal cell line
Polypeptide antibiotics	Ability to prevent growth of various indicator microorganisms

INITIAL RECOVERY OF PROTEIN

The initial step of any purification procedure involves recovery of the protein from its source. The complexity of this step depends largely upon whether the protein of interest is intracellular or extracellular. Many proteins produced by fermentation of microorganisms or by animal cell culture are secreted into the media. Initial product recovery in such cases involves the separation of the whole cells from the fermentation media by filtration or centrifugation. The protein of interest is present in the cell-free medium, often in very dilute form.

In the case of intracellular microbial proteins, cell harvesting from the culture medium is followed by resuspension of the cells in buffer or water with subsequent cell disruption. Resuspension of microbial cell pastes can often be achieved by simple stirring, although in some cases a more vigorous approach using mechanical mixing is required. Such cell pastes are resuspended in much smaller volumes than the original volume of fermentation broth from which they were prepared. Reduced volumes facilitate more efficient handling during subsequent purification steps.

Most proteins obtained from animal or plant tissue are intracellular in nature. The initial step in processing such material obviously involves collection of the appropriate tissue required. Specific examples include the collection of pituitary glands from which hormones such as follicle-stimulating hormone (FSH) and luteinizing hormone (LH) may be purified, collection of blood from which various blood proteins are obtained, or collection of internal organs such as liver and kidneys from which various enzymes and other proteins of interest may be obtained. If purification is not scheduled to begin directly after harvest of source material, the material can usually be stored frozen.

Cell disruption

If the protein required is an intracellular one, collection of the source cells or tissue is followed by their disruption. Most mammalian cells and tissues are relatively easily disrupted. Animal cells, unlike their bacterial or plant counterparts, are devoid of a protective cell wall. Most techniques rely on physical disruption of the cell membrane. A well-known example is that of the Potter homogenizer. Disruption is achieved in this case by shear forces generated between a rotating pestle and the inside wall of a test tube-like container.

Additional methods by which some animal cell types can be disrupted include osmotic shock and freeze – thaw cycles. Disruption via osmotic shock entails placing the cells in a buffer of high osmotic pressure (e.g. a buffer containing 15–25 per cent sucrose). This results in migration of water from inside the cell into the sucrose medium. After a suitable time (usually 1 h or less) the cells are then transferred into a weak buffer or distilled water (i.e. a solution of low osmotic pressure). As a result water floods back into the cell, often rupturing the plasma membrane. As the name suggests, freeze – thaw cycles entails freezing the cells, followed by thawing. This cycle is repeated if necessary. Cell rupture is usually prompted by damage to the plasma membrane as a result of intracellular ice crystal growth during the freezing process. This method is not widely used.

Efficient homogenization of plant tissue is more challenging, mainly because of the presence of the outer cell wall. Disruption is often achieved by physical means, in which the plant material is subjected to homogenization by rapidly rotating blades. The Waring blender (which is somewhat similar in design to a domestic food blender) is often used. This approach can also be used to disrupt animal tissue

Microbial cell disruption

Disruption of microbial cells is also rendered difficult due to the presence of the microbial cell wall. Despite this, a number of very efficient systems exist which are capable of disrupting large quantities of microbial biomass (Table 3.4). Disruption techniques, such as sonication or treatment with the enzyme lysozyme, are usually confined to laboratory-scale operations, due either to inadequate equipment or on economic grounds. Large-scale cell disruption by chemical means has been employed successfully in some instances. Chemicals utilized include a variety of detergents and antibiotics, solvents such as toluene or acetone, and treatment under alkaline conditions or with chaotropic agents such as urea or guanidine.

Protein extraction procedures employing detergents are effective in many instances, but suffer from a number of drawbacks. The mode of

Table 3.4 Some chemical, physical and enzyme-based techniques that may be employed to achieve microbial cell disruption

---

Treatment with chemicals:
    detergents
    antibiotics
    solvents (e.g. toluene, acetone)
    chaotropic agents (e.g. urea, guanidine)

Exposure to alkaline conditions

Sonication

Homogenization

Agitation in the presence of abrasives (usually glass beads)

Treatment with lysozyme

---

detergent action primarily involves solubilization of the cell's membrane. Ionic detergents such as sodium lauryl sulfate are more efficient than non-ionic detergents such as polysorbates. The major disadvantage associated with the utilization of any detergent system is that they often induce protein denaturation and precipitation. This obviously limits their usefulness. Many other chemicals, including various solvents or incubation under alkaline conditions, also suffer from this disadvantage. Furthermore, even if the chemicals employed do not adversely affect the protein, their presence may adversely affect a subsequent purification step (e.g. the presence of detergent can prevent proteins from binding to a hydrophobic interaction column). In addition, the presence of such materials in the final preparation, even in trace quantities, may be unacceptable for a number of reasons. Detergent-based cell disruption systems have been successfully employed in a number of specific cases. Triton for example has been used to render *Nocardia* cells permeable in the large-scale extraction and purification of cholesterol oxidase (Chapter 9).

Disruption of microbial cells (and some animal or plant tissue types) is most often achieved by mechanical methods such as homogenization, or by vigorous agitation with abrasives. During the homogenization process a cell suspension is forced through an orifice of very narrow internal diameter at extremely high pressures. This generates very high shear forces. As the microbial suspension passes through the outlet point, it experiences an almost instantaneous drop in pressure to normal atmospheric pressure. The high shear forces and subsequent rapid pressure drop act as very effective cellular disruption forces, and result in the rupture of most microbial cell types (Figure 3.2). In most cases a single pass through the homogenizer results in adequate cell breakage, but it is also possible to recirculate the material through the system for a second or third pass.

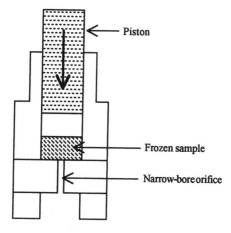

Figure 3.2   Diagrammatic representation of a cell homogenizer. This represents one of a number of instruments routinely used to rupture microbial cells, and in some cases animal or plant tissue

Although prototype homogenizers were developed and used as far back as the 1950s, many design improvements have since been incorporated. An efficient cooling system minimizes protein denaturation (denaturation would otherwise occur due to the considerable amount of heat generated during the homogenization process). Homogenizers capable of handling large quantities of cellular suspensions are now available, many of which can efficiently process several thousand litres per hour.

An additional method often employed to achieve microbial cell disruption, both at laboratory level and on an industrial scale, involves cellular agitation in the presence of glass beads. In such bead mills, the microorganisms are placed in a chamber together with a quantity of glass beads of 0.2–0.3 mm in diameter. This mixture is then shaken or agitated vigorously, resulting in numerous collisions between the microbial cells and the glass beads. It also results in the grinding of cells between the rotating beads. These forces promote efficient disruption of most microbial cell types. Operational parameters such as ratio of cells to beads, and the rate and duration of agitation may be adjusted to achieve optimum disruption of the particular cells in question. Laboratory systems can homogenize several grams of microbial cells in minutes. Industrial-scale bead-milling systems can process in excess of 1000 litres of cell suspension per hour. Cooling systems minimize protein inactivation by dissipating the considerable heat generated during this process.

REMOVAL OF WHOLE CELLS AND CELL DEBRIS

Upon completion of the homogenization step cellular debris and any remaining intact cells can be removed by centrifugation or by filtration. As previously mentioned these techniques are also used to remove whole cells from the medium during the initial stages of extracellular protein purification.

Centrifugation

Batch (fixed volume) centrifuges are capable of processing, at most, a few litres during any one spin cycle, and hence their use is usually restricted to laboratory-scale operations. Industrial-scale protein purification generally requires the processing of several hundred or thousands of litres of crude extract. Industrial scale centrifugation is normally achieved using continuous-flow centrifuges, through which homogenate is continuously pumped and the clarified solution continually collected. The deposited solids can be removed from the centrifuge bowl by periodically stopping the centrifuge and manually removing the pelleted material. Most modern continuous-flow centrifuges, however, are designed to allow continuous discharge of collected solids through a peripheral nozzle, or alternatively facilitate intermittent discharge of pelleted material via a suitable discharge valve. A number of different continuous flow centrifuge designs are commercially available. The three basic types supplied are the disc-type centrifuge, the hollow bowl centrifuge and the basket centrifuge. Most microbial cells are sedimented by batch centrifugation by applying a centrifugal force of approximately 5000 g for 15 min or less. Efficient removal of cell debris requires the application of higher centrifugal forces for longer time periods, typically 10 000 g for 45 min.

Although higher centrifugal forces are attained by batch centrifuges, modern continuous-flow centrifuges generate sufficient gravitational force to allow effective processing at high flow rates. Typically this may be of the order of several hundred litres per hour. Centrifugation remains the method of choice to effect cell and cellular debris separation, both at laboratory scale and on an industrial scale. However a number of factors, most notably the high capital and running costs have led many to investigate alternative means of collecting cells or cellular debris. The most popular alternative method utilized is filtration.

Filtration

Both whole cells and cell debris may be removed from solution by filtration. Either depth filters or (far more commonly) membrane filters may be used. Depth filters consist of randomly orientated fibres (usually manufactured from glass fibre or cellulose) which form an irregular network of channels or mesh-like structures. Such filters retain particles not only on their surface but also within the depth of the filter. Depth filtration is sometimes used to remove whole cells from fermentation media, as will be discussed later in this chapter. Such filters are also used to remove or reduce levels of cellular debris, denatured protein aggregates, or other precipitates from solution.

Membrane filtration, also termed microfiltration, is achieved using thin, membrane-like sheets of polymeric substances such as cellulose acetate or nitrate, nylon or polytetrafluoroethylene (PTFE), in which very small pores have been generated. Pore sizes available generally range between 10 and 0.02 μm. Membrane filters of pore diameter 0.2–0.45 μm will retain all microbial cells. The retention of particles or microbial cells occur only on or in the surface layer of the filter. The material to be filtered is applied to the filtration system under pressure in order to achieve satisfactory flow rates. Membrane filtration enjoys increasing popularity as a system of choice to remove cell and cellular debris from solutions. The method is efficient and requires relatively simple equipment. Filter configuration may be of the flat disc type although (particularly in the case of industrial systems) the filter is usually shaped into a cartridge configuration (Figure 3.3). This is achieved by placing the rectangular membrane filter sheet on a supporting mesh of the same size and subsequently folding it into a pleated structure. The two ends are then sealed together to form a cylinder which is placed between a

Figure 3.3 Photographs illustrating (a) a range of cartridge filters and (b) a range of filters and their stainless steel housings. In each case the pleated filter is protected by an outer plastic supporting mesh (photographs courtesy of Pall Process Filtration Ltd.)

plastic core and outer structure which physically protects and supports the filter material itself. Pleating allows a large filtration surface area to be accommodated in a compact area. Such filters are normally housed in stainless steel filter housing systems.

One of the main shortcomings of membrane filtration systems relates to their tendency to clog easily. This results in a sharp decrease in flow rate and the blockage can also lead to a pressure build-up in the filtration system, which could potentially destroy filter integrity. Most modern filters however withstand the application of relatively high pressures.

Incorporation of a suitable prefiltration system invariably increases effective filter life span and sustains higher flow rates through the main filter. Thus, a fermentation broth or a homogenate may first be passed through a depth filter, the eluate from which is then passed through a membrane filter. Another approach is to use two or three membrane filters of decreasing pore size, connected in series. Such systems are popular at industrial scale, where microfiltration is employed to yield a sterile flow of liquid.

Microfiltration is most often employed to sterilize a protein solution. Removal of all microbial cells is achieved by use of an 'absolute' 0.2 μm filter. Membrane filters may be classified as 'absolute' or 'nominal'. Absolute filters are guaranteed to remove all particles larger than the indicated filter pore size, i.e. they are 100 per cent effective. Nominal filters on the other hand, while effective, may not be 100 per cent effective so these, although considerably cheaper than absolute filters, should not be used for critical operations such as sterilization. Sterile filtration on a laboratory scale is usually achieved by passing the protein solution directly through a 0.22 μm filter. On an industrial scale it is often achieved by passing a solution or suspension through a filtration series consisting of a 1 μm filter followed by a 0.45 μm filter, followed by a 0.2 μm filter. Reduction in microbial levels (bioburden), may be achieved by using a 5 μm and 1 μm filter connected in series, often followed by a 0.45 μm filter. Stepwise reduction in filter pore size prolongs the lifetime of the filter with the smaller pore sizes

Many modern filtration systems incorporate several filters of different pore sizes into a single cartridge system (Figure 3.4). For example, sheets of a prefilter, two membrane filters of differing pore size and a supporting

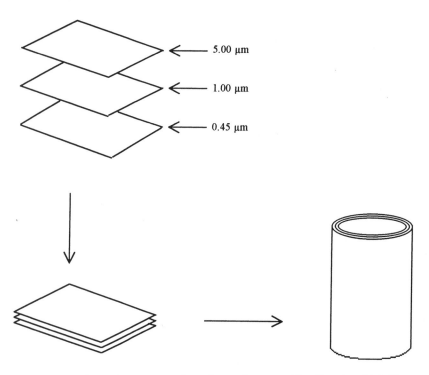

Figure 3.4 Schematic representation of a modern cartridge filter in which filter sheets of three different pore sizes are housed

mesh may be placed on top of each other in series, folded into pleats and formed into a cylindrical cartridge by joining both ends. One such filter can thus effectively replace a series of two or three filters of monopore sizes in a filtration system. Most modern cartridge filters can be repeatedly sterilized by autoclaving or on line steaming and they may also be operated for considerable periods of time at elevated temperatures (70–75°C).

To ensure their effectiveness, filters are often integrity tested, both before and after use. Membrane integrity tests most commonly employed include the bubble point test and the pressure hold method. Both tests rely on monitoring the effect of pressurized air on the filter system.

Although most filtration media are relatively inert, the possibility exists that certain proteins or other biomolecules may adhere to the filter matrix due to electrostatic, hydrophobic or other interactions. This possibility must usually be determined by direct experimentation. In summary, filtration techniques may be used at a number of stages during a protein purification process:

- to achieve separation of whole cells from fermentation media;

- to remove whole cells and cell debris after cell disruption;

- to achieve a reduction or totally eliminate microbial species from the product stream at later stages in a purification process.

In some instances, such separations may also be achieved by centrifugation. Although centrifugation represents the traditional method used, filtration is increasingly being used, particularly on an industrial scale. The increasing popularity of filtration is largely due to reduced capital and operating costs, as well as reduced inactivation of particularly labile proteins. Such proteins may be damaged by shear forces or frothing generated during continuous-flow centrifugation processes. Most filtration systems also offer a reduction in actual processing time when compared to centrifugation. As will be discussed later in this chapter, another form of filtration, ultrafiltration, can be used to concentrate protein solutions.

Aqueous two-phase partitioning

The technique of aqueous two-phase partitioning may be used to separate whole cells or cell debris from soluble protein. It can also be used to achieve a limited degree of protein purification and concentration. Although studied at laboratory- and pilot-plant scale for many years, aqueous two-phase partitioning does not enjoy widespread industrial use. This is mainly due to a lack of understanding as to how the technique works at a molecular level.

Aqueous two-phase partitioning is based on the fact that many water-soluble (aqueous) polymers are incompatible with each other, or with salt solutions that are of high ionic strength. Thus, if one such polymer is mixed with a salt solution or with a second (incompatible) polymer, two phases are formed upon standing. Such partitioning systems may be employed to separate proteins from cell debris or other impurities. The debris partitions to the lower, more polar and more dense phase, while soluble proteins tend to partition into the top, less polar and less dense phase. Subsequent separation of the two phases achieves effective separation of cellular debris from soluble protein.

The most commonly employed polymers are polyethylene glycol (PEG, a polymer consisting of ethylene molecules linked by ether bonds), and dextran (a polymer consisting of repeating glucose residues, linked by $\alpha 1 \rightarrow 6$ bonds, Figure 3.5). The most common polymer-salt system is based on polyethylene glycol and a phosphate salt (normally sodium or potassium phosphate). If solutions of PEG and dextran are mixed with a cell homogenate under appropriate conditions, the protein components partition into the upper PEG phase whereas cellular debris accumulate in the lower dextran phase. In this way, effective separations can be achieved upon prolonged standing. If the mixture is one of PEG and phosphate, the proteins tend to accumulate in the upper PEG-rich phase whereas the debris accumulates in the lower phosphate-rich phase (Figure 3.6).

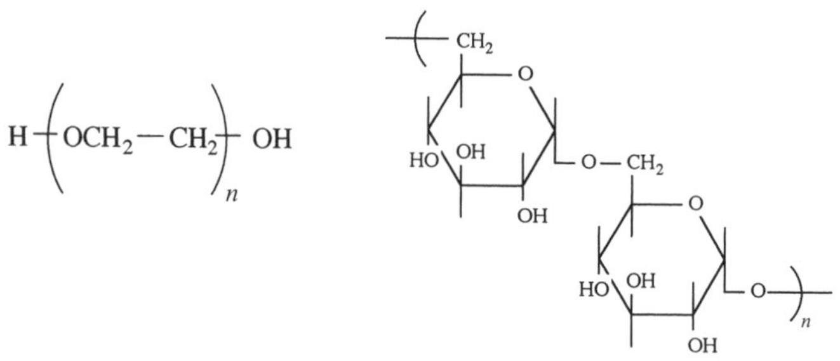

Figure 3.5 Molecular structure of (a) polyethylene glycol (*n* is usually between 4 and 200) and (b) a fragment of the dextran backbone

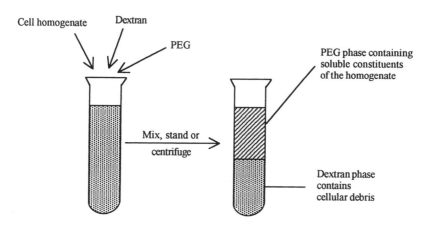

Figure 3.6 Principle of an aqueous two-phase purification system as applied to separation of cellular debris from soluble proteins

When two such incompatible water-soluble moieties are mixed subsequent phase separation occurs at a slow rate. However, the rate of phase separation is increased by applying a centrifugal force, so phase separation is accelerated by employing a subsequent centrifugation step. Once the phase separation has been completed, the phase containing the protein is subjected to further processing. On a laboratory scale, batch centrifuges are normally used to accelerate phase separation. Continuous centrifuges are more appropriate to industrial-scale operations. Application of the phase containing the protein of interest to an ultrafiltration system (see next section) results in an immediate concentration of the protein product. Alternatively, further purification may be achieved by employing a second-phase extraction step, in which the partition coefficient of the (new) system is altered. This may be achieved by, for example, altering the molecular mass of the polymers used (various PEG preparations may be obtained with molecular masses ranging from 200–20 000 Da), or by changing the ionic strength of the solutions. During separation of this second-phase system, some of the protein species remain in the PEG phase while others will partition to the lower phase, depending upon the conditions employed. In this way, a limited purification of the protein of interest may be attained.

A high degree of protein localization in one phase may be obtained by attaching a ligand which specifically binds the protein of interest to the least polar polymer of the two phase system (normally PEG). This technique is termed affinity partitioning. In this case conditions are employed which promote partitioning of most contaminant proteins in addition to nucleic acids and cell debris to the more polar phase. The protein of interest will, however, partition into the less polar phase due to the presence of the ligand in this phase. Affinity ligands used can be specific, for example a substrate or inhibitor of an enzyme of interest, or non-specific, such as various dyes that bind a number of protein types. Affinity ligands are discussed in more detail later in this chapter.

Aqueous two-phase partitioning is a potentially useful clarification or partial purification technique for a number of reasons:

- it is a gentle method, having little or no adverse effect on the biological activity of most proteins;

- many of the polymers used exhibit protein-stabilizing properties;

- the yield of protein activity recovered is generally high;

- little if any technical difficulties arise during process scale-up.

The major disadvantage associated with this technique relates to the lack of understanding of the molecular mechanisms involved in the partitioning process. Development of a two-phase process to achieve effective partitioning is wholly empirical. Without an intimate understanding of the underlining principles involved, a rational approach to designing such systems cannot be undertaken. Technical grade dextran is also quite expensive. This may be overcome by employing crude dextran preparations, or substituting the dextran with other polymers such as polyvinyl alcohol (PVA) or polyvinyl pyrrolidone (PVP). The disadvantages outlined above, coupled with the availability of alternative, established separation techniques such as centrifugation or filtration, has thus far limited the utilization of this technique.

Removal of nucleic acid and lipid

In some cases it is desirable and necessary to remove or destroy the nucleic acid content of a cell homogenate prior to subsequent purification. Liberation of large amounts of nucleic acids often significantly increases the viscosity of the cellular homogenate. This generally renders the homogenate more difficult to process, particularly on an industrial scale. Significant increases in viscosity place additional demands upon the method of cell debris removal that is used. Increased centrifugal forces for longer time periods may be required to efficiently pellet cell debris in such solutions. If a filtration system is employed to remove cellular debris, increased viscosity will also adversely affect flow rate and filter performance. Increased viscosity due to liberation of nucleic acids during homogenization is often most noticeable when prokaryotic organisms are used, as the DNA in such organisms is not bounded by an intracellular protective membrane, the nuclear membrane, as is the case in eukaryotic cells.

The inclusion of a specific nucleic acid removal step is not required for all purification procedures. Inclusion or exclusion of such a step depends upon the extent to which liberated nucleic acids affect the viscosity of particular suspensions and on the intended use of the final protein product. Effective nucleic acid removal is particularly important when purifying any protein destined for therapeutic use. Regulatory authorities generally insist that the nucleic acid content present in the final preparation be, at most, a few picograms per therapeutic dose (see Chapter 4).

Effective removal of nucleic acids during protein purification may be achieved by precipitation, or by treatment with nucleases. A number of cationic (positively charged) molecules are effective precipitants of DNA and RNA; they complex with, and precipitate, the negatively charged nucleic acids. The most commonly employed precipitant is polyethylenimine, a long-chain cationic polymer (Figure 3.7). The precipitate is then removed, together with cellular debris, by centrifugation or filtration. The use of polyethylenimine during purification of proteins destined for therapeutic applications is often discouraged, as small quantities of unreacted monomer may be present in the polyethylenimine preparation. Such monomeric species may be carcinogenic. If polyethylenimine is utilized in such cases the subsequent processing steps must be shown to be

$$-\left(CH_2-CH_2-NH\right)-$$

Figure 3.7   Molecular structure of the repeat unit of polyethylenimine

capable of effectively and completely removing any of the polymer or its monomeric units that may remain in solution.

Nucleic acids may also be removed by treatment with nucleases, which catalyse the enzymatic degradation of these biomolecules. Indeed nuclease treatment is quickly becoming the most popular method of nucleic acid removal during protein purification. This treatment is efficient, inexpensive and, unlike many of the chemical precipitants used, nuclease preparations themselves are innocuous and do not compromise the final protein product.

Some cell or tissue types (most notably of animal origin) contain an appreciable level of lipid. Removal of the lipid layer from the crude protein solution before further purification is desirable as (a) it is a contaminant, and (b) it can interfere with subsequent purification steps (e.g. clog chromatographic columns). The lipid layer can be removed by passage of the solution through glass wool or a cloth of very fine mesh size.

CONCENTRATION AND PRIMARY PURIFICATION

During the initial stages of many protein purification procedures the protein of interest is present in dilute solution, thus large volumes of process liquid must be handled. This is particularly true in the case of proteins secreted into the medium during microbial fermentation or during animal cell culture. It thus becomes necessary to concentrate such solutions in order to render the extract volume more manageable for subsequent purification steps. Methods used to achieve concentration on a laboratory scale are listed below:

- ultrafiltration;

- precipitation (salt, solvent, etc.);

- ion-exchange chromatography;

- vacuum dialysis;

- freeze drying;

- addition of dry Sephadex G-25 (see below).

Addition of dry Sephadex G-25 (or analogous beads) to a dilute protein solution results in bead swelling. During hydration of the beads, water enters the internal bead structure. As protein molecules are too large to enter the gel matrix, they remain in the decreased volume of liquid surrounding the beads, and hence are effectively concentrated. Unlike most of the other techniques mentioned, this method of concentration is not amenable to scale-up, mainly on economic grounds.

For large-scale applications, concentration of extracts is normally achieved by precipitation, ion exchange chromatography or ultrafiltration.

Each of these techniques has its own inherent advantages and disadvantages. All can effectively concentrate protein solutions and may also result in a limited degree of purification

Concentration by precipitation

Protein precipitation can be promoted by agents such as neutral salts, organic solvents, high molecular mass polymers, or by appropriate pH adjustments (Table 3.5). Concentration by precipitation is one of the oldest concentration methods known. Ammonium sulfate is likely the most common protein precipitant utilized. This neutral salt is particularly popular due to its high solubility, inexpensiveness, lack of denaturing properties towards most proteins, and its stabilizing effect on many proteins.

The addition of small quantities of neutral salts to a protein solution often increases protein solubility; the 'salting in' effect. However, increasing salt concentrations above an optimal level leads to destabilization of proteins in solution and eventually promotes their precipitation. This is known as 'salting out' (Figure 3.8). At high concentrations, such salts effectively compete with the protein molecules for water of hydration. This promotes increased protein – protein interactions, predominantly interactions between hydrophobic patches on the surface of adjacent protein molecules. Such increased protein – protein interactions eventually result in protein precipitation.

Addition of various organic solvents to a protein solution can also promote protein precipitation. Added organic solvents lower the dielectric constant of an aqueous solution. This in turn promotes increased electrostatic attraction between bodies of opposite charge in the solution, in this case proteins. Such increasing interactions between proteins of opposite charges eventually leads to their precipitation. Organic solvents

Table 3.5 Various methods and techniques which may be used to precipitate proteins from solution

Addition of neutral salts	(e.g. ammonium sulphate)
Addition of organic solvents	(e.g. ethanol or acetone)
Addition of organic polymers	(e.g. polyethylene glycol)
Affinity precipitation	(addition of a ligand – often an antibody – which precipitates the target protein from solution on the basis of bio-specific molecular interactions)
Adjustment of solution pH	(some proteins precipitate out of solution at their pI values)
Selective denaturation	(approach can be used to precipitate contaminant proteins from the protein of interest, if the latter is more stable than the former in the presence of some denaturing influence (e.g. extremes of pH or elevated temperature)

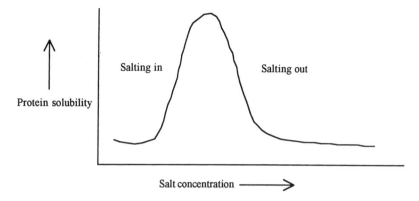

Figure 3.8 Effect of salt concentration on protein solubility. Increases in salt concentration from low initial values often increase protein solubility (salting in). Further increases above an optimal value will result in destabilization of the protein, and eventually its precipitation from solution (salting out)

frequently used to promote precipitation include ethanol, isopropanol, acetone, and diethylether (Figure 3.9). Protein precipitation by utilizing organic solvents must be carried out at temperatures at or below 0°C in order to prevent protein denaturation. As such solvents depress the freezing point of aqueous solutions, it is usually feasible to maintain the solution temperature several degrees below 0°C.

Precipitation may also be promoted by the addition of organic polymers such as polyethylene glycol. The addition of such polymers however, often dramatically increases the viscosity of the resultant solution, making recovery of the precipitate more difficult. Recovery is normally achieved by centrifugation or by filtration. The precipitate can subsequently be redissolved in a smaller volume of resuspending liquid and, in this way, it is effectively concentrated.

Concentration by precipitation has also played an important role in the downstream processing of many industrially important proteins. The technique is relatively straight forward to perform and requires only a

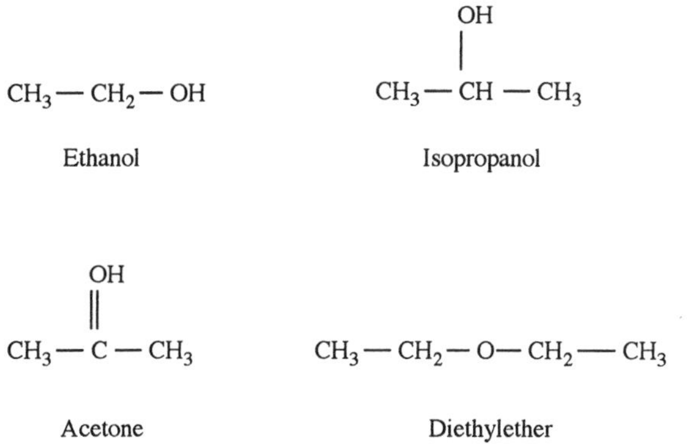

$$CH_3 - CH_2 - OH$$

Ethanol

$$CH_3 - \overset{\overset{\displaystyle OH}{|}}{CH} - CH_3$$

Isopropanol

$$CH_3 - \overset{\overset{\displaystyle OH}{\|}}{C} - CH_3$$

Acetone

$$CH_3 - CH_2 - O - CH_2 - CH_3$$

Diethylether

Figure 3.9 Structure of the organic solvents most commonly used as protein precipitants

limited amount of equipment. In many cases, precipitation also achieves some degree of protein purification (different protein types generally require different concentrations of a precipitant to effect their removal from solution). High recoveries of biological activity are also usually recorded. There are also a number of disadvantages associated with this technique, particularly at industrial scale:

- Many of the precipitants used are highly corrosive, in particular towards stainless steel equipment such as centrifuge rotors.

- Precipitation is often quite inefficient if employed under circumstances where the initial protein concentration is low. In such cases, low recoveries of protein may be recorded.

- Some precipitants, such as acetone and diethylether, are highly inflammable and hence are hazardous to work with. Others, such as ethanol, are quite expensive.

- Many precipitants (e.g. most organic solvents) must be disposed of carefully after use.

- In many cases all traces of the precipitant present in the precipitate must be removed before further processing. For example, ammonium sulfate must be removed from resuspended salt mediated precipitates before the resultant protein solution can be applied to an ion exchange column. This can be achieved by dialysis or diafiltration (discussed later).

Purification of serum proteins is one example of an industrially important process which traditionally included several protein precipitation steps. Precipitants normally used includes both ethanol and ammonium sulfate. Most modern processes, however, rely on methods other than precipitation to achieve initial concentration of the proteins.

Concentration by ion exchange

At any given pH value proteins display either a positive, negative, or zero nett charge. Using this parameter of molecular distinction, different protein molecules can be separated from one another by judicious choice of pH, ionic strength and ion exchange materials. Positively charged proteins will bind to cation exchangers whereas negatively charged proteins will bind to anion exchangers. Elution of such bound protein may be easily achieved by subsequent irrigation with a solution of high ionic strength (e.g. buffer containing a salt such as NaCl or KCl added to a final concentration of up to $0.5\,\text{M}$). Ion exchange offers an effective and relatively inexpensive method of achieving initial concentration. This may also result in a limited degree of protein purification.

Batchwise adsorbtion of proteins present in dilute solutions is easily achieved by the direct addition of the ion exchanger to the solution in question. Examples of such 'dilute' protein solutions include: (a) fermentation broths or cell culture media containing extracellular proteins from which the whole cells have been removed; or (b) cell homogenates from which cell debris has been removed. The ion exchange material may be recovered by centrifugation, and is then generally placed in a filter funnel or stainless steel vessel. Such holding vessels contain an outlet covered by a mesh, which ensures retention of the ion exchange material. Bound proteins are then eluted from the ion exchange by addition of a suitable solution at high ionic strength. The protein containing eluate is collected and subject to further processing. The ion exchange material can be regenerated and reused for the next purification cycle.

If the protein of interest is negatively charged, anion exchangers containing functional groups such as aminoethyl- or diethyl-aminoethyl-moieties may be employed. In the case of positively charged proteins, cation exchangers containing functional groups such as carboxymethyl moieties are used (Table 3.6). These functional groups are covalently linked to porous beads usually made from cross-linked dextran (e.g. sephadex exchangers), agarose (e.g. sepharose exchangers) or cellulose (e.g. sephacel exchangers).

Ion exchange techniques are moderately popular as most are relatively inexpensive, robust and can easily be regenerated. Very high levels of protein recovery are generally recorded. Batch adsorbtion – elution also results in considerable clarification of the resultant protein solution. Many impurities that have adverse effects on solution characteristics either do not bind to the ion exchanger, or do not subsequently elute from such exchangers under the conditions employed to elute the target protein. Undesirable impurities include particulate material that previous clarification steps failed to remove, various lipid and/or carbohydrate

Table 3.6 Functional groups commonly attached to chromatographic beads in order to generate cation or anion exchangers

Group name	Group structure	Exchanger type
Diethylaminoethyl (DEAE)	$-O-(CH_2)_2-\overset{\overset{\displaystyle H}{\mid}}{N^+}-(CH_2-CH_3)_2$	Anion exchanger
Quaternary ammonium (Q)	$-CH_2-N^+-(CH_3)_3$	Anion exchanger
Quaternary aminoethyl (QAE)	$-O-(CH_2)_2-\overset{\overset{\displaystyle (C_2H_5)_2}{\mid}}{N^+}-CH_2-\overset{\overset{\displaystyle OH}{\mid}}{CH}-CH_3$	Anion exchanger
Carboxymethyl (CM)	$-O-CH_2-COO^-$	Cation exchanger
Methyl sulphonate (S)	$-CH_2-SO_3^-$	Cation exchanger
Sulphopropyl (SP)	$-CH_2-CH_2-CH_2-SO_3^-$	Cation exchanger

molecules, in addition to partially denatured or aggregated protein. Effective clarification of such protein solutions is desirable as it prevents fouling of columns during the subsequent purification steps. Such initial ion exchange treatments also lead to some degree of protein purification, as only other molecules with similar charge characteristics will bind and subsequently co-elute with the protein of interest.

Concentration by ultrafiltration

Protein solutions may be quickly and conveniently concentrated by ultrafiltration (Figure 3.10) and this method of concentration is the one most widely applied, both on a laboratory and industrial scale. As previously discussed, the technique of microfiltration is effectively utilized to remove whole cells or cell debris from solution. Membrane filters employed in the

Figure 3.10   Ultrafilter used for laboratory-scale molecular separations (photograph courtesy of Amicon Ltd)

microfiltration process generally have pore diameters ranging from 0.1 to 10 μm. Such pores, while retaining whole cells and large particulate matter, fail to retain most macromolecular components such as proteins. In the case of ultrafiltration membranes, pore diameters normally range from 1 to 20 nm. These pores are sufficiently small to retain proteins of low molecular mass. Ultrafiltration membranes with molecular mass cut-off points ranging from 1–300 kDa are commercially available. Membranes with molecular mass cut-off points of 3, 10, 30, 50, and 100 kDa are most commonly used.

Traditionally, ultrafilters have been manufactured from cellulose acet-ate or cellulose nitrate. Several other materials such as polyvinyl chloride and polycarbonate are now also used in membrane manufacture. Such plastic-type membranes exhibit enhanced chemical and physical stability when compared to cellulose-based ultrafiltration membranes. An important prerequisite in manufacturing ultrafilters is that the material utilized exhibits low protein adsorbtive properties. No matter which material is utilized the pore size obtained is not uniform, a point which must be emphasized. A range of pore sizes of wide deviation from the mean pore size are generally observed. The molecular mass cut-off point quoted for any such filter is thus best regarded as being a nominal figure. For protein work it is advisable to use a membrane whose stated molecular mass cut-off point is 5 kDa or more lower than the molecular mass of the protein of interest. It is also important to realize that the molecular mass cut-off point quoted applies to globular proteins. The overall shape of the protein of interest affects its ultrafiltration characteristics. If the protein is somewhat elongated, it may not be retained by an ultrafilter whose cut-off point is significantly lower. Extensive post-translational modifications, in particular glycosylation, may also affect ultrafiltration behaviour.

Ultrafiltration is generally carried out on a laboratory scale using a stirred cell system. The flat membrane is placed on a supporting mesh at the bottom of the cell chamber, and the material to be concentrated is then transferred into the cell. Application of pressure, usually nitrogen gas, ensures adequate flow through the ultrafilter. Molecules of lower molecular mass than the filter cut-off pore size (e.g. water, salt and low molecular weight compounds) all pass through the ultrafilter, thus concentrating the molecular species present whose molecular mass is significantly greater than the nominal molecular mass cut-off point. Concentration polarization (the build-up of a concentrated layer of mol-ecules directly over the membrane surface which are unable to pass through the membrane) is minimized by a stirring mechanism operating close to the membrane surface. If unchecked, concentration polarization would result in a lowering of the flow rate. Additional ultrafilter formats used on a laboratory scale include cartridge systems, within which the ultrafiltration membrane is present in a highly folded format. In such cases the pressure required to maintain a satisfactory flow rate through the membrane is usually generated by a peristaltic pump.

Large-scale ultrafiltration systems invariably employ cartridge-type filters (Figure 3.11). This allows a large filtration surface area to be accommodated in a compact area. Concentration polarization is avoided by allowing the incoming liquid to flow across the membrane surface at right angles, i.e. tangential flow. The ultrafiltration membrane may be pleated, with subsequent joining of the two ends to form a cylindrical cartridge. Alternatively, the membrane may be laid on a spacer mesh and this may then be wrapped spirally around a central collection tube, into which the filtrate can flow.

Another widely used membrane configuration is that of the hollow fibres. In this case, the hollow cylindrical cartridge casing is loaded with bundles of hollow fibres. Hollow fibres have an outward appearance somewhat similar to a drinking straw, although their internal diameters

Figure 3.11   Example of an ultrafiltration system used on an industrial scale. This system (SPM 180) comprises of 16.7 m² of spiral wound membrane. It can be used in ultrafiltration or diafiltration mode (photograph courtesy of Amicon Ltd)

may be considerably smaller. In this configuration, the liquid to be filtered is pumped through the central core of the hollow fibres. Molecules of lower molecular mass than the membrane rated cut-off point pass through the walls of the hollow fibre. The permeate, which emerges from the hollow fibres along all of their length, is drained from the cartridge via a valve. The concentrate emerges from the other end of the hollow fibre and is collected by an outlet pipe – this is referred to as the retentate.

The permeate is then normally discarded while the retentate, containing the protein of interest, is processed further. The retentate may be recycled through the system if further concentration is required.

Ultrafiltration has become prominent as a method of protein concentration for a variety of reasons:

- the method is very gentle, having little adverse effect on bioactivity of the protein molecules;

- high recovery rates are usually recorded – some manufacturers claim recoveries of over 99 per cent;

- processing times are rapid when compared to alternative methods of concentration;

- little ancillary equipment is required;

- ultrafiltration may also achieve some degree of protein purification (on the basis of differences in molecular mass).

One drawback relating to this filtration technique is its susceptibility to rapid membrane clogging. Viscous solutions also lead to rapid decreases in flow rates and prolonged processing times.

Ultrafiltration may also be utilized to achieve a number of other objectives. As discussed above, it may yield a limited degree of protein purification and may also be effective in depyrogenating solutions. This will be discussed further in Chapter 4. The technique is also widely used to remove low molecular mass molecules from protein solutions by diafiltration.

Diafiltration

Diafiltration is a process whereby an ultrafiltration system is utilized to reduce or eliminate low molecular mass molecules from a solution. In practice, this normally entails the removal of salts, ethanol and other solvents, buffer components, amino acids, peptides, added protein stabilizers or other molecules, from a protein solution. Diafiltration is generally preceded by an ultrafiltration step to initially reduce process volumes. The actual diafiltration process is identical to that of ultrafiltration except for the fact that the level of reservoir is maintained at a constant volume.

This is achieved by the continual addition of solvent lacking the low molecular mass molecules which are to be removed. By recycling the concentrated material and adding sufficient fresh solvent to the system such that five times the original volume has emerged from the system as permeate, over 99 per cent of all molecules which freely cross the membrane will have been removed from the solution. Removal of low molecular mass contaminants from protein solutions may also be achieved by dialysis or by gel filtration chromatography. Diafiltration, however, is emerging as the method of choice, as it is quick, efficient and utilizes the same equipment as used in ultrafiltration.

COLUMN CHROMATOGRAPHY

The initial steps in any purification process are designed to (a) liberate and concentrate the protein of interest, and (b) to remove particularly undesirable contaminants such as particulate matter, lipids, or other substances which may subsequently result in fouling or clogging of high-precision chromatographic columns. The degree of purification achieved by such preliminary techniques is often marginal. Such initial steps should yield a high percentage recovery of the desired protein product. Further protein purification is generally achieved by column chromatography.

As applied to protein purification, column chromatography refers to the separation of different protein types from each other according to their differential partitioning between two phases: a solid stationary phase (the chromatographic beads, usually packed into a cylindrical column), and a mobile phase (usually a buffer). With the exception of gel filtration, all forms of chromatography used in protein purification protocols are adsorptive in nature. The protein mix is applied to the column (usually) under conditions which promote selective retention of the target protein. Ideally this target protein should be the only one retained on the column, but this is rarely attained in practice. After sample application the column is washed (irrigated) with mobile phase in order to flush out all unbound material. The composition of the mobile phase is then altered in order to promote desorption of the bound protein. Fractions of eluate are collected in test tubes, which are then assayed for both total protein and for the protein of interest (Figure 3.12). The fractions containing the target protein are then pooled and subjected to the next step in the purification process.

Individual protein types possess a variety of characteristics that distinguish them from other protein molecules. Such characteristics include size and shape, overall charge, the presence of surface hydrophobic groups, and the ability to bind various ligands. Quite a number of protein molecules may be similar to one another if compared on the basis of any one such characteristic. All protein types, however, present their own unique combination of characteristics, a protein chromatographic 'finger print'. Various chromatographic techniques have been developed which separate

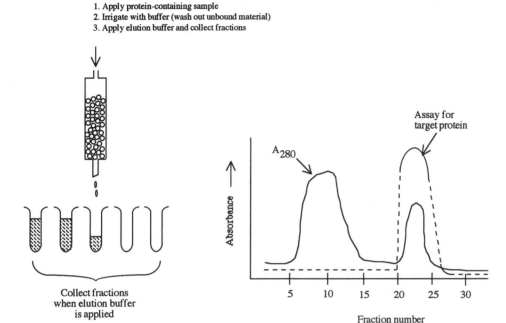

1. Apply protein-containing sample
2. Irrigate with buffer (wash out unbound material)
3. Apply elution buffer and collect fractions

Collect fractions
when elution buffer
is applied

(a)

Assay for
target protein

A_{280}

Absorbance

5   10   15   20   25   30

Fraction number

(b)

Figure 3.12   Typical sequence of events undertaken during an (adsorption based) protein purification chromatographic step (a). Note that the chromatographic beads are not drawn to scale, and in reality these display diameters <0.1 mm. Fractions collected during protein desorption are assayed for (i) total protein, usually by measuring absorbance at 280 nm, and (ii) target protein activity (b). In the case illustrated, two major protein peaks are evident only one of which contains the protein of interest. Thus desorption as well as adsorption steps can result in selective purification

proteins from each other on the basis of differences in such characteristics (Table 3.7). Utilization of any one of these methods to exploit the molecular distinctiveness usually results in a dramatic increase in the purity of the protein of interest. A combination of methods may be employed to yield highly purified protein preparations.

While chromatographic techniques can effectively separate various proteins from one another, the percentage recovery of the desired protein may be significantly less than 100 per cent. Typically, recovery values can range from 25 to 95 per cent. It is therefore desirable to design a purification scheme which yields the desired level of protein purity, using as few chromatographic steps as possible. Optimization of each step employed is critical if overall yields are to be maximized. Most purification systems employ three to five high-resolution chromatographic steps. The chromatographic techniques most commonly used to purify proteins are discussed below.

Table 3.7 Chromatographic techniques most commonly used in protein purification protocols with their basis of separation

Technique	Basis of separation
Ion exchange chromatography	Differences in protein surface charge at a given pH
Gel filtration chromatography	Differences in mass or shape of different proteins
Affinity chromatography	Based upon biospecific interaction between a protein and an appropriate ligand
Hydrophobic interaction chromatography	Differences in surface hydrophobicity of proteins
Chromatofocusing	Separates proteins on the basis of their isolectric points
Hydroxyapatite chromatography	Complex interactions between proteins and the calcium phosphate-based media. Not fully understood.

Size exclusion chromatography (gel filtration)

Size exclusion chromatography, also termed gel permeation or gel filtration chromatography, separates proteins on the basis of their size and shape. As most proteins fractionated by this technique are considered to have approximately similar molecular shape, separation is often described as being on the basis of molecular mass, although such a description is somewhat simplistic.

Fractionation of proteins by size exclusion chromatography is achieved by percolating the protein-containing solution through a column packed with a porous gel matrix in bead form (Figure 3.13). As the sample travels down the column, large proteins cannot enter the gel beads and hence are quickly eluted. The progress of smaller proteins through the column is retarded as such molecules are capable of entering the gel beads. The internal structure of the matrix beads could be visualized as a maze, through which proteins small enough to enter the gel must pass. Various possible routes through this maze are of varied distances. All proteins capable of entering the gel are thus not retained within the gel matrix for equal time periods. The smaller the protein, the more potential internal routes open to it and thus, generally, the longer it is retained within the bead structure. Protein molecules are therefore usually eluted from a gel filtration column in order of decreasing molecular size.

In most cases the gel matrices utilized are prepared by chemically cross-linking polymeric molecules such as dextran, agarose, acrylamide and vinyl polymers. The degree of cross-linking controls the average pore size of the gel prepared. Most gels synthesized from any one polymer type are thus available in a variety of pore sizes. The higher the degree of cross-linking introduced, the smaller the average pore size, and the more rigid the resultant gel bead. Various highly cross-linked gel matrices such as

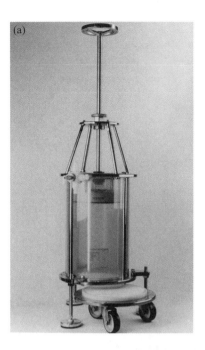

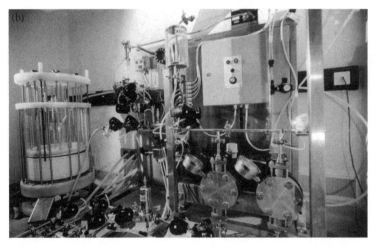

Figure 3.13   Chromatographic columns. The glass column illustrated in (a) is manufactured by Merck. A wide variety of columns (ranging in size from one ml to several litres, and constructed from glass, plastic or stainless steel) are available from this and a number of other manufacturers (e.g. Bio-Rad and Pharmacia Biotech). (b) Process scale chromatographic system. This particular system is utilized by a UK-based biotech company in the manufacture of a (protein) drug for clinical trials. The actual column is positioned to the left of picture

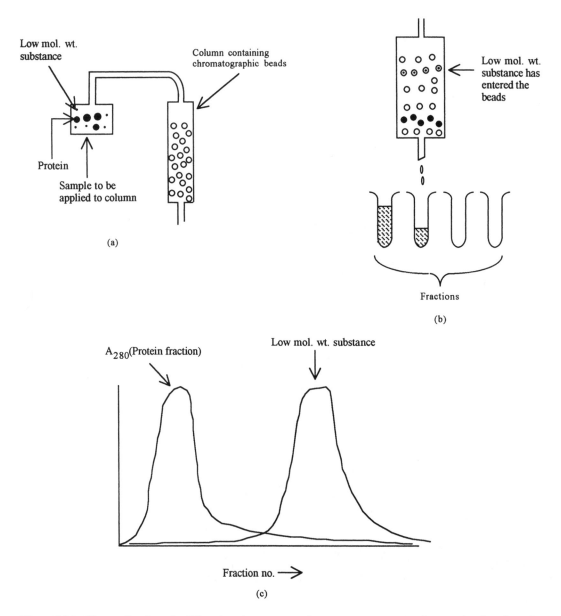

Figure 3.14 The application of gel filtration chromatography to separate proteins from molecules of much lower molecular weight. The mobile phase (the 'running buffer') will be devoid of the molecular species to be removed from the protein. Highly cross-linked porous beads are used, which exclude all protein molecules. The lower molecular weight substances, however, can enter the beads and their progress down through the column will therefore be retarded (a and b). The earlier fractions collected will contain the proteins while the latter fractions will contain the low molecular weight contaminants (c). In practice this 'group separations' application of gel filtration chromatography is mainly used to separate proteins from salt (e.g. after an ammonium sulfate precipitation step) or for buffer exchange

Sephadex G-25 or Bio Gel P2, have pore sizes that exclude all proteins from entering the gel matrix. Such gels may be used to separate proteins from other molecules which are orders of magnitude smaller, and are often used to remove low molecular weight buffer components and salts from protein solutions (Figure 3.14)

Gels of larger pore size are used to separate proteins from each other. The gel of most appropriate pore size for chromatographic application in any given circumstance depends largely on (a) the molecular mass of the protein of interest, and (b) the molecular mass of the major protein-aceous contaminants. All companies which produce gel filtration media publish the most effective fractionation range of each gel type (Table 3.8). Most gel filtration media are also available in a number of different particle sizes, typically course, medium, fine and superfine grades. Although smaller beads (fine/superfine) yield higher resolution, flow rates are lower.

The Sephadex range of fractionation gels were among the first to be developed. These gels are prepared by cross-linking dextran with epi-chlorohydrin. Dextran is a polysaccharide composed of glucose mono-mers, linked predominantly by $\alpha 1 \rightarrow 6$ glycosidic bonds. Some Sephadex gels, such as Sephadex G-25 or G-50 are quite rigid due to a high degree of cross-linking. Dextran-based gels of larger pore size, (Sephadex G-100 or G-200), are not as rigid, and tend to compress very easily. For this reason, the latter types of Sephadex are suitable only for small scale (i.e. laboratory-based) applications.

The Sephacryl range of gel filtration media are more rigid and physic-ally stable when compared to the Sephadex range. Hence they are more applicable to industrial situations. The Sephacryl based gels are prepared by cross-linking allyl dextran with *N-N*-methylene bisacrylamide. The Sepharose range of gels are prepared from the polysaccharide agarose. While these gels are particularly suited to fractionation of proteins of high molecular mass, their industrial usefulness is limited, mainly due to lack of physical stability. This gel exhibits a very open pore structure which is stabilized not by covalent linkages, but by hydrogen bonding between adjacent agarose molecules. The Sepharose gel structure disintegrates at temperatures above 40°C and thus, unlike the Sephadex or Sephacryl range of gels, may not be sterilized by autoclaving. The chemical and physical stability of Sepharose gels may be enhanced by covalently cross-linking the agarose moieties in the gel structure. Such cross-linking is achieved by incubation of the Sepharose with the cross-linking agent 2,3-dibromopropanol under alkaline conditions.

The Bio-Gel P range of media are prepared by cross-linking acrylamide with *NN'*-methylene-bis-acrylamide. Bio-Gel P is normally prepared in

---

Note: In practice the chromatographic beads are tightly packed in the column. They are separated from each other in this diagram only for the purpose of clarity. Also, the drawing is not to scale; protein molecules are considerably smaller than individual beads.

Table 3.8 Fractionation range and bead diameter of selected commercially available gel filtration media

Gel type	Fractionation range (kDa; globular proteins)	Bead diameter (μm)
Sephadex G-10	0–0.7	40–120
Sephadex G-25	1–5	50–150
Sephadex G-100	4–150	40–120
Sephadex G-200	5–600	40–120
Sephacryl S-100	1–100	25–75
Sephacryl S-300	10–1 500	25–75
Sepharose 6-B	10–4 000	45–165
Sepharose 2-B	70–40 000	60–200
Bio-Gel P-2	0.1–1.8	45–90
Bio-Gel P-6	1–6	90–180
Bio-Gel P-30	2.5–40	90–180
Bio-Gel P 100	5–100	90–180
Bio-Gel A 0.5 m	10–500	75–150
Bio-Gel A 5 m	10–5 000	75–150
Bio-Gel A 15 m	40–15 000	75–150

beaded form, and is especially useful when fractionating samples containing enzymes that degrade gels prepared from biological substances such as dextran. Bio-Gel A beads are made from agarose, with the pore size being controlled by the percentage agarose present.

Fractogel, produced by Merck, is a more recently developed gel material. This gel is a copolymer of oligoethyleneglycol, glycidylmethacrylate and pentacrythrol-dimethacrylate. The internal pore walls of fractogel are formed from intertwined polymer agglomerates which result in a very high degree of mechanical stability. Such mechanical strength facilitates use of these gels for large-scale preparative purposes.

Protein fractionation by size exclusion chromatography demands the use of long chromatographic columns. Generally, columns are 25–40 times greater in length than in width. Such dimensions are required in order to achieve adequate resolution of protein mixtures which separate into discrete protein bands on the basis of molecular size and shape, as they migrate down through the column.

Industrial-scale gel filtration columns may be several metres in length. Gel utilized for preparative purposes must be mechanically rigid so as to avoid gel compression and reduced flow rates. An alternative approach to industrial-scale column design involves the use of stacking gel systems, which entail packing the gel matrix into a number of identical short, wide or fat columns with the subsequent vertical connection of a number of such stacking columns in series; see industrial production of insulin, Chapter 7). The connecting distance between each column is kept to a minimum. The overall stack system behaves similarly to a single chromatographic column of similar dimensions. The chromatographic gel in such systems, however, experiences lower differential pressure. Furthermore, if one particular column (usually the first in series) becomes fouled it may be easily disconnected and replaced by a fresh column.

Size exclusion chromatography is rarely employed during the initial stages of protein purification. Small sample volumes must be applied to the column in order to achieve effective resolution. Application volumes are usually in the range of 2–5 per cent of the column volume. Furthermore, columns are easily fouled by a variety of sample impurities. Size exclusion chromatography is thus often employed towards the end of a purification sequence, when the protein of interest is already relatively pure and is present in a small, concentrated volume. After sample application, the protein components are progressively eluted from the column by flushing with an appropriate buffer. In many cases, the eluate from the column passes through a detector. This facilitates immediate detection of protein-containing bands as they elute from the column. The eluate is normally collected as a series of fractions. On a preparative scale each fraction may be a number of litres in volume. While size exclusion chromatography is an effective fractionation technique, it generally results in a significant dilution of the protein solution relative to the starting volume applied to the column. Column flow rates are also often considerably lower than flow rates employed with other chromatographic media. This results in long processing times which, for industrial applications, has adverse process cost implications.

Ion exchange chromatography

Several of the 20 amino acids that constitute the building blocks of proteins exhibit charged side chains. At pH 7.0, aspartic and glutamic acid have overall negatively charged acidic side groups, while lysine, arginine and histidine have positively charged basic side groups (Figure 3.15). Protein molecules therefore possess both positive and negative charges – largely due to the presence of varying amounts of these seven amino acids. (N-terminal amino groups and the C terminal carboxy groups also contribute to overall protein charge characteristics.) The nett charge exhibited by any protein depends on the relative quantities of these amino acids present in the protein, and on the pH of the protein

LIVERPOOL
JOHN MOORES UNIVERSITY
AVRIL ROBARTS LRC
TITHEBARN STREET
LIVERPOOL L2 2ER
TEL. 0151 231 4022

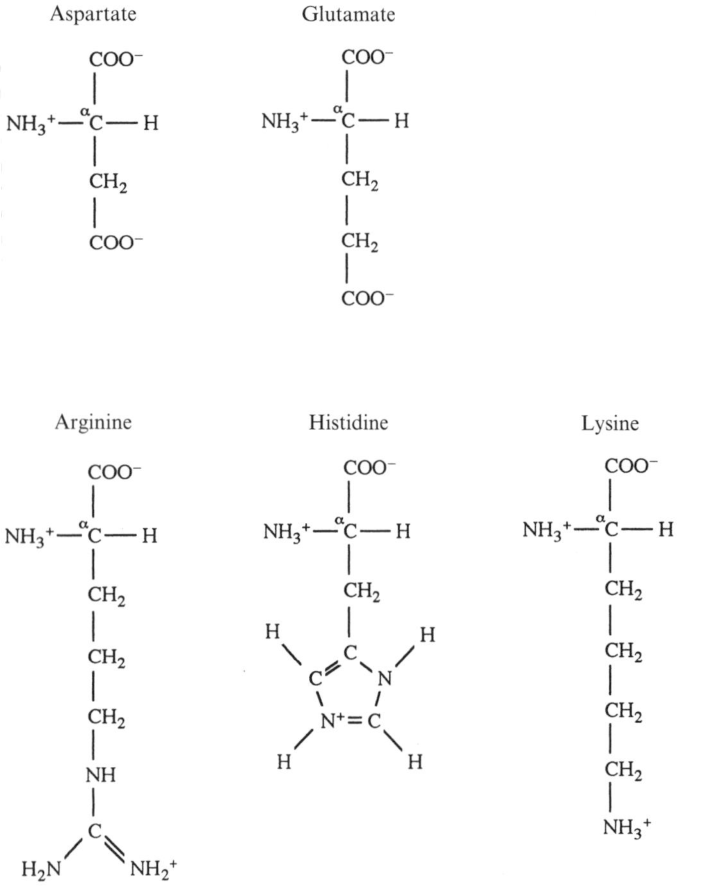

Aspartate

$$COO^-$$
$$NH_3{}^+ — \overset{\alpha}{C} — H$$
$$CH_2$$
$$COO^-$$

Glutamate

$$COO^-$$
$$NH_3{}^+ — \overset{\alpha}{C} — H$$
$$CH_2$$
$$CH_2$$
$$COO^-$$

Arginine

$$COO^-$$
$$NH_3{}^+ — \overset{\alpha}{C} — H$$
$$CH_2$$
$$CH_2$$
$$CH_2$$
$$NH$$
$$C$$
$$H_2N \qquad NH_2{}^+$$

Histidine

$$COO^-$$
$$NH_3{}^+ — \overset{\alpha}{C} — H$$
$$CH_2$$
$$C$$
$$C \quad N$$
$$N^+ = C$$

Lysine

$$COO^-$$
$$NH_3{}^+ — \overset{\alpha}{C} — H$$
$$CH_2$$
$$CH_2$$
$$CH_2$$
$$CH_2$$
$$NH_3{}^+$$

Figure 3.15 Structure of amino acids having overall nett charges at pH 7.0 In proteins the charges associated with the α-amino and α-carboxyl groups in all but the terminal amino acids are not present, as these groups are directly involved in the formation of peptide bonds

solution. The pH value at which a protein molecule possesses zero overall charge is termed its isoelectric point (pI). At pH values above its pI, a protein will exhibit a nett negative charge whereas at pH values below the pI, proteins will exhibit a nett positive charge.

Ion exchange chromatography is based upon the principle of reversible electrostatic attraction of a charged molecule to a solid matrix which contains covalently attached side groups of opposite charge (Figure 3.16). Proteins may subsequently be eluted by altering the pH, or by increasing the salt concentration of the irrigating buffer. Ion exchange matrices that contain covalently attached positive groups are termed anion exchangers. These will adsorb anionic proteins; i.e. proteins with a nett negative charge. Matrices to which negatively charged groups are covalently attached are termed cation exchangers, adsorbing cationic

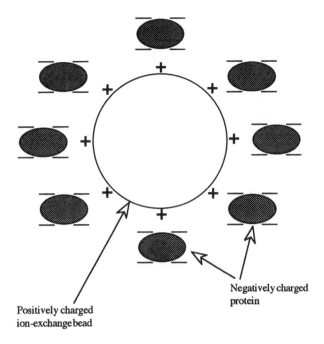

Positively charged
ion-exchange bead

Negatively charged
protein

Figure 3.16 Principle of ion exchange chromatography, in this case anion exchange chromatography. The chromatographic beads exhibit an overall positive charge. Proteins displaying a nett negative charge at the pH selected for the chromatography will bind to the beads due to electrostatic interactions

proteins; e.g. positively charged proteins. Positively charged functional groups (anion exchangers) include species such as aminoethyl and diethylaminoethyl groups. Negatively charged groups attached to suitable matrices forming cation exchangers include sulfo- and carboxymethyl groups (see Table 3.6). Ion exchangers may also be described as 'strong' or 'weak'. Strong ion exchange resins remain ionized over a wide pH range whereas weak ion exchange resins are ionized within a narrow pH range.

During the cation exchange process positively charged proteins bind to the negatively charged ion exchange matrix by displacing the counter ion (often H^+) which is initially bound to the resin by electrostatic attraction. Elution may be achieved using a salt-containing irrigation buffer. The salt cation, often Na^+, of NaCl, in turn displaces the protein from the ion exchange matrix. In the case of negatively charged proteins, an anion exchanger is obviously employed, with the protein adsorbing to the column by replacing a negatively charged counter ion.

The vast majority of purification procedures employ at least one ion exchange step; it represents the single most popular chromatographic technique in the context of protein purification. Its popularity is based upon the high level of resolution achievable, its straight forward scale-up (for industrial application), together with its ease of use and ease of column regeneration. In addition, it leads to a concentration of the protein of interest. It is also one of the least expensive chromatographic methods available. At physiological pH values most proteins exhibit a nett negative charge. Anion exchange chromatography is therefore most commonly used.

A wide range of support matrices are utilized in the manufacture of ion exchange media. The ideal matrix should be inert, structurally rigid and highly porous. Rigidity is required in order to facilitate scale-up and operation at high flow rates as soft matrices often compact, even under their own weight when packed in large, industrial scale columns. Gels exhibiting pore sizes large enough to allow free entry of proteins into the internal matrix cavities is desirable, as this dramatically increases binding capacity (i.e. proteins can be retained in internal bead cavities, and not just on the bead surface).

Cellulose-based ion exchangers were among the first to be widely employed in protein purification systems. Cellulose-based cation and anion exchangers are both readily available. Due to ease of compression, such exchangers are not widely utilized in industrial-scale chromatography, though they are used to concentrate proteins by batchwise adsorbtion. Improved cellulose-based ion exchangers, such as diethylaminoethyl (DEAE) Sephacel (Pharmacia) have been developed, although these are also somewhat confined to laboratory-scale applications. Sephadex ion exchangers are prepared by covalent attachment of charged groups to Sephadex G-25 or G-50 (refer to previous section). G-50 types are relatively easily compressed. G-25-based exchangers, although quite sturdy, have low binding capacities as the bead pore diameters are too small to allow penetration by proteins.

A number of additional ion exchange resins based upon polymers such as agarose are also available (e.g. Bio Gel A ion exchangers from Bio-Rad and sepharose based exchangers from Pharmacia). Such ion exchange media are generally suitable for both laboratory and large-scale applications, as they combine high capacity with rigidity, and hence high flow rates with no compression. The high capacity of such exchangers reflects their large open pore nature, which allows proteins free access to binding sites within the gel beads.

Fractogel (Merck) represents yet another ion exchange material suitable for use in laboratory and large-scale separations. Its suitability derives from its mechanical and chemical stability. To facilitate its use as an ion exchange media, the basic gel structure is chemically modified by the covalent introduction of appropriate charged groups, both on external and internal bead surfaces. The gel has a high protein binding capacity, as the pore size is sufficiently large to allow most proteins free entry into the gel matrix.

An alternative ion exchange resin design, termed 'tentacle type' (Merck), has also been developed. This is suitable for both analytical and preparative applications. Traditional ion exchangers consist of charged (ionic) groups directly attached to the ion exchange matrix.

This obviously presents a very rigid array of binding sites to which the protein is adsorbed. This in turn implies that binding may distort the conformational structure of the protein in question. It is not clear to what extent such distortion may lead to protein denaturation and inactivation. It must be stated, however, that the high recovery of biologically active proteins recorded when such traditional exchangers are employed suggest that such effects are minimal for most proteins. Tentacle type exchangers consist of a bead support matrix, to which linear charged polymers (the tentacles) are attached (Figure 3.17a). Such tentacle structures are flexible and may automatically adopt a configuration which allows maximum binding contact with the adsorbed proteins. This flexibility minimizes or eliminates any induced distortion of bound proteins. The binding capacity of such ion exchangers may be controlled by altering the tentacle length. Suitable support materials must carry primary or secondary hydroxyl groups on their surface. So far, matrices utilized are based on silica gel or organic polymers. The tentacle polymer usually consists of

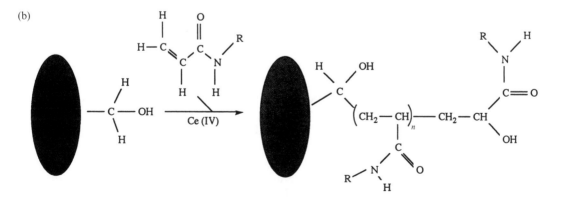

Figure 3.17 Tentacle-type ion exchangers as manufactured by Merck. (a) The interaction between a protein and a tentacle-type exchanger of opposite charge. (b) The principle of chemical modification by graft polymerization of acrylamide derivatives onto hydrophilic support beads

repeating acrylamide groups, which have been substituted with suitable charged groups. Suitable groups include DEAE groups in the case of anion exchangers or sulpho groups in the case of cation exchangers. These are covalently attached to the support matrix by a process known as graft polymerization. In this process, the matrix support is incubated with the substituted acrylamide monomers and Cs (IV) ions (Figure 3.17b). The caesium ions bind the hydroxyl groups on the matrix surface and promote radical formation. This initiates a chemical chain reaction resulting in the covalent attachment of a substituted acrylamide group to the support matrix, with subsequent polyacrylamide chain growth. A linear polyacrylamide chain, usually consisting of 15–25 acrylamide monomers, is then synthesized which contains charged groups spaced along its length.

Hydrophobic interaction chromatography

Of the 20 amino acids commonly found in proteins eight are classified as hydrophobic, because of the non-polar nature of their side chains (R groups, Figure 3.18). Most proteins are folded such that the majority of their hydrophobic amino acid residues are buried internally in the molecule, and hence are shielded from the surrounding aqueous environment (Chapter 1). Internalized hydrophobic groups normally associate with adjacent hydrophobic groups. A minority of hydrophobic amino acids are, however, present on the protein surface, and hence are exposed to the outer aqueous environment. Different protein molecules differ in the number and type of hydrophobic amino acids on their surface, and hence their degree of surface hydrophobicity. Hydrophobic amino acids tend to be arranged in clusters or patches on the protein surface. Hydrophobic interaction chromatography fractionates proteins by exploiting their differing degrees of surface hydrophobicity. It depends on the occurrence of hydrophobic interactions between the hydrophobic patches on the protein surface and hydrophobic groups covalently attached to a suitable matrix.

The most popular hydrophobic interaction chromatographic resins are cross-linked agarose gels to which hydrophobic groups have been covalently linked. Specific examples include octyl- and phenyl-Sepharose gels from Pharmacia, which contain phenyl and octyl hydrophobic groups, respectively (Figure 3.19). Attachment of octyl groups to a suitable matrix yields chromatographic beads, which are very hydrophobic in character. Chromatography of highly hydrophobic proteins using octyl-Sepharose should be avoided, as the protein may bind so strongly to the gel matrix that elution becomes impossible without inducing protein denaturation. A variety of hydrophobic interaction chromatography resins are also commercially available from companies such as Bio-Rad and Merck.

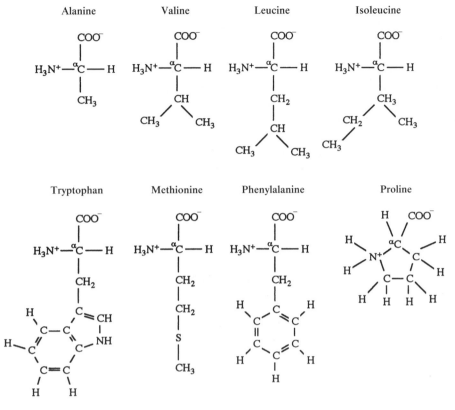

Figure 3.18 Structural formulae of the eight commonly occurring amino acids that display hydrophobic characteristics

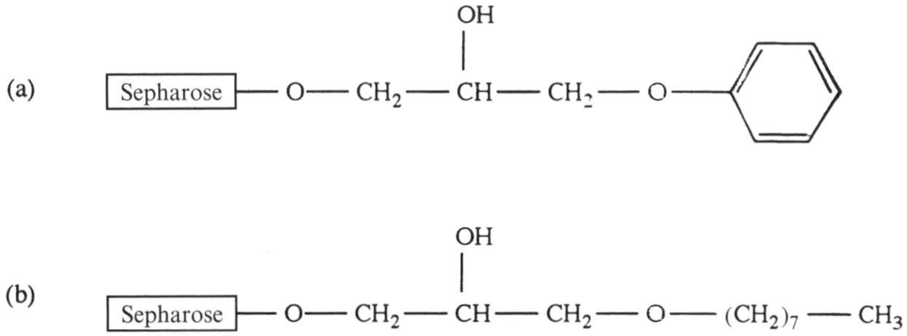

(a) [Sepharose] ─ O ─ CH$_2$ ─ CH ─ CH$_2$ ─ O ─ ⬡

with OH above the central CH

(b) [Sepharose] ─ O ─ CH$_2$ ─ CH ─ CH$_2$ ─ O ─ (CH$_2$)$_7$ ─ CH$_3$

with OH above the central CH

Figure 3.19 Chemical structure of (a) phenyl and (b) octyl Sepharose, widely used in hydrophobic interaction chromatography

Protein separation by hydrophobic interaction chromatography is dependent upon interactions between the protein itself, the gel matrix and the surrounding solvent – which is usually aqueous. Increasing the ionic strength of a solution by the addition of a neutral salt (e.g. ammonium sulfate or sodium chloride) increases the hydrophobicity of protein molecules. This may be explained somewhat simplistically on the basis that the hydration of salt ions in solution results in an ordered shell of water molecules forming around each ion. This attracts water molecules away from protein molecules, which in turn helps to unmask hydrophobic domains on the surface of the protein. As such increases in ionic strength enhance the surface hydrophobicity of protein molecules hydrophobic interaction chromatography is often preformed following a salt precipitation step, or after ion exchange chromatography.

Not all ionic solutions are equally effective in promoting hydrophobic interactions. Anions may be arranged in order of increasing salting-out effect, that is in the order of increased ability to promote hydrophobic interactions.

$$\text{increasing salting-out effect} \rightarrow$$

$$\text{Anions: } SCN^-, I^-, ClO_4^-, NO_3^-, Br^-, Cl^-, CH_3COO^-, SO_4^{2-}, PO_4^{3-}$$

Cations are often arranged in order of increasing chaotropic effect, the increasing tendency to disrupt the structure of water. Increases of such chaotropic effects leads to a decrease in the strength of hydrophobic interactions.

$$\text{increasing chaotropic effect} \rightarrow$$

$$\text{Cations: } NH_4^+, Rb^+, K^+, Na^+, Cs^+, Li^+, Mg_2^+, Ca_2^+, Ba_2^+$$

Protein samples are therefore best applied to hydrophobic interaction columns under conditions of high ionic strength. As they percolate through the column, proteins may be retained via hydrophobic interactions. The more hydrophobic the protein, the tighter the binding. After a washing step, bound protein may be eluted utilizing conditions that promote a decrease in hydrophobic interactions. This may be achieved by irrigation with a buffer of decreased ionic strength, inclusion of a suitable detergent, or lowering the polarity of the buffer by including agents such as ethanol or ethylene glycol. Subsequent to protein elution hydrophobic interaction chromatographic resins must be washed extensively to remove any tightly bound proteins. An initial washing step with a detergent such as sodium dodecyl sulfate (SDS) may be required to remove very tightly bound protein. Subsequent cleaning steps often involve the use of ethanol, butanol and water.

Reverse phase chromatography may also be used to separate proteins on the basis of differential hydrophobicity. This technique involves applying the protein sample to a highly hydrophobic column to which most proteins will bind. Elution is promoted by decreasing the polarity of the mobile phase. This is normally achieved by the introduction of an organic solvent. Elution conditions are harsh, and generally result in denaturation of many proteins.

Affinity chromatography

Affinity chromatography is often described as the most powerful highly selective method of protein purification available. This technique relies on the ability of most proteins to bind specifically and reversibly to other compounds, often termed ligands (Figure 3.20). A wide variety of ligands may be covalently attached to an inert support matrix, and subsequently packed into a chromatographic column. In such a system, only the protein molecules which selectively bind to the immobilized ligand will be retained on the column. Washing the column with a suitable buffer will flush out all unbound molecules. An appropriate change in buffer composition, such as inclusion of a competing ligand, will result in desorbtion of the retained proteins.

A wide variety of ligands have been employed to selectively purify proteins (Table 3.9). Biospecific affinity ligands, as the name suggests, interact in a specific manner with the target protein. The exact degree of specificity achieved allows sub-classification into a 'specific ligand' or 'general ligand' approach. Pseudoaffinity chromatography, as typlified by dye affinity systems, utilizes ligands which interact with the proteins in a far less biospecific manner; indeed the actual basis of interaction in such instances is often poorly understood.

Biospecific affinity chromatography   The general ligand approach is exemplified by the immobilization of ATP or cofactors such as nicotinamide adenine dinucleotide (NAD^+) in affinity systems designed to purify various ATP or cofactor-binding enzymes, or the use of immobilized lectins to purify glycoproteins (see below). One advantage of such systems

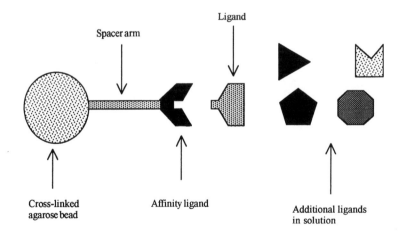

Figure 3.20 Schematic representation of the principle of biospecific affinity chromatography. The chosen affinity ligand is chemically attached to the support matrix (agarose bead) via a suitable spacer arm. Only those ligands in solution that exhibit biospecific affinity for the immobilized species will be retained

Table 3.9   The various forms of affinity chromatography which may be used to purify a protein

Classification	Description
Biospecific affinity chromatography; specific ligand approach	Use of a ligand displaying a high degree of biospecificity for the target protein, e.g. the use of a highly specific enzyme substrate/ substrate analogue to purify an enzyme or the use of an antibody raised against the target protein (immunoaffinity chromatography)
Biospecific affinity chromatography; general ligand approach	Use of a ligand capable of binding a specific category / family of proteins; e.g. immobilized NAD^+ for the purification of NAD^+-dependant dehydrogenases, or immobilized lectins to purify certain glycoproteins.
Pseudoaffinity chromatography	Use of a ligand known to bind various proteins, but not in a biospecific matter; e.g. dye affinity chromatography

is that the same affinity gel may be used to purify a number of different proteins but the degree of selectivity achieved falls well below that of the specific ligand approach. In this case ligands used include enzyme substrates, substrate analogues or inhibitors, in addition to antibodies raised specifically against the protein of interest.

One particularly elegant example illustrating the concept of the specific ligand approach involves the purification of lactate dehydrogenase (LDH) using an immobilized oxamate affinity column (Figure 3.21). Oxamate is a structural analogue of pyruvate, a substrate for LDH. The reaction catalysed by LDH may be represented as

$$\text{Pyruvate} + \text{NADH} + \text{H}^+ \overset{\text{LDH}}{\rightleftharpoons} \text{Lactate} + \text{NAD}^+$$

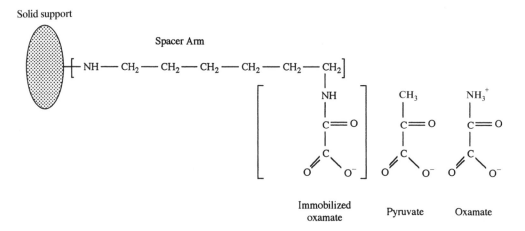

Figure 3.21   Structural relationship of pyruvate, oxamate and the immobilized oxamate employed in the affinity purification of lactate dehydrogenase

Lactate dehydrogenase exhibits an ordered kinetic mechanism in which the enzyme must first bind reduced NAD (NADH) before binding pyruvate or alternatively oxamate. If NADH is removed the enzyme can no longer bind the substrate.

Inclusion of NADH in a crude enzyme preparation containing LDH followed by percolation through an oxamate column, promotes binding of LDH to the immobilized oxamate. All other protein molecules pass through unbound. The column may subsequently be washed with buffer, to ensure that all unbound molecules are washed out. The LDH molecule remains adsorbed to the column as long as this washing buffer contains NADH. Omission of the NADH in the elution buffer results in spontaneous elution of the enzyme. Using such an affinity column LDH from crude extracts of human placenta was purified to homogeniety in a single step, with a yield in excess of 98 per cent.

Synthesis of any affinity gel may be discussed under three main headings;

- choice of affinity ligand;

- choice of support matrix;

- choice of chemical coupling technology.

The chosen ligand should ideally exhibit high specificity towards the protein of interest and binding to it should be reversible, such that desorption may be achieved under relatively mild conditions. The ligand should also be quite stable, as conditions required for their chemical coupling to the support matrix are often harsh. As already mentioned, many biological macromolecules have been successfully employed as affinity ligands. Most such ligands, however, suffer from a number of limitations, including high cost and low stability. The development of synthetic dye ligands which can bind a variety of proteins has overcome many such disadvantages.

The support matrix to be used should be;

- physically and chemically stable;

- sufficiently rigid to allow high working flow rates;

- chemically inert so as not to contribute to non-specific binding of proteins;

- exhibit a high degree of porousity, with pores sufficiently large to allow free entry of protein molecules to internally bound ligands, which greatly increases the binding capacity of the column;

- it must also be derivatizable, inexpensive and reusable.

Support matrices employed include agarose, cellulose, silica and various organic polymers. Cross-linked agarose is probably the most popular matrix employed.

A wide variety of chemical immobilization techniques may be used to couple the ligand to the matrix. In many cases, the ligand is not coupled directly to the support, but is attached via a spacer arm. A variety of spacer arms may be used, the most popular of which consists simply of a number of methylene groups (Figure 3.21). Spacer arms are required in many instances in order to overcome steric difficulties associated with the binding of large protein molecules. Matrices may be activated by a number of chemical means which generates a chemically unstable and therefore reactive support matrix. This is subsequently derivitized with the chosen ligand, or initially by a spacer arm followed by coupling of the ligand. Activation with cyanogen bromide was one of the first matrix activation methods developed. Cyanogen-bromide activated agarose is still widely used in the immobilization of a wide variety of ligands containing primary amino groups. The resultant ligand, however, exhibits a positive charge at neutral pH which confers ion exchange properties on the gel. The coupling bond is also susceptable to hydrolysis under certain conditions, which may result in leakage of ligand from the column. Many other methods of activation have subsequently been developed, some of which yield extremely stable derivitized gels.

Elution of bound protein from an affinity column is achieved by altering the composition of the elution buffer, such that the affinity of the protein for the immobilized ligand is greatly reduced. A variety of non-covalent interactions contribute to protein–ligand interaction. In many cases, changes in buffer pH, ionic strength, inclusion of a detergent or agents such as ethylene glycol which reduce solution polarity, may suffice to elute the protein. In other cases, inclusion of a competing ligand promotes desorbtion. Competing ligands often employed include free substrates, substrate analogues or cofactors. Use of a competing ligand generally results in more selective protein desorbtion than does the generalized approach such as alteration of buffer pH or ionic strength. In some cases, a combination of such elution conditions may be required. Identification of optimal desorbtion conditions often requires considerable empirical study.

Affinity chromatography offers many advantages over conventional chromatographic techniques. The specificity and selectivity of biospecific affinity chromatography cannot be matched by other chromatographic procedures. Increases in purity of over 1000-fold, with almost 100 per cent yields are often reported, at least on a laboratory scale. Incorporation of an affinity step could thus drastically reduce the number of subsequent steps required to achieve protein purification. This in turn could result in dramatic time and cost savings which would be particularly significant in an industrial setting. Despite such promise, biospecific affinity chromatography does not enjoy widespread industrial use. This fact reflects a number of disadvantages associated with this approach:

- as mentioned earlier, many biospecific ligands are extremely expensive and often exhibit poor stability;

- many of the ligand coupling techniques are chemically complex, hazardous, time-consuming and costly;

- any leaching of coupled ligands from the matrix also gives cause for concern for two reasons: (a) it effectively reduces the capacity of the system and (b) leaching of what are often noxious chemicals into the protein products is undesirable, particularly if the protein is being purified for commercial application.

It is not normally prudent to employ biospecific affinity chromatography as an initial purification step, as various enzymatic activities present in the crude fractions may modify or degrade the expensive gels. It should however, be utilized as early as possible in the purification procedure in order to accrue the full benefit afforded by its high specificity. For all the above reasons classical affinity systems, despite their great promise, have found limited use outside research laboratories.

Immunoaffinity purifications Immobilized antibodies may be used as affinity adsorbants for the antigens that stimulated their production (Figure 3.22). Antibodies, like many other biomolecules may be immobilized on a suitable support matrix by a variety of chemical coupling procedures. Many of the initial immunoaffinity columns utilized mammalian polyclonal antibodies raised against purified preparations of antigen. Thus, a prerequisite designing an immunoaffinity column was that an effective purification scheme for the antigen in question must already exist. Many immunoaffinity columns employing polyclonal antibody preparations exhibited low binding capacity for the protein of interest.

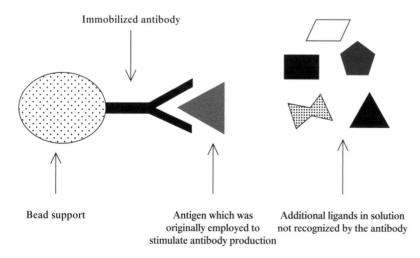

Figure 3.22 Principle of immunoaffinity chromatography. Only antigen that is specifically recognized by the immobilized antibody will be retained on the column

In addition, they also adsorbed some other proteins as the antibodies specific for the protein of interest constituted only a small proportion of the total polyclonal antibody preparation.

Monoclonal antibody technology has resulted in an increased application of this technique, as it makes possible the production of large quantities of antibody which are monospecific towards the antigen of interest. Furthermore, it is not necessary to fully purify the protein antigen of interest in order to generate monoclonal antibodies.

Although immunoaffinity chromatography is probably the most highly specific of all forms of biospecific chromatography its general applicability as a protein purification method has been delayed for a number of reasons.

- Like other forms of affinity chromatography, it is a relatively high cost technique. Generation and immobilization of appropriate monoclonal antibodies is certainly not an inexpensive exercise.

- Ligand (antibody) leakage from the column may occur, and is undesirable.

- In many cases, elution of the bound protein from the immobilized antibody is not readily achieved. Desorbtion generally requires conditions which result in partial denaturation of the bound protein. This is often achieved by alteration of buffer pH or by employing agents such as urea or guanidine. One of the most popular elution methods employed involves irrigation with a glycine-HCl buffer at pH 2.2–2.8. In some cases, elution is more readily attainable at alkaline pH values. Specific examples have been documented in which protein elution was performed under relatively mild conditions, such as a change of buffer system or an increase in ionic strength, however, such examples are exceptional.

Despite such drawbacks, immunoaffinity chromatography does offer extreme biospecificity and has found some application on both a laboratory and industrial scale. The inclusion of an immunoaffinity step in the purification of recombinant blood factor VIII used to treat haemophilia (Chapter 5) is one example of the industrial use of this technique.

Protein A chromatography   Most species of *Staphylococcus aureus* produce a protein known simply as protein A. This protein consists of a single polypeptide chain of molecular mass 42 kDa. Protein A binds the Fc region (the constant region) of immunoglobulin G (IgG) obtained from human and many other mammalian species with high specificity and affinity. Immobilization of protein A on Sephadex or agarose beads provides a powerful affinity system which may be used to purify IgG. There is however a considerable variation in the binding affinity of protein A for various IgG subclasses obtained from different mammalian sources. In some cases another protein, protein G, may be used instead of protein A. Most immunoglobulin molecules that bind to immobilized protein A do so under alkaline conditions, and may subsequently be eluted at acidic pH values. Protein A affinity chromatography has significant industrial potential in terms of the purification of high value antibodies destined for diagnostic or therapeutic use.

Lectin affinity chromatography   Lectin affinity chromatography may be used to purify a range of glycoproteins. Initially, lectins were studied because of their ability to promote agglutination of erythrocytes and a number of other cell types. Lectins are a group of proteins synthesized by plants, vertebrates and a number of invertebrate species. Especially high levels of lectins are produced by a variety of plant seeds. Plant lectins are often termed phytohaemagglutinins. All lectins have the ability to bind certain monosaccharides (such as α-D-mannose, α-D-glucose, D-*N*-acetyl galactosamine) and the sugar specificity for many are known (Table 3.10). Among the best known and most widely used lectins are concanavalin A (Con A), soybean lectin (SBL) and wheat germ agglutinin (WGA).

Lectin affinity chromatography may be utlilised to purify a variety of proteins, including various hormones, growth factors and cytokines. Glycoproteins generally bind to lectin affinity columns at pH values close to neutrality. Desorbtion may be achieved in some cases by alteration of the pH of the eluting buffer. The most common method of desorbtion however involves inclusion of free sugar molecules for which the lectin exhibits a high affinity in this elution buffer, i.e. the inclusion of a competing ligand.

Table 3.10   Some lectins commonly used in immobilized format for the purification of glycoproteins

Lectin	Source	Sugar specificity	Eluting sugar
Con A	Jack bean seeds	α-D-mannose, α-D-glucose	α-D-methyl mannose
WG A	Wheat germ	*N*-acetyl-β-D-glucosamine	*N*-acetyl-β-D-glucosamine
PSA	Peas	α-D-mannose	α-D-methyl mannose
LEL	Tomato	*N*-acetyl-β-D-glucosamine	*N*-acetyl-β-D-glucosamine
STL	Potato tubers	*N*-acetyl-β-D-glucosamine	*N*-acetyl-β-D-glucosamine
PHA	Red kidney bean	*N*-acetyl-D-galactosamine	*N*-acetyl-D-galactosamine
ELB	Elderberry bark	Sialic acid or *N*-acetyl-D-galactosamine	Lactose
GNL	Snowdrop bulbs	$\alpha - 1 \rightarrow 3$ Mannose	α-Methyl mannose
AAA	Freshwater eel	α-L-fucose	L-Fucose

Although lectin affinity chromatography may be utilised to purify a variety of glycoproteins it has not been widely employed for a number of reasons:

- Most lectins are quite expensive.

- Crude protein sources containing one glycoprotein usually contain multiple glycoproteins. In most such instances lectin based affinity systems will result in the co-purification of several such glycoproteins.

- Limited application of this approach means it has little track record, particularly on an industrial scale.

Dye affinity chromatography   The development of dye affinity chromatography may be attributed to the observation that some proteins exhibited anomalous elution characteristics when fractionated on gel filtration columns in the presence of blue dextran. Blue dextran consists of a triazine dye (cibacron blue F3G-A) covalently linked to a high molecular mass dextran. It is often used to determine void volumes of gel filtration columns. The discovery that some proteins bind the triazine dye soon led to its use as an affinity adsorbant by immobilization on an agarose matrix. A variety of other triazine dyes (Figure 3.23) also bind certain proteins and hence have also been used as affinity adsorbants. Dye affinity chromatography has become popular for a number of reasons:

- The dyes are readily available in bulk and are relatively inexpensive.

- Chemical coupling of the dyes to the matrix is usually straightforward, often requiring no more than incubation under alkaline conditions at elevated temperature. The use of noxious chemicals such as cyanogen bromide is avoided and in many cases incorporation of spacer arms is not required.

- The dye – matrix bead linkage is relatively resistant to chemical, physical and enzymatic degradation. In this way ligand leakage from the column is minimized and is easily recognizable if it does occur – due to the dye colour.

- The protein binding capacity of immobilized dye adsorbants is also high and exceeds the binding capacity normally exhibited by natural biospecific adsorbtion ligands.

- Elution of bound protein is also relatively easily achieved.

One problem is that textile dyes are produced in bulk and preparations often contain varying amounts of impurities, which may adversely affect the columns chromatographic properties. It is not possible to accurately predict if a specific protein will be retained on a dye affinity column, or what conditions will allow optimum binding and elution. Such information must be derived by empirical study. Many companies now supply

Procion Blue MX 3G

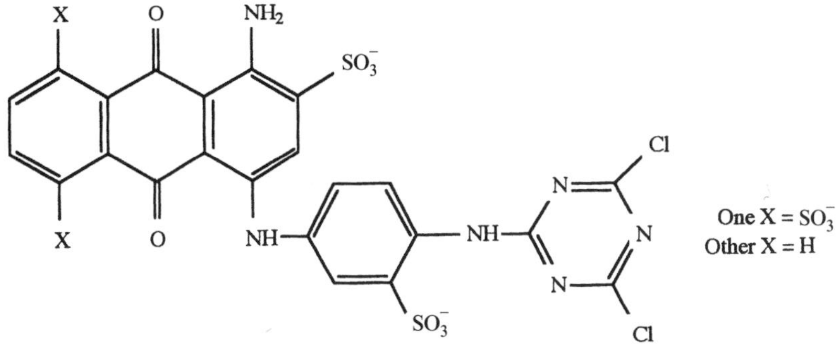

Procion Red H-3B

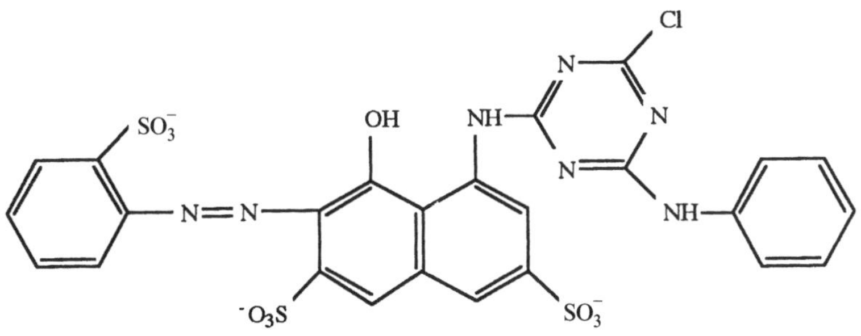

Procion Yellow H-A

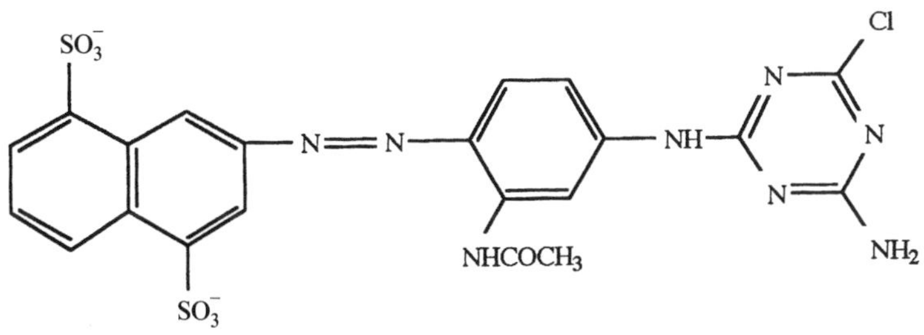

Figure 3.23 Some mono- and dichloro triazine dyes commonly used as affinity ligands in dye affinity chromatography

kits containing a variety of different dye affinity ligands which may be employed in such initial empirical studies. An understanding of the specific interactions which allow many apparently unrelated proteins to bind to dye affinity ligands, while other proteins are not retained, is often lacking. As reactive dyes do not occur in nature truly 'biospecific' interactions with proteins are not possible. Several such dyes however are structurally similar to nucleotide cofactors (e.g. NADH), and can often bind dehydrogenase enzymes. X-ray crystallographic studies confirm that binding occurs at the nucleotide fold of such nucleotide dependant enzymes. Not unsurprisingly reactive dyes have therefore been used to purify a wide range of such dehydrogenases. These dyes however also bind a range of proteins unrelated to nucleotide dependant enzymes. The basis of binding specificity in such cases is less clear. The presence of negatively charged sulfonate groups lends triazine dyes an ion exchange character. These dyes also contain aromatic groups which can lend them some degree of hydrophobicity. Hydrophobic interactions therefore may play some role in protein adsorption. The dyes can also hydrogen bond with proteins.

Metal chelate affinity chromatography   Metal chelate affinity chromatography is a pseudoaffinity protein purification technique first developed in the 1970s. The mode of adsorption relies upon the formation of weak coordinate bonds between basic groups on a protein surface with metal ions immobilized on chromatographic beads (Figure 3.24). The affinity media is synthesized by covalent attachment of a metal chelator

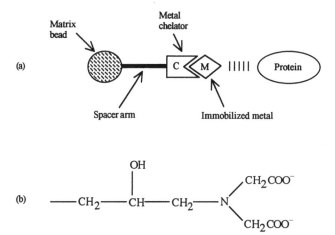

Figure 3.24   Schematic representation of the basic principles of metal chelate affinity chromatography. Certain proteins are retained on the column via the formation of co-ordinate bonds with the immobilized metal ion (a). The actual structure of the most commonly used metal chelator, iminodiacetic acid (IDA) is presented in (b)

to the chromatographic matrix (e.g. cross-linked agarose) via a spacer arm. Chelating agents such as iminodiacetate (IDA) are capable of binding a number of metal ions (e.g. Fe, Co, Ni, Cu, Zn, Al), and binding effectively immobilizes the ion on the bead. The affinity gel is normally supplied without bound metal, so the gel can be 'charged' with the metal of choice (by flushing the column with a solution containing a salt of that metal, e.g. $CuSo_4$ in the case of copper). The metal ions most commonly used are Zn^{2+}, Ni^{2+} and Cu^{2+}. Basic groups on protein surfaces, most notably the side chain of histidine residues, are attracted to the metal ions, forming the weak coordinate bonds. Elution of bound proteins is undertaken by lowering the buffer pH (this causes protonation of the histidine residues, which are then unable to coordinate with the metal ion). Alternatively a strong competitor complexing agent (e.g. the chelating agent ethylenediaminetetraceticacid; EDTA) can be added to the elution buffer.

Metal chelate affinity chromatography finds most prominent application in the affinity purification of recombinant proteins to which a histidine tag has been attached (described later). As protein binding occurs via the histidine residues this technique is no more inherently useful for the purification of metalloproteins than for the purification of non-metallo-proteins (a common misconception, given its name).

Chromatography on hydroxyapatite

Hydroxyapatite occurs naturally as a mineral in phosphate rock and also constitutes the mineral portion of bone. It also may be used to fractionate protein by batch or column chromatography.

Hydroxyapatite is prepared by mixing a solution of sodium phosphate (Na_2HPO_4) with calcium chloride ($CaCl_2$). A white precipitate known as brushite is formed. Brushite is then converted to hydroxyapatite by heating to $100°C$ in the presence of ammonia. Although hydroxyapatite may be obtained commercially, it is relatively easily prepared in house.

$$Ca_2HPO_4.2H_2O \xrightarrow{\text{heat} 100°C} Ca_{10}(PO_4)_6(OH)_2$$
$$\text{Brushite} \qquad\qquad\qquad \text{Hydroxyapatite}$$

The underlying mechanism by which this substance binds and fractionates proteins is poorly understood. Protein adsorbtion is believed to involve interaction with both calcium and phosphate moieties of the hydroxyapatite matrix. Elution of bound species from such columns is normally achieved by irrigation with a potassium phosphate gradient.

Calcium phosphate–cellulose gels represent a variation of hydroxyapatite chromatography. These gels are prepared by incubation of a calcium phosphate slurry with a slurry of cellulose powder, which seems to result in coating of the cellulose matrix with the calcium phosphate.

Chromatography on hydroxyapatite is not regularly encountered, most likely because of the availability of other chromatographic methods whose mode of fractionation is understood. Preparation of hydroxyapatite, although straightforward, represents an additional inconvenience. In some cases, variation in batch to batch chromatographic characteristics have been observed.

Chromatofocusing

Fractionation by chromatofocusing represents a relatively new chromatographic technique, which separates proteins on the basis of their isoelectric points. This technique basically involves percolating a buffer of one pH through an ion exchange column which is pre-equilibrated at a different pH. Due to the natural buffering capacity of the exchanger, a continuous pH gradient may be set up along the length of the column. In order to achieve maximum resolution a linear pH gradient must be constructed. This necessitates the use of an eluent buffer and exchanger that exhibit even buffering capacity over a wide range of pH values. The range of the pH gradient achieved will obviously depend on the pH at which the ion exchanger is pre-equilibrated and the pH of the eluent buffer. The matrix most often used in chromatofocusing is a weak anion exchanger, which exhibits a high buffering capacity. This anion exchanger is pre-equilibrated at a high pH value. The sample is then applied, usually in the running buffer, whose pH is lower than that of the pre-equilibrated column. After sample application, the column is constantly percolated with a specially formulated (commercially available) buffer containing a large number of buffering species. Because of its lower pH value relative to the initial column pH, this percolation results in the establishment of an increasing pH gradient down the length of the column.

Upon sample application, negatively charged proteins immediately adsorb to the anion exchanger, while positively charged proteins flow down the column. Because of the increasing pH gradient formed such positively charged proteins will eventually reach a point within the column where the column pH equals their own pI values, (their isoelectric points). Immediately upon further migration down the column such proteins become negatively charged, as the surrounding pH values increase above their pI values – hence they bind to the column. Overall therefore, upon initial application of the elution buffer, all protein species will migrate down the column until they reach a point where the column pH is marginally above their isoelectric points. At this stage they bind to the anion exchanger. Proteins of differing isoelectric points are thus fractionated on the basis of this parameter of molecular distinction. The pH gradient formed is not a static one. As more elution buffer is applied, the pH value at any given point along the column is continually increasing. Thus, any protein which binds to the column will be almost immediately desorbed as once again it experiences a surrounding pH value above its pI, and becomes positively charged. Any such desorbed protein flows down the column until it reaches a further point where the pH value is marginally above its pI value, and it again rebinds. This process is repeated until the protein emerges from the column at its isoelectric point. To achieve best results, the isoelectric point (pI value) of the required protein should ideally be in the middle of the pH gradient generated. Chromatofocusing can result in a high degree of protein resolution, with protein bands being eluted as tight peaks. This technique is particularly effective when used in conjunction with other chromatographic methods during protein purification. Most documented applications of this method still pertain to laboratory-scale procedures. Scale-up to industrial level is somewhat discouraged by economic factors, most notably the cost of the eluent required.

Protein chromatography based on aqueous two-phase separations

As previously outlined, protein partitioning in aqueous two-phase systems can represent a convenient method of achieving a certain degree of protein fractionation. This technique may also be adapted such that it may be used in the form of column chromatography. Construction of a liquid – liquid partition column requires the selective binding of one or other of the two incompatible aqueous phases to a suitable chromatographic support, forming the stationary phase which is then packed into a chromotographic column. The second (mobile) phase is then percolated through the column. Fractionation of proteins applied to such a column depends upon their relative affinity for the two phases used. Proteins exhibiting a higher affinity for the mobile phase than for the stationary phase elute more rapidly than proteins which exhibit greater interactions with the stationary phase.

Matrix supports generally used are incompatible with one or other of the phase forming polymers. The support material will then repel the incompatible polymer and is coated by the second phase, effectively shielding the support from the incompatible phase. Cellulose is incompatible with PEG and is thus coated by dextran in PEG – dextran phases. However, due to its dense fibrous structure, the capacity of this system is relatively low. A dextran-rich phase will also coat cross-linked polyacrylamide supports. Polyacrylamide beads, however, tend to swell in the presence of dextran. The resultant swollen particles exhibit poor mechanical stability and collapse under low pressures. This limits the use of polyacrylamide-based systems for such purposes.

An alternative approach to designing suitable stable support matrices for use in aqueous two-phase partition chromatography, involves the fixing of linear polyacrylamide chains on the surface of a suitable support matrix bead by the process of graft polymerization. The resultant support bead exhibits high mechanical stability and are readily coated by the dextran-rich phase. Support materials onto which polyacrylamide may be successfully grafted include silica and polyacrylated vinyl polymer. Liquid – liquid partition chromatography in two-phase systems may be used to separate nucleic acids as well as proteins.

HPLC of proteins

Most of the chromatographic techniques described thus far are performed under relatively low pressures, generated by gravity flow or by low pressure pumps (low pressure liquid chromatography or LPLC). Fractionation of a single sample on such chromatographic columns typically requires a minimum of several hours to complete as relatively slow flow rates must be maintained. Low flow rates are required because as the protein sample flows through the column, the proteins are brought into contact with the surface of the chromatographic beads by direct (convective) flow. The protein molecules then rely entirely upon molecular diffusion to enter the porous gel beads. This is a slow process, especially when compared to the direct transfer of proteins past the outside surface of the gel beads by liquid flow. If a flow rate significantly higher than the diffusional rate is used band spreading (and hence loss of resolution) will result. This occurs due to the fact that any protein molecules which have not entered the bead will flow through the column at a faster rate than the (identical) molecules which have entered into the bead particles. Such high flow rates will also result in a lowering of adsorption capacity as many molecules will not have the opportunity to diffuse into the beads as they pass through the column.

One approach which allows increased chromatographic flow rates without loss of resolution entails the use of microparticulate stationary-phase media of very narrow diameter. This effectively reduces the time required for molecules to diffuse in and out of the porous particles. Any reduction in particle diameter dramatically increases the pressure required to maintain a given flow rate. Such high flow rates may be achieved by utilizing high pressure liquid chromatographic systems. By employing such methods, sample fractionation times may be reduced from hours to minutes. Preparative HPLC has attracted considerable industrial interest in recent years due to the reductions in processing times. Although significant advances have been made in this regard, the use of preparative HPLC systems for the purification of proteins is still relatively limited.

The successful application of HPLC was made possible largely by (a) the development of pump systems which can provide constant flow rates at high pressure and (b) the identification of suitable pressure resistant chromatographic media. Traditional soft gel media utilized in low pressure applications are totally unsuited to high pressure systems due to their compressability.

An ideal support material should:

- be mechanically and chemically stable;
- exhibit a low degree of non-specific adsorbtion;
- be readily reuseable;
- be inexpensive;
- be available in small particle sizes with a narrow range of particle and size distribution;
- exhibit a high degree of porosity.

Obviously no one material satisfies all of the above criteria. Silica gel fulfils some of the more important criteria, most notably its high mechanical stability. As such, it has found widespread application in high pressure chromatographic systems. Silica gel, however, is unstable under alkaline conditions, and does display some non-specific adsorbtion characteristics. Its relative instability at alkaline pH values limits its usefulness in the purification of therapeutically important proteins (a major potential use) as the current preferred method of depyrogenation (Chapter 4) of chromatographic columns involves prolonged standing in solutions of 1 M NaOH. This disadvantage may be overcome by coating the silica particles with a suitable hydrophilic layer.

Various organic polymers such as crosslinked polystyrene have been developed as alternatives to silica. Although many such substances are chemically more stable, some are difficult to produce with diameters as low as the silica particles normally used. Typically, stationary phase particle diameters for HPLC range between 5–55 μm, tenfold smaller than diameters of particles utilized in low pressure liquid chromatographic applications.

In the context of protein purification and characterization HPLC may be used for analytical or preparative purposes. Most analytical HPLC columns available have diameters ranging from 4 to 4.6 mm and lengths ranging from 10 to 30 cm. Preparative HPLC columns currently available have much wider diameters, typically up to 80 cm, and can be longer than a metre (Figure 3.25). Various chemical groups may be incorporated into the matrix beads; thus, techniques such as ion exchange, gel filtration, affinity, hydrophobic interaction and reverse phase chromatography are all applicable to HPLC.

Many small proteins, in particular those that function extracellularly (e.g. insulin, growth hormone and various cytokines) are quite stable and may be fractionated on a variety of HPLC columns without significant denaturation or decrease in bioactivity. Preparative HPLC may be used in industrial scale purification of insulin and of interleukin-2 (IL-2). In contrast, many larger proteins (e.g. blood factor VIII) are relatively labile and loss of activity due to protein denaturation may be observed upon high pressure fractionation.

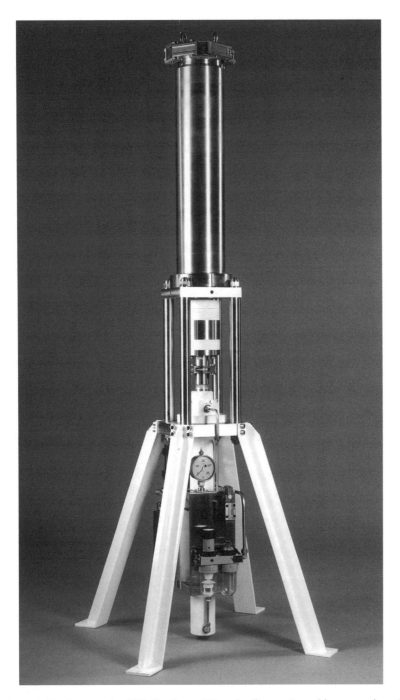

Figure 3.25   Preparative HPLC column (15 cm in diameter) used in processing of proteins required for therapeutic or diagnostic purposes. Column manufactured by Prochrom, Nancy, France (photograph courtesy of Affinity Chromatography Ltd)

At both preparative and analytical levels HPLC exhibits several important advantages as compared to low pressure chromatographic techniques:

- HPLC offers superior resolution due to the reduction in bead particle size. The diffusional distance inside the matrix particles is minimized, resulting in sharper peaks than those obtained when low pressure systems are employed.

- Because of increased flow rates HPLC systems also offer much improved fractionation speeds, typically in the order of minutes rather than hours.

- HPLC is amenable to a high degree of automation.

The major disadvantages associated with HPLC include cost and to a lesser extent, capacity. Even the most basic HPLC systems are expensive compared to low pressure systems. Solvents and other disposables required for the day to day operation of such systems are also quite expensive. At a preparative level, the installation and maintenance costs of a HPLC system must be balanced against the financial savings associated with the overall reduction in processing times achieved. In some instances HPLC column capacity may also be an important deciding factor, although some modern preparative systems may fractionate 100 g or more of protein per production cycle.

For both technical and economic reasons preparative HPLC is employed almost exclusively in downstream processing of low volume, extremely high value proteins, mostly intended for therapeutic use (Figure 3.26). HPLC systems will find wider biotechnological applications at an analytical level. The high resolution, sensitivity and in particular speed afforded by this technique renders the method suitable for routine in-process purity assessments, and in finished product quality control checks. HPLC systems are already widely used in such applications, most notably in the purification of human insulin produced semisynthetically from porcine insulin, in addition to recombinant human insulin produced in microbial systems (Chapter 7).

An alternative chromatographic system to HPLC has been introduced relatively recently. Termed FPLC or fast protein liquid chromatography, this technique employs operating pressures significantly lower than those used in conventional HPLC systems. Lower pressures allow use of matrix beads based on polymers such as agarose. FPLC chromatographic columns are constructed of glass or inert plastic materials. Conventional HPLC columns are manufactured from high grade stainless steel. In many cases, FPLC systems are economically more attractive than HPLC. Despite their operation at lower pressures they still combine high resolution with enhanced speed of operation when compared to low pressure systems. For larger scale applications (e.g. pilot scale studies or small scale production) FPLC-based chromatographic systems such as Pharmacia's BioPilot system are most appropriate. The BioPilot system is a

Figure 3.26 A preparative high-performance liquid chromatography system, manufactured by Prochrom, Nancy, France (photograph courtesy of Affinity Chromatography, Ltd)

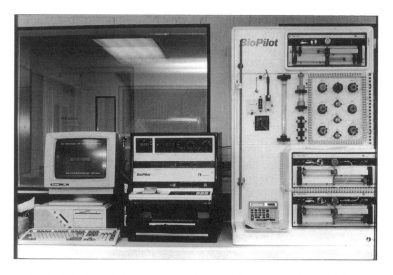

Figure 3.27 The BioPilot system manufactured by Pharmacia Ltd (photograph courtesy of British Biotechnology Ltd)

fully automated liquid chromatography system based on FPLC technology (Figure 3.27). The system consists of control and fractionation units. The computerized control unit allows control of parameters such as flow rates and eluate composition and controls the UV monitor and chart recorder, which detect and record protein elution profiles from the column. The separation unit consists of all the buffer reservoir systems in addition to the chromatographic columns. Several different chromatographic columns may be housed in the BioPilot system and are interconnected by a series of connection tubes and valves. The system may be operated with column volumes ranging from 1 ml to 30 l. A wide variety of column types, such as ion exchange, gel filtration, hydrophobic interaction and affinity columns, may be used in the system. Integrated high precision pumps allow maximum flow rates of up to 100 ml/min.

Larger automated chromatographic systems have been designed to cater for large scale production of high value proteins. Pharmacia's Bio-Process system consists of an integrated control panel, process pump and separation unit, which may consist of several different types of chromatographic columns. Flow rates of up to 400 l h are achievable and this system can purify proteins in multigram to kilogram quantities per production cycle. As in the case of the BioPilot system, the BioProcess system is designed to allow rigorous sanitizing, which is required to successfully produce proteins for *in vivo* diagnostic or therapeutic applications (Chapter 4). The high degree of process automation to which such systems lend themselves contributes enormously to overall process economy.

Expanded bed chromatography

Conventional protein purification procedures require removal of particulate matter from the protein sample prior to chromatographic fractionation. As the stationary phase (the beads) are tightly packed in the chromatographic column, application of a protein stream containing particulates (e.g. whole cells or cell debris) would quickly result in the formation of a plug of trapped solids near the bed surface, with consequent reductions in flow rate, and increased back pressure.

Expanded bed chromatography aims to overcome this difficulty, allowing direct application of whole cell or cellular debris containing protein solutions to an adsorption chromatographic column. As this would render redundant initial clarification steps, the protein purification process would be more convenient and of shorter duration (Figure 3.28). While the approach is equally applicable on the laboratory and industrial scale its benefits would be most apparent for large-scale industrial operations, when the consequent reduction is equipment (e.g. centrifuges, filters) and processing time required would translate into considerable financial savings. As such, expanded bed chromatography has generated most interest in the industrial sector. Expanded bed chromatography is characterized by upward flow of mobile phase through the column, which in turn allows the beads to rise from their settled state, as illustrated in Figure 3.29. Particulate matter present in the applied solution can therefore pass straight through the column without clogging it.

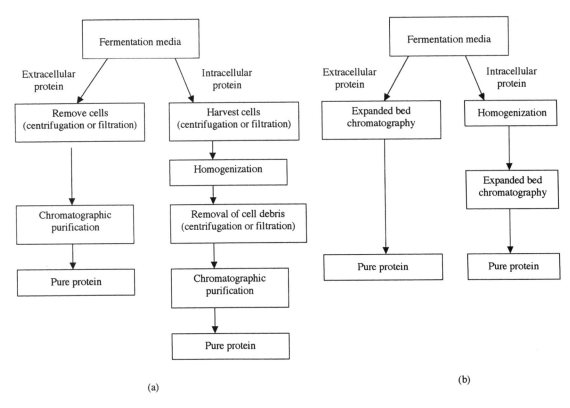

Figure 3.28 Comparison of a generalized protein purification scheme for a microbial protein using (a) conventional column chromatography and (b) expanded bed chromatography. Refer to text for details

The generation of a stable expanded bed (Figure 3.29 c) is achieved by judicious choice of:

- bead density;
- flow rate of mobile phase;
- bead size distribution.

Obviously the greater the bead density the greater the upward flow rate required to maintain the appropriately expanded state. Beads such as cross-linked agarose can be used but only at very low flow rates. Beads of higher density have thus been developed, allowing the use of higher flow rates. Examples include porous glass and perfluoropolymer based beads. In addition, composite beads have been developed, consisting of a core dense material (e.g. quartz or various metals) coated with an outer layer of agarose. Empirical study reveals that beads displaying densities of 1.1–1.3 g/ml are best suited to expanded bed format. An appropriate

Mobile phase

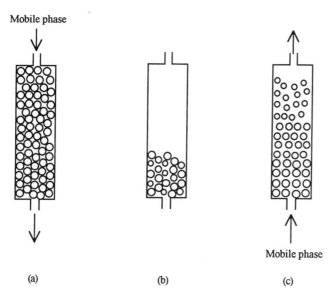

Mobile phase

(a)   (b)   (c)

Figure 3.29 Simplified schematic comparison of packed bed versus expanded bed chromatography. In packed bed mode the chromatographic beads are closely packed and mobile phase flow is in a downward direction (a). In the case of expanded bed operations the beads, when settled, only take up a proportion (usually half or less) of the column volume (b). The mobile phase is applied in an upward direction which promotes lifting of the beads as shown in (c). This ensures that particles in the mobile phase can pass through the column without blocking it. Specialized beads are used in expanded bed operations, and these display a broad range of diameters, as discussed in the main text

capture ligand is then attached to the bead and expanded bed chromatography may therefore be run in ion exchange, hydrophobic interaction or affinity mode.

The use of beads displaying an appropriate range of diameters is also central to the generation of a stable expanded bed. Typically these beads will display a particle size distribution from approximately 100 to 300 μm in diameter. As the other parameters (bead density and flow rate) are constant, generation of the expanded bead will result in the beads of greatest diameter remaining at the column base, while the smaller beads will be found near the top (Figure 3.29 c). Once it has reached equilibration this size distribution will ensure the beads will not move vertically (to any great extent) within the column as it operates in expanded bed mode. This is obviously important in terms of chromatographic performance. (Chromatographic beads used in traditional packed bed systems also display some variation in diameter, but as the beads are tightly packed in this case, there is no scope for vertical movement in any case).

The equipment required for expanded bed chromatography is similar to that required for packed bed operations. The column design is modified slightly to ensure the liquid inlet achieves an even distribution of flow over the entire cross-sectional area of the beads. The adapter at the top of the column must also be movable in order to facilitate bed expansion.

Operation of expanded bed columns is straightforward. After washing with the initial equilibrating buffer (which also creates the stabilised expanded bed), the particulate containing feedstock is applied. Selective protein adsorption occurs as usual. This is followed by a washing step, which removes unbound material, including the remaining particulate matter. Elution can be carried out in expanded bed format, but invariably the beads are allowed to settle before the elution buffer is applied (i.e. the column reverts to packed bed format). This facilitates protein elution in a smaller overall volume. The scientific literature contains increasing numbers of papers reporting the successful operation of column chromatography in expanded bed mode. The technique however remains to be widely used, at laboratory and industrial scale.

Membrane chromatography

Membrane chromatography resembles expanded bed chromatography in that it represents an alternative chromatographic format to that of the packed bead approach. In this case, however, the capture ligand is attached to the inner walls of the pores running through a microfiltration membrane (Figure 3.30). The capture ligand can be charged, hydrophobic or biospecific for a particular protein, allowing operation in ion exchange, hydrophobic interaction or affinity modes (Table 3.11).

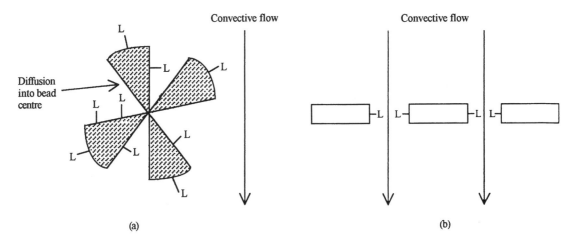

Figure 3.30   Adsorption chromatography using bead versus membrane-based stationary phases. (a) illustrates a trans-section of a chromatographic bead. Solutes (e.g. protein) in the mobile phase are brought into contact with the bead surface by direct convective flow, but rely on diffusion to enter the bead pores. (b) illustrates a trans-section of a chromatographic membrane. Here solutes are brought into contact with the capture ligand by convective flow alone

Table 3.11 A representative sample of some proteins purified using a protocol in which at least one chromatographic step was carried out in membrane mode

Protein purified	Capture ligand in membrane
Bovine serum albumin	Ion exchanger
β-Galactosidease	Ion exchanger
α-Chymotrypsinogen	Ion exchanger
Lysozyme	Ion exchanger
Interleukin-1	Ion exchanger
Bovine serum albumin	Hydrophobic ligand
Chick egg albumin	Hydrophobic ligand
Tumour necrosis factor	Hydrophobic ligand
Soybean trypsin inhibitor	Hydrophobic ligand
Bovine serum albumin	Anti-BSA-antibody (immunoaffinity)
Human IgG	Protein A (affinity)
Interleukin-2	Anti-IL-2 antibody (immunoaffinity)
Bovine catalase	Cibacron blue F3GA (dye affinity)
Cytochrome C	IDA -Cu^{2+} (metal affinity)

The major advantage of membrane chromatography over more conventional systems is increased process speed. In the case of porous beads the protein is brought in contact with the bead surface by direct convective flow. However, the protein molecules must then rely on the much slower process of diffusion to travel down bead pores, where the majority of capture ligands are actually based (Figure 3.30). In practice this diffusional constraint places an upper limit on flow rates appropriate for use with column chromatography. In the case of membrane chromatography the protein-containing mobile phase is brought in contact with the capture ligand only by convective flow, which allows operation of the system at much higher flow rates (typically up to ten times faster than in the case of bead columns). Increased processing speeds are beneficial in terms of e.g. decreasing the duration during which the protein of interest may be in contact with particularly detrimental contaminants such as proteases, and minimizing the inactivation of particularly labile proteins. At an industrial scale faster processing times will save time and hence money.

Membranes manufactured from various different substances, including cellulose, modified cellulose, polyethylene, nylon and acrylic composites are available. The membranes may be configured in a variety of ways, mainly as flat sheets, radial flow or hollow fibre systems.

Purification of recombinant proteins

Proteins produced by recombinant DNA technology are usually purified by means identical to those available for purification of non-recombinant proteins. In fact purification of recombinant proteins can be somewhat more straightforward as high expression levels of the target protein can be attained. This increases the ratio of target protein to contaminants, and can eliminate or reduce the requirement for an initial concentration step. Purification can also be simplified if the recombinant protein displays some pronounced physicochemical characteristic, which is not displayed by native host proteins. This is exemplified by the recombinant production of extremely thermostable proteins (derived from hyperthermophiles; chapter 10) in mesophilic host organisms such as *E. coli.* Heating of a crude extract to 70–80°C for several minutes' results in the denaturation and precipitation of native *E. coli* proteins. The precipitate can be removed by centrifugation, leaving behind a highly purified recombinant protein in solution.

Two features of recombinant production, however, can impact very significantly upon the approach subsequently taken to purify the recombinant product; inclusion body formation and the incorporation of purification tags. The processes of inclusion body formation, recovery and recombinant protein renaturation have been considered in Chapter 2, to which the reader is referred. Once the recombinant protein has been refolded additional purification (if required) follows traditional lines.

Genetic engineering techniques also facilitate the incorporation of specific peptide or protein tags to the protein of interest. A tag is chosen which confers on the resultant hybrid protein some pronounced physicochemical characteristic, facilitating its subsequent purification. Such a molecule is normally produced by fusing a DNA sequence which codes for the tag to one end of the genetic information encoding the protein of interest. Tags that allow rapid and straightforward purification of the hybrid protein by techniques such as ion exchange, hydrophobic interaction or affinity chromatography have been designed and successfully employed.

Addition of a polyarganine (or polylysine) tag to the C terminus of a protein confers on it a strong positive charge. The protein may then be more readily purified by cation exchange chromatography. This approach has been used in the purification of various interferons and urogastrone. Addition of a tag containing a number of hydrophobic amino acids confers on the resultant molecule a strongly hydrophobic character, which allows its effective purification by hydrophobic interaction chromatography. A purification tag consisting of polyhistidine may be employed to purify proteins by metal chelate chromatography.

Tags that facilitate protein purification by affinity chromatography have also been developed. The gene coding for protein A may be fused to the gene or cDNA encoding the protein of interest. The resultant hybrid may be purified using a column containing immobilized IgG. Immunoaffinity purification may be employed if antibodies have been raised against the tag utilized. One synthetic peptide known as Flag has been developed specifically for use as an immunoaffinity tag. The Flag peptide includes a cleavage site recognized by the protease enterokinase. Monoclonal antibodies have been generated which specifically bind the Flag peptide. One such monoclonal antibody has been identified which, when immobilized on a chromatographic matrix, allows desorbtion of the Flag – protein hybrid under mild conditions. Various enzymes may also be used as purification tags. The use of β-galactosidase is particularly popular, as it facilitates purification by readily prepared affinity chromatographic systems.

Upon purification of the hybrid protein it is generally desirable to remove the tag. This is especially critical if the protein is destined for therapeutic use, as the tag itself will be immunogenic. Removal of the tag is generally carried out by chemical or enzymatic means. This is achieved by designing the tag sequence such that it contains a cleavage point for a specific protease or chemical cleavage method at the protein – tag fusion junction. Sequence-specific proteases often employed to achieve tag removal include the endopeptidases trypsin, factor Xa and enterokinase. Exopeptidases such as carboxypeptidase A are also sometimes utilized. Generally speaking endopeptidases, which cleave internal protein peptide bonds, are used to remove long tags whereas exopeptidases are used most often to remove short tags. The exopeptidase carboxypeptidase A, for example, sequentially removes amino acids from the C-terminus of a protein until it encounters a lysine, arginine or proline residue. Chemical cleavage of specific peptide bonds relies on the use of chemicals such as cyanogen bromide or hydroxylamine.

Although several methods exist which can achieve tag removal, most such methods suffer from some inherent drawbacks. One essential prerequisite for any method is that the protein of interest must remain intact after the cleavage treatment. The required protein therefore should not contain any exposed (surface) peptide bonds susceptible to cleavage by the specific method chosen. Chemical methods, for example, must generally be carried out under harsh conditions, often requiring high temperatures or extremes of pH. Such conditions can have a detrimental effect on normal protein functioning. Proteolytic removal of tags is also often less than 100 per cent efficient. Selective cleavage of the tag must be followed by subsequent separation of the tag from the protein of interest. This may require a further chromatographic step.

PROTEIN INACTIVATION AND STABILIZATION

Proteins may be subjected to a wide range of influences that results in loss of their biological activity. Influences can be chemical, physical or biological in nature and can destroy biological activity by either inducing denaturation, by covalently modifying the protein, or by partially degrading it (Table 3.12). In other words, any influence that alters the protein's native structure will likely influence its biological activity (in the vast majority of cases it's biological activity will be decreased or abolished).

Loss of biological activity can occur during (a) protein recovery from its producer source, (b) during the protein purification procedure or (c) subsequent to purification, during protein storage. Proteins vary widely in terms of their susceptibility to any given inactivating influence, and their relative stability or lability in any given circumstance ultimately depends upon their structure.

Although exceptions exist, as a general principle, extracellular proteins tend to be more inherently stable than do intracellular proteins. Extracellular proteins have evolved to function in a largely uncontrolled extracellular environment. Intracellular proteins on the other hand have evolved to function in a far more regulated environment:

- Intracellular pH is generally maintained within narrow parameters (usually between pH 6–7).

- The cell membrane regulates entry of substances into the cell, and intracellular compartmentalisation (eucaryotic cells) also helps keep incompatible biomolecules physically separated. Lyzosomes for example house many hydrolytic enzymes including proteases while in plant cells many chemicals capable of denaturing or modifying proteins are found in vacuoles.

Table 3.12 Various chemical, biological and physical influences which can lead to protein inactivation

Inactivating influence	Example	Comments
Chemical influences	Detergents, organic solvents, chaotrophic agents, oxidising agents, heavy metals	Most induce inactivation by interfering with covalent/non–covalent bonds which stabilize protein structure, and hence lead to protein denaturation. Some (e.g. oxidizing agents) can induce loss of activity by covalently modifying an amino acid residue essential to the protein's biological activity
Biological influences	Proteolytic enzymes, Carbohydrases (glycoproteins) Phosphatases (phosphoproteins) Microbial contaminants	All induce protein inactivation via hydrolysis of covalent bonds. Microbial contaminants (of e.g. a stored protein solution) can induce loss of activity via production of degradative enzymes
Physical influences	Extremes of temperature or pH, Freeze and thawing, vigorous agitation	Most induce protein inactivation by causing denaturation

The negative impact of many such influences is time and concentration dependent. Refer to text for details.

- The intracellular environment is invariably a reducing one, which prevents inactivation of oxygen – sensitive proteins.

- Protein concentration within cells is generally high (up to 400 mg/ml). Proteins are generally more stable when present as concentrated solutions.

- Many chemical metabolites found within cells may specifically or non-specifically stabilize proteins.

Liberation of intracellular proteins by cellular homogenization reverses many of the above conditions. Particular attention to maximizing and maintaining protein stability is thus advisable if working with an intracellular protein.

Chemical inactivation

A wide range of chemical substances can inactivate proteins. Detergents such as triton and SDS, chaotrophic agents such as urea and guanidinium chloride (Figure 3.31) as well as a wide range of organic solvents can lead to protein denaturation by interrupting the non-covalent forces stabilizing the protein's native structure. Detergents and organic solvents interrupt hydrogen bonding. Heavy metals can also promote protein modification and hence loss of biological activity. The metal ions can complex directly with selected amino acid residues or enhance the reactivity of additional substances with amino acid residues. Various heavy metals for example can complex directly with cysteine thiol groups, or can enhance oxygen mediated oxidation of such thiol groups.

Inactivation of proteins by chemical agents is usually concentration and contact time dependent. Inactivation by such means can obviously be avoided by ensuring such substances are not present in the protein

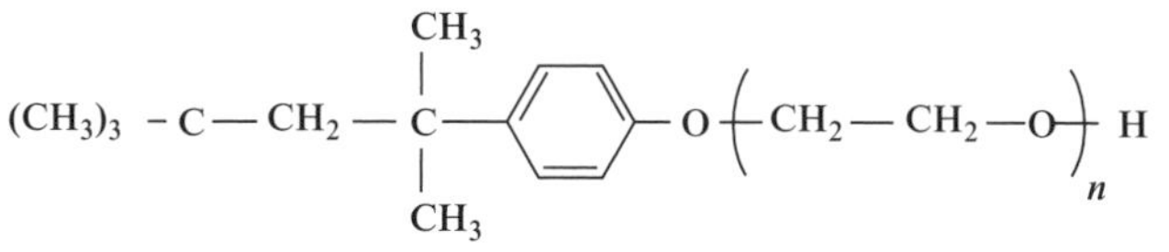

Triton

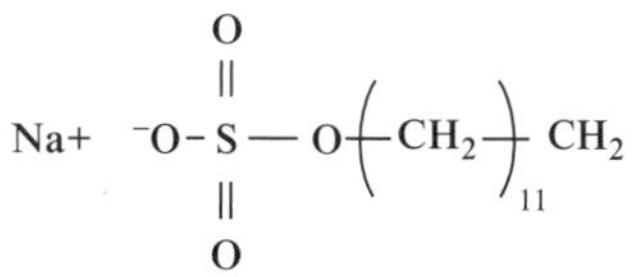

Sodium deodecyl sulfate

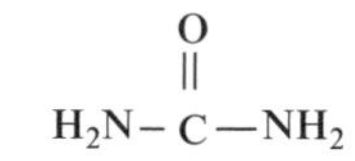

Urea

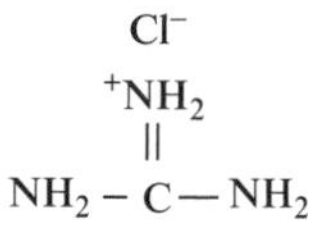

Guanidinium chloride

Figure 3.31 Molecular structure of Triton, sodium dodecylsulfate (SDS) urea, and guanidinium chloride

solution in the first place. However, sometimes these substances are deliberately added to the protein (to achieve e.g. solubilization of inclusion bodies or solvent-mediated protein precipitation). Subsequent removal of such substances is therefore advisable. Protein precipitation using organic solvents does not result in protein denaturation at low temperatures (very close to or below 0°C). However at higher temperatures protein 'flexing' or 'breathing' occurs, which allows entry of organic solvent into the internal protein structure. By interacting with hydrophobic amino acid residues, the solvent molecules disrupt intramolecular hydrophobic interactions, so favouring protein denaturation.

Detergents are not normally deliberately added to a protein stream (except in special circumstances, e.g. if a highly hydrophobic protein is being purified). Detergent molecules, however, can contaminate the protein if purification equipment is used which was not sufficiently rinsed with water after a prior cleaning step. Heavy metals are also rarely deliberately added to protein solutions (unless inhibitor studies are being carried out). However, they may sometimes leach into the protein stream from processing equipment, or be present as trace contaminants in chemicals (e.g. buffer components, salts, etc.) added to the protein stream. Chemical reagents used for such purposed therefore should be of the highest grade of purity available.

The side chains of various amino acids are susceptible to oxidation, either by molecular oxygen or various additional oxidizing agents (e.g. peroxide). The sulfur atoms present in the side chain of methionine and cysteine are particularly susceptible to oxidation, and this process is often accelerated in the presence of metal ions. In general intracellular proteins are more prone to inactivation via oxidation. Oxidation of methionine yields a sulfoxide or a sulfone derivitive (Figure 3.32). Oxidation of cysteine usually results in disulfide formation although cysteic acid can be generated in the presence of strongly oxidizing substances (Figure 3.32). In some instances oxidation of methionine or cysteine residues have little effect on a protein's biological activity. However, if the residue plays a central role (either structurally or functionally) its oxidation can result in complete loss of activity. Protein oxidization, if a problem, can be minimized by the addition of selected reducing agents to the protein solution. The ones most commonly used include dithiothreitol, as well as β-mercaptoethanol, ascorbic acid and free cysteine (Figure 3.33).

Inactivation by biological or physical influences

By far the most significant biological influence which can lead to protein inactivation is the presence of proteolytic enzymes, (proteases, Chapter 11). Different proteases exhibit different specificities in terms of the exact peptide bond(s) they hydrolyse. If susceptible peptide bonds are present on the protein surface then proteolysis is likely to occur, a process which usually leads to protein inactivation (in some instances partial proteolysis

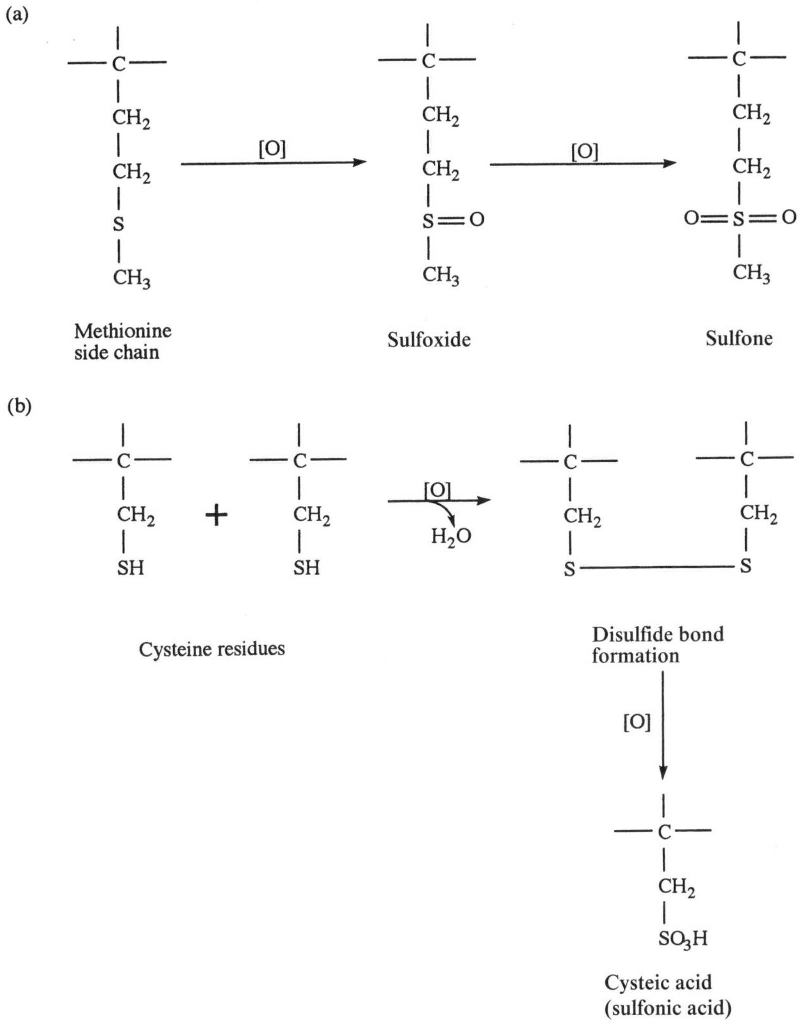

Figure 3.32 Oxidation of (a) methionine and (b) cysteine side chains, as can occur upon exposure to air or more potent oxidizing agents (e.g. peroxide, super-oxide, hydroxyl radicals). Refer to text for details. Reproduced with permission from Walsh, G. (1998) *Biopharmaceuticals, Biochemistry and Biotechnology*, John Wiley & Sons Ltd, Chichester, UK

can have little effect, or even can have a stimulating effect on protein activity). Most proteins resistant to proteolytic cleavage become suscep-tible when denatured (susceptible bonds being exposed upon protein unfolding). Again in general, intracellular enzymes are likely to be most susceptible to proteolytic degradation.

Proteolysis can be minimized by undertaking the initial stages of pro-tein purification as quickly as possible (thereby quickly separating the

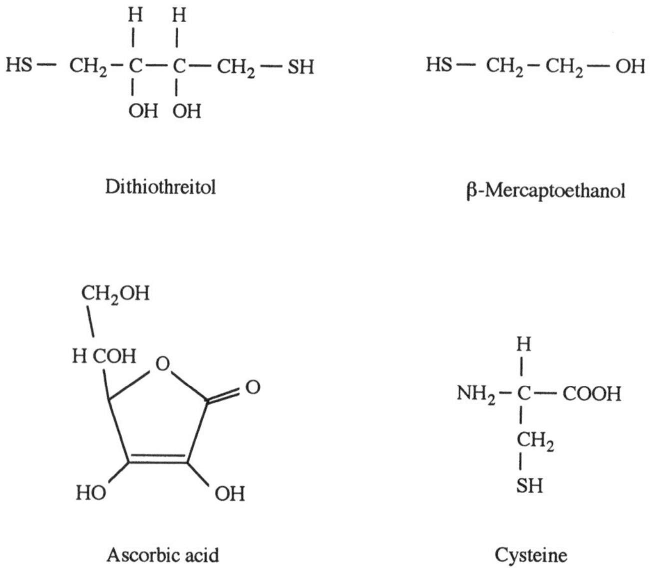

Figure 3.33 Molecular structure of some reducing agents commonly/potentially used in order to maintain a protein in a reducing environment

proteases from the protein of interest), by maintaining low processing temperatures (the lower the temperature the lower the proteolytic activity) or by including one or more proteolytic inhibitors in the processing buffers (Chapter 11).

Carbohydrases and phosphatases can also sometimes influence the biological activity of glycoproteins or phosphoproteins, although this problem is not nearly as widespread as proteolytic inactivation. Modulation or removal of the carbohydrate component of a glycoprotein can influence its solubility and sometimes its biological activity (Chapter 1). Significant deglycosylation can also potentially render the protein more prone to proteolysis if susceptible peptide bonds are unmasked. Dephosphorylation of phosphoproteins can potentially seriously affect the latter's biological activity, as the state of phosphorylation can directly influence the activity of these proteins. As in the case of proteases, minimization of unwanted carbohydrase or phosphatase activity can be achieved by rapid processing times, maintaining low temperatures and the use of inhibitors.

Vigorous agitation is amongst the most common physical influence that can lead to loss of protein activity. Agitation usually results in the incorporation of a gaseous phase (usually air) into the protein solution. The presence of air bubbles obviously greatly increases the total liquid – gas interface area. Proteins tend to align themselves along such interfaces, a process which often results in their partial or complete denaturation (Chapter 13). Agitation is usually a feature of cellular disruption, but can also occur during the actual protein purification process, (e.g. by over-vigorous stirring during a precipitation step or during mechanical pumping operations characteristic of large-scale protein purification schemes).

Extremes of temperature and pH can also lead to protein inactivation, usually by promoting denaturation. The term 'extreme' must be taken in the context of the 'normal' environment of the protein at which it is usually maximally or near maximally stable. For example, a temperature of 80°C would be extreme in the case of animal proteins, but would be close to the optimal operation temperature of most hyperthermophile-derived proteins (Chapters 1 and 10). Extremes of pH can promote protein unfolding, as this will affect the ionization status of the side chains of (ionizable) amino acid residues. This in turn will affect the range and distribution of ionic attractive and repulsive forces, which play an important role in stabilizing the native conformation of most proteins. Exposure of a protein to extremes of pH can be avoided by keeping the latter in an appropriate buffer at all times during extraction, purification and storage. Buffers used for such a purpose are listed in Table 3.13.

Table 3.13 Buffers *commonly used to maintain the pH of a protein solution at a pre-specified value

Buffer	Effective pH range
Clark and Lubs	1.0–2.2
Glycine–HCl	2.2–3.6
Citric acid–sodium citrate	3.0–6.2
Sodium acetate–acetic acid	3.7–5.6
Tris–Maleate–NaOH	5.4–8.4
Na_2HPO_4–NaH_2PO_4	5.8–8.0
Tris–HCl	7.1–8.9
HEPES–NaOH	7.2–8.2
Glycine–NaOH	8.6–10.6
Carbonate	9.7–10.9
Hydroxide–chloride	12.0–13.0

*A buffer is a solution which resists changes in its pH when small quantities of an acid or base are added. It consists of a weak acid (H^+ donor) and its conjugate base (H^+ acceptor). Any buffer will maintain its buffering capacity only over a specific pH range, as indicated below. Exact buffer composition can be obtained from *Data for Biochemical Research*, Oxford Science Publications, 1986.

Extremes of pH can also induce protein destabilization by promoting covalent modification. For example, the amide group of asparagine and glutamine residues are labile at extremes of pH (and at high temperatures), and often deamidate under such conditions (Figure 3.34).

Elevated temperatures can promote protein unfolding, by disrupting the non-covalent forces that stabilize a protein's conformation. Again, temperatures can be maintained at low levels by refrigeration, keeping the protein solution on ice, etc. Although maintaining a protein solution at low temperatures is generally stabilizing, low-temperature operation can slow down the purification process (e.g. low temperatures slow down flow rates through chromatographic columns if the protein is present in a viscous buffer). This in turn can be detrimental by, for example prolonging exposure times of the protein with any potential denaturing influences present.

While the process of freezing and subsequent thawing has no detrimental effect on most proteins, it can inactivate some proteins. Inactivation is often dependent upon the exact solute composition of the protein solution. As described later, freezing effectively concentrates all solutes present in the aqueous solution. If any 'contaminating' solutes are potential denaturants, this concentration effect could result in protein inactivation. Also, as the temperature continues to decrease during the freezing process some solutes will selectively crystallize out of solution. This can be harmful if, for example one buffer component crystallizes before the other.

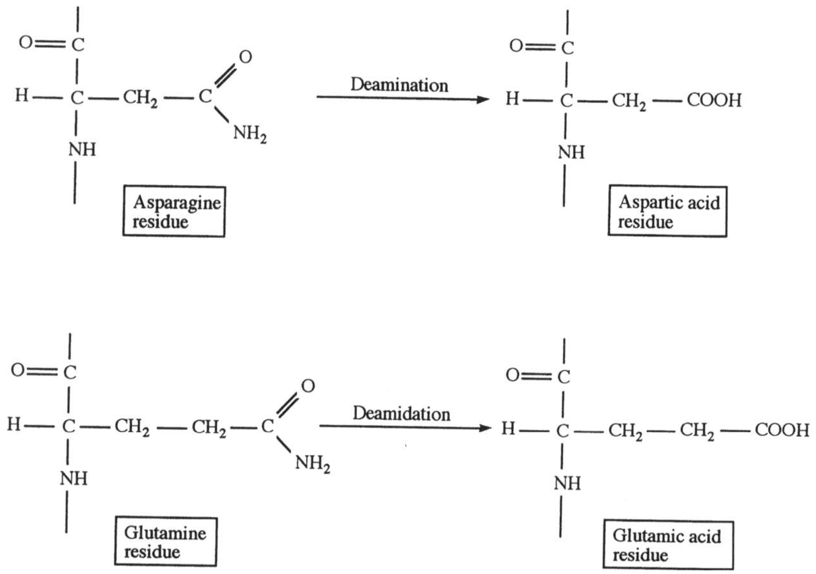

Figure 3.34 Deamidation of asparagine and glutamine residues, yielding aspartic acid and glutamic acid, respectively. Reproduced from Walsh, G. (1998) *Biopharmaceuticals, Biochemistry and Biotechnology*. John Wiley & Sons, Ltd, Chichester, UK

Such an event would lead to a huge swing in the pH value of the remaining uncrystallized liquid, which contains the protein.

Approaches to protein stabilization

For some proteins no specific stabilization steps need be undertaken during extraction, purification or storage. However, a number of general steps can be undertaken in order to maximize protein stability. These usually include:

- Ensure the protein is always maintained in a buffered solution set at a pH value in which it is maximally stable.

- Ensure extraction and processing temperatures are controlled such that the protein is not exposed to elevated temperature values.

- Minimize processing times, to ensure protein contact time with any denaturing influence encountered is minimized.

- Avoid processes such as vigorous agitation or the addition of chemicals known to promote denaturation of the target protein.

- Include substances in all processing buffers which might counteract or inactivate known inactivators of the target protein (e.g. proteolytic inhibitors for protease sensitive proteins and reducing agents in the case of oxygen-sensitive proteins).

- Include stabilizing agents of the target protein (if known) in all extraction and processing buffers.

Protein stabilizing agents can be of two types: specific or general. As the name suggests, specific stabilizers achieve their effect by interacting biospecifically with the target protein. Examples include substrates (or competitive inhibitors) for many enzymes, and in some cases, antibodies that specifically bind to the protein of interest. General stabilizers act in a non-biospecific manner. Examples include glycerol, various carbohydrates, polymers such as polyethylene glycol, amino acids such as glycine and bulk proteins such as bovine serum albumin (BSA).

Substances such as glycerol, sugars and polyethylene glycol are often used as general stabilization agents, particularly for intracellular-derived proteins. These substances likely achieve their intended effect mainly by reducing water activity. Most of the water within intact cells is not freely mobile, but is bound or loosely associated with proteins and other cellular molecules (i.e. 'free water' activity is low). Cellular homogenization (usually using several volumes of buffer per unit weight of cells) liberates intracellular proteins into a dilute solution, displaying greatly increased free water levels. Glycerol, sugars and various polymers can reduce free water levels by hydrogen bonding with bulk water. This more closely mimics normal intracellular conditions, under which proteins are generally more stable.

Various amino acids also stabilize some proteins in solution. Glycine is most commonly used, although alanine, lysine and threonine have also found application in this regard. The molecular mechanism of stabilization is not fully understood. Direct interaction with some proteins is probably a factor but they may also effectively stabilize the protein by reducing its adsorption to internal walls of containers.

Although stable at higher concentrations, some proteins become considerably labile when diluted. Dilute proteins may be stabilized in solution by the addition of 'bulking' proteins such as BSA. In some cases this may exert a stabilizing influence by direct interaction with the protein, although they may also function by providing alternative 'targets' for any inactivating agents present, and by decreasing levels of surface absorption of the target protein to the container's surface. Addition of any stabilizing agent is appropriate only if it does not interfere with the subsequent purpose(s) for which the protein is being purified.

Once purified to an acceptable level most proteins are stored for a period of time before being used for their intended academic or applied purpose. The storage stability of proteins can vary widely, and depends upon (a) the inherent stability of the protein and (b) the storage conditions chosen. Optimization of storage conditions is invariably a trial and error process, but generally the following options are considered:

- The protein must be stored under conditions of temperature and pH at which it is maximally stable. Many proteins are stable for months when stored at room temperature, others are only stable when stored at 4°C or frozen.

- If stored in liquid format, the addition of stabilizing agents (as previously discussed) should be considered. The protein solution should also ideally be filter sterilised and preservatives added before storage.

- If stored frozen the protein solution should be quickly frozen, and subsequently maintained at −20°C, or preferably −70°C.

- The protein may be more stable if stored in a dry format.

Lyophilization

Lyophilization involves the drying of protein (or other materials) directly from the frozen state. This is achieved by firstly freezing the protein solution in a suitable, unstoppered, container. A vacuum is then applied and the temperature is increased in order to promote sublimation of the ice, which occurs under conditions of reduced pressure. The ice is drawn off directly as water vapour. The containers can be sealed following completion of the freeze drying process. Many freeze-dried proteins may be stored at room temperature for prolonged periods with little or no loss of biological activity. Some freeze-dried products, however, exhibit significant loss of activity if stored under such conditions, and thus must be stored at lower temperatures. Some advantages or disadvantages associated with the freeze drying technique are listed in Table 3.14.

Many proteins sold commercially, in particular high value, low volume products such as vaccines, therapeutic enzymes, hormones, antibodies and diagnostic reagents, are often marketed in freeze-dried form. The technique is also routinely used at a research laboratory scale. The average moisture content of a freeze-dried protein preparation is in the order of 3 per cent. However, domains often exist within the product which contains a much higher moisture content. This can contribute to product inconsistency.

The first step in the freeze-drying process involves freezing the protein solution in suitable containers, generally glass vials or flasks. As the temperature decreases, ice crystals begin to form. Such crystals contain only pure water molecules. As the ice crystals grow, the protein concentration, and the concentration of all other solutes present in the remaining liquid phase, steadily increases. Any solute species, such as salts, buffer components, other chemical additives or proteases present in the product are concentrated many-fold. The greater the solute concentration, the greater the reaction rate between such solutes. Proteins that are damaged by high concentrations of these solutes may be inactivated at this. Such inactivation may be due to chemical or biological modification of the protein, or may be caused by protein aggregation.

As cooling continues, some of the solutes present in the concentrated solution may also crystallise and hence are removed from the solution. As the temperature decreases still further, the viscosity of the unfrozen solution increases dramatically, and becomes more and more 'rubber-like'. Eventually, the unfrozen solution will change from a rubber consistency to that of glass. The temperature at which this occurs is known as the glass transition temperature (T_g). The glass consists of all the uncrystallized solute

Table 3.14 Some advantages and disadvantages associated with the freeze-drying of proteins; some points are more relevant in the context of industrial scale operations

Advantages	Disadvantages
Freeze-drying represents one of the least harsh methods of protein drying	Equipment required at an industrial scale is extremely expensive
It yields a lightweight product which reduces shipping and distribution costs	Running costs are high
Freeze-dried proteins can be rapidly reconstituted (rehydrated) prior to use	Freeze-drying entails long processing times (typically 3–5 days)
Freeze-drying is accepted by regulatory authorities as a technique suitable for the preservation of finished products destined for parenteral administration	Some proteins exhibit an irreversible decrease in biological activity upon freeze-drying

molecules, including the protein, as well as all uncrystallized water molecules present in association with these solutes. Structurally, this glass is not a solid but actually a liquid which exhibits a very slow flow rate, of the order of micrometres per year. The same is true of glass formed from any material – this includes window glass or spectacle glass. Mechanically, however, it may be regarded as a solid. Molecular mobility of all solutes within the glass, which contains up to 50 per cent water, is to all intents and purposes, non-existent. Chemical mobility, and hence reactivity, within the glass phase all but ceases. Upon initial formation of ice crystals it is therefore desirable to reduce the solution temperature below the glass transition temperature as quickly as possible, in order to minimize protein inactivation.

The T_g of any given solution depends upon its composition and may be determined experimentally by a technique known as differential scanning calorimetry. This involves heating the glassy product and subsequently plotting temperature versus its specific heat value. A sharp increase in heat flow is observed at the glass transition temperature. Determination of the T_g value of a protein solution facilitates the development of a more rational freeze-drying protocol for that particular protein preparation.

During the freeze-drying process a vacuum is applied to the system once a temperature below the glass transition temperature has been attained. For most protein solutions this involves decreasing the temperature to between -40 and $-60°C$. The temperature may then be allowed to increase, in order to promote sublimation of the crystalline water. This requires an input of energy. In most industrial freeze drying systems, this is achieved by simply increasing the shelf temperatures. Adequate temperature control is required to ensure maximum efficiency during the primary drying process. The shelf temperature must normally be maintained at values above $0°C$ in order to promote efficient sublimation. The internal vial temperature will still be well below $0°C$. If the temperature is maintained below the T_g value during primary drying, removal of crystallized water will leave behind a very well defined protein cake. If, however, the product temperature increases above the T_g, which could occur if the shelf temperature were too high, the viscous solution left behind as the crystallized water sublimes can 'flow'. This results in the formation not of a 'light fluffy cake' but of a very dense cake in which the protein may be inactivated and which will rehydrate slowly, if at all. The glass transition temperature of any solution depends upon its composition. During the drying process the T_g value is continually changing, increasing as water vapour is removed from the product.

Upon completion of primary drying, the protein cake still retains appreciable quantities of moisture, the uncrystallized water molecules associated with the glass. This water is removed during secondary drying by allowing the internal vial temperature to increase still further, such that it too is removed by sublimation. Upon completion of secondary drying, a vacuum is released and the vials are then sealed.

Although several different models of freeze-dryers are available, all consist of a number of essential features. All possess a drying chamber into which the product is loaded and subsequently frozen. In the case of basic laboratory-scale systems, this often consists of a round-bottom flask which may be connected to the freeze-dryer via a suitable port. A vacuum pump is required in order to evacuate the drying chamber and maintain a vacuum during drying. A suitable condenser is also required. During drying, the water vapour emanating from the product is collected in the condensing chamber. Due to the low internal temperature (typically $-50°-60°C$), the water accumulates in the form of ice. Upon completion of the drying cycle, the chamber temperature is increased so that the ice melts. The water is then drained from the chamber before the next freeze drying cycle begins.

Industrial scale freeze dryers usually cater for thousands of product vials during each drying cycle. Upon completion of the product fill, rubber stoppers are partially inserted into the mouth of each vial in such a way as to allow water vapour to flow freely from within the vial during the drying process. The vials are placed on a series of trays which are loaded onto shelves in the drying chamber. Each shelf may be electrically heated or cooled. After loading is completed, the drying chamber door is closed. The shelf temperature is then decreased to temperatures in the order of $-40°-60°C$ in order to freeze the product. Various probes are inserted into test vials such that the actual product temperature may be accurately monitored.

Upon reaching the predetermined temperature, which should be below the product's glass transition temperature, chamber evacuation is initiated. Appropriate increases in shelf temperature promotes efficient primary and secondary drying under vacuum. Upon completion of the freeze-drying cycle, the vacuum is released and the vials are sealed *in situ*. Shelf design in modern freeze-drying systems allows hydraulic upward, or downward, movement of the individual shelves. As any one particular shelf moves upwards, the partially inserted rubber seals will be inserted fully into the vials when they come in contact with the shelf immediately above. In modern freeze-drying systems all variable parameters, such as temperature and reduced pressure levels, may be pre-programmed on a central control console. The freeze-drying cycle may thus be fully automated. Parameters such as individual shelf temperature, product temperature and vacuum level are continually monitored and recorded during the process. Most lyophilization systems also allow *in situ* heat sterilization of the drying chamber. This greatly facilitates its subsequent use when freeze-drying proteins, which must be processed under sterile conditions.

Micropurification of proteins

The scale of protein purification pursued depends upon the quantity of purified product required (Table 3.1). The general purification sequences described in this chapter are those applied to yield from mgs of pure protein up. Smaller quantities (ng–μg level) may sometimes be purified by alternative means. Such small quantities of a previously uncharacterized protein are often sufficient to allow partial sequence and mass spectrum analysis. The information retrieved can be used, for example to isolate the gene coding for the protein, allowing its large-scale recombinant production. Micropurification often simply entails separation of crude protein preparations using analytical scale equipment (e.g. two-dimensional electrophoresis, capillary electrophoresis or analytical HPLC columns).

PROTEIN CHARACTERIZATION

Once purified, most proteins are subjected to a battery of characterization studies. Table 3.15 summarizes the most important of these. The exact range of studies undertaken will very much depend upon the ultimate project aim. Proteins purified as part of academic research projects are usually extensively characterized, particularly if the protein has not been previously studied. The level of purification and characterization pursued in the case of proteins produced for commercial purposes vary depending upon the intended product use (Chapter 4). Proteins destined for therapeutic applications are most extensively characterized.

Table 3.15   Major studies potentially undertaken to fully characterize a protein, along with the associated analytical techniques by which these are achieved

Characteristic studied	Methods/techniques by which achieved
Functional characteristics	Depends upon the biological activity of the protein. Examples include determination of specific activity, the effects of e.g. temperature and pH on biological activity, ligand interaction studies, etc.
Evidence of purity	SDS–PAGE, two-dimensional electrophoresis, isoelectric focusing, HPLC analysis, mass spectrometry, capillary electrophoresis
Structural studies	
Molecular mass determination	SDS–PAGE, non-denaturing electrophoresis (Ferguson plots). Gel filtration, analytical ultracentrifugation, mass spectrometry
Analysis of primary structure	Amino acid composition, peptide mapping, N-terminal sequencing, complete amino acid sequencing
Analysis of secondary/tertiary/quaternary structure	Circular dichroism, fluorescence spectroscopy NMR, X-ray crystallography, analytical ultracentrifugation
Analysis of post-transitional modifications	Various, depending upon the exact modification

Characterization of recombinant proteins is undertaken using the same techniques and procedures as used for non-recombinant proteins. Depending upon the expression system and purification procedures used, however, some characterization studies may adopt a special significance:

- conformational studies in the case of proteins produced as inclusion bodies, which required denaturation and refolding steps;

- N- or C-terminal sequencing in the case of proteins produced as fusion products and whose tag was removed during purification;

- post-translational modification analysis in the case of proteins which are glycosylated in the native state.

Characterization studies may be undertaken on the target protein alone, or may be undertaken on the purified protein and a 'standard', for comparative purposes. Purified recombinant proteins for example are sometimes characterized in conjunction with the protein obtained by direct extraction and purification from its native source. In this way direct comparisons between the recombinant and native protein can be made. If the nucleotide sequence of the gene coding for the protein under study is known, characterization studies may be undertaken simply to confirm identity of the protein or to identify any post-translational modifications present.

A summary of the major characterization studies usually employed is presented below. The theory behind many of these techniques is not presented in detail as comprehensive coverage is outside the scope of this text. Note also that several of the techniques used to structurally characterize a protein have been discussed in Chapter 1, to which the reader is referred.

Functional studies

Functional studies, as the name suggests, aim to investigate the biological function and activities of the purified proteins, and how these activities are affected by various pertinent influences. The type and range of studies undertaken very much depend upon (a) the protein type and (b) its intended application. For enzymes, basic functional studies usually encompass issues such as:

- determination of specific activity (number of activity units per mg protein);

- determination of substrate range and specificity;

- kinetic characteristics (e.g. determination of K_m V_{max} and K_{cat} values);

- effects of various influences (e.g. temperature, pH, inhibitors) on enzyme activity.

If the enzyme is to be used for a specific applied purpose, functional studies of particular significance in the context of that purpose will be investigated in great detail (e.g. the temperature versus activity profile and thermal stability of enzymes destined for application in high temperature industrial processes).

Evidence of purity

Purity refers to the absence of contaminants in the final purified protein. Contaminants can represent different things to different people and the contaminant profile studied often depends upon the ultimate purpose for which the protein is being purified. In the case of those purified for academic characterization the main contaminant of interest (i.e. for which evidence of its absence from the purified protein is sought) are proteins. Protein contaminants may be unrelated to the target protein or may be modified or partially degraded forms of that protein. In the case of proteins purified for therapeutic application the 'contaminant list' is longer, including not only other, modified protein forms but also pyrogens, nucleic acids, microorganisms and viral particles (detailed in Chapter 4). In the context of this chapter the discussion relating to evidence of purity is restricted to determination of the presence/absence of contaminant proteins. One point to remember is that the detection of any contaminant type is inherently limited by the sensitivity of the detection technique used.

The most common method used to investigate protein purity is that of one-dimensional polyacrylamide gel electrophoresis run in the presence of the negatively charged detergent sodium dodecyl sulfate (SDS–PAGE; Figure 3.31). Any charged species, including proteins, will migrate when placed in an electric field. The rate of migration depends upon the ratio of charge to mass (i.e. the charge density) of the protein. In the case of PAGE migration occurs through a polyacrylamide gel, the average pore size of which is largely dependent upon the concentration of polyacrylamide present. A sieving effect therefore also occurs during PAGE so the rate of protein migration is influenced by its size and shape as well as charge density.

Incubation of the protein with SDS has two notable effects: (a) it denatures most proteins, giving them all approximately the same shape; (b) it binds directly to the protein at the constant rate of approximately one SDS molecule per two amino acid residues. In practice this confers essentially the same (negative) charge density to all proteins. Separation of proteins by SDS–PAGE therefore occurs by a sieving effect, with the smaller proteins moving fastest towards the anode (Figure 3.35). SDS–PAGE can be used for two purposes: (a) determination of purity and

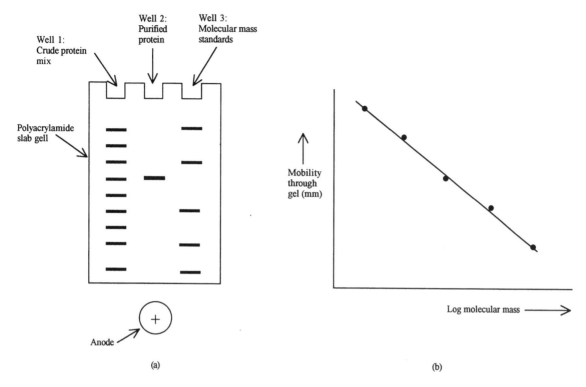

Well 1:
Crude protein
mix

Well 2:
Purified
protein

Well 3:
Molecular mass
standards

Polyacrylamide
slab gell

Mobility
through
gel (mm)

Log molecular mass

Anode

+

(a)

(b)

Figure 3.35 Separation of proteins by SDS–PAGE. Protein samples are incubated with SDS (as well as reducing agents, which disrupt disulfide linkages). The electric field is applied across the gel after the protein sample(s) to be analyzed are loaded into the gel wells. The rate of protein migration towards the anode is dependant upon protein size. After electrophoresis is complete, individual protein bands may be visualized by staining with a protein binding dye (a). If one well is loaded with a mixture of proteins each of known molecular mass, a standard curve relating distance migrated to molecular mass can be constructed (b). This allows estimation of the molecular mass of the purified protein

(b) determination of subunit molecular mass. Purity is indicated by the presence of a single band in the case of single polypeptide proteins (or proteins containing multiple identical polypeptides). Purified proteins containing two or more dissimilar polypeptide subunits (e.g. antibodies) will yield two or more bands on SDS gels. Protein bands can be visualized by staining the gel with protein binding dyes such as Coomassie blue, or silver-based stains (which are 10–100 times more sensitive than Coomassie blue).

One-dimensional PAGE of proteins can also be undertaken under non-denaturing conditions. Again a pure protein will usually yield a single band, visualized by staining with Coomassie blue. As the protein has retained its native conformation it is also possible to confirm its identity by using some sort of activity stain or by probing with fluorescent labelled antibodies.

Isoelectric focusing is an additional form of electrophoresis which can be used to analyse (or sometimes micropurify) proteins. Again, the technique involves electrophoresis through a polyacrylamide gel, but this time in the presence of a mixture of low molecular mass organic acids and bases (ampholytes). The ampholytes distribute themselves in the gel under the influence of the electric field, forming a pH gradient. The protein solution to be applied is normally first incubated with urea. This denatures it, exposing all amino acid side chains, which contribute to the protein's isoelectric point (pI).

Upon application of the protein sample to the gel the proteins present migrate until they reach a point at which the pH equals their pI, (at its pI a protein has a nett zero charge, and hence will not move under the influence of an electric field). A mixture of commercially available protein standards (of known pI values) are run concurrently. This facilitates the generation of a standard curve which equates distance migrated to the pI value. Thus, the pI value of the purified protein can also be ascertained.

A more stringent approach to assessing protein purity entails the application of a combination of SDS–PAGE and isoelectric focusing in so-called two-dimensional electrophoretic analysis (Figure 3.36). In addition to purity analysis this technique can be used to analyse complex protein solutions. Each separation mode can individually resolve about 100 protein bands, but when combined 1000–2000 bands can be resolved.

Capillary electrophoresis is yet another electrophoretic format, and separates molecules on the basis of charge density. In this case, however,

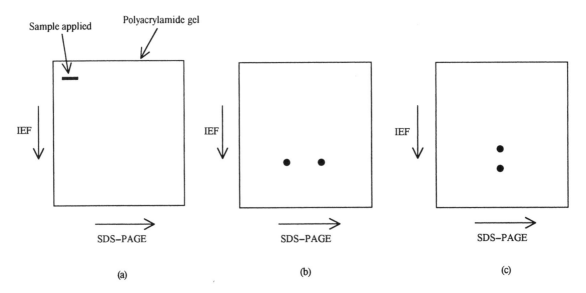

Figure 3.36 Principle of two-dimensional gel electrophoresis. The protein sample is applied to the polyacrylamide gel and firstly subjected to isoelectric focusing (IEF). After this is complete the protein bands are subjected to SDS–PAGE in the perpendicular direction (a). This combination has greater resolving power than either technique alone. The resolution of two proteins with with equal pI values but different molecular masses is illustrated in (b). Resolution of two proteins with of equal molecular mass but differing pI values is illustrated in (c)

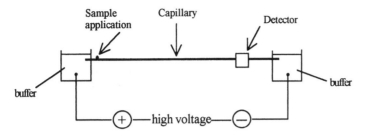

Figure 3.37 Schematic representation of capillary electrophoresis. After sample application a high voltage is applied, and the proteins migrate under the influence of the resultant electric field. Visualization of proteins is achieved using an in-line UV detector. The capillary tube, which can be coiled for convenience, is generally 30–60 cm in length, with an internal diameter in the region of 50 μm

electrophoretic separation occurs not in a polyacrylamide gel but along a narrow-bore capillary tube usually packed with a fused silica matrix (Figure 3.37). Although most frequently used for analysis of low molecular mass substances (drugs, metabolites, peptides), capillary electrophoresis may also be applied to protein analysis. Protein purity is also often tested using HPLC. As a technique HPLC is characterized by superior peak resolution and fast fractionation speeds. Use of sensitive protein in line detectors allows rapid and sensitive analysis. Assessment of purity entails application of the protein sample to at least two different HPLC column types, most often a reverse phase column and an ion-exchange column.

Mass spectrometry may also be used to investigate purity, although its major application relates to molecular weight determination, as detailed in the next section.

Molecular mass determination: mass spectrometry

The most recent (and sensitive) analytical technique adapted to the determination of protein molecular mass is that of mass spectrometry (MS). The basic features of this technique are presented in Figure 3.38. MS has for many years been a central technique used to identify and determine the molecular mass of small molecules. It's routine application to protein work has only recently been made possible by (a) the development of suitable ionization techniques that allow generation of gas phase ionized proteins and (b) the increased availability of user-friendly mass spectrometers.

Two approaches to generating gas phase ionized protein molecules have now come to the fore: electrospray ionization (ESI) and matrix-assisted laser desorption and ionization (MALDI). In the case of ESI the protein solution is passed through a narrow-bore needle held at a high electrical potential. A fine mist of droplets containing charged protein molecules is generated, followed by rapid evaporation of the solvent.

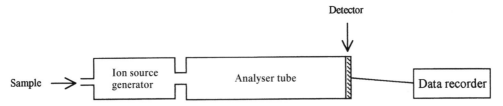

Figure 3.38 Basic principle of mass spectrometry, as applied to protein analysis. The system is composed of three essential components, an ion source which generates gas phase ionized proteins, an analyser tube along which ion separation occurs (on the basis of molecular mass) and a detector, which reveals protein ion arrival at the end of the tube. The data generated is transmitted to a suitable recording system. Refer to text for details of operation

The charged proteins, now in the gas phase, are fed directly into the mass analyser tube. The ionization process occurs at atmospheric pressure and hence ESI-based systems can be interfaced directly with chromatographic systems such as HPLC.

The MALDI method, which has become the method of choice for protein molecular mass determination, entails the initial mixing of the protein sample with a molar excess of 'matrix' compounds. These matrix compounds are low molecular mass organic acids, which strongly absorb UV radiation. Examples include sinapinic acid and 2,5-dihydroxy benzoic acid. The mixture is dried and then bombarded with laser generated UV photons. The matrix molecules absorb the UV, resulting in a rapid breakdown of the matrix – protein lattice, which sends both into the gas phase. Transfer of photons from the excited matrix molecules to the protein molecule results in ionization of the latter. The protein ions are now accelerated from the ionization point down the analyser tube under the influence of a strong electric field. Mass analysers separate ions on the basis of their mass – charge ratios (m/z).

Time-of-flight TOF analysers which as the name suggests, measure the time ion(s) require to reach the detector at the tubes end, are normally used in conjunction with MALDI-generated protein ions (MALDI–TOF). As they enter the analyser tube all the protein ions have essentially the same kinetic energy and charge. Because of this the time required for each protein ion to reach the detector reflects its molecular mass, with smaller proteins travelling fastest. MALDI–TOF systems are operated under high vacuum.

MS can provide an accurate estimation of mass for proteins up to about 500 kDa in size, with accuracy in the region of 0.01 per cent (SDS–PAGE molecular mass estimates are accurate to only 5–10 per cent). A sample size of only a few picomoles of protein is all that is required for analysis. Furthermore comparison of the actual mass recorded with theoretical mass (from e.g. gene/c DNA sequences) can reveal the presence of protein modifications and post–translational processing. For example if the actual mass is greater than theoretical by 80 Da, a phosphate or sulfate group is likely present.

In addition to mass analysis, MS can also be used to generate protein sequence data. As such the technique is complementary to the Edman degradation analysis, as discussed in Chapter 1. Sequence analysis requires the use of a tandem mass spectrometer (MS/MS). This consists of two mass analyser tubes linked sequentially and separated only by a collision cell. The protein to be sequenced is firstly fragmented, either chemically or enzymatically (Chapter 1). The fragments are separated along the first analyser tube. One peptide ion fragment is selected at a time and fed (alone) into the collision tube, where it collides with intert gas molecules (He or Ar). This promotes further fragmentation into a range of complementary peptides which are separated on the basis of mass in the second tube.

Computerized analysis of the mass of each fragment generated in the second tube can yield nearly complete or complete sequence data. Interpretation of the mass spectrum is a complex, computerized task, but MS /MS displays several potential advantages:

- it can provide sequence information from blocked or modified peptides;

- it is faster than Edman degradation;

- it is as sensitive as the Edman technique.

Additional methods by which the molecular mass of a protein can be determined include gel filtration analysis, non-denaturing electrophoresis and analytical ultracentrifugation. As previously outlined, gel filtration separates proteins on the basis of size and shape. As most globular proteins are roughly spherical in shape, their separation is essentially on the basis of mass. Mass analysis by gel filtration entails initial calibration of the gel filtration column using a series of protein standards (i.e. globular proteins of known molecular mass, Table 3.16). Log molecular mass, when plotted against the protein elution volume, yields a linear relationship. A standard curve can thus be generated which can be used to equate the elution volume of the unknown protein with its molecular mass.

Non-denaturing PAGE separates proteins on the basis of size, shape and charge. Non-denaturing systems can be used to generate Ferguson plots, which can in turn be used to determine protein molecular mass. For any given protein subjected to non-denaturing PAGE, the relative mobility (R_f) of the protein through the gel decreases if gels of increasing acrylamide concentration are used. For a given protein, by running several PAGE gels at different gel concentrations, a graph linking R_f to acrylamide concentration can be generated (a Ferguson plot). A straight line is observed, the slope of which is termed the K_r (retardation coefficient). Large proteins have higher K_r values than smaller proteins. By determining the K_r value of several proteins of known molecular mass, a standard curve equating K_r to molecular mass can be generated. The K_r value of the protein of unknown molecular mass is then determined, and by use of the standard curve generated its mass can be obtained.

Table 3.16 Molecular mass markers frequently used to calibrate gel filtration columns for determination of the mass of a globular protein

Protein standard	Molecular mass (kDa)
Horse heart cytochrome C	12.4
Bovine erythrocyte carbonic anhydrase	29.0
Bovine serum albumin	66.0
Yeast alcohol dehyrogenase	150.0
Potato β-amylase	200.0
Horse spleen apoferritin	443.0
Bovine thyroglobulin	669.0

Analytical ultracentrifugation can also be used to determine the molecular mass of a protein. This entails subjecting the protein to very high centrifugal forces (using rotor speeds of up to 60 000 r.p.m), while following the rate of migration of the protein through the centrifugal tube. This is achieved using specially designed sample cells. From this data the sedimentation coefficient(s) of the protein can be determined. This value in turn is related to the protein's molecular mass by a mathematical expression termed the Svedberg equation.

FURTHER READING

Books

Crabb, J. (1994) *Techniques in Protein Chemistry*. Academic Press, London.
Harris, E. (2000) *Protein Purification Applications*. Oxford University Press, Oxford.
Harris, E. (2001) *Protein Purification Techniques*. Oxford University Press, Oxford.
Matejtschuk, P. (1997) *Affinity Separations*. IRL Press, Oxford
Oliver, R. (1998) *HPLC of Macromolecules*. IRL Press, Oxford
Sadana, A. (1997) *Bioseperations of Proteins*. Academic Press, London.
Scopes, R. (1993) *Protein Purification*. Springer Verlag, Godalning.

Articles

Protein purification
Afeyan, S.P. *et al.* (1990) Perfusion chromatography, an approach to purifying biomolecules. *Bio/Technology* **8**, 203–206.

Charcosset, C. (1998) Purification of proteins by membrane chromatography. *J. Chem. Technol. Biotechnol.* **71**, 95–110.

Chase, H. (1994) Purification of proteins by adsorption chromatography in expanded beds. *Trends Biotechnol.* **12**, 296–303.

Dyr, J. & Suttnar, J. (1997) Separation used for purification of recombinant proteins. *J. Chromatogr. B* **699**, 383–401.

Geisow, M.J. (1991) Stabilizing protein products: coming in from the cold. *Trends Biotechnol.* **9**, 149–150.

Hjorth, R. (1997) Expanded bed adsorption in industrial bioprocessing: recent developments. *Trends Biotechnol.* **15**, 230–235.

Huddleston, J. *et al.* (1991) The molecular basis of partitioning in aqueous two-phase systems. *Trends Biotechnol.* **9**, 381–388.

Knight, P. (1990) Bioseparations: Media and modes. *Bio/Technology* **8**, 200–201.

Muller, C.J., (1990). Separation of proteins and nucleic acids using 'tentacle type' ion exchangers. *Mod. Methods Protein Nucl. Acid Res.* 99–115.

Muller, W. (1990). New ion exchangers for the chromatography of biopolymers. *J. Chromatogr.* **510**, 133–140.

Narayanan, S. (1994) Preparative affinity chromatography of proteins. *J. Chromatogr. A* **658**, 237–258.

O'Carra, P. & Barry, S. (1972) Affinity chromatography of lactate dehydrogenase. *FEBS Lett.* **21**, 281–285.

Ohlson, S. *et al.* (1989) High performance liquid affinity chromatography: a new tool in biotechnology. *Trends Biotechnol.* **7**, 179–185.

Pilak, M.J. (1991) Freeze drying of proteins, part I: process design. *Pharm. Technol. Int.* **3**, 37–43 (originally printed in *BioPharm* 1980 **3**(8), 18–27).

Pilak, M.J. (1991) Freeze drying of proteins, part II: formulation selection. *Pharm. Technol. Int.* **3**, 40–43 (originally printed in *BioPharm.* 1990 **3**(9), 26–30).

Sassenfeld, H.M. (1990) Engineering proteins for purification. *Trends Biotechnol.* **8**, 88–93.

Scopes, R. (1987). Dye ligands and multifunctional adsorbents: an empirical approach to affinity chromatography. *Anal. Biochem.* **165**, 235–246.

Scopes, R. (1996) Protein purification in the nineties. *Biotechnol. Appl. biochem.* **23**, 197–204.

Shamey, E.Y. *et al.* (1989) Stabilization of biologically active proteins. *Trends Biotechnol.* **7**, 186–190.

Thommes, J. & Kula, M. (1995) Membrane chromatography – an integrative concept in the downstream processing of proteins. *Biotechnol. Progr.* **11**, 357–367.

Protein characterization

Kemp, G. (1998) Capillary electrophoresis: a versatile family of analytical techniques. *Biotechnol. Appl. Biochem.* **27**, 9–17.

Novotny, M. (1996) Capillary electrophoresis. *Curr. Opin. Biotechnol.* **7**, 29–34.

Roepstorff, P. (1997) Mass spectromertry in protein studies from genome to function. *Curr. Opin. Biotechnol.* **8**, 6–13.

4

Large-scale protein purification

SOME GENERAL PRINCIPLES

The various sources from which proteins may be obtained have been discussed in Chapter 2. The approaches which can be adopted to purify and characterize these proteins have been outlined in Chapter 3. The general principles provided in these two chapters are, by enlarge, equally applicable to laboratory- and industrial-scale operations. Indeed, several issues pertinent specifically to large-scale protein purification have been discussed in Chapter 3.

At an industrial scale, the manufacture of a protein product is often divided into upstream and downstream processing (Figure 4.1). Upstream processing refers to the phases of production in which biosynthesis of the protein takes place. For most proteins therefore, this refers to microbial cell fermentation or animal cell culture. Discussion of how these operations are undertaken at an industrial scale is beyond the scope of this text. The interested reader is referred to the Further Reading section at the end of this chapter.

Downstream processing refers to the extraction of the protein from the producer source, and its subsequent purification. It also encompasses additional manufacturing activities such as quality control evaluation, end product stabilization, adjustment of product potency to within specified limits, filling of product into suitable final product containers, labelling of containers, etc. A number of guidelines may be followed when designing a downstream protein purification scheme. Such guidelines, which are little more than statements of common sense, are general in nature, and as such may not be applicable or appropriate to some purification schemes. These guidelines include:

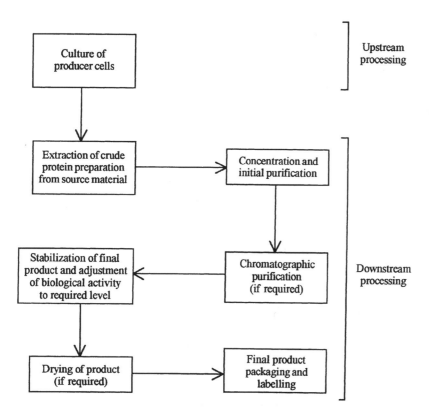

Figure 4.1 Flow diagram outlining the major steps constituting upstream and downstream processing of protein products. Refer to text for details

- Initial characterization of the starting material should be undertaken in detail, such that a rational purification scheme may be developed. The final product purity required should be clearly defined.

- The purification scheme should be kept as simple and as straightforward as possible, ensuring that the highest end product yield is obtained within the shortest time period practicable.

- Fractionation techniques (and the conditions under which the techniques are used) should be selected such that they exploit the greatest differences in physicochemical properties between the protein of interest and other impurities.

- A highly selective step should be used as early as possible in the purification scheme. This may reduce the number of subsequent steps which must be employed to achieve the desired level of purity.

- If possible, the most expensive processing step should be left until last.

When purifying a protein for research purposes technical issues are usually the sole issues considered during design of the purification procedure. In general the protein is purified to homogeneity, and less emphasis is placed upon factors such as percentage yield obtained or the duration/cost of the purification scheme. In the case of commercially produced proteins, economic and often regulatory issues must also be taken into account. Commercial logic dictates that the product be produced to the required specification as inexpensively as possible. Manufacturing guidelines issued by regulatory authorities such as the US Food and Drug Administration (FDA) must also be taken into account in the manufacture of proteins destined for therapeutic, diagnostic or food use.

The degree of purity required is an over-riding factor in the design of any downstream processing procedure. This is largely dependent upon the intended application of the final product. As a general principle, the extent of downstream processing of any protein is maintained at the minimum required to produce an acceptable final product. Unnecessary steps in the purification protocol often reduces the product's biological stability, and the yield of product decreases as the number of purification steps increases. Most proteins of industrial interest may be grouped into one of two broad categories:

- Proteins produced in bulk, often as relatively crude preparations. Such proteins are usually enzymes that have a wide variety of applications in the food and beverage processing industries (Chapters 10–12) or are milk-derived proteins used for food application (Chapter 13).

- Proteins destined for therapeutic and/or diagnostic applications. These proteins are generally produced in quantities which are orders of magnitude lower than bulk protein preparations, and to a very high degree of purity.

Proteins used for therapeutic or *in vivo* diagnostic purposes are subjected to the most stringent purification procedures, as the presence of molecular species other than the intended product may have an adverse clinical impact. Proteins used for *in vitro* diagnostic and analytical purposes are also usually highly purified. In such cases however, the level of purification required is generally not as high as for those proteins intended for *in vivo* administration. In some instances, design of a downstream purification procedure which removes specific contaminating proteins is more important than purification of the protein to homogeneity. While various general issues relating to large-scale protein purification are considered over the remainder of this chapter, specific examples of how various protein products are produced commercially are also provided in many of the subsequent chapters of this text.

Scale up of protein purification

Industrial-scale protein purification protocols are initially designed at laboratory level. Scale up studies are then undertaken in order to produce sufficient quantities of the protein to meet market demands as economically as possible. Generally speaking, the costs associated with the production of a unit quantity of any protein decline with increasing scale of production:

- most chemicals and other raw materials required may be purchased more cheaply in bulk;

- many overhead costs remain independent of the scale of production;

- labour costs (per unit of product produced) decrease sharply with increased scale of production.

Many of the techniques and procedures used for laboratory-scale purifications are not amenable to scale up. Many methods employed to disrupt bacterial cells in the laboratory are not suitable for large-scale application – either for technical or economic reasons. Disruption of the bacterial cell wall by sonication, while feasible on a small scale, is inefficient when applied to large-scale situations. Treatment with lysozyme on a large scale would be uneconomic. Other techniques routinely used in laboratory-scale purification procedures must be modified before they can be successfully used on a large scale. Laboratory-scale centrifugation for example is generally carried out in a batch (fixed volume) centrifuge, whereas continuous flow centrifuges are employed in industrial-scale purification systems (Chapter 3). Problems associated with scale up of specific techniques should always be borne in mind when designing a purification procedure at a laboratory level.

Rational design of a protein purification procedure, with all stages being amenable to direct scale up, is especially desirable when working with a protein of therapeutic interest. Such proteins are initially produced in small quantities, which are then subjected to animal trials. If encouraging results are obtained, limited clinical studies may then be initiated. Such trials, while requiring relatively little protein, could take several years to complete and are expensive to carry out. If they prove the protein is safe and therapeutically effective, scale up of production is normally initiated, in line with projected market demands. Any but the most minor of changes made to the original purification process during such scale up would invalidate the earlier clinical studies. In such cases therefore, it is critically important to ensure that the purification system initially developed in the laboratory can be scaled up without difficulty. The three stages involved in the scale up of protein purification are outlined in Figure 4.2.

It is difficult to define what exactly constitutes a small- or large-scale protein purification process. Laboratory-scale purification procedures generally yield microgram to milligram quantities of the protein product. Pilot-scale production often yields gram quantities while large-scale purification yields quantities in the order of kilograms. Initially it is often quite difficult to decide what level of scale up is desirable for any given protein product. It may not always be clear what the level of market demand for the protein in question will be. Although there are several hundred proteinaceous products currently on the market, most are produced in relatively small quantities. There are probably no more than about 100 proteins whose annual level of production is in the range of kilograms, while a very small number of proteins are marketed whose annual production levels exceeds 1000 kg (See Table 3.1).

Scale up of purification systems

Scale up of protein purification systems poses many difficulties and pitfalls, not only for scientific personnel but also for the process engineer. The process engineer is charged with designing and installing all of the equipment required to successfully process the protein product on a preparative scale. As can be seen from Figure 4.3 process design is far from simple.

Figure 4.2 Sequence of steps undertaken during the design and scale up of a protein purification protocol. Initial studies are undertaken at laboratory level. These identify the minimum steps required to yield a protein product of the desired purity. Pilot scale studies are then undertaken which serve to pinpoint and resolve difficulties associated with process scale up. Finally the process scale operation is undertaken

Figure 4.3 Bioreactor at a Boehringer Mannheim manufacturing facility (photograph courtesy of Boehringer Mannheim, UK)

Most large-scale process equipment such as holding vessels and transfer pumps are constructed from stainless steel or plastics such as polypropylene. Glass vessels, so commonly used on a laboratory scale, are seldom used for large-scale preparative work, mainly due to the lack of structural strength. The grade of stainless steel or the plastic type used in the manufacture of process vessels must be selected with care. Materials must be inert and resistant to the corrosive action of any chemical used during the process. They should not allow any leaching of potentially toxic metals or chemicals into the product stream.

Liquid transfer, so easily achieved on a laboratory scale by physically pouring the liquid from one vessel into another, is achieved on a preparative scale by pumping the liquid via a series of stainless steel or plastic pipes. Only certain pump designs are suitable for transfer of liquid protein solutions. Some pump systems tend to entrap air during their pumping action, which can lead to protein denaturation.

Preparative-scale chromatographic equipment also often differs in design and appearance from laboratory-scale systems. Most small-scale chromatographic columns are manufactured from glass or from transparent plastics and range in capacity from 1 ml to 200 or 300 ml. Preparative scale chromatographic columns range in capacity from some hundreds of ml to several litres and, in some cases, up to several hundred litres. While some such columns are manufactured from glass, the majority are manufactured from reinforced plastic or stainless steel. Scale up is normally achieved by increasing the column diameter as opposed to column height. Columns several metres in diameter are available commercially. In the case of some chromatographic systems (e.g. gel filtration) the degree of separation achieved is proportional to the column length. Scale up of such systems is made more difficult by the compressible nature of many gel beads. Such difficulties may be overcome by employing more recently developed gel types which exhibit enhanced structural characteristics, or by altering column design. Stack columns are often used for preparative scale gel filtration of many industrially important proteins such as insulin (Chapters 3 and 7).

Complete large-scale chromatographic systems are designed, manufactured and marketed by several companies. Most such purification systems consist not only of the process equipment required but also of a computerized control unit, which facilitates a high degree of system automation. In general, innovations in downstream processing techniques are introduced infrequently, largely for two reasons: First, regulatory authorities such as the FDA review licence applications to produce and market proteins designed for use in the food or healthcare industries. Such applications are more likely to be looked upon favourably if proposed downstream processing procedures are based upon established, validated methodologies. Second, many protein products produced in a highly purified form command a high sale price. Many manufacturers have therefore tended simply to add on the cost of downstream processing to the final product price. As the level of competition between rival product manufacturers increases downstream processing costs, which can account for up to 80 per cent of the overall production costs, will become more and more critical in determining product competitiveness.

Bulk protein production

Industrial-scale enzymes are produced in bulk, and are generally subjected to little downstream processing. The vast majority of such 'bulk' enzymes are biopolymer degrading enzymes such as amylases, proteases and pectinases (Chapter 10). These are produced extracellularly by submerged fermentation of various (recombinant and non-recombinant) microbial species – particularly species of *Bacillus* or selected species of

Aspergillus. Many such bulk enzyme preparations simply consist of concentrated, cell-free fermentation products.

A generalized downstream processing scheme utilized in the production of bulk enzyme preparations from microbial sources is outlined in Figure 4.4. Once maximum yield of the protein has been achieved in the fermentation process the fermentation media is cooled to under 5 °C. This stabilizes the protein product and discourages further microbial growth.

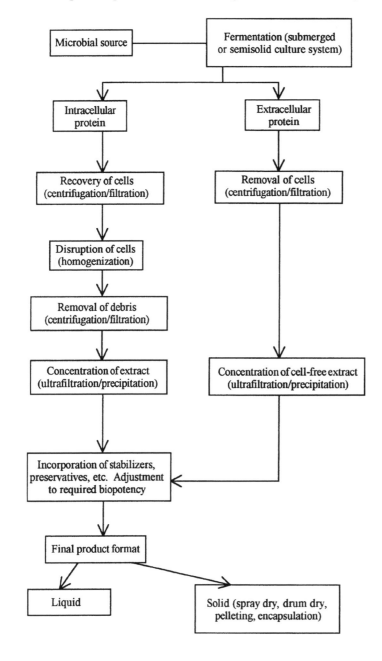

Figure 4.4 Bulk protein production from microbial sources, a generalized outline

The microbial cells must then be removed from the culture medium, typically by centrifugation or by filtration. The collected cells are usually washed several times in order to recover any enzyme trapped in the cell paste, and the wash is added to the cell-free media, which contains the bulk of the desired enzyme. If the enzyme is extracellular, as is usual, then filtration tends to be the method of choice. If the enzyme is intracellular, cell harvesting by continuous centrifugation is usually undertaken. After harvesting the cells are re-suspended in water or buffer and are disrupted in order to release their intracellular contents. The cell homogenate must then be centrifuged in order to remove any intact cells, in addition to cellular debris. The protein of interest should now be present in the supernatant. If the enzyme is produced by recombinant DNA technology, regulatory laws generally require the inclusion of a specific processing step that kills all the producer (genetically modified) organisms before their disposal.

In both cases a concentrating step is generally undertaken next. This is necessary to reduce the process volume to more manageable levels, especially in the case of extracellular enzymes present in large volumes of fermentation medium. Concentration is often achieved by methods of low-temperature evaporation or, more recently, ultrafiltration. In some cases further concentration is undertaken by vacuum evaporators. As the name suggests this entails applying a vacuum to the enzyme-containing liquid, with subsequent heating to around 40 °C. The concentrate may then be subjected to an additional filtration step, in order to remove any intact microorganisms still remaining in the extract. The most common filter employed at this stage is a cellulose fibre depth filter.

If the crude enzyme preparation is presented in liquid form, the final steps of the downstream processing procedure normally involves the addition of various stabilizers and preservatives, in order to increase the product shelf life. The most common preservatives and stabilizers used include propylene glycol, glycerol and sorbitol. The final product volume is then adjusted as appropriate, in order to ensure that the enzyme activity falls within its specified limits. The product is then drummed and stored at low temperatures. Some enzymes are also formulated as slurries. This entails addition of agents such as ammonium sulfate, sodium sulfate or sodium chloride to the enzyme concentrate. This promotes enzyme precipitation and crystallization. Various substances such as silicon dioxide can be added to the slurry in order to keep the enzyme in suspension. In many cases enzyme slurries exhibit longer shelf lives than enzymes when stored fully in solution.

Many bulk enzymes are sold in solid form. Perhaps the simplest method of producing a powdered enzyme preparation is by spray-drying the enzyme concentrate after stabilizer addition and a final filtration step. Spray-drying essentially involves the generation of an aerosol of tiny droplets from the enzyme-containing liquid or slurry, which are directed into a stream of hot gas. This results in evaporation of the water content of the droplets, leaving behind solid, enzyme-containing particles (Figure 4.5). Spray-dried enzyme preparations may subsequently be further dried

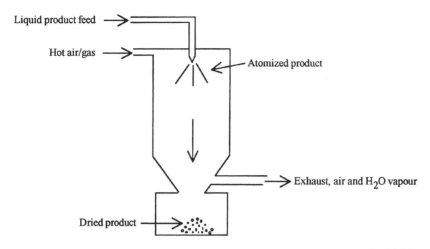

Liquid product feed →

Hot air/gas →

Atomized product

Exhaust, air and H_2O vapour

Dried product →

Figure 4.5 Schematic representation of the spray-drying process. The highly concentrated protein slurry/liquid is atomized, forming tiny product droplets. Hot air/gas is fed into the system as shown, and travels concurrently with the atomized product down the length of the drying chamber. In this way the hottest, driest air comes into contact with the wettest material. The product is dried by the air. The dried product collects at the base of the drying chamber, while the wet air/gas is removed as shown

in vacuum ovens if required. The dried preparation may then be ground into a fine powder and the activity level adjusted to the required bioactivity by the addition of suitable powder diluents.

Powdered enzyme preparations suffer from the disadvantage of a high level of dust formation during handling. Exposure of downstream processing personnel and/or end product users to such enzyme-containing dust may cause severe allergic reactions (Chapter 10). The problem can be overcome by encapsulating the enzymes. Initially encapsulating enzymes entailed their mixing with a liquefied wax and subsequent spray drying. This process, know as prilling, generated granules in which both the wax and enzyme were homogeneously distributed. An alternative process involved granulation of a pure enzyme-containing paste. These granules are subsequently coated with wax. The coating process ensures retention of granule integrity and thus minimizes subsequent dust formation. Granulation technology allowed the widespread re-incorporation of enzymes in detergents while essentially eliminating the health risks associated with dust formation.

Bulk enzyme preparations marketed in dried form include a variety of microbial amylases used in the starch processing industry, pectinases used in the clarification of fruit juice and proteases incorporated into detergent preparations (Chapters 11 and 12). These preparations may be quite impure and the desired enzymatic activity may represent no more than 2–5 per cent of the total protein content of the final product. All such enzyme products are now also normally tested for toxicity and allergenicity.

Milk-derived caseins and whey proteins as well as animal proteins such as gelatin are also produced commercially in large quantities. The approaches undertaken in the industrial-scale manufacture of these proteins are generally protein specific, and are outlined in Chapter 13.

Purification of proteins used for therapeutic or diagnostic purposes

Proteins destined for diagnostic or therapeutic applications must be purified to a very high degree. This is particularly true in the case of any protein preparation which is to be administered parenterally. Parenteral preparations are defined as sterile products which are intended for administration by injection, infusion or implantation. The high level of purity demanded is necessary in order to minimize or eliminate the occurrence of adverse clinical reactions against unwanted trace contaminants in the product. This will be discussed in detail later in this chapter. Many proteins used for *in vitro* diagnostic or analytical purposes (e.g. antibodies or enzymes) must also be partially purified in order to remove any contaminating substances which would otherwise interfere with their diagnostic applications.

The downstream processing of such proteins invariably entails not only preliminary treatments but also several high-resolution chromatography steps (Chapter 3). A typical overall purification process is outlined in Figure 4.6 If the protein of interest is secreted into the extracellular medium, the initial purification step involves the removal of whole cells by centrifugation or ultrafiltration. If the protein of interest is intracellular, cell harvesting is followed by homogenization to achieve cell disruption. If the intracellular protein has accumulated in soluble form, the next step involves the removal of any remaining whole cells and cellular debris, generally by centrifugation. If the protein has accumulated in the form of an inclusion body, homogenization is followed by a low speed centrifugation step which collects the inclusion bodies. After a washing step, the inclusion bodies are solubilized by a suitable method (Chapter 2).

A minimum of three chromatographic steps are usually utilized during the purification of a protein intended for parenteral administration. Although there are many different chromatographic techniques available, by far the most common techniques used industrially are those of ion exchange (most common), affinity chromatography, hydrophobic interaction chromatography and gel filtration chromatography. Gel filtration chromatography is often the last or 'polishing' step, undertaken for a number of reasons. It effectively fractionates the required protein molecules from dimeric or higher molecular mass aggregates, which may form during the earlier purification steps. It may also effectively remove molecules such as ligands, which may have leached from affinity columns. Gel filtration columns have, however, a low loading capacity. They also must be run at relatively low flow rates, and hence are often quite

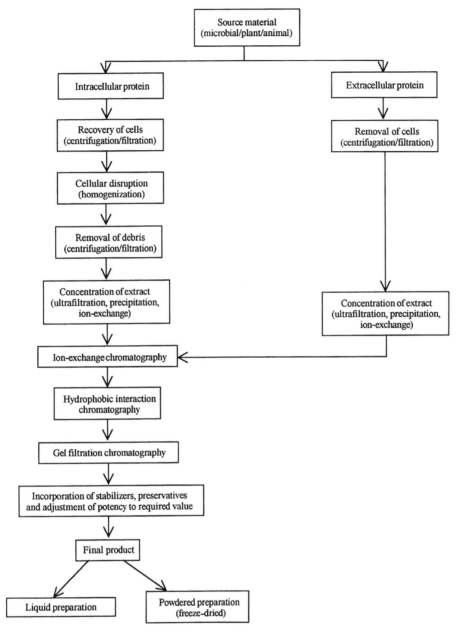

Figure 4.6   Production of protein products destined for therapeutic or diagnostic application (generalized outline). Additional steps may be required to assure complete removal of specific non-protein contaminants. Other chromatography types sometimes used include affinity chromatography or chromatofocusing. Final protein products are generally in the 98–99 per cent pure range if destined for therapeutic application. Numerous quality control steps are also undertaken

expensive in terms of their operation. The old maxim that 'time is money' is particularly true when applied to downstream processing. By employing the relevant irrigation buffer gel filtration may also be used to change the final purified protein into the buffer system most suitable for its final processing or storage.

As yet many chromatographic systems used are run under low pressure conditions, although preparative systems which operate under intermediate pressure, such as FPLC, and preparative high-pressure systems are assuming increasing industrial significance (Chapter 3). The degree of purification achieved after each chromatographic step may be monitored by assessing the quantity of protein of interest present relative to the total protein content.

Unlike traditional pharmaceutical preparations most therapeutic proteins are heat labile. They are inactivated at temperatures significantly above 40–50 °C. Therefore, terminal sterilization of product by heat (e.g. autoclaving) is not an option. In many cases, sterilization by irradiation is also impractical due to lack of appropriate irradiation facilities, in addition to the fact that many proteins are adversely affected by irradiation.

Therapeutic and diagnostic proteins are sterilized by filtration and subsequently introduced into their final (pre-sterilized) containers using strict aseptic techniques. Sterilization by filtration is achieved using a filter of 0.22 μm pore size or smaller. Such filters will remove bacteria and fungi, but are not guaranteed to remove all viruses or mycoplasmas. In-line 1.0 μm or 0.45 μm filters are usually utilized immediately prior to the sterilizing (0.22 μm) filter. Filters employed should in no way affect the product; thus fibre-shedding filters should be avoided and it must be verified that no product ingredient is altered or removed during the filtration step. The integrity of all filters should be tested upon completion of a fill.

The level of risk of contamination in relation to aseptically prepared products is dependant upon three parameters: (a) the concentration of airborne microorganisms; (b) the neck diameter of the container to be filled with product and (c) the time period that the open container is exposed to the environment. Reduction of any one of these parameters decreases the likelihood of accidental microbial contamination during an aseptic fill. The concentration of airborne microorganisms is minimized by carrying out all manipulations within a clean area environment (see below), and minimizing the number of personnel present during the aseptic operation. In this regard, design and installation of highly automated aseptic filling systems should greatly reduce the dangers of product contamination. A number of automated filling systems are available commercially and some may be cleaned in place and subsequently sterilized by steaming in place. The neck diameter of the container to be filled should be minimized and the filling operation should be operated at maximum speed in order to further reduce the likelihood of accidental microbial contamination. Sterile aqueous preparations which are filled aseptically may contain suitable antimicrobial preservatives added to an appropriate final concentration. Insulin injections for example, generally contains 0.1–0.25 per cent phenol or cresol as preservative. If the final product is sold in a powdered format, it is freeze-dried as described in Chapter 3.

THERAPEUTIC PROTEIN PRODUCTION: SOME SPECIAL ISSUES

Certain precautions must be observed when producing therapeutic proteins destined for parenteral administration. Such modes of administration can obviously introduce the protein or drug directly into the bloodstream. Body defence mechanisms associated with natural assimilation of molecular species through, for example the skin or gastrointestinal tract are thus bypassed. For this reason, every aspect of the manufacturing protocol employed in the production of such therapeutic proteins must attain high standards of safety. Manufacturing processes are undertaken in environmentally controlled areas (clean areas) and cleaning, decontamination and sanitization of equipment coming into direct contact with the product stream is particularly important. The solvent or diluent used during the manufacturing process is almost always highly purified water, termed water for injections (WFI). WFI is usually manufactured in-house in the biopharmaceutical plant by subjecting deionized water to distillation or reverse osmosis (Figure 4.7).

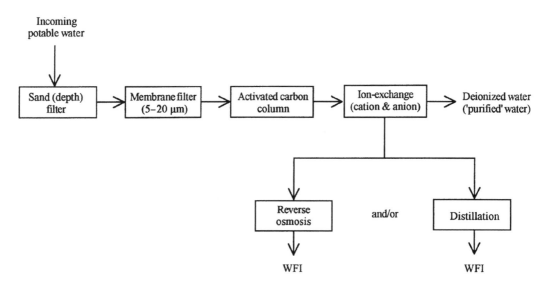

Figure 4.7   Generalized system by which potable water is purified in order to facilitate its use in biopharmaceutical processing. Water passed through cation and anion exchange columns is termed 'purified' or 'deionized' water. The generation of water for injections (WFI) entails a subsequent distillation or reverse osmosis step

Clean areas

Clean areas are specially designed rooms in which general environmental conditions are tightly controlled. High efficiency particulate air filters (HEPA filters) are installed in the ceilings of such rooms. All air entering the clean room area is filtered through the HEPA units. Filters may be classified according to their ability to remove particulate matter. Air normally exits a clean room via specialized outlets incorporated into the walls just above floor level. HEPA filters are evenly spaced over the ceiling area in order to generate a relatively uniform downward current of filtered air throughout the room (Figure 4.8).

Cleaning, decontamination and sanitation

Cleaning may be defined as the removal of 'dirt'. Dirt, as defined here, includes all organic and inorganic material which may accumulate in process equipment and processing areas during routine downstream processing. In this regard any protein or other molecules retained in a chromatographic column subsequent to elution of the protein of interest may be regarded as dirt. Decontamination generally refers to the inactivation and removal of undesirable substances such as endotoxins and other pyrogenic compounds (see later), in addition to other harmful substances such as viruses. Sanitation refers to the removal and inactivation of viable organisms. In many but not all instances cleaning procedures are also effective decontamination and sanitation methods. Such procedures are applied to both process equipment and to the area in which the manufacturing process is undertaken. Cleaning,

Figure 4.8 Generalized representation of HEPA-filtered air flow through a clean room

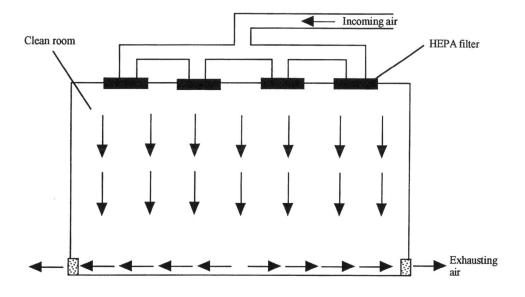

decontamination and sanitation of process equipment forms an integral part of any manufacturing procedure (Table 4.1).

Process equipment, which may be detached and/or dismantled, can be cleaned and sanitized relatively easily. Processing vessels may therefore be cleaned simply by rinsing or scrubbing thoroughly with high-purity water. Detergents may be required but if used, all traces must be subsequently washed away with water for injections. Detachable tubing may be cleaned in a somewhat similar fashion. Pumps and other equipment such as continuous flow centrifuges should be dismantled if possible to ensure a thorough cleaning of the constituent parts. Such cleaning procedures will remove virtually all 'dirt' and will significantly reduce the level of viable organisms (bioburden) present. It will not, however, guarantee sterility. Components may be sterilized by a number of methods, including moist or dry heat, irradiation or by chemical means. Sterilization of any process equipment with chemical agents such as formaldehyde, hypochlorite or peroxides must be undertaken with caution, due to the possibility that traces of such chemicals would remain in the sanitized equipment. This would contaminate products during subsequent production runs. Most processing equipment may be effectively sterilized by autoclaving, as can ancillary equipment such as flexible tubing and other plastic materials. Prior to their use, most filters are assembled in their final housings (usually stainless steel modules) and are also sterilized by autoclaving.

The internal surfaces of process fixtures such as metal pipework or some vessels may be sterilized by passing 'live' steam through for an appropriate period of time. Cleaning and sanitation of fixtures without their disassembly or removal from their normal functioning positions is usually termed cleaning in place (CIP). While all process equipment should be cleaned at regular intervals, only those items of equipment that come into direct contact with products are generally sterilized.

Chromatographic systems require regular cleaning, decontamination and sanitation. Such systems are generally cleaned in place. Prior to pouring or re-pouring a column, the column and gel matrix may be cleaned or sterilized separately. Before sterilization, the chromatographic column should be disassembled and scrupulously cleaned, firstly with an appropriate detergent, followed by rinsing thoroughly with high purity

Table 4.1   The range of substances that cleaning, decontamination and sanitation of therapeutic protein production equipment aims to remove

Cellular debris	Viral particles
Precipitated/aggregated materials	Microorganisms
General particulate material	Traces of protein or other molecules from previous production runs
Pyrogenic substances	Lipids and related substances

water. Depending upon the material from which it is manufactured, the column may subsequently be sterilized by autoclaving or by treatment with chemicals such as sodium hydroxide (NaOH). Most chromatographic media may be sterilized by autoclaving, though chemicals such as hypochlorite or peroxides may also be used under certain circumstances.

Upon packing columns with relevant gel materials it becomes necessary to devise a suitable protocol to achieve effective CIP of the chromatographic system. Several cleaning or sanitizing agents may be employed. Upon completion of a production run a thorough flushing of the column with several column volumes of water or suitable process buffer may flush out many column contaminants. Rinsing of the gel with a concentrated solution of neutral salts such as KCl of NaCl is often effective in removing precipitated or aggregated proteins, or other molecules which may be weakly bound to the gel. Inclusion of a chelating agent such as EDTA in the buffer may help remove any metal ions which remain associated with the gel. Subsequent washing with a solution of NaOH, or in certain cases with detergents, may be required in order to remove tightly bound material. In some instances, where column fouling by lipids is particularly problematic; increasing the column temperature to 40–60 °C may be effective. Most lipid material is solubilized at such temperatures. This approach may be adopted only if it has been shown that subjecting the chromatographic system to such changes in temperature will not adversely effect its fractionation characteristics.

Sodium hydroxide is extensively used as a CIP agent. It is effective in removing many contaminants that bind tightly to the gel matrix, in addition to removing or destroying pyrogenic material such as bacterial endotoxins, viral particles and bacteria. Sodium hydroxide is popular not only due to its efficiency as a cleaning agent, but also due to its ready availability and the fact that it is inexpensive.

The real possibility of the presence of pathogenic viral particles in biological source material is a matter of concern. Many viruses harbour oncogenes (cancer-causing genes) while others represent the causative agents of serious illnesses such as hepatitis or AIDS. NaOH will inactivate the HIV virus within minutes.

The effectiveness of NaOH as a cleaning and sanitizing agent is both time and concentration dependent. Generally speaking NaOH is applied at concentrations of 0.5–1.0 M, and should have a gel contact time of an hour or more. NaOH may be used only if it is compatible with both the chromatographic gel and all of the column components. Most chromatographic media may be exposed to NaOH for reasonable time periods without adverse effect. Silica gel is an exception as it quickly dissolves at pH values greater than 8.0. Following exposure to the NaOH solution, the chromatographic system should immediately be thoroughly rinsed with sterile, pyrogen-free water or an appropriate buffer. Efficient removal of NaOH subsequent to completing a CIP protocol is necessary in order to prevent column deterioration due to prolonged exposure to alkaline conditions and to ensure that residual NaOH does not contaminate the product during the next production run.

Routine application of an effective CIP procedure is greatly simplified if the overall chromatographic system has been designed with process hygiene in mind. The choice of gel and column type will determine what CIP agents may be utilized. Design of the chromatographic column and ancillary equipment will determine the susceptibility to microbial contamination. Valves and pipe connections represent danger points when considering the risk of introducing microbial contaminants. Pipe connections in particular should be designed such that no 'dead leg' zones are present.

RANGE AND MEDICAL SIGNIFICANCE OF IMPURITIES POTENTIALLY PRESENT IN PROTEIN-BASED THERAPEUTIC PRODUCTS

Proteins destined for parenteral administration must be free from all impurities which might have an adverse effect on the well being of the patients to whom they are administered. Impurities most commonly encountered are outlined in Table 4.2.

The majority of techniques utilized during downstream processing are designed to separate protein molecules from each other, i.e. to purify the protein of interest from the hundreds or in many cases thousands of other proteins present in the starting material. In many instances non-proteinaceous impurities may be efficiently removed by one or more of the protein fractionation steps of downstream processing. In other instances it may be necessary to include specific steps in order to remove certain impurities.

Table 4.2   The range and medical significance of potential impurities present in therapeutic protein products

Impurity	Medical consequence
Microorganisms	Potential establishment of severe microbial infections – septicaemia
Viral particles	Potential establishment of a severe viral infection
Pyrogenic substances	Fever response which in serious cases culminates in death
DNA	Significance is unclear – could bring about an immunological response
Contaminating proteins	Immunological reactions. Potential adverse effects if the contaminant exhibits an unwanted biological activity

Details of the various potential product impurities, their medical significance and methods to minimize or eliminate these from the final protein product forms the subject matter of the remainder of this chapter.

Microbial contaminants

Pharmaceutical products intended for parenteral administration must be sterile, with the exception of live bacterial vaccines. The presence of microbiological contaminants in the final product could result not only in microbial degradation of the product, but also in the establishment of a severe microbial infection in the recipient patient. Direct introduction of viable microorganisms into the bloodstream could induce a range of serious illnesses such as septicaemia. Septicaemia is characterized by widespread damage or destruction of various tissues, due to adsorption from the bloodstream of pathogens or toxins derived from such pathogens.

Low levels of microbial contaminants are frequently found in association with many protein products during downstream processing. In some cases such microorganisms may have produced the protein of interest. In others, the microbial contaminants are introduced into the product from the general environment, from non-sterile processing equipment or from downstream processing personnel.

In some circumstances the product may be subjected to one or several in-process filtration steps. Samples of product may be withdrawn at various junctures during the downstream processing and subjected to microbiological analysis in order to assess the microbial load or bioburden at that stage of processing. A decision to include a filtration step may be taken on the basis of this analysis. Agar plates are also usually placed at strategic working points during downstream processing operations and are exposed under normal working conditions. Detection of a high microbial count on any such plate(s) suggests the product may be subject to microbial contamination at that point. Appropriate action must then be taken to identify, reduce or eliminate the source of such microorganisms.

Viral contaminants

Traditional pharmaceuticals manufactured from chemical substances are usually considered free of viral contaminants. Therapeutic protein products can however potentially harbour such contaminants. It is conceivable that viral species may be introduced into the product during downstream processing, from infected personnel and contaminated equipment. However, contaminant viral particles usually originate from infected source material. Products obtained from human tissue, blood or urine may contain a number of human viral contaminants (Table 4.3). HIV and the hepatitis B virus are in many cases the two most likely contaminants.

Table 4.3 Viral contaminants that may potentially be present in biological raw materials obtained from human volunteers

Virus	Medical significance
Human immunodeficiency virus	Causative agent of AIDS
Hepatitis B virus	Causative agent of hepatitis B
Human T-cell leukaemia viruses (HTLV-I, HTLV-II)	Can cause leukaemia
Herpes simplex virus (HSV)	HSV-1 is the major causative agent of: herpetic stomatitis, herpes labialis (cold sore) keratoconjunctivitis. HSV-2 represents the major causative agent of genital herpes
Human papillomavirus	Causative agent of common wart. Also implicated in anal canal carcinoma
Cytomegalovirus (CMV)	Generally symptomless: however, represents a serious opportunistic infection in immunocompromised individuals
Epstein–Barr virus (EBV)	Causes infectious mononucleosis, linked to Burkitt's lymphoma and nasopharyngeal carcinoma

Some proteins destined for parenteral administration are obtained from bovine sources. Examples include insulin and follicle-stimulating hormone. In addition, bovine serum and fetal calf serum are often included in cell culture media. As in the case of proteins derived from human sources, proteins obtained from bovine tissue are usually purified from a large number of pooled samples. The presence of even one infected sample thus contaminates all of the starting material. Contaminant viruses most likely to be present in bovine-derived biological material include infectious bovine rhinotracheitis and bovine immunodeficiency virus.

Animal cell culture also represent an important source of some therapeutically beneficial proteins. A wide range of murine viruses such as lymphocytic choriomeningitis virus may therefore be potential contaminants of monoclonal antibody preparations derived from murine hybridomas. Chinese hamster ovary (CHO) or baby hamster kidney (BHK) cells, used in the recombinant production of many therapeutic proteins (Chapters 5–8) can also potentially harbour animal cell viruses.

Elimination of viral contaminants and the associated potential health risks to recipient patients is most effectively achieved by utilizing raw material sources free from viral contamination. Individual donor tissues should therefore be tested in order to eliminate any diseased samples before purification begins. Donor herds employed should have disease-free status. The development of animal cell culture which facilitates cellular growth in serum-free media represents an important break-through. This effectively eliminates the possibility of such cellular products being contaminated with bovine-derived viral particles. The production of proteins of therapeutic interest in recombinant microbial systems offers the greatest degree of assurance that the final therapeutic product will be free of clinically significant viral contaminants.

Two strategies are generally employed to reduce the risk of final product contamination by viruses. The first strategy involves the incorporation of specific steps capable of viral destruction in the downstream processing procedure. The second involves ensuring that one or more of the purification steps employed is effective in separating viral contaminants from the protein of interest.

It is not necessary to include specific viral inactivation steps in many purification procedures. Certain buffer components or other chemical additives inactivate a range of fragile virus particles. Specific inactivation steps which may be used include heat and irradiation steps. Maintenance of elevated temperatures (35–60 °C) for several hours has been shown to inactivate most viruses. Heat steps are used extensively to inactivate blood-borne viruses. Some more recent studies have investigated the effectiveness of high temperature–short time treatments. UV irradiation has also been used as a method of inactivation and several other methods continue to be evaluated. These include exposure to extremes of pH and multiple repeat filtration through a 0.1 μm or 0.2 μm filters. Some such methods may however have a detrimental effect on the protein of interest.

Many steps employed in protein purification systems effectively separate viral particles from proteins. This is particularly true with regard to chromatographic fractionation. Viral particles and protein molecules exhibit widely differing physicochemical properties. In some instances viral contaminants may adsorb to the chromatographic matrix and are removed from the product stream. Such a possibility renders regular column CIP operations essential. In other instances, viral particles percolate through the column at different rates to those of the proteins and hence are separated out. Gel filtration offers a particularly effective viral clearance step.

Central to successful viral detection studies is the availability of sensitive assay systems. Most assays currently available detect a specific virus or a group of related viruses. In many cases therefore, it becomes necessary to identify the viral particles whose potential presence is most likely in the source material and to assay specifically for those viral species.

A variety of assay types may be utilized. Some such assays employ a virus-specific DNA probe. DNA probes are often used to identify certain human retroviruses such as HTLV-1 or HTLV-2. Other viral assays rely on immunological-based diagnostic tests such as enzyme-linked immunosorbent assays (ELISA; Chapter 9).

Some *in vitro* viral assays involve the incubation of potentially infected samples with a detector cell line sensitive to a range of human or murine viruses. Upon incubation, the sensitive cell line is examined for any associated cytopathic effects for 14 days or longer.

A range of mouse, rat or hamster antibody production tests (MAPs, RAPs or HAPs) may also be used to detect the presence of various viruses. In such systems the mouse, rat or hamster is injected with the test sample. Approximately 4 weeks later serum samples are collected and screened for the presence of antibodies recognising a range of viral antigens.

Extensive validation studies must be carried out to ensure that protein purification protocols used during downstream processing effectively inactivates any contaminant viruses, or separates such viruses from the final protein product. Validation studies normally involve spiking a sample of raw material, from which the protein of interest is purified, with a known amount of viral particles. The sample is then subjected to the complete purification procedure. The purification process is normally scaled down to laboratory size during such studies. The level of viral inactivation or clearance may be monitored after each purification step, thus an associated viral reduction factor may be calculated.

Typically, a mixture of at least three different viral types are employed during such validation studies. The viral species chosen depends upon a number of factors, including knowledge of the viral types most likely to be present in the source material. The chosen viruses should exhibit a range of physicochemical properties (e.g. size & shape) and should all be easily detected.

Although the levels of viruses used in spiking vary, quantities of up to 1×10^{10} viral particles are commonly used. The cumulative inactivation and/or removal effects of the scaled down purification procedure should lead to a reduction in the viral population of up to 10^{16}. Thus, the likelihood of a single dose of final product containing a single viral particle should not be greater than 1 in 1 000 000.

Pyrogenic contaminants

Pyrogens represent another group of hazardous contaminants whose presence in parenteral products is highly undesirable. Pyrogens are substances which influence the hypothalmic regulation of body temperature, and usually cause a fever. Various miscellaneous chemicals and particulate matter can be pyrogenic. Contamination of finished products by such substances is avoided by implementation of comprehensive raw material and in-process quality control testing, and by strict adherence to the principles of good manufacturing practice. Most solutions destined for parenteral administration are passed through a 0.45 or 0.22 μm filter directly prior to filling into final product containers. Such filtration steps will remove particulate matter which otherwise could elicit a pyrogenic response in recipient patients.

By far the most common pyrogenic substances likely to contaminate products however, are bacterially derived endotoxins. The presence of endotoxins is difficult to control due to their ubiquitous nature. Endotoxins are lipopolysaccharide (LPS) substances originating from the cell wall of Gram-negative bacteria. Unlike the cell wall of Gram-positive organisms, which consists of a thick protective layer of peptidoglycan, the cell envelope of Gram-negative bacteria consists of an underlying layer of peptidoglycan covered by an outer membraneous structure. This outer membrane is composed mainly of phospholipids, LPS and proteins. The LPS, which is the endotoxin, consists of a core polysaccharide attached to a lipid A moiety (which serves to anchor the structure in the membrane) and polysaccharide O antigens, which project outward from the membrane surface. The core polysaccharide usually contains a variety of seven carbon sugars (heptoses) in addition to hexoses such as galactose, glucose and *N*-acetylglucosamine. The *O*-linked polysaccharide also contain a variety of hexoses such as glucose, galactose, mannose, in addition to more unusual dideoxy sugars such as abequose and paratose. The lipid structure does not contain glycerol and constituent fatty acids are linked (via ester linkages) to *N*-acetylglucosamine. It is believed that the lipid component is mainly responsible for the toxic properties of the LPS molecules while the polysaccharides render the molecule water soluble. It must also be stressed that not all LPS molecules found in association with Gram-negative bacteria exhibit toxic effects. Lysis of Gram-negative bacteria results in the liberation of free endotoxin molecules. The presence, and in particular lysis of Gram-negative bacteria during downstream processing may result in contamination of the product with large quantities of endotoxin.

One of the most clinically notable responses to endotoxin is that of fever, i.e. a pyrogenic response. This effect is indirect. Endotoxin serves as a potent stimulant to interleukin-1 (IL-1) production by macrophages. IL-1 is also known as endogenous pyrogen and initiates the fever response. Endotoxin is a potent activator of the complement system, and is considered to be instrumental in the development of septic shock. This may be triggered at least in part by injected endotoxin, though it is often initiated in response to the presence of high circulatory levels of endotoxin due to Gram-negative bacteraemia (the presence of Gram-negative bacteria in the blood). Septic shock is a particularly serious medical condition, which in severe cases results in patient death (Figure 4.9).

The presence of pyrogenic substances in parenteral preparations may be detected by a number of methods. Historically, the rabbit assay was the most widely used method. This test involves parenteral administration of a sample of the product, with subsequent monitoring for changes in rabbit temperature. While this test is capable of detecting a wide range of pyrogenic substances it suffers from a number of limitations. The process is laborious, space and time consuming and requires experienced animal technicians. Excitation of the rabbits employed can affect the experimental results obtained. Large-scale rabbit testing using different rabbit

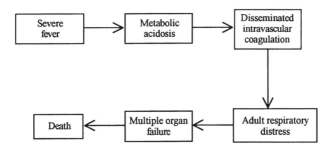

Figure 4.9 Clinical stages associated with septic shock

colonies are also subject to variations, which can lead to the possibility of variable standards. An alternative *in vitro* test based on endotoxin-stimulated coagulation of amoebocyte lysate obtained from horseshoe crabs is also available.

It has been known since the mid-1950s that Gram-negative bacterial infections causes intravascular coagulation (blood clotting) in the American horseshoe crab, *Limulus polyphemus*. The factor responsible for coagulation exists within the crab's circulating blood cells (termed amoebocytes). The presence of bacterial endotoxin results in activation of this clotting agent. Scientific investigations have revealed that endotoxin activates a coagulation cascade, not unlike the coagulation cascade system found in higher animals. Activation of the cascade is dependent not only upon the presence of endotoxin, but also on the presence of divalent cations such as magnesium or calcium. The terminal stage of the cascade system (Figure 4.10) involves the proteolytic conversion of an inactive proenzyme to an activated clotting enzyme. This enzyme catalyses the proteolytic cleavage of the clotting protein, coagulogen, forming coagulin and free peptide C. Free coagulin molecules interact noncovalently resulting in clot formation.

The *Limulus* amoebocyte lysate (LAL) assay is based upon this coagulation cascade. LAL reagent is prepared by extracting and washing *Limulus* amoebocytes, followed by induction of cellular lysis. The *Limulus* lysate, which is commercially available, is incubated with the test preparation in pyrogen-free test tubes, usually for a period of 1 h. Any endotoxin present in such test samples will result in activation of the coagulation cascade and therefore clot formation in the test tube.

The LAL assay is widely used to detect the presence of endotoxin in parenteral products, in bulk reagents such as WFI used in the manufacture of parenterals and in biological fluids such as serum or cerebrospinal fluids. The popularity of the test is a reflection of its sensitivity, speed, specificity and reproducibility. The LAL test has largely replaced the rabbit test as the method of choice for detection of endotoxin in water for injections and finished products parenteral solutions. The sensitivity of the LAL test remains unsurpassed as it can detect a few picograms of endotoxin per millilitre of test sample. Its extreme specificity however is cited by some as a disadvantage. The LAL assay will not detect non-endotoxin pyrogenic substances.

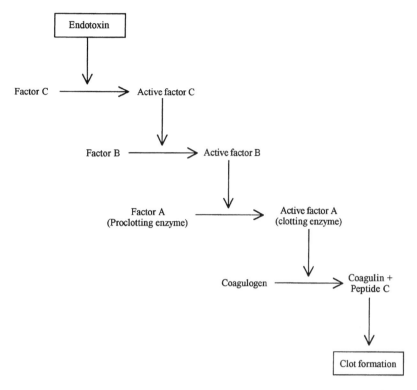

Figure 4.10 Endotoxin-mediated intravascular coagulation, as occurs in the American horseshoe crab, upon vascular exposure to endotoxin. The lysate of the crab's amoebocytes contains all the constituents of the coagulation cascade. Incubation of such a lysate with a test sample will result in gel formation (i.e. coagulation) if the sample contains endotoxin

Validation of the LAL assay system, as applied to any biopharmaceutical process, must also be undertaken. It is particularly important to show that gel formation is not inhibited by any constituents present in the sample being tested. This may be achieved by spiking the test sample with known quantities of endotoxin. The presence of inhibitory substances would prevent clotting. The LAL assay system has been adapted to a chromagenic test format. This involves incubation of the test sample with amoebocyte lysate and an additional suitable substrate. The substrate utilised is normally a short peptide to which a chromagen has been attached. Activation of the coagulation cascade by endotoxin results in the proteolytic degradation of the substrate, with release of free chromagen. Unlike the bound chromagen, the free chromagen absorbs specific wavelengths of visible light. The level of absorption at such wavelengths reflects the level of endotoxin present in the test sample. This method is therefore quantitative. Colour formation is rapid and the test may be completed in 15 min.

Endotoxin may be destroyed or removed from process equipment or product by a number of means. LPS is heat stable and is not easily destroyed by thermal treatment. Depyrogenation of heat resistant process equipment and test tubes used in LAL assays may be achieved using dry heat (180 °C for 3 h, or 240 °C for 1 h), or by moist heat (three consecutive autoclave cycles).

Chromatographic columns are normally depyrogenated by exposure to NaOH during CIP procedures, or by exhaustive rinsing with pyrogen-free water or buffer until the eluate is shown to be pyrogen free. Product exposure to bacterial contaminants during downstream processing should be minimized. This is most effectively achieved by: (a) ensuring that all relevant purification procedures are undertaken in areas of controlled environmental conditions, such as clean room areas; (b) that all equipment coming into contact with the product is cleaned, sanitized and, if possible, sterilized before commencement of the operations; and (c) by preventing bacterial build up in the process during various stages of downstream processing. Particular care must be taken with regard to endotoxin control if the protein of interest is sourced from a Gram-negative bacteria.

Endotoxin molecules exhibit physicochemical properties which differ greatly to the physicochemical characteristics of most proteins. For this reason many protein fractionation techniques result in effective separation of endotoxin from the protein of interest. Inclusion of a specific endotoxin removal step as part of a downstream processing protocol is therefore unnecessary in many cases. Specific removal steps, if required, normally take advantage of the charge or molecular mass characteristics of the endotoxin molecule. While individual molecules of lipopolysaccharide exhibit a molecular mass of less than 20 kDa, such molecules aggregate in aqueous conditions giving rise to a structure of molecular mass 100–1000 kDa. Gel filtration chromatography may thus effectively separate such endotoxin molecules from the protein of interest if the molecular mass of the protein in question is below 100 kDa. Conversely, if the protein of interest has a molecular mass greater than 100 kDa, chelating agents such as EDTA may be included in the buffers used, as such agents promote depolymerization of endotoxin molecules to the monomeric form. Under such conditions, gel filtration may also be used to separate endotoxin from protein. Separation on this basis may also be achieved using an ultrafiltration membrane of appropriate pore size.

LPS possess a high negative charge, and thus most ion exchange steps will separate endotoxin from protein molecules. Endotoxin may also be removed from bulk solutions such as water by passing the solution through a membrane filter exhibiting a positive surface charge. This depyrogenation method is unsuitable if the solution contains other negatively charged important molecules in addition to endotoxin.

DNA contaminants

The significance of DNA-based contaminants in parenteral products remains unclear. Theoretically, entry of contaminant DNA into the genome of recipient cells, if such a process were to occur, could have serious clinical implication. These could include alteration of the level of expression of cellular genes, or expression of a foreign gene product. Many cell lines employed to produce certain therapeutic proteins (e.g. hybridomas, recombinant CHO and BHK cell lines) are known to contain active oncogenes. Although health risks associated with the presence of naked DNA in parenteral preparations is considered to be minimal, the potential presence of oncogenes in an injectable product is deemed inappropriate.

In many cases there is little need to incorporate specific DNA removal steps in downstream processing procedures. Nucleic acids present in crude cell extracts are often degraded by endogenous nucleases. Exogenous nucleases can, however, be added if considered necessary. Nuclease treatment will result in degradation of nucleic acids, yielding nucleotides. Nucleotides are a much less potential hazard compared to DNA segments corresponding to intact genes. Most protein purification steps subsequently remove nucleotides from product-containing fractions. Nucleic acid molecules may also be precipitated from solution by addition of positively charged polymers such as polyethyleneimine (Chapter 3).

Due to major differences in their physicochemical properties, most chromatographic steps will result in effective separation of DNA from proteins. As in the case of endotoxin, DNA's highly negative charge ensures that an ion exchange step is particularly effective in its removal from the product stream. Levels of contaminating DNA are normally measured using a species-specific DNA hybridization assay. If the protein of interest is purified from *E. coli*, total *E. coli* DNA is radiolabelled and used as a probe to detect any *E. coli* DNA contaminating the product.

DNA validation studies are normally carried out in order to illustrate that the downstream purification process used is capable of reducing the level of DNA in the final product to within acceptable limits. The definition of what constitutes a suitable upper limit is somewhat arbitrary, but quantities of up to 10 pg of residual DNA per therapeutic dose are generally considered acceptable.

DNA validation studies are carried out by methods very similar to those of viral validation studies. This involves spiking a raw material sample with a known amount of DNA and applying this to a scaled-down version of the proposed purification system. The DNA distribution profile may be monitored after each purification step, together with the fractionation behaviour of the protein of interest. In this way, a DNA reduction factor may be calculated. DNA obtained from the same source as the protein of interest is normally used in validation studies. The DNA employed to spike samples may itself be radiolabelled, allowing direct detection. Alternatively, the DNA applied may be unlabelled, and in such circumstances, a specific labelled DNA probe must be used to detect the spiking material. The actual quantity of DNA used must be carefully considered as excess DNA may itself adversely affect the purification procedure. Large quantities of exogenous DNA added to the product sample could for example, bind anion exchange resins and therefore significantly reduce their capacity to bind other molecules. The physical characteristics of the DNA employed should also receive some thought as DNA molecules present in any given protein starting sample may vary widely in terms of molecular mass. It may thus be prudent to carry out validation studies using DNA spikes, which exhibit a wide range of molecular masses.

Protein contaminants

The majority of purification steps included in downstream processing protocols are designed to specifically fractionate differing protein molecules from one another. Despite the availability of a wide range of fractionation methods it remains a difficult task to purify a specific protein to homogeneity, while obtaining an economically viable yield. The range of potential protein contaminants is dependant upon the source of the protein, its method of production and the downstream processing procedures used.

Modified forms of the protein product itself may also be considered as impurities. All proteins are susceptible to a variety of structural modifications which may alter their biological activity or immunological characteristics (See also chapter 3).

- aggregation;
- oxidation of methionine residues;
- incorrect disulfide bond formation;
- proteolysis by endo- or exo-acting proteases;
- enzymatic alteration of post-translational modifications (glycosylation).

In many instances, such altered products are generated during downstream processing and their effective separation from the intact parental molecules can prove quite difficult. Aggregated or extensively degraded molecules may be removed from the intact product by gel filtration chromatography, as can some molecules with extensively altered glycosylation patterns. Oxidized or deaminated proteins may be separated by techniques such as ion exchange chromatography or isoelectric focusing.

A major clinical significance of protein impurities relates to their antigenicity. Some protein contaminants may also display a biological activity deleterious either to the protein product or to the recipient patient. The potential for adverse immunological reactions depends not only upon the immunogenicity of the product or contaminants, but also upon the route of administration and in particular upon the frequency of administration. Contaminating proteins also act as adjuvant-like materials, thereby further increasing the immunogenicity of the protein. Immunological complications are less likely to feature if a particular therapeutic product is administered on a once-off basis, in contrast to a product which must be repeatedly administered to the recipient patient.

Upon their administration, all injected proteins run the risk of eliciting an immune response. Some protein vaccines or toxoids are administered specifically for this purpose. Many therapeutically important proteins are derived from human sources and as such, are non-immunogenic. Other proteins often share extensive structural homology with the natural human product. Porcine insulin for example differs from human insulin by only one amino acid residue. Bovine insulin differs from human insulin by three amino acids. As such, administration of bovine, and in particular, porcine insulin normally does not elicit a strong immunological response.

Different protein molecules differ in their intrinsic ability to stimulate an immune response. It is not yet possible to pre-determine if any particular protein will initiate a strong immunological reaction. In general however the less homology exhibited between the product and the analogous human protein, the greater the possibility that an immunological reaction will be observed. Larger polypeptides or proteins also tend to exhibit increased immunogenicity as do proteins exhibiting extensive post-translational modifications which differ from modifications observed in their human counterparts.

Immunological responses to administered proteins are clinically undesirable for a number of reasons (unless the product is used specifically as an immunogen). In cases where repeat administration is required, antibodies may be raised against the 'foreign' protein. These antigen-specific antibodies may decrease or nullify the potency of the product. This effectively means that the patient develops clinical resistance to the protein drug or that increasing quantities of the protein must be administered in order to sustain a particular level of clinical responsiveness. Binding of antibody to the therapeutic protein may also distort the dose–response curve. While initial binding of antigen by antibody may decrease the products apparent potency, the reversibility of antibody–antigen binding usually results in a slow, sustained release of free protein in the blood stream. Binding of the therapeutic protein by antibodies may also effect normal degradation of the product. Many hormones such as insulin are removed from general circulation by a process of receptor-mediated endocytosis, and binding of antibody may severely restrict this uptake mechanism. Patients may also develop allergic responses to the particular product or to impurities consistently found in association with the products. Such allergic reactions or hypersensitivity generally result from the action of IgE antibodies or from T-cell mediated cellular toxicity.

The various approaches to detecting protein based impurities in a purified protein product have been reviewed in Chapter 3. One point sometimes overlooked is that many protein contaminants, be they modified versions of the protein of interest or different protein molecules, may be completely innocuous. Many would suggest that the presence of trace amounts of protein contaminants does not by itself, automatically justify introduction of additional purification steps in order to achieve their removal. Pre-clinical and clinical trials are perhaps the only real means by which the medical significance of any such trace contaminants may be properly assessed.

Chemical and miscellaneous contaminants

A variety of miscellaneous contaminants are often found in the product stream at various stages of purification. Some such contaminants, such as minor levels of lipid or polysaccharide, may be derived from the producing cells. The majority of miscellaneous contaminants, however, are introduced from exogenous sources. Examples of such potential exogenous contaminants are listed in Table 4.4. The nature of the purification procedure largely dictates which if any such contaminants might be present in the product stream. In some instances contaminants introduced during the initial stages of purification may subsequently be removed by one or more of the latter purification steps. In other cases contaminants may co-purify with the protein of interest. Under such circumstances it may be necessary to incorporate an additional purification step in order to achieve effective removal of the contaminant(s) concerned. It is

Table 4.4 Miscellaneous contaminants which may be introduced into the product stream during downstream processing

Buffer components	Detergents
Chaotrophic agents	Ethanol
Proteolytic inhibitors	Stabilizers
Salts	Ligands and other chromatographic breakdown products
Glycerol	Chemicals and ions leached from process equipment and pipe work
Antifoaming agents	

particularly important to ensure that all traces of potentially toxic, carcinogenic or otherwise unsafe contaminants are removed from the finished product. Final product containers of suitable quality should be utilized in order to reduce or eliminate the risk of leaching of chemical or other substances into the product during storage.

Many chemical contaminants (e.g. buffer components and ligands) are low molecular mass compounds. In such instances, gel filtration or ultrafiltration may effectively remove this material. Higher molecular mass contaminants are often more difficult to separate from the product. Methods utilized to remove such contaminants may be logically chosen if the physiochemical properties of both contaminant and protein product are known.

In some instances (e.g. if a particular likely contaminant is known to be toxic) it may become necessary to validate the purification system with regard to its ability to effectively remove that contaminant. This is best achieved by spiking a sample of the material at the stage of purification where the putative contaminant is introduced, and subjecting this spike sample to the remainder of the purification steps. An appropriate assay capable of detecting the contaminant in question must be available.

As an additional safety measure finished products are often subjected to 'abnormal toxicity' or 'general safety' tests. Safety tests normally entail the intravenous administration of a dose of up to 0.5 ml of product to at least five healthy mice. The animals are then observed for a period of 48 h, and should exhibit no adverse symptoms other than the symptoms expected. Death or illness of one or more of the test animals signals further analysis and in such instances the test may be repeated using a larger number of animals. Safety tests are undertaken in order to detect any unexpected or unacceptable biological activities associated with the product concerned.

LABELLING AND PACKING OF FINISHED PRODUCTS

Upon filling and sealing in their final containers all protein products are subsequently labelled and packed. Such operations are generally highly automated and do not require significant technical input. Labelling, however, is a critical operation in its own right. Mislabelling remains one of the most frequent causes of product recall. Information present on a product label should include the name and strength or potency of the product, batch number, date of manufacture and expiry date, in addition to the storage conditions to be employed. Information detailing the presence of any preservatives or other excipients may also be included, in addition to a brief summary of the correct mode of product usage.

FURTHER READING

Books

Avis, K & Wu, V. (1998) *Biotechnology and Biopharmaceutical Manufacturing, Processing and Preservation*. Interpharm Press, Chicago.

Butler, M. (1996) *Animal Cell Culture and Technology*. IRL Press, Oxford.

Desai, M. (2000) *Downstream Protein Processing Methods*. Humana Press, Totowa, NJ.

Doran, P. (1995) *Bioprocess Engineering Principles*. Academic Press.

El-Mansi, M. (1999) *Fermentation Microbiology and Biotechnology*. Taylor & Francis, London.

Freshney, R. (2000) *Animal Cell Culture*. Oxford University Press, Oxford.

Godfrey, T. (1996) *Industrial Enzymology*. MacMillan, Basingstoke.

Goldberg, E. (1995) *Handbook of Downstream Processing*. Blackie Academic and Professional, Glasgow.

Harrison, R. (1993) *Protein Purification Process Engineering*. Marcel Dekker, New York.

Shanbury, P. (1995) *Principles of Fermentation Technology*. Pergamon Press, Oxford.

Shuler, M. (2000) *Bioprocess Engineering*. Prentice-Hall, Englewood Cliffs, NJ.

Spier, R. (1994) *Animal Cell Biotechnology*. Academic Press, London.

Uhlig, H. (1998) *Industrial Enzymes and their Applications*. John Wiley & Sons, Chichester.

Walsh, G. (1998) *Biopharmaceuticals; Biochemistry and Biotechnology*. John Wiley & Sons, Chichester.

Wiseman, A. (1995) *Handbook of Enzyme Biotechnology*. Ellis Horwood, Chichester.

Articles

Helmrich, A. & Barnes, D. (1998) Animal cell culture equipment and techniques. *Methods Cell Biol.* **57**, 3–17.

Hjorth, R. (1997) Expanded bed adsorption in industrial bioprocessing: recent developments. *Trends Biotechnol.* **15**, 230–235.

Hu, W.-S. & Aunins, J. (1997) Large scale mammalian cell culture, *Curr. Opin. Biotechnol.* **8**, 148–153.

Konrad, M. (1989) Immunogeneity of proteins administered to humans for therapeutic purposes. *Trends Biotechnol.* **7**, 175–178.

Middelberg, A. (1996) Large-scale recovery of recombinant protein inclusion bodies expressed in *E. coli. J. Microbiol. Biotechnol.* **6** (4), 225–231.

Narayanan, S. (1994) Preparative affinity chromatography of proteins. *J. Chromatogr. A.* **658**, 237–258.

Rech, M. *et al.* (1991) Current trends in facilities and equipment for aseptic processing. *Pharm. Technol. Int.* **3**, 48–52.

Rhodes, M. & Birch, J. (1988) Large-scale production of proteins from mammalian cells. *Bio/Technology* **6**, 518–523.

Schugerl, K. (2000) Integrated processing of biotechnology products. *Biotechnol. adv.* **18** (7), 581–599.

Tai, J. & Liu, T.Y. (1977) Studies on *Limulus* amoebocyte lysate. *J. Biol. Chem.* **252**, 2178–2181.

Thornton, R.M. (1990) Pharmaceutically sterile clean rooms. *Pharm. Technol. Int.* **2**, 26–29.

Varley, J. & Birch, J. (1999) Reactor design for large scale suspension animal cell culture. *Cytotechnology* **29** (3), 177–205.

5

Therapeutic proteins: blood products and vaccines

INTRODUCTION

The human vascular system contains between 5 and 6 l of blood. On average, this accounts for 8.5–9.0 per cent of total body weight. Whole blood consists of red blood cells (which constitute 99 per cent of all blood cells), white blood cells and platelets. These are suspended in a fluid called plasma. Plasma has a characteristic straw colour, due largely to the presence of bilirubin, a breakdown product of haemoglobin. It may be obtained by centrifugation of whole blood, following the addition of a suitable anticoagulant to freshly drawn blood. The anticoagulant prevents clotting of the blood while the centrifugation step removes the cells suspended in the plasma. If blood is allowed to clot, the clot exudes a fluid called serum. The clot consists of suspended cellular elements and platelets entrapped or enmeshed in an extensive cross-linked network of fibrin molecules. Fibrin is derived from fibrinogen, a plasma protein. Plasma is essentially serum in which fibrinogen is present.

The various components of blood serve a wide range of physiological functions within the body. Blood is ideally suited to play a transportational role, due to the extensive network formed by the vascular system. Blood functions to transport a wide range of substances within the body, such as nutrients, waste products, gases, antibodies, enzymes, parenterally administered substances, hormones and other regulatory factors. Blood also plays a vital role in several additional physiological processes, such as maintenance of tissue hydration levels and regulation of body temperature.

Many specific functions of blood are carried out by proteins found in plasma. Electrophoretic separation of plasma proteins reveals five major bands, referred to as albumin, α_1-globulins, α_2-globulins, β-globulins and γ-globulins. Serum albumin represents the most abundant protein, accounting for more than 50 per cent of total plasma protein. All globulin protein fractions contain a variety of different protein molecules.

BLOOD PRODUCTS

Whole blood and blood plasma

Whole blood is aseptically collected from human donors and is immediately mixed with an anticoagulant to prevent clotting. Suitable anticoagulants include heparin and sodium citrate. Whole blood may be used as a source of a variety of blood constituents:

- red blood cells;
- platelets;
- various clotting factors;

- immunoglobulins;

- Additional plasma constituents.

The use of whole blood for such purposes would be considered wasteful if the specific purified component required is already available. Furthermore, fractionation procedures used to produce specific purified blood products considerably reduce the risks of accidental transmission of disease from contaminated blood donations.

Blood obtained from donors must be screened for the presence of a variety of likely pathogenic contaminants, in particular, Hepatitis B and C as well as HIV viral species. Donated blood samples are therefore usually screened individually for the presence of hepatitis B surface antigen and the presence of anti-HIV antibodies. Whole blood is normally administered to patients following severe blood loss. Concentrated red blood cell fraction and plasma-reduced blood is also available, and may be administered for certain clinical conditions.

Blood plasma is prepared from whole blood by centrifugation and is then normally stored frozen until use. Plasma is normally employed clinically as a source of therapeutically important plasma proteins such as clotting factors, when the purified products are unavailable. Yet another widely employed derivative of whole blood is termed plasma protein solution or plasma protein fraction. This aqueous solution is normally prepared by limited fractionation of serum or plasma and consists predominantly of serum albumin with minor quantities of globulin proteins. Plasma protein solution is administered in cases of shock caused by a large decrease in the volume of blood. Such sudden blood loss may be as a result of internal or external bleeding, extensive burns or dehydration.

Blood-derived proteins

Specific blood proteins that are of therapeutic use include a range of factors involved in the blood clotting process, fibrinolytic agents that degrade clots, serum albumin and immunoglobulin preparations. Most such proteins, in addition to whole blood and blood plasma preparations, have been commercially available for many years. While these products have traditionally been obtained from blood donated by human volunteers, some are now produced by recombinant DNA technology, as described later.

Blood coagulation factors: biochemistry and function

A variety of plasma proteins form an integral part of the blood clotting process. A genetic or induced deficiency in any one blood factor results in severely impaired coagulation ability, with serious medical consequences. The vast majority of hereditary diseases characterized by poor coagulation responses result from a deficiency of blood factors VIII and IX. A variety of other non-hereditary clinical conditions, such as vitamin K deficiency, may also result in the impairment of the blood coagulation process.

When a blood vessel is damaged or cut, specific elements in blood initiate the process of haemostasis – the curtailment and eventual cessation of blood loss. Haemostasis ultimately depends upon two interdependent physiological processes: (a) the formation of a platelet plug and (b) the blood coagulation process. Haemostasis is characterized by the rapid attachment of platelets to the damaged area. Platelets also adhere to each other, and in this way often stem blood flow. The congregated platelets also secrete a variety of amines such as adrenaline which stimulate localised constriction of the blood vessels. The process of blood coagulation is also initiated, resulting in the formation of a blood clot (thrombus) at the site of damage.

The process of blood coagulation is dependant upon a number of associated blood clotting factors. Such factors are designated by Roman numerals, although each is also known by a common name. Blood factors are listed in Table 5.1. With the exception of factor VI (calcium ions), all other factors are proteinaceous in nature.

Many of the blood factors, (e.g. factor II, VII, IX, X, XI and XII) exhibit proteolytic activity upon activation. The unactivated factors are protease zymogens. Activated factors catalyse the proteolytic cleavage of another factor in the clotting sequence, thus resulting in activation of this next factor. The clotting sequence may thus be regarded as a molecular cascade system in which a sequential activation of clotting factors is observed. Each single activated molecule will in turn, catalyse the activation of numerous molecules of the next factor in the sequence. This results in considerable stepwise amplification of the initial signal.

Not all of the clotting factors listed in Table 5.1 exhibit proteolytic activity. Those non-proteolytic accessory factors (factors III, IV, V, and VIII), however, form essential components of the coagulation system. They are generally activated by one of the proteolytic factors, and upon activation, serve to enhance the rate of activation of other blood factor zymogens. The presence of certain phospholipid components released from damaged tissue or from platelets also serves to accelerate the rate of coagulation.

Two distinct blood coagulation pathways exist: the intrinsic and extrinsic pathways. The extrinsic pathway relies upon factors which are normally present in plasma. This pathway is initiated when factor VII is activated through contact with injured surfaces. Functioning of the extrinsic pathway requires, in addition to blood factors, the presence of tissue factor (factor III). Tissue factor, along with calcium (factor IV), factor VII and phospholipid, greatly stimulates activation of factor X. Tissue factor is an accessory protein present in a wide range of tissue

Table 5.1 Blood clotting factors

Factor number	Factor name
Factor I	Fibrinogen
Factor II	Prothrombin
Factor III	Thromboplastin (tissue factor)
Factor IV	Calcium
Factor V	Labile factor (proaccelerin)
Factor VI	*
Factor VII	Proconvertin
Factor VIII	Antihaemophilic factor
Factor IX	Christmas factor (plasma thromboplastin component)
Factor X	Stuart factor
Factor XI	Plasma thromboplastin anticedent
Factor XII	Hageman factor
Factor XIII	Fibrin stabilizing factor

Activated forms of the above factors are designated by the addition of the letter 'a' to the factor number, e.g. factor VIIa = activated factor VII.
*The protein originally termed factor VI was later discovered to be factor Va, thus factor VI is now unassigned.

types. It is particularly abundant in saliva, in addition to lung and brain, and is an integral membrane glycoprotein. Tissue factor is released upon tissue damage, along with phospholipid components of the membranes. Hence, it can initiate the extrinsic coagulation cascade at the site of damage. The clotting process occurs most rapidly if initiated via this pathway. While the initial activation sequence of these pathways differ, the terminal sequence of events is identical in both cases (Figure 5.1). These final (common) steps of the coagulation cascade involve the conversion of prothrombin (factor II) to thrombin (factor IIa). This proteolytic reaction is catalysed by activated Stuart factor (Xa).

Thrombin catalyses the proteolytic cleavage of soluble fibrinogen yielding insoluble fibrin. Fibrin monomers subsequently interact forming a fibrin clot. Initially this interaction is of a non-covalent nature, yielding a large aggregated 'soft' clot. This soft clot is subsequently converted into

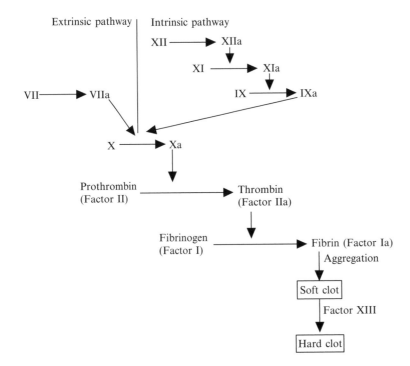

Figure 5.1 Simplified version of the intrinsic and extrinsic blood coagulation pathways

a hard clot by factor XIII, also known as fibrin stabilizing factor (FSF). Activated factor XIII catalyses the formation of covalent cross links between a lysine residue of one fibrin monomer and a glutaminyl residue of an adjacent fibrin molecule. Factor XIII is itself activated by thrombin in the presence of calcium.

HAEMOPHILIA A AND B

Genetic defects, which significantly decrease the level of production or alter the amino acid sequence of any blood factor, may result in serious illness. Such illness is characterized by poor coagulational ability, with resulting prolonged haemorrhage. Defects in all clotting factors with the exception of tissue factor, calcium and phospholipid have been documented and characterized. Well in excess of 90 per cent of all such defects relate to a deficiency of factor VIII. Many of the remaining cases are due to a deficiency of factor IX. The clinical disorders associated with deficiencies of factors VIII or IX include haemophilia A, von Willebrand's disease (vWD) and haemophilia B.

Factor VIII complex consists of two separate gene products. The smaller (170 kDa) polypeptide exhibits coagulant activity and is often designated VIII:C. This polypeptide is coded for by the factor VIII gene. The larger polypeptide, designated von Willebrand factor (VIII:vWB), is predominantly associated with platelet adhesion. This factor is coded for by the vWB gene. Upon synthesis, individual von Willebrand factor polypeptides polymerize forming large multimeric structures. The product of the factor VIII gene, (VIII:C polypeptide) then associates with the multimeric VIII:vWB protein, forming the overall complex structure, (VIII:C and VIII:vWB which may be co-purified from plasma). This overall structure displays a molecular mass in excess of 1 million Da, approximately 15 per cent of which is carbohydrate.

Failure to synthesize VIII:C results in classic haemophilia (haemophilia A), while failure to synthesize VIII:vWB results in von Willebrand's disease. In the case of haemophilia A, the VIII:vWB gene product is synthesized as usual; however, Von Willebrand's disease is characterized by absence of both factors VIII:C and VIII:vWB. Patients suffering from von Willebrand's disease actually synthesize normal factor VIII:C; however, this polypeptide is rapidly degraded as stabilization of this factor requires its association with the VIII:vWB polypeptide (Figure 5.2). Haemophilia B, also known as Christmas disease, results from a deficiency of factor IX. Its clinical consequences are identical to those of classic haemophilia but it does not occur as frequently as the latter disease (see also Box 5.1). The nature and severity of the clinical features of haemophilia depend upon the level of the factor in the plasma. Patients with very low levels (e.g. <1 per cent of normal quantity) of factor VIII:C or factor IX are likely to experience frequent and spontaneous bouts of

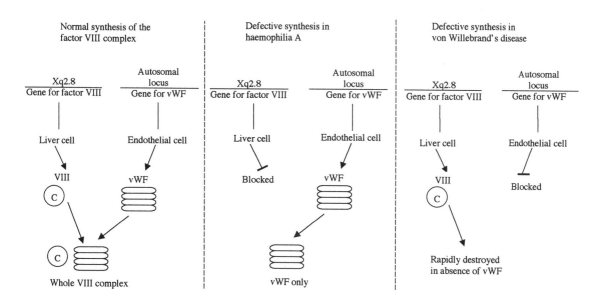

Figure 5.2 Normal factor VIII:c and vWF synthesis and defective synthesis in haemophilia A and von Willebrand's disease (reproduced with permission from Kumar, P. & Clark, M. (1990) *Clinical Medicine*, 2nd edn. Baillière Tindall, London)

bleeding. Persons with higher levels of active factor (3–5 per cent or above) experience less severe clinical symptoms.

Production of clotting factors for medical use

Management of the above bleeding disorders is normally attained by administration of concentrates of the relevant deficient factor. Factors VIII and IX are normally purified by suitable fractionation techniques from plasma obtained from healthy human donors. Factor IX, when purified by traditional fractionation procedures (mainly precipitation steps), usually contains appreciable quantities of factors II, VII, X and XI. This preparation therefore may also be used in clinical cases where deficiency of one of these additional factors is observed. More recently plasma-derived factor VIII preparations have been purified to a greater extent using high resolution chromatographic techniques, including immunoaffinity chromatography. However produced, the final product is normally sterilized by filtration, filled into final containers and freeze-dried. The containers are sealed under vacuum or under an oxygen free nitrogen atmosphere in order to minimize the possibility of oxidative deterioration of the product. Anticoagulants such as heparin or sodium citrate are usually present in the final preparation, although such preparations do not contain antibacterial agents or other preservatives.

The importance of utilizing plasma free of detectable viral contaminants in the preparation of blood products cannot be over-emphasized. It is estimated that over 60 per cent of haemophiliacs have, at some stage, been infected by blood-borne pathogens from contaminated plasma or blood factor products. In addition to screening blood donations, several other approaches may be adopted in order to further reduce the likelihood of accidental transmission of infectious agents. Such approaches include addition of an antiviral substance to the final product preparation or heat inactivation treatment. Purified or partially purified preparations are also less likely to contain potential pathogens as compared to whole plasma, due to chromatographic resolution.

Recombinant blood factors

Recombinant DNA techniques facilitate the production of recombinant blood factors. To date several such products have gained regulatory approval for medical use (Table 5.2). Recombinant production overcomes any potential problems of source availability (due to inadequate supplies of blood donations). However, the major advantage of recombinant production is the virtual elimination of the risk of accidental transmission of blood-borne diseases. As most blood factors display post-translational modifications (glycosylation, proteolytic processing and, in some cases, γ-carboxylation) eukaryotic expression systems are

Table 5.2 Recombinant blood clotting factors which have gained regulatory approval for general medical use. Most products are now approved in several world regions

Product	Company	Therapeutic indication	Approved
Bioclate (rhFactor VIII produced in CHO cells)	Centeon	Haemophilia A	1993 (USA)
Benefix (rhFactor IX produced in CHO cells)	Genetics Institute	Haemophilia B	1997 (USA, EU)
Kogenate (rhFactor VIII produced in BHK cells. Also sold as Helixate by Centeon via a licence agreement)	Bayer	Haemophilia A	1993 (USA)
NovoSeven (rhFactor VIIa produced in BHK cells)	Novo-Nordisk	Some forms of haemophilia	1995 (EU) 1999 (USA)
Recombinate (rhFactor VIII produced in an animal cell line)	Baxter Healthcare/ Genetics Institute	Haemophilia A	1992 (USA)
ReFacto (Moroctocog-alfa, i.e. B-domain deleted rhFactor VIII produced in CHO cells)	Genetics Institute	Haemophilia A	1999 (EU) 2000 (USA)

used. Products thus far approved are produced in engineered CHO cells (e.g. Benefix and ReFacto) or BHK cells (e.g. NovoSeven). Recombinant blood factors can also be produced in the milk of transgenic animals (Chapter 2), and the first cloned sheep ('Dolly') harbours the gene for human factor IX.

After initial recovery from the production system, recombinant blood factors are subject to a number of chromatographic purification steps in order to yield a purified product, which is usually marketed in lyophilized format. An overview of the production on one such product (Benefix) is presented in Box 5.1.

Non-hereditary coagulation disorders

Not all coagulation disorders are due to hereditary deficiencies in clotting factors. Most such acquired coagulation disorders stem from disorders in liver function and/or vitamin K deficiency. Many blood factors, such as fibrinogen, prothrombin, and factors VII, IX and X, are synthesized in the liver. Vitamin K serves as an essential cofactor for a carboxylase enzyme. This enzyme catalyses the γ-carboxylation of certain glutamate residues on factors II, VII, IX and X (Figure 5.3) This vitamin-K-dependent post-translational modification must take place if these factors are to successfully bind calcium ions.

Box 5.1   Case study: Production of 'Benefix', a recombinant coagulation factor IX

Haemophilia B is caused by a mutation in the factor IX gene, which very significantly reduces expression of biologically active factor IX protein. The condition occurs almost exclusively in males as the factor IX gene is located on the X chromosome. Its incidence ranges between 1 in 25 000 and 1 in 30 000 live male births. Factor IX is a single chain, 415 amino acid 55 kDa glycoprotein, composed of five structural domains (a 'γ-carboxyglutamate' (Gla) domain, two 'epidermal growth factor-like' domains, an 'activation peptide' (AP) domain and a 'serine protease' domain). The mature protein displays multiple post-translational modifications. The Gla domain houses 12 glutamate residues which are normally γ-carboxylated, and the protein also houses a total of seven potential glycosylation sites. Factor IX is a zymogen, converted *in vivo* into its enzymatically active form by proteolytic cleavage at two sites (Arg 145 and Arg 180). This cleavage yields activated factor IX (IXa), a heterodimer held together by a disulfide linkage, and a 35 residue peptide.

Benefix is a recombinant factor IX product developed by Genetics Institute (Cambridge, MA, USA). It is produced in an engineered CHO cell line containing multiple copies of an expression vector which houses a nucleotide sequence coding for the human protein. The manufacturing process (summarized schematically below) involves culture of the CHO cells in a 2500 l bioreactor for 3 days (until high cell densities are attained). Intact cells are then removed from the (product-containing) extracellular media by microfiltration. The media is next subjected to an ultrafiltration (concentration) step, followed by diafiltration (to ensure the product's presence in a suitable processing buffer). High resolution purification is achieved using a combination of four successive chromatographic steps: ion exchange (Q Sepharose), an affinity step using Matrex cellufine sulfate (a heparin analogue used for affinity purification of proteins with a heparin binding domain and which also partially acts via an ion exchange mechanism, due to the negatively charged sulfate groups), a hydroxyapatite chromatographic step, and finally an immobilized metal affinity step, using copper (II) as the immobilized metal ion.

A nanofiltration step is also undertaken using a membrane filter displaying a 70 kDa cut off point. This step is included largely as an added layer of viral safety. The 55 kDa product passes through the membrane, which would retain any potential viruses present in the product stream. A final ultrafiltration–diafiltration step is undertaken in order to concentrate the purified product and achieve buffer exchange (i.e. place the product in its final formulation buffer) before product fill. Lyophilization is then undertaken. The formulation buffer contains as excipients: Histidine (as a buffer), polysorbate-80 (protects the protein from damage during freezing), sucrose (stabilizes and protects the protein in the freeze-dried state), and glycine (helps produce a high quality freeze-dried cake). The final product displays a shelf life of at least 2 years when stored at 2–8 °C.

Final product purity is assessed (confirmed) by SDS–PAGE, size exclusion HPLC, reverse phase HPLC and N-terminal sequencing. Product identity is confirmed by bioassay, degree of electrophoretic mobility, peptide mapping, carbohydrate fingerprinting and determination of the molecule's Gla content. The bioassay, which also is used to determine final product potency, is based on a clotting assay using factor IX deficient plasma. In total over 150 tests are performed on each batch of product before its release for sale.

Box Figure 5.1

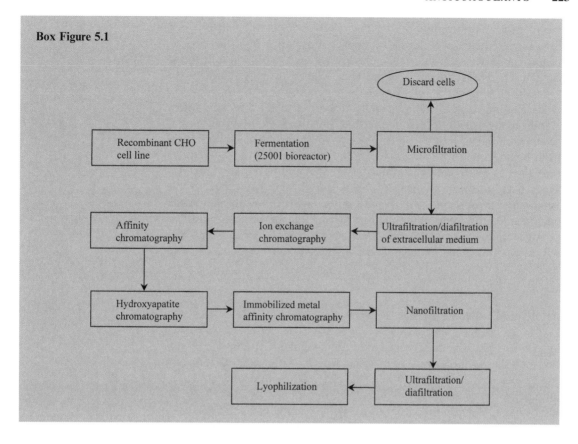

Figure 5.3 Reaction catalysed by the vitamin K dependent carboxylase. Refer to text for details

ANTICOAGULANTS

Anticoagulants function to prevent blood from clotting. They are used clinically in cases where a high risk of blood clot formation is diagnosed and are also utilised to prevent the formation of further clots. Anticoagulants are thus often administered to patients who have suffered heart attacks, strokes or deep vein thrombosis, in an effort to prevent recurrent episodes.

Thrombosis refers to the formation of blood clots. The blood clot itself is termed a thrombus. Thrombosis will most readily occur within diseased blood vessels. The formation of a thrombus in an artery will obstruct blood flow to the tissue it normally supplies. Formation of a thrombus in the coronary artery (coronary thrombosis) will obstruct blood flow to the heart muscle. This results in a heart attack, which is usually characterised by death, or 'infarction', of part of the heart muscle – hence the term myocardial infarction. The presence of thrombi which arrest blood flow to brain tissue usually results in a stroke. Furthermore, upon its formation, a thrombus or part of a thrombus may become detached and travel in the blood, only to be lodged in another blood vessel. This obstructs blood flow at that point. This process is termed embolism, and may also induce heart attacks or strokes. Deep vein thrombosis refers to the formation of thrombi in the deep veins of the leg. It can occur in normal, apparently healthy veins, although risk factors include surgery, immobility, advanced age, and pregnancy. Deep vein thrombi display a tendency to form emboli. Some anticoagulants of therapeutic value are listed in Table 5.3.

Traditional anticoagulants

Heparin is a glycosaminoglycan-based anticoagulant. It consists of sulfated polysaccharide chains of varying length, with molecular masses ranging between 6000 and 30 000 Da. It is synthesized and stored in many body tissues, especially in lung, liver, intestinal cells and cells lining blood vessels. Heparin preparations available commercially are normally prepared from porcine intestinal mucosa or from beef lung. This glycosaminoglycan exerts its anticoagulant activity by binding and thus activating antithrombin III, an α_2-globulin of molecular mass 60 kDa which is found in plasma. Activated antithrombin III inhibits a variety of activated blood factors, including IIa, IXa, Xa, XIa and XIIa. Thus heparin, released naturally *in vivo* or administered therapeutically, inhibits the blood coagulation cascade. Administration of inappropriately large doses of heparin may result in severe haemorrhage. The anticoagulant

Table 5.3 Some anticoagulants used therapeutically

Anticoagulant	Structure
Heparin	Glycosaminoglycan
Dicoumarol	Coumarin-based vitamin K antimetabolite
Warfarin	Coumarin-based vitamin K antimetabolite
Ancrod	Serine protease
Hirudin	Polypeptide capable of binding and inactivating thrombin

activity of heparin is related to the molecular mass of the polysaccharide molecules. This fact has facilitated the development of low molecular mass heparins which retain anticoagulant ability while exhibiting decreased haemorrhagic effects.

Dicoumarol and warfarin are coumarin-based anticoagulants. Both may be administered orally. As previously outlined, these vitamin K antimetabolites exert their anticoagulant effect by inhibiting the vitamin K dependant γ-carboxylation of coagulation factors II, VII, IX and X. γ-Carboxylation of such factors is essential if they are to bind calcium ions, as is required during the normal coagulation process. The major adverse effects of administration of dicoumarol or warfarin is, predictably, the possibility of severe haemorrhage. Lethal doses of warfarin are employed as the active ingredient in many rat poisons.

Hirudin

The buccal secretion of leeches contain an anticoagulant termed hirudin. Components present in the saliva of leeches do not participate in the digestive process but function primarily to interact with, and inhibit, the host animals haemostatic mechanism. Leech bites are thus characterized by subsequent prolonged bleeding, often lasting several hours.

Leeches have been used medically for centuries as vehicles to promote blood letting, as well as in instances where localized anticoagulant activity was required. Indeed the leech is enjoying somewhat of a comeback in medicine where it is sometimes employed to remove blood from inflamed areas and in procedures associated with plastic surgery.

The saliva of leeches contains a variety of peptides of low molecular mass. Hirudin is the major anticoagulant present in the saliva of the European leech (*Hirudo medicinalis*). Hirudin was first reported in the 1880s, though its characterization was not undertaken until the 1950s. The hirudin gene was cloned in the mid-1980s and has subsequently been expressed in a number of host systems. The polypeptide consists of 65 amino acids and has a molecular mass of 7 kDa. The molecule contains a sulfated tyrosine residue at position 63 and is also characterized by a high content of acidic amino acids towards the C-terminal end. The overall conformation of the molecule consists of a globular domain which is stabilised by three sites of intramolecular disulfide bridges, and an elongated C-terminal region.

The anticoagulant activity of hirudin stems from its ability to bind thrombin, i.e. factor IIa, tightly. This results in inactivation of the thrombin molecule. Thrombin not only catalyses the proteolytic cleavage of fibrinogen thus forming fibrin and hence promoting clot formation, but also plays a role in activation of factors V, VIII and XIII). Binding of hirudin to thrombin masks both thrombin's fibrinogen binding site and its catalytic site. Although the natural anticoagulant activity of hirudin had been recognized for quite some time, its lack in adequate quantities rendered its widespread clinical use impractical. Hirudin appears to have a number of therapeutic advantages over some other anticoagulants:

- Hirudin acts directly on thrombin.

- It does not require a cofactor to exert its inhibitory effect.

- High doses are less likely to promote haemorrhage.

- It is a particularly weak immunogen.

Over the past number of years recombinant hirudin has become available and two hirudin based products (trade names Revasc and Refludan) are now approved for general medical use. Both products are produced in *Saccharomyces cerevisiae* strains transformed with an expression plasmid housing a synthetic nucleotide sequence coding for hirudin. The products are secreted by the yeasts into the fermentation medium, from where they are recovered and purified. Both products display identical biological activity to that of native hirudin. Their structures differ from the native molecule only by the absence of a sulfate group on tyrosine 63.

The saliva of some species of leech contain polypeptides other than hirudin which exhibit anticoagulant activity. One such anticoagulant is the protein antistasin. Antistasin has a molecular mass of approximately 15 kDa and exerts its anticoagulant effect by inhibiting factor Xa. This protein is currently the subject of study as it also displays antitumour activity. A number of other proteins obtained from various species of leech may also be of potential clinical significance. Specific examples include destabilase, an enzyme which catalyses the depolymerization of fibrin clots, and decorsin which seems to inhibit platelet aggregation by interacting with a platelet surface glycoprotein.

Ancrod

Ancrod is a serine protease with anticoagulant activity. This enzyme has a molecular mass of approximately 35 kDa and is highly glycosylated. It is purified from the venom of the Malaysian pit-viper. Ancrod catalyses the proteolytic cleavage of microparticulate fibrin molecules prior to clot formation. However, the enzyme has no effect on blood clots once they are formed. It continues to be evaluated in clinical trials.

THROMBOLYTIC AGENTS

Tissue plasminogen activator, urokinase and streptokinase are proteins utilized therapeutically as thrombolytic (clot-degrading) agents. As such they are administered in a variety of situations including treatment of myocardial infarction, embolisms, strokes, and deep vein thrombosis. Many such products are administered over relatively short time periods subsequent to thrombus formation. Such treatment is then followed by administration of a suitable anticoagulant for longer periods of time. On a global basis thrombolytic agents are administered to in excess of 600 000 patients annually.

The fibrinolytic system

Fibrinolysis forms an intrinsic part of the natural wound healing process (Figure 5.4). This process refers to the enzymatic degradation and removal of blood clots from the circulatory system. The process is largely mediated by the serine protease plasmin. Plasmin catalyses the enzymatic degradation of the fibrin strands present in the clot. Plasmin is derived from plasminogen – its circulating zymogen. Human plasminogen is a 90 kDa glycoprotein synthesized in the kidney. Plasminogen consists of a single polypeptide chain and contains numerous intrachain disulfide linkages. Two natural forms exist which differ in carbohydrate content.

Plasminogen may be activated by a variety of specific serine proteases, yielding active plasmin. Tissue plasminogen activator (t-PA) represents the most important physiological activator of plasminogen (Figure 5.5). t-PA, also referred to as fibrinokinase, is a 527 amino acid, 70 kDa glycoprotein displaying serine protease activity. It activates plasminogen by cleaving a single Arg–Val bond. t-PA found in human plasma is predominantly formed in the vascular endothelium. Two forms of t-PA may be purified. Type 1 t-PA is a single chain polypeptide whereas type II consists of two polypeptide chains connected by a disulfide linkage. Type II t-PA is derived from type I by proteolytic cleavage.

Fibrin contains binding sites for both plasminogen and t-PA. Activation of plasminogen by t-PA forming plasmin thus occurs most efficiently on the surface of blood clots. Active plasmin found free in blood is quickly inactivated by another plasma protein termed α_2-antiplasmin. This glycoprotein plays an important role in regulation of plasmin activity.

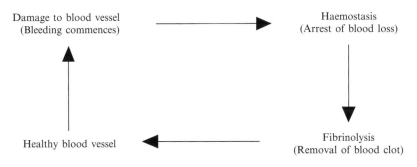

Figure 5.4 Simplified representation of wound healing

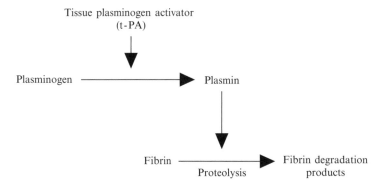

Figure 5.5   The fibrinolytic system, showing action of tissue plasminogen activator (t-PA). Other plasminogen activators include urokinase, kallikrein and bacterial streptokinase

t-PA-based products

Human t-PA preparations produced by recombinant DNA methods have been available commercially since the late 1980s (Table 5.4). While the efficacy of these products are beyond doubt they are quite expensive, especially when compared to streptokinase (discussed below).

Activase was the first recombinant t-PA product to be approved (in 1987) for therapeutic use. It is produced in engineered CHO cells, which harbour a cDNA sequence coding for natural human t-PA. The additional products listed in Table 5.4 are all engineered forms of t-PA, in which three of the five natural t-PA domains are removed. The resultant smaller (355 amino acid) molecule consists only of t-PA's catalytic

Table 5.4   Recombinant t-PA-based products which have gained regulatory approval for general medical use. Most products are now approved in several world regions

Product	Company	Therapeutic indication	Approved
Activase (Alteplase, rh-t-PA produced in CHO cells)	Genentech	Acute myocardial infarction	1987 (USA)
Ecokinase (Reteplase, rtPA; differs from human t-PA in that 3 of its 5 domains have been deleted. Produced in *E. coli*)	Galenus Mannheim	Acute myocardial infarction	1996 (EU)
Retavase (Reteplase, rt-PA; see Ecokinase)	Boehringer Manheim/ Centocor	Acute myocardial infarction	1996 (USA)
Rapilysin (Reteplase, rt-PA; see Ecokinase)	Boehringer Manheim	Acute myocardial infarction	1996 (EU)

domain and kringle 2 domain (the latter is involved in fibrin binding selectivity). These engineered forms of t-PA are produced in recombinant *E. coli*, and accumulate intracellularly as inclusion bodies. Downstream processing thus includes denaturation and refolding steps, subsequent to cellular homogenization (Figure 5.6). As *E. coli* cannot undertake post-translational modifications, the products are unglycosylated. Despite the molecular differences between the native and engineered forms of t-PA, all exhibit the same clinical efficacy as thrombolytic agents.

Several different t-PA preparations are currently in development or undergoing clinical trials. Recombinant t-PA has most recently been produced in the milk of transgenic animals (Chapter 2). t-PA produced in transgenic animals is predominantly the two-chain form of the enzyme. The protein may be purified by a combination of techniques such as acid fractionation, hydrophobic interaction chromatography and immunoaffinity chromatography. As expected, the glycosylation pattern observed on the purified product differs somewhat from the patterns obtained when the protein is produced by the more conventional method of CHO cell culture.

Additional thrombolytic agents

Additional thrombolytic agents include urokinase, streptokinase and staphylokinase. Urokinase is a serine protease produced in the kidney. It also activates plasminogen, forming plasmin. Urokinase is produced as prourokinase and is found in both plasma and in urine. Two variants of the enzyme have been purified. High molecular mass urokinase has a molecular mass of 54 kDa while the low molecular mass form is approximately 30 kDa. Both forms exhibit similar biological activity. The low

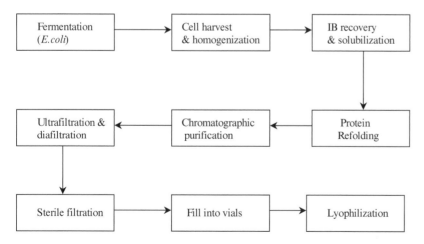

Figure 5.6 Overview of the production of 'Rapilysin', a modified t-PA produced in recombinant *E. coli* cells. Note: IB, inclusion body

molecular mass form of the enzyme is produced from the high molecular mass form by proteolysis. Urokinase may be purified from human urine by a variety of techniques. Chromatographic steps utilized include affinity chromatography and gel filtration. However, urokinase used therapeutically is usually produced by primary cell culture of kidney cells harvested post mortem from human neonates.

Bacterial streptokinase may also be employed to activate plasminogen. Streptokinase is produced by a variety of haemolytic streptococci and is generally purified from the filtrate obtained from cultures of several such organisms. The molecular mass of the purified protein is in the region of 45 kDa. Because of its microbial source, streptokinase is immunogenic and its administration sometimes results in adverse allergic reactions. However, it is less expensive than other thrombolytics, particularly t-PA.

Unlike the plasminogen activators already discussed, streptokinase is devoid of proteolytic activity. This protein binds tightly to plasminogen, which results in a conformational change in the plasminogen molecule rendering it proteolytically active. The activated plasminogen molecules are now capable of catalysing the proteolytic activation of other plasminogen molecules, thereby yielding active plasmin.

Staphylokinase is a 136 amino acid protein produced by *Staphylococcus aureus*. It displays thrombolytic activity and as such is currently being evaluated in clinical trials. The staphylokinase gene has been cloned and expressed in recombinant *E. coli* cells, and its tertiary structure has been elucidated. Staphylokinase displays a mechanism of action broadly similar to that of streptokinase. It binds to plasminogen or plasmin in a 1:1 stoichiometric complex, and this complex can subsequently activate additional plasminogen molecules. In contrast to streptokinase, however, staphylokinase reacts poorly with free plasminogen in plasma, but will react with high affinity to traces of plasmin at the clot surface. This resultant staphylokinase–plasmin complex will thus activate additional plasminogen molecules specifically at the clot surface.

ADDITIONAL BLOOD-RELATED PRODUCTS

In addition to proteins involved in the promotion or dissolution of blood clots, several other serum proteins are of considerable actual or potential therapeutic value. Examples include human serum albumin, α_1-antitrypsin, haemoglobin and of course immunoglobulins.

Human serum albumin

Serum albumin constitutes the most abundant protein present in serum, representing approximately 60 per cent of total plasma protein. It is also one of the smallest known plasma proteins with a molecular mass of approximately 69 kDa. It is one of a small number of plasma proteins that is devoid of a carbohydrate moiety. The protein is synthesized in the liver as preproalbumin. Removal of several amino acid residues from its amino terminus during passage through the endoplasmic reticulum yields mature albumin, which consists of 585 amino acids. The albumin molecule exhibits several intrachain disulfide linkages. It consists of three similar domains (I, II and III), and is 67 per cent α-helical in structure.

A major function of albumin is to provide most of the natural osmotic pressure of plasma. It also plays an important transportational function and is especially important in transporting sparingly soluble substances in aqueous media within the body. Most free fatty acids bind tightly to albumin and are transported in plasma in this manner.

Very little albumin is found in the urine of healthy individuals. Certain medical conditions, especially some forms of kidney disease, however are characterized by secretion of large quantities of serum albumin into the urine. Some forms of liver disease may also result in significant decreases of hepatic protein synthesis, with a resultant marked reduction in the concentration of several plasma proteins, most notably albumin.

Aqueous solutions of human serum albumin are available commercially. These range in concentration from 5 to 25 per cent. Human serum albumin preparations are administered to patients suffering from some forms of kidney or liver disease and is also used as a plasma volume expander for patients suffering from shock resulting from a decrease in the volume of blood. Such decreases are often associated with surgery or occur subsequent to serious injury.

Albumin is normally purified from serum, plasma or from placentae obtained from healthy donors. All raw materials are screened for the presence of viral or other potentially pathogenic organisms prior to purification. The methods of purification utilised, precipitation and chromatography, must yield product which is 95–96 per cent pure. Like most proteinaceous preparations the purified albumin is sterilized by filtration and subsequently aseptically filled into sterile containers. Although no preservatives are added, stabilizers such as sodium caprylate, which protect the product, particularly against the effects of heat, are normally included in the final product. Subsequent to final filling, the product is subjected to a heat step which promotes inactivation of certain pathogens which may be present. This normally involves heating to 60 °C for a period of up to 10 h.

While commercially available human serum albumin is currently obtained from blood or placentae, the protein has been produced as a heterologous product in a number of recombinant systems. Systems employed have included bacteria such as *Bacillus subtilis* and *E. coli*, yeast such as *Saccharomyces cerevisiae* and plants such as the potato. While albumin can be produced inexpensively by direct extraction from its native source, the possibility of accidental transmission of blood-borne disease via infected product continues to be the main rationale for producing a recombinant form of the protein. Human serum albumin is also sometimes used as a stabilizing excipient in biopharmaceutical products.

α_1-Antitrypsin

α_1-Antitrypsin is a 52 kDa serum glycoprotein. The protein consists of 394 amino acid residues and contains three glycosylation sites. The α_1-antitrypsin gene (*PI*) is located on the distal long arm of chromosome 14. α_1-Antitrypsin is synthesized in the liver and constitutes over 90 per cent of the α_1-globulin band, observed upon electrophoresis of serum. It is normally present in serum at a concentration of 2 g/l. A variety of genetic variants of the protein have been described.

α_1-Antitrypsin constitutes the major serine protease inhibitor present in mammalian serum. It serves as a potent inhibitor of the protease elastase and as such prevents damage to lung tissue by neutrophil elastase. Neutrophils are a particular type of granulocyte, essentially a type of white blood cell. Genetic deficiencies resulting in the absence of α_1-antitrypsin in human plasma have been described. Such deficiencies are particularly prevalent in persons of northern European descent. Persons suffering from such deficiencies often develop life-threatening emphysema. This results from unchecked damage to lung tissue by neutrophil elastase. Replacement therapy of α_1-antitrypsin delivered by intravenous infusion may arrest progression of this disease.

α_1-Antitrypsin preparations administered to patients suffering from congenital α_1-antitrypsin deficiency are normally obtained from pooled plasma fraction. The large quantities of inhibitor required (approx. 200 g/patient/year), and the possibility of accidental transmission of disease via infected source material has hastened the development of alternative recombinant sources. α_1-Antitrypsin has more recently been successfully produced in the milk of both transgenic mice and sheep.

Blood substitutes

Much work has been carried out over the past number of years with regard to development of a blood substitute or perhaps more correctly, a red blood cell substitute. Development of such a substitute would decrease or eliminate concerns over viral contamination of blood cell preparations. It is hoped that any such substitute would also exhibit a longer shelf life than red cell preparations, which currently must be used within 42 days of collection. Furthermore, as antigens determining blood group type reside on the surface of red blood cells, the development of a cell-free substitute would allow its universal administration. Red blood cell concentrates are currently used clinically in the treatment of certain forms of anaemia and haemolytic disease.

A number of substances have been investigated as potential blood substitutes. Fluorinated hydrocarbons (fluorocarbons), have been assessed for this purpose, and have been used clinically in a limited number of cases. Such fluorocarbons can successfully transport both oxygen and carbon dioxide, and may be used to replace whole blood in some animal models. Inhalation of high oxygen concentrations is required if these compounds are to bind and transport clinically significant quantities of oxygen. In addition, some clinical investigations suggest that fluorocarbons may result in compromised immune function. Such disadvantages militate against their widespread use as suitable substitute substances.

Purified haemoglobin has also been investigated as a potential red blood cell substitute. Haemoglobin is the principal protein found in circulating red blood cells or erythrocytes. It is a tetramer, consisting of two different polypeptide chains, α and β. The native molecule may be represented as $\alpha_2\beta_2$ with an overall molecular mass of approximately 64 kDa. Each of the four polypeptide subunits also contain a haem prosthetic group which confers upon the molecules their oxygen binding properties (Figure 5.7). Binding of one oxygen molecule to one haem group greatly increases the affinity for oxygen of the remaining groups.

Such oxygen binding kinetics renders haemoglobin ideally suited to its oxygen transporting role. Haemoglobin molecules, however, are not themselves suitable as blood substitutes due to (a) instability of the protein outside the environment of the red blood cell and (b) its high affinity for oxygen. Additionally, haemoglobin molecules, when free in solution, quickly dissociate into $\alpha\beta$ dimers. Such dimers are rapidly removed from the circulatory system. Normal human erythrocytes also contain large quantities of 2,3-diphosphoglycerate. Binding of 2,3-diphosphoglycerate to haemoglobin reduces the protein's affinity for oxygen. This promotes the release of bound oxygen, upon reaching oxygen-

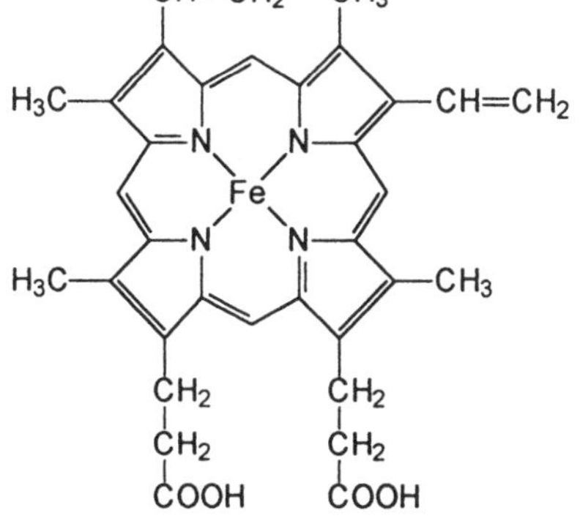

Figure 5.7 Chemical structure of the haem prosthetic group. The haem group is present not only in haemoglobin but also in myoglobin and a variety of other haem proteins

requiring tissue. The absence of 2,3-diphosphoglycerate renders the oxygen affinity of purified haemoglobin preparations too great to allow efficient oxygen release in such a manner.

Many of the problems relating to the stability of free haemoglobin and its high affinity for oxygen may be overcome, at least in part, by chemical modification. Treatment with gluteraldehyde or other agents which promote polymerization will prevent dissociation of the tetrameric molecule. Covalent attachment of pyridoxal 5′-phosphate groups reduces its affinity for oxygen. The introduction of specific cross-links between subunits will both stabilize individual molecules of haemoglobin, and will also reduce its affinity for oxygen. All such modified haemoglobin molecules may be described as haemoglobin-based oxygen carriers (HBOC).

The vast majority of such HBOC preparations are derived from haemoglobin purified from human blood donations which have surpassed their designated useful shelf life. Some studies have also been undertaken which use preparations of bovine haemoglobin. Bovine haemoglobin exhibits a lower affinity for oxygen than does its human counterpart. Administration of such a 'non-self' protein, however, may result in immunological or allergic complications. Porcine haemoglobin may be of interest in this regard as it more closely resembles the human molecule.

Like many other clinically interesting proteins, haemoglobin has been produced by recombinant methods. The haemoglobin α and β genes may be expressed in separate systems and intact haemoglobin may be reconstituted from the purified recombinant products. Alternatively, both genes may be expressed within a single host. This usually results in the automatic production of mature tetrameric haemoglobin molecules. Haemoglobin has been successfully produced in both recombinant *E. coli* and yeast systems and has been produced in the blood of transgenic animals.

Although production of recombinant human haemoglobin is technically feasible, outdated stocks of donated human blood still remain the major source from which it is extracted. This remains the most inexpensive method of production. Protein contaminants present in recombinant haemoglobin preparations would be non-human in nature, and thus immunogenic. This difficulty should not arise when haemoglobin is obtained from human red blood cells. Recombinant DNA technologies may however allow researchers to rationally develop and express engineered haemoglobins with desirable characteristics such as improved stability or decreased oxygen affinity. Research in this area is ongoing.

VACCINE TECHNOLOGY

Traditional vaccines

If an antigen is introduced into the bloodstream, either naturally or artificially, the host's immune system launches an immunological response to the challenge. This response involves the production of anti-bodies which specifically recognizes and binds the antigenic material that elicited their production. Antibodies are synthesized by a subset of white blood cells termed B lymphocytes. Binding of antibody to antigen should inactivate or neutralise the offending antigen, and mark it for destruction by additional elements of the immune system. In addition, a second type of white blood cell – the T-lymphocyte – may be activated, and these will mount an immunological response. Such cells play a central role in the destruction of foreign antigenic material, such as bacteria and some viral species. After the immune system has successfully defeated the antigenic challenge, long-lived B and T cells, termed memory cells, remain in circulation. Memory cells are capable of recognizing the antigenic sub-stance that elicited their initial production. Thus, if this antigenic material re-enters the body the memory cells will trigger a full-blown immuno-logical response. Unlike the primary response which builds up somewhat slowly, the repeat response is immediate.

The process of vaccination is designed to exploit the natural defence mechanism conferred upon us by our immune system, and involves artificially exposing the immune system to antigenic preparations. Such processes will, by the process described above, induce active immunity against specific pathogenic organisms. If man subsequently comes into contact with such a pathogen, an immediate and specific immunological response is initiated. This should quickly destroy the offending organism before it has the chance to establish a full-blown infection.

A vaccine therefore is a preparation of antigenic components usually consisting of, or derived from, or related to, pathogens. When the vaccine is administered, it stimulates an immune reaction and thereby will confer active immunity on the recipient. This helps prevent subsequent establish-ment of an infection by the same, or antigenically similar pathogens. Vaccines are thus used prophylactically to prevent the future occurrence of diseases which the recipient is likely to encounter. Most vaccination protocols involve administration of an initial dose followed by one or more subsequent doses over a suitable time period. Administration of booster shots ensures a maximal immunological response, with associated production of high levels of circulating antibody. Vaccine production represents a significant niche of biotechnological endeavours. Most coun-tries implement a systematic vaccination programme against key infec-tious diseases.

Diseases against which vaccine preparations have been developed are caused by a variety of biological entities, including viruses, rickettsiae, and various microorganisms. Vaccines have been prepared which protect against other harmful microbial substances such as toxins. Perhaps the most effective method of inducing artificial immunization is to expose the recipient to a vaccine containing small quantities of the actual pathogen against which immunity is sought. This approach however, is rarely adopted in practice as there is a risk that the recipient will develop the uncontrolled disease. Dead, inactivated or attenuated bacteria and viruses are often used. Such antigenic preparations exhibit little or no virulence but retain immunological characteristics similar to the wild-type pathogen. Organisms may be killed or inactivated by a variety of chemical or physical means. Attenuation is normally achieved by growing the organisms in an unnatural host. Safe vaccine preparations against a variety of bacterial toxins are produced by inactivating the toxin such that its toxic properties are eliminated while retaining its immunogenicity.

Many such preparations suffer from one or more disadvantages. Dead or inactivated particles may be significantly less immunogenic compared to their wild type counterparts. If methods used to kill or inactivate the organisms are not consistently 100 per cent efficient a possibility exists that live virulent organisms may be accidentally employed as vaccines. There is also a danger, albeit slight, that some attenuated species might revert to the pathogenic state. Such theoretical dangers may be reduced by using a purified antigenic component of the pathogens. Surface polysaccharide–protein antigens have been purified and utilized to successfully induce active immunity in several cases. Table 5.5 lists most of the major conventional vaccine preparations commercially available.

Recombinant vaccines

Recent advances in vaccine development centre around recombinant DNA technologies. Such technologies allow the production and synthesis of specific protein antigens found on the pathogenic organisms, in recombinant organisms such as *E. coli*, yeast cells or mammalian cell lines. The desired protein product, which is identical to the antigen sourced from the wild-type pathogen, is then purified and employed as a vaccine. This method of vaccine production exhibits several advantages over conventional vaccine production methodologies:

- clinically safe;
- unlimited supply;
- defined product;
- less likely to cause unexpected side effects.

The recombinant vaccine is extremely safe and should consist of a single antigenic constituent of the pathogen against which vaccination is required. It is thus impossible to accidentally or otherwise induce the diseased state with such vaccine preparations. Furthermore, there is little possibility that the final preparation can be contaminated with additional pathogenic organisms. Administration of such a defined and structurally less complex vaccine is also less likely to induce unexpected adverse clinical reactions. Vaccine production by recombinant methods also ensures a continuous and convenient supply of material from a safe source.

Table 5.5 Vaccine preparations most commonly used to induce active immunity

Vaccine	Vaccine preparation	Use
BCG vaccine (Bacillus Calmette–Guérin)	Live attenuated strain of mycobacterium tuberculosis	Active immunization against tuberculosis
Measles vaccine	Live attenuated strain of the measles virus, usually propagated in culture cells from chick embryo	Active immunization against measles
Mumps vaccine	Live attenuated strain of *Paramyxovirus parotitidis*, the causative agent of mumps – usually propagated in cultures of chick embryo cells	Active immunization against mumps
Rubella vaccine	Live attenuated strain of rubella virus, grown in cell culture	Active immunization against German measles
Poliomyelitis vaccine (Sabin vaccine – oral)	Live attenuated strain of poliomyelitis virus propagated in cell culture	Active immunization against polio
Poliomyelitis vaccine (Salk vaccine – injection)	Inactivated poliomyelitis virus propagated in cell culture	Active immunization against polio
Yellow fever vaccine	Attenuated strain of live yellow fever virus cultured in chick embryos	Active immunization against yellow fever
Cholera vaccine	Dead strain(s) of *Vibrio cholera*	Active immunization against cholera
Influenza vaccines	Inactivated strains of influenza virus, either individual or mixed or suspension of influenza surface antigens (e.g. haemaglutinin & neuraminidase)	Active immunization against influenza
Pertussis vaccine	Killed strains of *Bordetella pertussis*	Active immunization against whooping cough
Plague vaccine	Formaldehyde killed *Yersinia pestis*	Active immunization against the plague
Rabies vaccine	Inactivated strain of rabies virus	Active immunization against rabies
Typhoid vaccine	Killed *Salmonella typhi*	Active immunization against typhoid fever

continues overleaf

Table 5.5 (*continued*)

Vaccine	Vaccine preparation	Use
Typhus vaccine	Killed *Typhus rickettsiae* (*Rickettsia prowazeki*)	Active immunization against typhus
Varicella-zoster vaccine	Live attenuated varicella virus	Active immunization against chicken pox
Diptheria vaccine	Formaldehyde treated toxin (i.e. a toxoid) derived from *Corynebacterium diptheria*	Active immunization against diptheria
Tetanus vaccine	Formaldehyde treated toxin (e.g. a toxoid) derived from *Clostridium tetani*	Active immunization against tetanus
Hepatitis B vaccines	Hepatitis B surface antigen isolated from plasma of carriers or produced by genetic engineering	Active immunization against hepatitis B
Haemophilus influenzae	Purified capsular polysaccharide of *Haemophilus influenzae* type B	Active immunization against *H. influenzae* – the major cause of infant meningitis
Meningococcal vaccines	Purified polysaccharide antigens obtained from one or more serotypes of *Neisseria meningitidis*	Active immunization against *N. meningitidis* – causes epidemics of meningitis
Pneumococcal vaccines	Purified polysaccharide antigens from different serotypes of *Streptococcus pneumoniae*	Active immunization against pneumococcal disease

In addition to such single constituent vaccines many vaccine preparations of mixed specificity are also available. Examples of the latter include measles, mumps and rubella vaccines, as well as vaccines for diptheria, tetanus and pertussis.

This should help to keep production costs down and prevent accidental transmission of disease to production personnel.

Recombinant hepatitis B-based vaccines

The first vaccine product produced by recombinant DNA technology to be approved for human use was that of hepatitis B. Traditionally hepatitis B vaccine was prepared by purifying hepatitis B surface antigen (HBsAg) from the plasma of infected individuals. When isolated from plasma sources, HBsAg polypeptides are usually not found free, but characteristically as polymeric particles of 22 μm in diameter. Production of hepatitis B vaccine from such sources suffers from several drawbacks. There is a risk that the final preparation may be contaminated with active hepatitis B virions, or indeed other pathogenic viruses such as HIV. Furthermore, production of the vaccine depends upon a constant supply of plasma from infected individuals.

Over the past number of years the HBsAg has been cloned and expressed in a number of recombinant systems (Table 5.6). rHBsAg produced in engineered *Saccharomyces cerevisiae* was the first recombinant subunit vaccine to be approved for general medical use (1986 in the USA). Since then a number of additional rHBsAg products, again produced in *Saccharomyces cerevisiae*, have come on the market. Purification of the recombinant molecules often features an immunoaffinity step using anti-HBsAg antibodies. rHBsAg may be formulated on its own as a single component vaccine. Additionally it may be included as one component of combination vaccines, used to simultaneously vaccinate against several pathogens. Tritanrix-HB for example is one such combination vaccine. Produced by SmithKline Beecham, this preparation is used to immunize against diphtheria, tetanus and pertussis toxin, as well as hepatitis B. Other hepatitis B antigens such as its core antigen (HBcAg) have also been successfully produced in recombinant systems.

A number of additional protein antigens derived from various pathogens have been produced by recombinant means (Table 5.7) and many are now in clinical trials.

Table 5.6 Some recombinant systems in which recombinant hepatitis B surface antigen has been produced

Bacterial systems	*E. coli*
Yeast systems	*S. cerevisiae*
Mammalian systems	Monkey kidney cells
	Human hepatoma cell line
	Monkey kidney cells

Table 5.7 Some additional protein-based antigens derived from pathogens which have been produced in recombinant systems

Antigen expressed	Potential vaccine against:
Bordetella pertussis surface antigen	Whooping cough
Tetanus toxin fragment C	Tetanus
Haemaglutinin	Influenza
Hepatitis A caspid proteins	Hepatitis A
Poliovirus caspid proteins	Polio
Rabies virus glycoprotein	Rabies
Surface glycoprotein D	Herpes

The gene for a surface antigen (p69) of *Bordetella pertussis* has been cloned and expressed at high levels in *E. coli*. Two members of the *Bordetella* family (*Bordetella pertussis* and *Bordetella parapertussis*) are the causative agents of whopping cough. Traditional pertussis vaccines are composed of killed strains of *Bordetella pertussis*. Surface antigens such as p69 purified from recombinant sources are likely to form the next generation of pertussis vaccine preparations.

Fragments of tetanus toxin have also been produced in *E. coli*, with a view to development of an alternative tetanus vaccine. Production of current tetanus vaccines involve treatment of tetanus toxin with formaldehyde, thus forming toxoid molecules. Native tetanus toxin consists of a single polypeptide chain of molecular mass 150 kDa. Proteolytic cleavage of the intact toxin yields two fragments. Fragment AB represents the 100 kDa N-terminal portion of the toxin while fragment C represents the 50 kDa carboxyl terminal end. Fragment C is non-toxic and is capable of inducing active immunity against tetanus in experimental animals. Recombinant fragment C preparations are likely to form the next generation of tetanus vaccines.

A number of other pathogen-derived antigens have been expressed in recombinant systems with a view to assessing their efficacy as subunit vaccines. In some instances, such recombinant proteins must be subject to certain post-translational modifications, which may best be achieved by using mammalian cells such as a CHO cell line as the recombinant host.

Additional approaches to vaccine design

Several other approaches have been adopted in the development of modern vaccine preparations. One such approach involves isolating or synthesizing a gene or cDNA which codes for a particular surface antigen obtained from

a pathogen of interest. This may then be introduced into the genetic complement of a clinically harmless virus, often fused to a viral surface protein.
Vaccinia virus is often employed for this purpose. Recipients immunized
with preparations of the live recombinant virus invariably develop immunity not only against the viral host, but also against the pathogen from which
the inserted genetic information was obtained (Figure 5.8). Genes derived
from hepatitis B virus and from HIV have been inserted into vaccinia viral
genomes in an attempt to develop live recombinant vaccines against these
pathogens. Subsequent animal trials confirmed the ability of these vaccine
vectors to elicit an immune response against the target antigen.

Another research avenue entails the synthesis of specific peptide sequences identical to a portion of a protein antigen found in the pathogen.
This method is attractive insofar as it also results in the generation of
safe vaccine preparations, free from harmful contaminants. The conformational structure adopted by any protein helps determine its level of
antigenicity. In many instances treatments that destroy its native conformation will adversely effect the antigen's ability to elicit a strong
immunological response. Many epitopes on the antigen's surface against
which antibodies are directed are created by the specific conformational
structure of the protein. Short peptide sequences may not accurately
reproduce such epitopes and, as such, may be of little use in vaccinations.

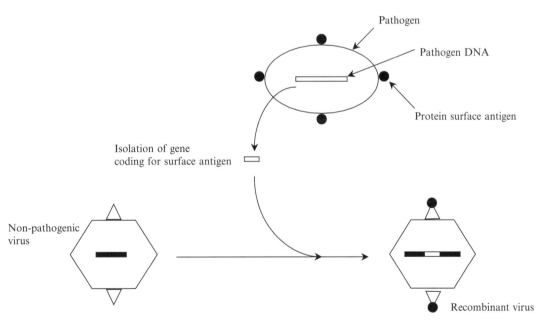

Figure 5.8 Basic approach to developing recombinant vaccine vectors. The nucleotide sequence coding for
most or all of a selected surface antigen on the target pathogen is identified and isolated. This is then
introduced into the genome of a clinically safe virus (e.g. Vaccinia). It is modified (e.g. by fusion with a viral
gene coding for a viral surface protein) to ensure that the transferred gene product will be expressed on the
viral cell surface

LIVERPOOL JOHN MOORES UNIVERSITY
LEARNING SERVICES

Anti-idiotype antibodies as vaccines

Anti-idiotype antibodies may also potentially be used as vaccines. Upon encountering any antigen, an individual's immune system sets in motion a specific train of immunological events. One consequence of immune stimulation relates to the production of B lymphocytes that produce antibodies which specifically recognize and bind epitopes on the foreign antigen. The antigen-binding sites of such antibodies are located in their hypervariable regions (Chapter 6). The hypervariable regions themselves contain antigenic determinants which can stimulate the production of yet another series of antibodies. The collection of antigenic determinants present on the hypervariable region are often termed idiotype. The second set of antibodies raised against the hypervariable region of the first set are termed anti-idiotype antibodies.

Some anti-idiotype antibodies will recognize epitopes in the primary antibody hypervariable regions which are not directly involved in binding to the foreign antigenic substance. Other anti-idiotype antibodies, however, are raised against the specific epitopes present in the hypervariable region, which constitutes the primary antibody's antigen-binding site. Such anti-idiotypic antibodies carry what may be termed as an internal image of the original antigen, and can mimic the antigen. They can therefore elicit the production of antibodies that will recognize and bind this original antigen.

Anti-idiotype antibodies could be used to vaccinate against a wide range of diseases caused by various pathogens. Such antibodies have been used in the laboratory to vaccinate against HBsAg, polio virus, rabies virus, *Streptococcus pneumoniae* and *Trypanosoma rhodesiense*, the causative agent of sleeping sickness. Thus far however, the vast majority of anti-idiotypic vaccination experiments have been confined to animal studies.

Production of subunit vaccines by recombinant DNA technology is likely to have the greatest impact on vaccine technology for the foreseeable future. Anti-idiotypic antibodies could find important applications as some specialist vaccine preparations. Many antigenic determinants are non-proteinaceous in nature and thus cannot be produced by recombinant DNA technology. Anti-idiotype antibodies can be prepared against such antigens and therefore employed as effective vaccines.

VACCINES FOR AIDS

The development of an effective vaccine against HIV constitutes an intensely active area of vaccine research. HIV is the causative agent of AIDS. While the rate of transmission of this incurable disease appears to be decelerating in the Western world, its proliferation within Africa and other poorer areas of the world has reached crisis proportions. According to statistics compiled by the World Health Organization, over 70 million people are currently infected with HIV, with an annual death rate of in excess of 2 million.

HIV is a retrovirus as its genetic material consists of RNA which, along with enzymes such as reverse transcriptase and integrase, are located in the viral core. This core material is in turn surrounded by the core protein, p24. The outer viral envelope consists of a membraneous structure originally derived from the host cell membrane in which the viral particle was synthesised. Protruding from the membrane is the glycoprotein, gp120. An additional glycoprotein, gp41, is found embedded in the membrane coat. gp120 and gp41 are both derived from the viral glycoprotein, gp160, by proteolytic cleavage of the latter during viral replication. An additional protein, p17, is located beneath the viral envelope.

HIV is capable of binding to, and thus infecting, certain cell types. Sensitive cells must display a protein, termed CD4, on their surface in order to facilitate viral attachment. The CD4 marker is located only on selected cell types, most notably T helper cells. These cells play a critical role in the generation of cell-mediated immunity. The virus binds to CD4 via its envelope glycoprotein, gp120. Upon entry into the cell, the viral RNA is used as a template by reverse transcriptase, thus generating DNA. The viral DNA is then integrated into the host's genome, a process facilitated by the enzyme integrase. Once integrated, the viral genes may remain dormant for an extended period of time and will be replicated along with the host chromosomal DNA should the infected cells divide. However, activation of the DNA results in the production of numerous viral particles. These particles eventually bud from the host cell membrane and hence are released into the bloodstream. Particularly rapid intracellular replication may even result in lysis of the host cells.

Several different approaches continue to be evaluated with a view to developing an effective AIDS vaccine. Administration of small quantities of live wild-type virus in order to illicit a protective immunological response is obviously out of the question. Development of a vaccine consisting of a live attenuated strain of the virus also remains unlikely, due to the possibility of reversion to virulent form. Furthermore, it is difficult to predict what constitutes a 'safe' attenuated HIV virus, as the molecular basis by which the virus destroys the immune system is still poorly understood.

Inactivated viral particles may constitute a potentially effective vaccine, and viral preparations inactivated by a combination of chemical (formaldehyde) and irradiation treatments are currently being evaluated clinically. Wild-type viral strains are firstly inactivated by heat. Anti-idiotype antibodies and synthetic peptide technologies are under investigation as potential vaccine preparations. The majority of vaccine preparations currently undergoing clinical trials consist of specific viral antigens produced by recombinant DNA technology. Somewhat mixed success has been reported in this regard. Most such antigens consist of recombinant viral gp160, gp120, or fragments of these molecules. These recombinant molecules are normally produced in mammalian cell lines to facilitate their post-translational glycosylation. While gp160, and therefore gp120, which is derived from gp160, are subject to a high rate of mutation, it is considered likely that particular sequences will remain highly conserved. The presence of such preserved antigenic determinants in vaccine preparations should, in theory, help confer immunity against a wide range of HIV strains.

Several additional vaccine preparations currently being evaluated consist of recombinant core antigens such as p24 and p17. Such proteins are subject to a lesser degree of genetic drift when compared to the envelope proteins. Hence, these offer the hope of providing a vaccine preparation of broad specificity. It has also been established that, as the immune function of the AIDS patient decreases, so do the circulatory levels of antibodies which recognize viral core proteins. In contrast, the level of circulatory antibodies that recognize the envelope glycoproteins remain elevated. It is therefore hoped that vaccines which specifically stimulate production of anti-core protein antibodies may be particularly effective in suppressing the development of full-blown AIDS.

A vaccine that could successfully induce an immunological response targeted against the wild type virus could be employed as a preventative and/or a therapeutic agent. An effective preventative vaccine should enable an immune response to be initiated against the AIDS virus immediately upon its entry into the body, thus destroying the virus before an infection can be established. The rationale behind therapeutic vaccine development is based on the fact that, subsequent to infection by HIV, the host's immune response prevents extensive viral replication for prolonged periods. This prevents progression of the disease and the patient remains clinically asymptomatic. It is hoped that vaccines used therapeutically would stimulate the immune function sufficiently to further delay development of the clinical symptoms of the disease.

Development of an effective vaccine against HIV, however, remains elusive for a number of reasons. HIV mutates at a higher rate than almost any other virus. Thus a vaccine which elicits the production of protective antibodies against one strain of the virus, may not induce immunological protection against subsequent genetic variants. Furthermore, many patients suffering from AIDS exhibit high levels of circulating anti-HIV antibodies, which seemingly fail to halt progression of the disease. It seems therefore, that a humoral response (the production of antibodies by B lymphocytes), alone is not sufficient to successfully combat the infection.

It is likely that an effective vaccine must elicit both a strong humoral and cell mediated immune response. T-cell mediated responses would play a crucial role in destroying cells infected with virus. Only T lymphocytes are capable of destroying cells that exhibit foreign viral antigen on their surface. Many, if not most, traditional vaccine preparations effectively stimulate only an antibody-mediated immune response.

The fact that HIV often enters a new host not as free viral particles but in the form of infected cells also renders difficult the development of an effective vaccine. Furthermore, immune surveillance may not be capable of recognizing and destroying many infected cells in which the integrated viral DNA remains quiescent. Additionally, the lack of a good animal model in which the progression of AIDS can be studied renders difficult the pre-clinical testing of potential vaccines. Chimpanzees probably represent the best such model system. These, however, are a protected species and are currently in short supply.

In addition to vaccine development several other research strategies are being pursued with a view to defeating this viral infection. The antiviral activity of drugs known to inhibit viral enzymes such as integrase or reverse transcriptase are currently being evaluated. Many such drugs, however, are characterized by a lack of specificity and many thus exhibit serious side effects.

Additional research efforts are targeted against the integrated viral DNA, in an effort to prevent activation of the viral genes. Other strategies under active consideration involve the production of recombinant soluble CD4 protein molecules. As previously discussed, viral infection is initiated by its binding to the CD4 protein found on the surface of susceptible cells. Parenteral administration of large quantities of soluble CD4 molecules may well retard or prevent viral attachment to healthy cells by acting as decoy viral receptors.

Cancer vaccines

It is estimated that some 80 million people suffer from cancer, and that the annual cancer mortality rate stands in excess of 7 million. Survival rates continue to increase due to surgical and non surgical advances in cancer therapies. An alternative approach to combating cancer would be to develop effective cancer vaccines, and in excess of 40 such vaccines are currently in clinical trials.

Scientists have known for decades that cellular transformation often results in altered cell surface properties, and early attempts at developing cancer vaccines centred around the use of mechanically disrupted or irradiated tumour cells. The identification of specific tumour surface antigens (TSA) associated with certain cancer types (Chapter 6) facilitates a more rational approach to cancer vaccine development. Many such experimental vaccines consists of recombinant soluble forms of specific TSA. In other cases a partial or complete nucleotide sequence coding for TSA have been expressed in viral vectors (e.g. vaccinia) fused to a viral surface protein. While such approaches are promising, the development of truly effective cancer vaccines is complicated by a number of factors. As described in more detail in Chapter 6, TSA expression is often not entirely unique to cancer cells. In the context of vaccine development, even low level expression of target TSA on untransformed cells could result in the development of autoimmune type conditions in vaccine recipients.

An alternative but related approach to cancer vaccine development entails the use of DNA-based vaccines. Experiments in the 1990s revealed that a proportion of foreign plasmid DNA administered to animals by intramuscular injection could be assimilated by muscle cells. Furthermore, it was demonstrated that once inside in the cells, the foreign DNA passed through the nuclear membrane and persisted in the nucleus as a non replicating episomal molecule. Low-level but long-lived expression of the foreign DNA was usually observed. The introduction (and subsequent expression) of nucleotide sequences coding for TSA into body cells in such a manner may prove to be an effective means of vaccinating against cancer.

Adjuvant technology

Many antigens, when administered by themselves, elicit a poor immunological response. This is also true of recombinant subunit vaccines, which usually induce a good B-cell but poor T-cell response. Antigen preparations are thus routinely administered in conjunction with substances termed adjuvants. Adjuvants function to enhance the immunological response against the injected antigenic substances by a number of means. For example they generally enhance the immunogenicity of antigens. They also tend to protect the antigen against rapid inactivation and removal. Adjuvants may also increase antibody production by enhancing antigen presentation to a variety of immune cell types. The only adjuvant preparations currently employed for use in humans are salts of aluminium such as aluminium hydroxide $(Al(OH)_3)$ or aluminium phosphate $(AlPO_4)$. Such adjuvants increase the humoral response against injected antigen by as much as 100-fold. They also exhibit little if any adverse side effects.

A variety of even more potent adjuvant preparations are also available. These include: (a) Freund's complete and Freund's incomplete adjuvants (FCA/FIA); (b) suspensions of dead *Bordetella pertussis* cells; and (c) immunostimulatory components of mycobacterial and other cell walls. Freund's complete adjuvant consists of a mixture of dead *Mycobacterium tuberculosis* in a water–oil emulsion. The emulsifying agent normally used is known as Arlacel A while the oil employed is usually mineral oil. Freund's incomplete adjuvant lacks the mycobacterial component. Freund's-based adjuvants have not been approved for use in humans and are restricted to use in animals because of side effects upon administration. Such side effects include severe local irritation and inflammation at the site of injection, as well as granuloma formation.

The active immunostimulatory components present in mycobacterial preparations appear to be muramyl dipeptides (MDP), in particular *N*-acetyl muramyl L-alamyl-D-isoglutamine, and trehalose dimycolates (TDM). These molecules are found in association with the mycobacterial cell wall. *N*-acetyl muramic acid is a base component of peptidoglycan, the structural polymer found in bacterial cell walls. Trehalose dimycolate consists of a molecule of trehalose to which two molecules of mycolic acid are covalently linked. Trehalose is a disaccharide consisting of two glucose residues linked via a glycosidic bond (Figure 5.9). Mycolic acid is a hydroxy lipid molecule, found almost exclusively on the surface of mycobacteria (Figure 5.9).

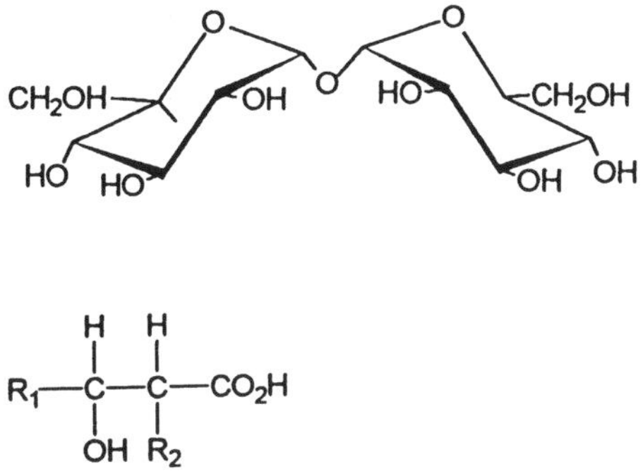

Figure 5.9 Structural formulae of (a) trehalose (α-D-glucopyranosyl $1 \rightarrow 1\alpha$-D glucopyranose) and (b) mycolic acid. R_1 and R_2 groups in mycolic acid represent long-chain aliphatic hydrocarbons

Purified extracts from mycobacterial cell walls containing such immunostimulants may also constitute valuable adjuvant material. Their administration results in marked stimulation of antibody production and an enhanced cell-mediated immune function. These stimulants are pyrogenic insofar as they potentiate release of interleukin-1, (IL-1, endogenous pyrogen), from macrophages and monocytes. A variety of muramyl dipeptides have been chemically synthesized, some of which are capable of inducing a marked immunostimulatory response.

A suspension of dead *Bordetella pertussis* cells is sometimes utilized as adjuvant material. The pertussis toxin is known to stimulate antibody production. LPS derived from the dead cells may also stimulate the immune function by inducing IL-1 secretion.

FURTHER READING

Books

Aledort, L. (1996) *Inhibitors to Coagulation Factors*. Plenum Publication Co., New York.

Becker, R. (2000) *Fibrinolytic and Antithrombotic Therapy*. Oxford University Press, Oxford.

Chung, Y.L. (1991) *Fibrinogen, Thrombosis, Coagulation and Fibrinolysis*. Plenum Publishing Co., New York.

Cohen, S. (1996) Novel strategies in the design and production of vaccines. Plenum Publishing Co., New York.

Institute of medicine (1996) *Blood and Blood Products*. National Academy Press, Washington, DC.

O'Hagan, D. (2000) *Vaccine Adjuvants*. Humana Press, New York.

Peters, T. (1995) *All About Albumin*. Academic Press, London.

Plotkin, S. (1999) *Vaccines*. W.B. Saunders, Philadelphia.

Poller, L. (1996) *Oral Anticoagulants*. Arnold, London.

Powell, M. (1995) *Vaccine Design*. Plenum Publishing Co., New York.

Ratnoff, O. (1996) *Disorders of Haemostasis*. W.B. Saunders, Philadelphia.

Rizza, C. (1997) *Haemophilia and Other Inherited Bleeding Disorders*. W.B. Saunders, Philadelphia.

Rudolph, A. et. al. (1997) *Red Blood Cell Substitutes*. Marcel Dekker, New York.

Stern, P. (2000) *Cancer Vaccines and Immunotherapy*. Cambridge University Press, Cambridge.

Tsuchida, E. (1998) *Blood Substitutes, Present and Future Prospects*. Elsevier, Amsterdam.

Articles

Clotting factors

Foster, P. (2000) Prions in blood products. *Ann. Med.* **32** (7), 501–513.

Fox, J.L. (1992) FDA panel okays two factor VIII's with conditions. *Bio/Technology* **10**, 15.

Legaz, M. *et al.* (1973) Isolation and characterization of human factor VIII (antihemophilic factor). *J. Biol. Chem.* **248**, 3946–3955.

Lowe, K. (1999) Perflourinated blood substitutes and artificial oxygen carriers. *Blood Rev.* **13** (3), 171–184.

Lusher, J. (2000) Recombinant clotting factors – A review of current clinical status. *Biodrugs* 13 (4), 289–298.

Perry, D. (1997) Factor X and its deficiency states. *Haemophilia* **3** (3), 159–172.

Sallah, S. (1997) Inhibitors to clotting factors. *Ann. Hematol.* **75** (1–2) 1–7.

Anticoagulants, thrombolytic and related blood products

Albers, C. *et al.* (2001) Antithrombolytic and thrombolytic therapy for ischemic stroke. *Chest* 119 (1), 300S–320S.

Becker, R. (1999) New thrombolytic, anticoagulants and platelet antagonists: the future of clinical practice. *J. Thromb. Thrombol.* **7** (2), 195–220.

Collen, D. (1998) Staphylokinase: a potent, uniquely fibrin-selective thrombolytic agent. *Nat. Med.* **4** (3), 279–284.

Datar, R. *et al.* (1993). Process economics of animal cell and bacterial fermentations: a case study analysis of tissue plasminogen activator. *Bio/Technology* **11**, 349–357.

Denman, J. *et al.* (1991). Transgenic expression of a varient of human tissue type plasminogen activator in goat milk: purification and characterization of the recombinant enzyme. *Bio/Technology* **9**, 839–842.

Ebert, K.M. *et al.* (1991) Transgenic production of a variant of human tissue type plasminogen activator in goat milk: generation of transgenic goats and analysis of expression. *Bio/Technology* **9**, 835–838.

Lubenow, N. & Greinacher, A. (2000) Management of patients with heparin-induced thrombocytopenia; focus on recombinant hirudin. *J. Thromb. Thrombol.* **10**, S47–S57.

Markwardt, F. (1991) Hirudin and derivatives as anticoagulant agents. *Thromb. Haemost.* **66**, 141–152.

Ogden, J. (1992) Recombinant haemoglobin in the development of red-blood-cell substitutes. *Trends Biotechnol.* **10**, 91–95.

Rydel, T.J. & Tulinsky, A. (1991) Refined structure of the hirudin-thrombin complex. *J. Mol. Biol.* **221**, 583–601.

Sawyer, R.T. (1991) Thrombolytics and anticoagulants from leeches. *Bio/Technology* **9**, 513–518.

Symons, P. *et al.* (1990) Production of correctly processed human serum albumin in transgenic plants. *Bio/Technology* **9**, 217–221.

Vandegriff, K. (1992) Blood substitutes; engineering the haemoglobin molecule. *Biotechnol. Gen. Eng. Rev.* **10**, 403–453.

Verstaete, M. (2000) Third generation thrombolytic drugs. *Am. J. Med.* **109** (1), 52–58.

Wright, G. *et al.* (1991) High level expression of active human alpha-1-antitrypsin in the milk of transgenic sheep. *Bio/Technology* **9**, 830–834.

Vaccine technology

Allison, A.C. & Bayrs, N. (1987a). Vaccine technology: developmental strategies. *Bio/Technology* **5**, 1038–1040.

Allison, A.C. & Bayrs, N. (1987b). Vaccine technology: adjuvants for increased efficacy. *Bio/Technology* **5**, 1041–1045.

Bolognesi, D.P. (1990) Approaches to HIV vaccine design. *Trends Biotechnol.* **8**, 40.

Cohen, J. (1993). Naked DNA points way to vaccines. *Science* **259**, 1691–1692.

Hansson, M. *et al.* (2000) Design and production of recombinant subunit vaccines *Biotechnol. Appl. Biochem.* **32**, 95–107.

Herlyn, D. & Birebent, B. (1999) Advances in cancer vaccine development. *Ann. Med.* **31**, 66–78.

Horig, H. & Kaufman, H. (1999) Current issues in cancer vaccine development. *Clin. Immunol.* **92** (3), 211–223.

Kohler, G. & Milstein, C. (1975). Continuous culture of fused cells secreting antibody of predefined specificity. *Nature* **256**, 495–497.

Mahon, B. *et al.* (1998) Approaches to new vaccines. *Crit. Rev. Biotechnol.* **18** (4), 257–282.

Markoff, A.J. *et al.* (1990) Protective surface antigen P69 of *Bordetella pertussis*: its characterization and very high level expression in *Escherichia coli*. *Bio/Technology* **8**, 1030.

McAleer, W.J. *et al.* (1984) Human hepatitis B vaccine from recombinant yeast. *Nature* **307**, 178–180.

Moingeon, P. (2001) Cancer vaccines. *Vaccine* **19**, (11–12), 1305–1326.

Peters, B. (2000) HIV immunitherapeutic vaccines. *Antiviral Chem. Chemother.* **11** (5), 311–320.

Salk, J. *et al.* (1993), A strategy for prophylactic vaccination against HIV. *Science* **260**, 1270–1272.

Wang, R. & Rosenberg, S. (1999) Human tumor antigens for cancer vaccine development. *Immunol. Rev.* **170**, 85–100.

6

Therapeutic antibodies and enzymes

INTRODUCTION

Antibodies and enzymes are widely used for *in vitro* diagnostic purposes (Chapter 9). Polyclonal antibody preparations have also been used therapeutically for many decades to induce passive immunity. More recently monoclonal antibody preparations, as well as antibody fragments produced by rDNA technology have found therapeutic use. This chapter reviews the *in vivo* medical applications of antibody and enzyme preparations.

ANTIBODIES USED FOR *IN VIVO* APPLICATIONS

Polyclonal antibody preparations

A wide variety of polyclonal antibody preparations are used clinically. They usually function to afford immediate immunological protection against specific pathogens or other harmful antigenic substances. Administration of such specific antibody preparations is termed passive immunization. The purified antibody preparations administered are normally termed antisera. A distinction is often made between antisera and immunoglobulin preparations. The former term refers to antibodies isolated from animals while the latter term refers to antibody preparations specifically obtained from human sources. In both cases the purified preparations consist predominantly of immunoglobulin G (IgG) molecules. Some of the more commonly used antiserum and immunoglobulin preparations are listed in Table 6.1

Passive immunization is generally used as a therapeutic measure if the patient is already suffering from a harmful condition caused or exacerbated by the presence of a known antigenic substance. Such antigenic substances include viruses, microorganisms or toxins/venom produced by certain spiders and snakes. The antibody preparations specifically recognize and bind the offending antigenic substances, thereby neutralizing or inactivating them. The exogenous antibody also helps to initiate a full immunological response against the foreign substance.

In contrast to this, the process of active immunization (the administration of a specific antigen, a vaccine, which stimulates the immune system to generate its own immunological response, Chapter 5), is normally used as a prophylactic measure. The term prophylactic or prophylaxis refers to measures taken to prevent the future occurrence of specific diseases. Under certain circumstances, passive immunization may also be employed as a prophylactic measure. For example if a person is likely to come in contact with pathogens because of work or travel, prior administration of antibodies capable of recognizing the pathogenic agent will afford transient immunological protection.

Specific antiserum preparations are raised in large animals, of which horses are the most popular. This is achieved by injecting the antigen of interest into the animal, thus initiating an immunological response (Figure 6.1). Booster shots of the antigen in question may be administered subsequently in order to further heighten the antibody response. Samples of blood are withdrawn from the immunised animal at regular intervals and the antibody titre measured by an appropriate assay. When large quantities of antibodies which specifically recognise the antigen of interest are detected, the animal is normally bled.

When large animals such as horses are used, several litres of blood may be withdrawn during each bleed. The blood is then allowed to clot and the resulting antiserum subsequently recovered. Alternatively, the blood may be withdrawn directly into containers containing suitable anticoagulants

Table 6.1 Polyclonal antibody preparations most commonly used to induce passive immunity

Antibody preparation	Usual source	Antibody specificity
Normal immunoglobulin	Human	Exhibits a wide range of specificities against pathogens which are prevalent in the general population
Hepatitis B immunoglobulin	Human	Antibodies which exhibit a specificity for hepatitis B surface antigen
Measles immunoglobulin	Human	Antibodies which exhibit a specificity for measles virus
Rabies immunoglobulin	Human	Antibodies which exhibit a specificity for rabies virus
Cytomegalovirus immunoglobulin	Human	Antibodies exhibiting a specificity for cytomegalovirus
Varicella zoster immunoglobulin	Human	Antibodies exhibiting a specificity for the causative agent of chicken pox
Tetanus immunoglobulin	Human	Antibodies which exhibit a specificity for the toxin of *Clostridium tetani*
Tetanus antitoxin	Horse	Antibodies raised against the toxin of *Clostridium tetani*
Botulism antitoxin	Horse	Antibodies raised against toxins formed by type A, B or E *Clostridium botulinum*
Diptheria antitoxin	Horse	Antibodies raised against diptheria toxin or toxoid
Gas-gangrene antitoxins	Horse	Antibodies raised against the alpha toxin of *Clostridium novyi*, *C. perfringens* or *C. septicum*
Scorpion venom antisera	Horse	Antibodies raised against venom of one or more species of scorpion
Snake venom antisera	Horse	Antibodies raised against venom of various poisonous snakes
Spider antivenins	Horse	Antibodies raised against venom of various spiders, in particular the Black Widow Spider

and centrifuged to remove blood cells, thus producing plasma. The serum or plasma is then subjected to fractionation in order to recover a purified antibody preparation. Traditional purification protocols employ several sequential precipitation steps, usually using ethanol and/or ammonium sulphate as precipitants. More recent purification protocols employ various chromatographic steps (Figure 6.2) including ion exchange, hydrophobic interaction chromatography or protein A chromatography. All procedures used in the collection, manipulation and purification of antiserum preparations are carried out in accordance with the principles of good manufacturing practice.

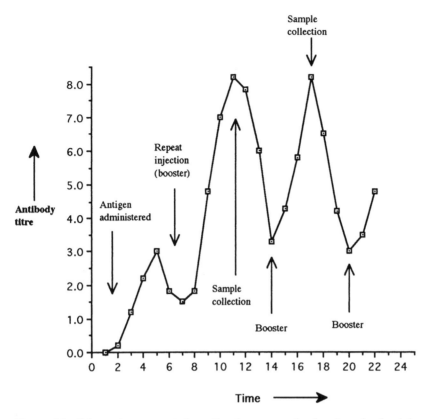

Figure 6.1   Schematic representation of antiserum production in animals. Administration of the antigen of interest results in an immune response. Serum IgG levels increase. Repeat injections of antigen (booster doses) maintain elevated levels of anti-antigen antibody. Careful monitoring of these levels facilitates bleeding (antibody collection) at the most appropriate junctures. Frequency of bleeding may be once every 7–14 days

Following antibody purification, the potency of the product is determined and adjusted to a suitable strength by dilution or concentration, as appropriate. Stabilizing agents such as NaCl or glycine are often added, as are antimicrobial agents such as phenol or thiomersal. The antiserum preparation is sterilized by filtration and filled aseptically into sterile containers, which are subsequently sealed to preserve sterility. The antiserum preparation may also be lyophilized. The antibody solution is often filled under an oxygen-free nitrogen atmosphere in order to prevent oxidative degradation of the product during long-term storage. The product is normally stored at 2–8°C. Under such conditions it should have a shelf life of anything up to 3–5 years.

Immunoglobulin preparations are purified from donated bloods obtained from human volunteers. The methods of purification used are broadly similar to those used in the purification of antibodies from

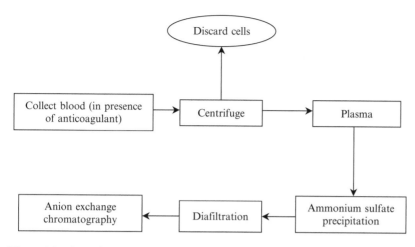

Figure 6.2 Overview of one method of purifying human IgG from plasma. In this case the IgG does not bind to the anion exchange column, whereas other plasma proteins do. Note that alternative procedures or variations of the above procedure may also be used

animal sources. Immunoglobulins purified from blood donations obtained from normal healthy individuals usually contain a wide variety of antibody specificities. The production of such specificities was elicited over the years as the donor came into contact, either naturally or artificially, with a variety of antigens. Alternatively, high antibody titres, which recognized specific antigens, may be purified from donated blood obtained from individuals who have been immunized with that antigen or who have recently recovered from an illness caused by the antigenic substance. Persons recently vaccinated against hepatitis B or who have suffered from hepatitis B infection for example would generally exhibit high titres of anti-hepatitis B antibody.

Administration of antibody preparations may sometimes result in adverse clinical reactions. This is especially true if the antiserum preparation used is of animal origin. Adverse reactions potentially associated with administration of animal serum include serum sickness and, in severe cases, anaphylactic shock. For this reason it is preferable to use immunoglobulin preparations of human origin whenever possible. Serum from which antibody preparations are purified should only be obtained from healthy animals or humans, and should be screened for the presence of potential pathogens prior to processing.

The range of pathogens for which donated blood is screened before being used for purposes of direct transfusion or as a raw material for the purification of various blood products is listed:

- HIV-I & HIV-II (A);
- Hepatitis B virus (A);

- Hepatitis C virus (A);

- *Treponema pallidum* (syphilis) (R);

- Cytomegalovirus (R);

- Epstein–Barr virus (R);

- HTLV-I & HTLV-II (F);

- Creutzfeld–Jakob disease (F).

Note that the pathogen screening list may vary somewhat in different world regions. All blood however will be screened for evidence of the presence of HIV-I and -II as well as hepatitis B and C viruses. This is indicated below by the appearance of an (A) after the pathogens mentioned. Screening regularly or often carried out are indicated by an (R), while screening tests that may be included in the future are indicated by an (F).

Monoclonal antibodies

Monoclonal antibodies are used in an ever-increasing range of biotechnological processes:

- *in vitro* diagnostic reagents;

- investigative tools for *in vivo* diagnostic purposes;

- therapeutic agents;

- vaccines;

- ligands in immunoaffinity chromatography;

- catalysts.

This is due in part to their specificity, selectivity, high antigen binding affinity and to the inexhaustible supply of identical antibody afforded by hybridoma technology. Recent advances which allow recombinant production of monoclonal antibodies will extend even further the availability and usefulness of these highly sophisticated biological reagents.

Most antigenic substances contain numerous epitopes or regions on the surface of the antigen which antibodies recognize. Therefore, when the immune system encounters such antigens, the synthesis of a variety of antibody molecules which recognize and bind such epitopes is initiated (antibody structure is illustrated in Figure 6.3) Each antibody type which binds to any particular epitope is produced by a clone derived from a specific B lymphocyte. If one such antibody-secreting B lymphocyte or plasma cell could be isolated and cultured successfully, antibody molecules of a single specificity (monoclonal antibodies) could be produced. Such antibody-producing cells, however, cannot be cultured in this manner over long periods of time.

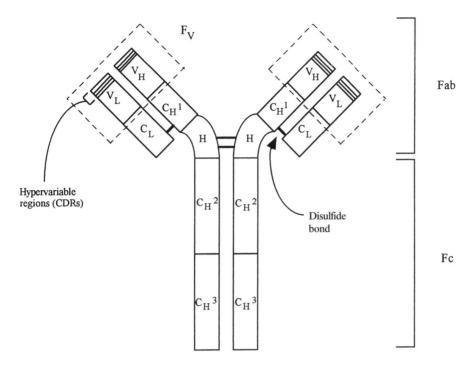

Figure 6.3 Antibody structure (immunoglobulin G) Immunoglobulin G consists of 4 polypeptide chains, 2 heavy (identical) and 2 light (identical). The four polypeptide units assemble to form a Y-shaped molecule. The overall structure is stabilized by both interchain and intrachain disulphide bonds, and by non-covalent interactions. Treatment with certain proteolytic enzymes results in cleavage of the antibody at the flexible hinge (H) region, yielding two antigen binding fragments (Fab), and a constant fragment (Fc). The constant fragment (Fc) normally mediates various effector functions. Both H chains and L chains contain variable (V) regions and constant (C) regions. Variable regions contain the antigen binding site. Variable regions of antibodies of different specificity differ in amino acid sequence. Constant regions on the other hand, do not. Structurally each H and L chain consists of a number of domains. One such domain forms the variable region of the H chain (V_H). The constant region of the H chain consists of three such domains, C_H1, C_H2 and C_H3. L chains on the other hand consist only of two domains; a variable one (V_L) and a constant one (C_L). An antibody fragment consisting of V_H and V_L domains is sometimes referred to as the F_V fragment. Each domain exhibits strikingly similar underlying architectural features, consisting of β-pleated sheets joined by loop regions. These loops in the variable domains (V_H and V_L) exhibit hypervariable sequences and form the antigen binding sites of the antibody. These hypervariable sequences are referred to as complementarity determining regions (CDRs). The remaining areas of the variable domains are often termed framework regions. Framework regions exhibit reduced variability when compared to CDR sequences. Immunoglobulins are glycoproteins. The carbohydrate moiety is normally associated with the constant domains of the heavy chains. Removal of the carbohydrate groups has no effect on the antigen binding ability of the antibody but does effect various antibody effector functions

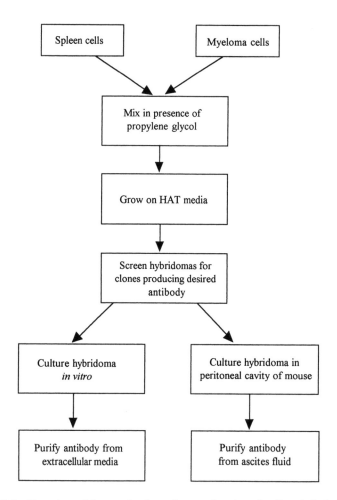

Figure 6.4 Overview of the production of monoclonal antibodies via hybridoma technology. The process begins with the immunization of a mouse with the antigen against which antibodies are required. The mouse is subsequently sacrificed (usually 1–4 weeks after initial immunization), and its spleen is removed. Lymphoid cells are washed out of the spleen. Typically 10^8 spleen-derived lymphocytes are fused with myeloma cells by co-incubation in the presence of polyethylene glycol. The myeloma cells used in the fusion protocol do not produce the enzyme hypoxanthine–guanine phosphoribosyl transferase (HGPRT). Lack of this enzyme means they cannot synthesize DNA if grown in a media containing hypoxanthine, aminopterin and thymidine (HAT media). A proportion of the daughter cells formed upon incubation will be derived from the fusion of a single lymphocyte and a single myeloma cell. These daughter cells (hybridoma cells) often retain the antibody producing characteristics of the parent lymphocyte and the immortal characteristics of the parent myeloma cell. Screening of such hybridoma cells post-fusion is possible by culture in HAT medium. Undfused myeloma cells cannot grow on this media, but fused cells (by inheriting the HGPRT enzyme from the parent lymphocyte) can. Unfused lymphocytes do not grow well on this media and either die or are rapidly overgrown by hybridoma cells, which double every

Hybridoma technology

In the mid-1970s, a technique was developed that facilitates the production of monospecific antibodies derived from a single antibody-producing cell. This process is termed hybridoma technology. (Figure 6.4) It stems from the observation that, if a population of antibody-producing cells are fused with immortal myeloma cells, a certain fraction of the resultant hybrid cells will retain the immortal characteristics of the myeloma cell while secreting large quantities of monospecific antibody.

Monoclonal antibody production initially involves the collection of lymphocytes from the spleen of a laboratory mouse immunized against the antigen of interest. A certain proportion of such lymphocytes produce antibodies that react with specific epitopes present on the antigen's surface. The lymphocytes are incubated together with mouse myeloma cells in the presence of a reagent that promotes cell fusion, such as propylene glycol. Hybrid cells derived from the fusion of myeloma cells and lymphocytes are termed hybridomas. Hybridomas may be selected from unfused cells by culture in a specific selection medium. The antibody-producing hybridomas may be separated from each other by dilution and are readily grown in tissue culture. The resultant clones must be screened in order to determine which one(s) produce antibodies recognizing the antigen of interest. The most appropriate positive clones may subsequently be cultured in order to produce large quantities of monospecific antibody.

The specificity, selectivity and high binding affinity exhibited by monoclonal antibodies renders them attractive biochemical tools. Moreover, once a suitable hybridoma has been identified and isolated, it may be used to produce large quantities of defined antibodies on an ongoing basis. In contrast to this, the range and quantity of antibody specificities present in polyclonal antibody preparations may vary from bleed to bleed and may not be exactly reproduced by replacement animals, when the initial producing animal dies.

18–48 h. Hybridoma cells growing on HAT medium can be separated from each other by serial dilution, with subsequent culture of individual cells, to form clones. Screening of these clones pinpoints those producing the antibody of interest. Clones can be stored deep frozen under liquid nitrogen if required. Growth of these cells to produce monoclonal antibodies can be undertaken by two means. Direct cell culture *in vitro* results in monoclonal antibody production to levels of μg/ml of media. Alternatively, the hybridoma cells can be injected into the peritoneal cavity of live mice, where the hybridomas grow as an ascitic tumour. The ascitic fluid produced contains the monoclonal antibodies, usually at levels 50–100 times those achievable by *in vitro* cell culture. However, large-scale production of monoclonal antibodies by this route raised ethical issues and is disadvantaged by the fact that the ascitic fluid is also contaminated by mouse immunoglobulins

Hybridoma technology represents the traditional method by which monoclonal antibodies are produced. However, the advent of gene technology has facilitated the production of antibodies or antibody fragments in recombinant systems, as described later. Many monoclonal antibody based products approved recently (as well as most products currently in clinical trials) are produced via the recombinant route.

Antibodies approved for *in vivo* application

In 1986 OKT3 became the first monoclonal antibody to be approved for an *in vivo* therapeutic purpose. This antibody is used to promote a reversal of acute kidney transplant rejection. OKT3 specifically recognizes a cell-surface antigen known as CD3, also referred to as the cluster of differentiation, which is associated with virtually all T cells. Binding of the antibody to CD3 can induce destruction of T cells – these cells usually mediate rejection of transplanted tissue. Treatment with OKT3 appears more effective in preventing transplant rejection as compared to treatment with more conventional immunosuppressive drugs. Administration of OKT3, however, often induces the release of a variety of cytokines, which may mediate a number of unwanted clinical responses.

By 2000 a total of 18 monoclonal antibody based products had been approved as either *in vivo* diagnostic or therapeutic agents (Table 6.2). Of these seven products serve to detect or diagnose various cancer types, while three aim to treat specific cancers. Some 90 monoclonal antibody based products are undergoing clinical evaluation (Table 6.3). This makes antibodies (together with vaccines) the largest categories of pharmaceutical biotechnology products currently in clinical trials.

Antibody-based tumour detection and destruction

A prerequisite to developing an antibody destined for *in vivo* cancer cell detection and destruction is the identification of a surface antigen unique to the target cell. Antibodies raised against a unique surface antigen (USA) will therefore bind only to that cell type (Figure 6.5).

When a healthy cell turns cancerous it begins to express a number of genes which are either unexpressed (or expressed at extremely low levels) in the pre-cancer state. Some of these newly synthesized proteins are retained on the surface of the transformed cell. Such tumour surface antigens (TSA) can therefore represent a USA on the cancer cell surface, and antibodies raised against TSA can target these cells. Identification of TSA remains an active area of research. Carcinoembryonic antigen (CEA) is one of the best characterized TSA. This 180 kDa transmembrane glycoprotein is often expressed at high levels on the surface of colorectal tumour cells. TAG-72 and CO17–1A represent additional

Table 6.2   Monoclonal antibody based products approved (by 2000) for *in vivo* use

Trade name	Indication
CEA-SCAN	Detection of recurrent colorectal cancer
OncoScint	Detection of colorectal & ovarian cancers
ProstaScint	Detection of prostate cancer
Verluma	Detection of small cell lung cancer
Indimacis 125	Diagnosis of ovarian adenocarcinoma
Humaspect	Detection of colorectal cancer
Rituxan	Treatment of non-Hodgkin's lymphoma
Tecnemab K1	Diagnosis of melanoma
Herceptin	Treatment of some forms of breast cancer
Mabthera	Treatment of non-Hodgkin's lymphoma
Remicade	Treatment of Crohn's disease
OKT-3	Reversal of acute kidney transplant rejection
Zenapax	Prevention of acute kidney transplant rejection
Simulect	Prevention of acute kidney transplant rejection
MyoScint	Myocardial infarction imaging agent
LeukoScan	Imaging of bone infections in patients with osteomyelitis
ReoPro	Prevention of blood clots
Synagis	Treatment of lower respiratory tract disease

colorectal TSA. TSA associated with other cancer types include G250 (renal cell carcinomas) and folate binding protein (ovarian tumours).

Four strategies have emerged by which anti-TSA monoclonals have been used to detect or destroy human tumours *in vivo*: (a) the administration of unmodified antibody, or the administration of antibody/antigen-binding antibody fragment to which (b) a radioactive tag, (c) a toxin or (d) a pro-drug activating enzyme has been attached (Figure 6.6).

Unconjugated antibodies   The administration of unmodified anti-TSA antibodies can potentially induce tumour cell destruction via immunological effector functions associated with the antibody's Fc region

Table 6.3 Range of medical conditions that monoclonal antibody based products currently in clinical trials aim to detect/treat

Medical condition	No. products in trials
AIDS	3
Autoimmune disorders	8
Blood disorders	1
Cancer	37
Heart disease	6
Digestive disorders	5
Infectious diseases	5
Neurological disorders	7
Respiratory disease	6
Transplantation	8
Other conditions	2

Source: Pharmaceutical Research and manufacturers of America.

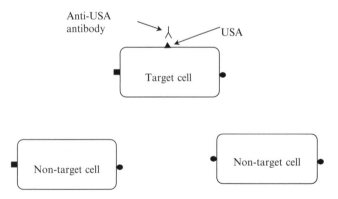

Figure 6.5 Basis of specificity of antibody products used for *in vivo* targeting of specific cell types. Antibodies are raised against an antigen uniquely found on the target cell surface (unique surface antigen, USA). Antibodies injected intravenously will bind only to the surface of the target cell type

(e.g. activation of complement and attraction of phagocytes). In general, clinical trials utilizing unmodified monoclonal have yielded disappointing results. Lack of efficacy could be caused by a number of factors, including:

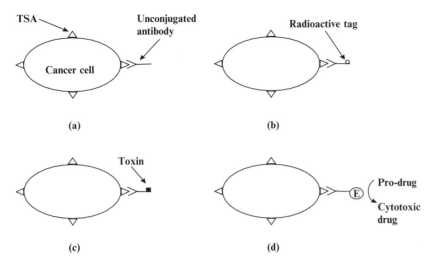

Figure 6.6 Approaches taken to the development of monoclonal antibody mediated detection/destruction of tumours. Refer to text for specific details

- Monoclonals used (initially at least) were of murine origin, produced by standard hybridoma techniques. The Fc region of mouse antibodies are poor activators of human immune function. Moreover, the mouse antibody itself would become the target of a human immunological response.

- In most instances TSA expression is neither static nor uniform. At any given time a proportion of tumour cells may not be expressing the TSA at all. The administered antibody would be blind with respect to this subset of transformed cells.

Radioactively tagged antibodies Radiolabelled antibodies can be used to either detect or destroy tumour cells. Various chemical techniques have been developed that allow coupling of radioactive atoms to antibodies or other proteins. Antibody-based tumour detection requires the use of radionuclides which emit γ-rays (to allow outward penetration from the body). Metastable technetium (^{99m}Tc) or indium (^{111}In) are most often used for this purpose. γ-Rays can be detected using a planar γ camera, and hence the presence and location of any tumours can be established.

Antibodies used to destroy cancer cells are most often labelled with β-emitting (e.g. ^{90}Y) or α-emitting (e.g. ^{212}Bi or ^{211}At) isotopes. β-Particles can penetrate several layers of cells, and will kill these cells by the ionizing effects of radiation. α-Particles have a shorter effective path length, but each α-emission has a greater likelihood of killing all cells in its path. Antibody-mediated targeted radiotherapy has proven effective in promoting regression of various tumours, although obviously a proportion of healthy cells in the immediate vicinity of the tumour will also be irradiated.

Antibody–toxin conjugates   Conjugation of toxins to anti-TSA antibodies will result in targeted toxin delivery to the tumour surface. In some instances binding of an antibody–toxin conjugate results in the subsequent internalization of the conjugate, thus delivering the toxin inside the cell. A variety of toxins derived from plants or microorganisms have been conjugated to anti-TSA antibodies, and these continue to be evaluated in clinical trials. The most popular toxins used to date include *Pseudomonas* exotoxin and the ricin A chain. All function by halting protein synthesis in the target cell, thus killing it.

Antibody–enzyme conjugates   Antibodies to which pro-drug activating enzymes are conjugated are also being clinically assessed (Figure 6.6). This approach to targeted cancer chemotherapy is termed antibody-directed catalysis (ADC) or antibody directed enzyme pro-drug therapy (ADEPT). The enzyme used is chosen on the basis of its ability to catalytically convert a (harmless) pro-drug into a cytocidal agent. Alkaline phosphatase is one such example. It activates etoposide phosphate by dephosphorylating it. Clinically, the conjugate is first administered and allow time to congregate at the tumour surface. The prodrug is then administered with its ensuing activation at the tumour surface.

Therapeutic antibodies, an initial disappointment

Despite initial enthusiasm, many of the monoclonal antibody based therapeutic products initially developed were a clinical disappointment. The disappointing results recorded with these intact murine monoclonals, manufactured by traditional hybridoma technology, may be attributed to a number of important factors (Table 6.4). Over the past number of years technical advances have negated many of these initial difficulties.

Chimaeric and humanized antibodies

The immunogenic nature of murine monoclonal antibodies, when administered to humans, has presented the greatest obstacle to their successful clinical application. The administration of murine antibodies stimulates the production of neutralizing antibodies by the host (human) immune system. These antibodies quickly negate the therapeutic effect of the administered product.

One obvious solution to such problems is to develop and employ monoclonal antibodies of human origin. This task has thus far proven extremely difficult. Human lymphocytes represent poor fusion partners for mouse myeloma cell lines. Upon fusion, a preferential loss of human genetic elements is often observed, and the resultant hybrids are highly unstable. Furthermore, it is often not possible to immunize human subjects using toxic or otherwise deleterious antigenic substances. Attempts

Table 6.4 Factors which limited the therapeutic efficacy of many first generation monoclonal antibody based therapeutic products

---

Murine monoclonal antibodies are recognized as foreign by the host human immune system

It is extremely difficult to generate monoclonal antibodies of human origin

Murine monoclonal antibodies, when administered to humans, fail to trigger a number of effector functions normally associated with the constant region of native antibody

Insufficient information existed regarding appropriate target cell surface markers which might be selectively found on the surface of certain tumour types

Short half life: Human antibodies (IgG) display a serum half life of up to 23 days. Murine monoclonals, when administered to man, exhibit a half life of 30–40 h

Poor penetration of tumour mass by whole antibodies is normally observed

Monoclonal antibodies are time consuming and expensive to produce

---

to immortalize human antibody producing cells by methods such as infection with Epstein–Barr virus have thus far proved disappointing.

Genetic engineering technologies have provided the most effective means of reducing the innate immunogenicity of murine monoclonal antibodies. Initial genetic manipulations centred around production of hybrid or chimaeric antibody molecules consisting of mouse variable regions and human constant regions. It was considered likely that such partially 'humanized' antibodies would be significantly less immunogenic when compared to unaltered murine monoclonal antibodies. This was found to be the case. However, such chimaeric antibodies still retain significant portions of mouse sequence (the entire variable region) (Figure 6.3) and these sequences can still induce substantial immunological responses.

The immunogenicity of murine monoclonal antibodies may be further reduced by 'grafting' the DNA sequences coding for the antigen-binding complementarity-determining regions (CDR) of the parent monoclonal antibody into the DNA sequences coding for a human antibody. The resultant humanized monoclonal antibody contains mouse sequences which code only for the hypervariable (antigen-binding) regions of the antibody (Figure 6.7). As the remainder of the molecule is entirely

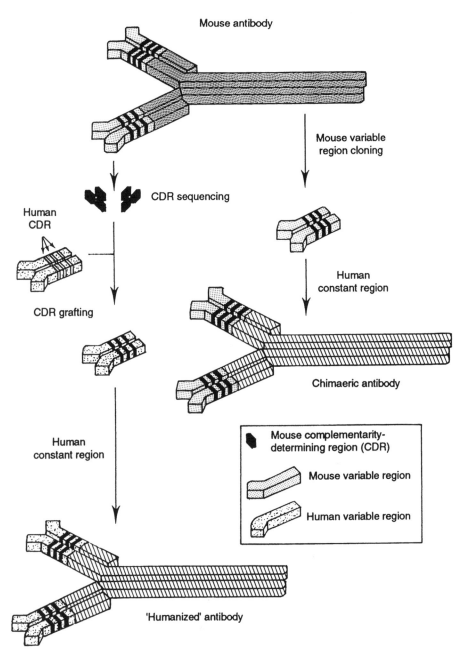

Figure 6.7 Recombinant DNA technology makes possible the production of 'chimaeric' and 'humanized' antibodies. Chimaeric antibodies consist of variable regions of murine origin and constant regions of human origin. Such antibodies retain significant proportions of mouse sequences, and generally remain immunogenic when administered to humans. Humanized antibodies are less immunogenic as the only sequences of murine origin they contain are those that constitute the complementary determining regions. This first appeared in *New Scientist*, London

human, it should not elicit an antigenic response when administered to humans. Many such humanized antibodies, however, exhibit decreased antigen-binding affinity when compared to the parent murine monoclonal antibody. Slight alteration of the human framework sequences which surround the hypervariable regions can usually restore antigen-binding affinity. Although the framework regions of the variable chains do not seem to engage directly in binding antigen, they nevertheless influence such binding.

The successful production of humanized monoclonal antibodies also overcomes the inability of murine monoclonal antibodies to mediate a variety of antibody effector functions in recipient human patients. Binding of antibody to antigen facilitates the destruction and removal of the antigen by a number of means. The binding of antibody renders more efficient the process of antigen ingestion and destruction by phagocytic cells. The Fc region of bound antibody also mediates complement. Complement is composed of a number of serum proteins which, when activated, facilitates the destruction of antigen by phagocytes in addition to promoting direct lysis of invading microorganisms. The Fc region of murine antibodies are poor mediators of such effector functions in humans. Humanized antibodies whose Fc domains are entirely human in origin do not suffer from such disadvantages.

TSA specificity

The specificity of antibody-mediated tumour detection and destruction is ultimately dependent on the identification of a TSA which is uniquely expressed on the target tumour cells. This absolute tissue specificity is rarely attained in practice. Many TSAs are also expressed (all be it at much lower levels) on the surface of one or more healthy cell types. Such cells will also be targeted by the TSA antibody. In other instances tumours secrete a portion of the TSA they produce in soluble form into the bloodstream. These soluble TSA will mop up some of the antibody administered. In other cases antibody cross-reactivity with host antigens bearing a structural resemblance to the TSA can occur.

CEA (previously described as a TSA associated with colorectal cancer cells) illustrates these difficulties. CEA is also expressed by some breast and lung cancer cell types, and at low levels by some untransformed cells in the lactating breast, as well as colonic mucosal cells. In addition cells which synthesize CEA also secrete a proportion of this antigen into the blood in soluble form. However, significantly elevated serum CEA levels are not necessarily indicative of cancer, as such levels are also increased by a host of unrelated medical conditions, including inflammatory bowel disease, cirrhosis and hepatitis. The biology of TSA is far from simple.

Despite such complications, anti-TSA antibody-based products have in some cases proven themselves effective diagnostic or therapeutic agents. Clinically however, they are rarely used alone for cancer diagnosis or therapy. Most antibody-based cancer diagnostic products are used in combination with other diagnostic indicators such as ultrasonography

Production of antibodies for clinical application

First generation monoclonal antibody products were produced by hybridoma technology. However, a large proportion of more recently approved products are engineered antibodies, produced by recombinant means. While hybridoma production immortalizes antibody-producing cells, DNA technology may be utilized to immortalize antibody genes. The development of chimaeric or humanized antibodies thus far approved for medical use generally entailed initial production of the antibody of interest by classic hybridoma technology. Genetic analysis of the hybridoma cells allowed determination of the exact nucleotide sequence coding for the heavy (H) and light (L) chain variable regions. Chimaeric antibody generation is achieved by splicing these sequences in front of sequences coding for (human) H and L chain constant regions. Humanized antibody production entails replacing nucleotide sequences coding for the CDR regions in a human antibody with those coding for the CDR of the (murine) monoclonal displaying the desired antigen-binding specificity.

The resultant gene constructs may be introduced and expressed in a recombinant animal cell line, thus facilitating glycosylation. Most antibody molecules are glycosylated in their CH_2 domains (certain antibody types exhibit additional glycosylation sites in other constant domains). The presence or absence of bound carbohydrate has no apparent influence on the antigen-binding capability of the antibody. However, most antibody effector functions are reduced or abolished if normal antibody glycosylation is prevented. Apart from influencing various effector functions, the glycosylation status of an antibody may also effect additional properties such as its solubility and its rate of degradation *in vivo*.

In some instances the use of an intact antibody is not necessary to achieve its *in vivo* diagnostic/therapeutic purpose. For example radiolabeled antibodies used for *in vivo* cancer diagnostic purposes are simply required to congregate at the tumour site. Binding is via the antibody's Fab region (Figure 6.3), and the antibody constant regions are redundant in such a context. Thus, antibody fragments containing the antigen binding sites can be (and are) used for such purposes. Antibody fragments can be generated by one of two approaches. Incubation of the intact antibody with certain proteolytic enzymes can generate Fab or $F(ab)_2$ fragments (Figure 6.3). These fragments are then resolved from the Fc fragment (by e.g. gel filtration chromatography).

Even smaller antigen-binding fragments can be generated by recombinant DNA technology. Nucleotide sequences coding for V_H and V_L domains of an antibody displaying the desired binding specificity can be joined by short linker sequences. Subsequent expression of the resultant full length nucleotide sequence results in the production of an Fv fragment, consisting of V_H and V_L domains, linked by a short peptide sequence. Fv fragments can be produced in *E. coli*, as the expressed product does not require post-translational modification.

Antibody fragments can be particularly useful in antibody mediated diagnosis/therapy of solid tumours. Due to their smaller size, these fragments can more effectively penetrate the tumour mass than can intact antibodies.

After their initial production by whichever means, intact antibodies or antibody fragments are next purified to homogeneity via suitable downstream processing techniques. In general, the purified products are marketed in lyophilized format. As an illustrative example, the production of one such monoclonal product ('Humaspect') is outlined in Box 6.1.

THERAPEUTIC ENZYMES

A variety of enzymes are used clinically in the treatment of various medical conditions (Table 6.5). Several enzymes (ancrod, t-PA, urokinase

Table 6.5 Some enzymes that may be used for therapeutic purposes

Enzyme	Therapeutic application
Ancrod (serine protease)	Anticoagulant
Tissue plasminogen activator	Thrombolytic agent
Urokinase	Thrombolytic agent
(Activated) Factors IIV and IX	Treatment of clotting disorders
Asparaginase	Treatment of some types of cancer
DNase	Treatment of cystic fibrosis
Glucocerebrosidase	Treatment of Gaucher's disease
Trypsin Papain Collagenase	Debriding/anti-inflammatory agents
Lactase Pepsin Pancrelipase Papain	Digestive aids
Superoxide dismutase	Prevention of oxygen toxicity

Box 6.1   Case study: production of 'Humaspect'

Humaspect is an intact human monoclonal antibody. It is unusual in that it is produced directly from a human lymphoblastoid cell line that was transformed by Epstein–Barr virus. The antibody is labelled with technetium (^{99M}Tc) prior to clinical use.

Humaspect specifically binds a cytokeratine tumour associated complex of antigens known as CTA#1 or CTAA16.88. These antigens are associated with human large bowel adenocarcinoma. The product's approved therapeutic indication is for the imaging of recurrence and/or metastases in patients with histologically proven carcinoma of the colon or rectum, and as an adjunct to standard non-invasive imaging techniques such as ultrasonography or CT scan.

Humaspect's manufacturing procedure is summarized schematically below. The production process begins by growing the antibody-producing lymphoblastoid cell line in a bioreac-

tor of hollow fibre design. The harvested antibody can be stored at − 70°C if further processing is not scheduled to occur immediately.

Downstream processing includes initial incubation of the crude antibody preparation for one hour at ambient temperature in the presence of 1% Triton-X-100. This step is designed to inactivate any lipid-enveloped viral contaminants which could potentially be in the product stream. Purification is achieved using a combination of protein G affinity chromatography and an anion exchange step. Excipients are then added. These include a sodium phosphate buffer, NaCl and lactose. The product is sterile filtered and aseptically filled into the product vials and lyophilized. The product is marketed in kit form. In addition to the purified antibody, the kit also contains a vial of coupling reagent, which allows direct coupling of the radiolabel to the antibody immediately before its clinical use.

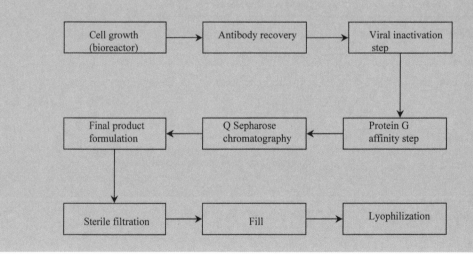

as well as (activated forms of) blood coagulation factors VII and IX) have been discussed in the previous chapter, and hence will not be considered further here. The majority of enzyme preparations described below are extracted directly from their native source material. In recent years however, two recombinant enzyme products (DNase and glucocerebrosidase) have been approved for medical use.

Asparaginase

Asparaginase catalyses the hydrolysis of the amino acid asparagine, yielding aspartic acid and ammonia (Figure 6.8). The enzyme is a tetramer with a molecular mass in the region of 130 kDa. It may be purified from a wide variety of microorganisms including yeast, fungi and bacteria such as *E. coli.*

All cells require asparagine to sustain normal metabolic activity. Although most human cells are themselves capable of synthesizing this amino acid, certain malignant cells lack this ability. Intravenous administration of asparaginase results in the rapid depletion of serum asparagine levels, which normally range between 0.5 and 1.5 mg/100 ml. Protein synthesis in malignant cells incapable of synthesizing asparagine is thus severely compromised. In contrast, untransformed body cells begin to synthesis their own asparagine (Figure 6.8).

Asparaginase preparations used clinically are normally purified from *E. coli* or from *Erwinia chrysanthemi. E. coli* produces two asparaginase isoenzymes, of which only one is clinically effective. Recombinant DNA technology now facilitates the recombinant synthesis of asparaginase from a range of other sources, some of which may display clinical potential. This enzyme is mostly used in the treatment of certain forms of childhood leukaemia. Side effects can include severe allergic reaction, as well as nausea, vomiting, fever, and compromised kidney as well as liver function. The latter effects may be attributable to the fact that most untransformed body cells do not themselves synthesize asparagine when it is available from the blood. The rapid depletion in of serum asparagine upon administration of asparaginase temporarily deprives such healthy cells of asparagine thus disrupting protein synthesis. Ironically, this slows the synthesis of the asparagine synthatase enzyme required by these cells to manufacture their own asparagine (Figure 6.8).

Given its microbial source it is not surprising that L-asparaginase elicits an immune response when administered to humans. This imposes obvious limitations on the enzyme's long-term clinical efficacy. Coupling the L-asparaginase to polyethylene glycol (PEG) has been shown to greatly

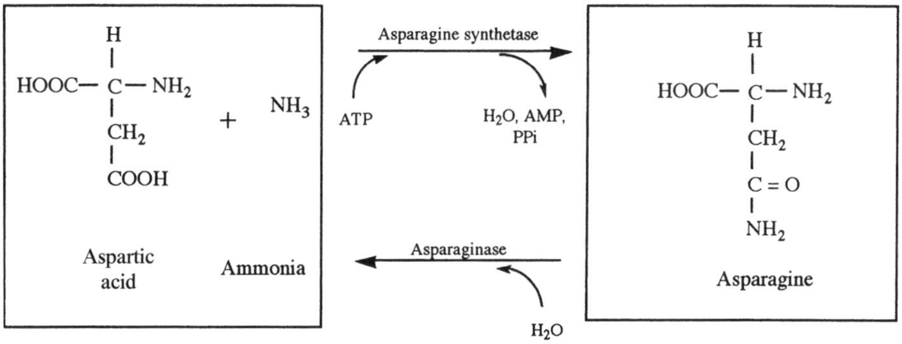

Figure 6.8 The biosynthesis and degradation of L-asparagine

reduce or eliminate the enzyme's immunogenicity, and clinical studies utilizing PEGylated asparaginase have been initiated.

Debriding and anti-inflammatory agents

Certain enzymes are employed as debriding agents. These effectively clean open wounds by removal of foreign matter and any surrounding dead tissue. This allows for rapid healing of the wound. Enzymes such as trypsin, papain and collagenase have often been used as debriding agents. Such preparations are normally applied topically to the affected areas.

Trypsin is a proteolytic enzyme synthesized by the mammalian pancreas. It has a molecular mass of 24 kDa and hydrolyses peptide bonds in which the carboxyl group has been contributed by either an arginine or lysine residue.

Papain is a proteolytic enzyme isolated from the leaves and the unripe fruit of the papaya tree. It catalyses the hydrolysis of peptide bonds involving basic amino acids such as lysine, arginine or histidine. In addition to its use as a debriding agent, papain has also been utilized as a meat tenderiser and for clearing beverages (Chapter 11).

Chymopapain, a second proteolytic enzyme produced by the papaya tree has also found medical application. Chemonucleolysis is the term used for the treatment of sciatica by injecting chymopapain into damaged intervertebral discs. Sciatica, a medical condition characterized by back and leg pain, is caused by the degradation of an intervertebral disc. The disc tends to protrude laterally, compressing spinal root neurones. The injection of chymopapain appears to speed up the underlining process of disc degradation, thereby more quickly reaching an end stage where it is stable and asymptomatic.

Collagenase is an enzyme which catalyses the proteolytic destruction of collagen. Although it may be isolated from culture extracts of various animal cells, it is normally obtained from the culture supernatant of various species of *Clostridium*. Some clostridial species are pathogenic, causing diseases such as gas-gangrene. The ability of such microorganisms to produce tissue degrading enzymes such as collagenase facilitates their rapid spread throughout the body.

Administration of certain enzymes has proven helpful in the reduction of various inflammatory responses. Such enzymes include chymotrypsin and bromelains. Chymotrypsin is a proteolytic enzyme produced in zymogen form, (chymotrypsinogen), by the mammalian pancreas. Chymotrypsinogen is converted to chymotrypsin in the small intestine, where it functions to catalyse the proteolytic degradation of dietary proteins. Bromelains are plant proteases purified from the fruit or stem of the pineapple. The molecular mechanisms by which chymotrypsin and bromelains achieve an anti-inflammatory effect remains to be properly elucidated. However, it is likely that the mode of action of these proteases centres around their ability to degrade protein-based inflammatory mediators, or proteins involved in promoting the synthesis of such mediators.

Enzymes as digestive aids

Various enzymatic preparations may used as digestive aids. Most such enzymes are depolymerases, catalysing the enzymatic breakdown of a number of dietary components including polysaccharides, proteins and lipids. Some such enzyme preparations consist of a single enzyme that catalyses the degradation of a specific dietary substance. Others contain multiple enzymatic activities that exhibit broad digestive ability.

α-Amylase catalyses the hydrolysis of the α1 → 4 glycosidic bonds, which form the main covalent linkage between glucose monomers in carbohydrates such as starch or glycogen. α-Amylase produced by microorganisms such as *Bacillus subtilis* or species of *Aspergillus* have found widespread industrial application, as will be discussed in Chapter 11. Amylase activity also plays an important digestive role in higher animals. The enzyme may be isolated from saliva or from pancreatic tissue, and various amylase preparations have been administered orally to aid the digestion of dietary carbohydrate.

Lactase catalyses the hydrolysis of the disaccharide lactose, the principal sugar of milk, yielding glucose and galactose. Preparations of lactase may be employed as a digestive aid, in particular to alleviate symptoms associated with lactose intolerance. While most infants exhibit high levels of intestinal lactase activity, the adult population of many geographical regions has greatly reduced lactase activity. Such persons are unable to digest lactose, and often suffer from intestinal upset upon consumption of milk.

Various proteolytic enzymes may also be employed as digestive aids. Such enzymes include papain, bromelains and pepsin. Pepsin is secreted naturally in the stomach of most animals, where it catalyses the proteolytic degradation of dietary protein. Pepsin preparations employed as digestive aids are obtained by extraction from the mucous membranes of the stomach of various slaughterhouse animals.

Some enzymatic preparations employed clinically contain multiple enzymatic activities. Pancreatin, for example, is a proteinaceous preparation extracted from the pancreas. It contains amylase, protease, lipase and nuclease activities, and may be administered orally to patients suffering from conditions caused by deficient secretion of pancreatic enzymes. These include chronic pancreatitis (pancreatic inflammation and failure), pancreatic carcinomas, and cystic fibrosis (characterized in part by the blockage of pancreatic secretion ducts by a thick mucous). Pancreatin (and many other enzymatic digestive aids) displays a pH versus activity profile which renders it maximally active in the upper portion of the small intestine (the duodenum). Such enzymes, when administered orally, thus must pass through the stomach in order to reach their site of action.

Inactivation of a large proportion of these enzymes can be witnessed in the stomach, due to the low pH values encountered therein, and also possibly due to their proteolytic degradation by stomach pepsin. (The stomach secretes approximately 2 l of HCl per day, and pH values can be below 2). A number of strategies can be adopted in order to minimise such gastric inactivation. The simplest approach entails co-administration of inhibitors of gastric acid secretion (e.g. cimetidine). However, by diminishing free acid levels in the stomach these agents can reduce the activity of pepsin and more importantly, can compromise the natural bacteriocidal effects of low stomach pH (a low pH value normally kills dietary-derived microorganisms in the stomach). An alternative entails enzyme administration in enteric coated tablet or capsule format. (Enteric coatings are impervious to acid but will dissolve at more neutral pH values, such as those characteristic of the duodenum.) Yet another approach under investigation is the use of microbial (mainly fungal) proteases, amylases and lipases. Many such microbial enzymes display superior acid stability in comparison to pancreatin derived enzymes. Pancrelipase is another enzymatic preparation derived from pancreatic tissue, which may be employed as a digestive aid, the principal enzymatic activity of which is lipase.

Superoxide dismutase

Superoxide dismutase catalyses the conversion of the highly reactive superoxide ion (O_2^-) to hydrogen peroxide (H_2O_2). This enzyme plays an important role in all aerobic organisms, as the superoxide ion is highly reactive and capable of causing serious cellular damage. During normal metabolic activity, aerobic organisms usually reduce oxygen molecules directly to water. This reaction as shown below, requires four electrons.

$$O_2 + 4e^- + 4H^+ \rightarrow 2H_2O$$

Incomplete oxidation of the oxygen molecule invariably generates molecules such as superoxide (O_2^-), hydrogen peroxide (H_2O_2) or hydroxyl (O^-) ions, all of which are extremely reactive. Hydrogen peroxide and superoxide radicals cause extensive cellular damage by reacting with unsaturated fatty acids in membranes. O_2^- and H_2O_2 may be generated by several cellular enzymes. Furthermore, the spontaneous oxidation of iron atoms present in various molecules such as haemoglobin also generate such reactive molecules. Two specific enzymes protect cells from the harmful effects of superoxide and hydrogen peroxide. The enzyme superoxide dismutase catalyses the conversion of superoxide to hydrogen peroxide as shown below:

$$O_2^- + O_2^- + 2H^+ \xrightarrow{\text{superoxide dismutase}} H_2O_2 + O_2$$

Catalase then catalyses the conversion of hydrogen peroxide into oxygen and water as follows:

$$H_2O_2 + H_2O_2 \xrightarrow{\text{catalase}} 2H_2O + O_2$$

Two forms of superoxide dismutase are found in eukaryotic cells. The cytoplasmic enzyme is a dimer consisting of two identical subunits and with a molecular mass of 31 kDa. Each enzyme subunit contains an atom of zinc and an atom of copper. Both metal atoms participate directly in the enzymatic process. The dismutase associated with mithocondria has a molecular mass of 75 kDa and contains two atoms of manganese. Two highly homologous yet distinct superoxide dismutases are also found in aerobic prokaryotes. One form contains manganese while the other contains Iron.

Superoxide dismutase isolated from bovine liver or erythrocytes has been available for a number of years, and to date it has been used clinically as an anti-inflammatory agent. Human superoxide dismutase may be produced as a heterologous protein in a number of recombinant systems. Such preparations are currently undergoing clinical trials in order to assess the enzyme's ability to prevent damage to tissue resulting from exposure to excessively oxygen-rich blood.

Nuclease treatment of cystic fibrosis

Cystic fibrosis (CF) represents one of the most commonly occurring genetic diseases. The frequency of occurrence varies among populations, with persons of northern European extraction being most at risk. Within such populations approximately 1 in 2500 newborns are affected. It has been recognized for many years that excessive salt loss occurs in the sweat of persons suffering from CF. More recently the gene, which is defective in persons suffering from this disease, has been identified. This gene codes for a polypeptide product which ordinarily functions as a chloride channel. It thus seems that the underlying cause of CF may be traced to a malfunction in ion transport. About 5 per cent of white people carry this defective gene. However, the trait is recessive so the condition only occurs in cases where one inherits defective gene copies from both parents. In such individuals expression of the aberrant gene results in compromised function of a number of tissue types, including the pancreas and sweat glands. The major clinical symptom of CF however, is undoubtedly the production of extremely viscous mucus in the respiratory tract. This compromises lung function.

It seems that alterations in ion transport in CF sufferers results in subtle changes in lung physiology, which is conducive to the establishment of recurrent bacterial infections. Such infections trigger an immune response in which large numbers of neutrophils are attracted to the site of infection. Ingestion and destruction of bacterial populations by such white blood cells results in the liberation of large amounts of DNA. This released DNA interacts with a variety of additional extracellular substances present in the infected lung, thus generating a highly viscous mucus.

In order to prevent build-up of such mucus CF patients suffering from bacterial infections are normally subjected to percussion therapy. This basically involves physically pounding on the patient's chest for extended time periods in order to dislodge the mucus, and hence allow the sufferer to expel it. It was postulated a number of years ago that treatment of the infected lung with a DNase preparation might alleviate such respiratory symptoms, by catalysing the degradation of the extracellular DNA and hence promoting a reduction in mucus viscosity.

DNase I is one of the most prominent and well-characterized members of the DNase family. This endonuclease catalyses the hydrolysis of internal phosphodiester linkages at the $P-O(3)$ bond (Figure 6.9). Its preferred substrate is double stranded DNA, although it will also degrade single stranded DNA, albeit more slowly. Bovine DNase I was first (partially) purified from pancreatic extracts in the late 1940s. Experiments over the subsequent decades clearly illustrated the enzyme's ability to significantly reduce the viscosity of the CF patient's lung mucus *in vitro*, and bovine pancreatic DNase was approved for the treatment of CF in the USA in the 1950s. While initially well tolerated, adverse reactions were reported upon its prolonged administration. In part this was due to the enzyme's bovine origin (and hence its immunogenicity in man). However, the major cause of such adverse reactions was most likely the presence of proteolytic contaminants in the product (these preparations were subsequently shown to contain up to 2 per cent trypsin and chymotrypsin).

The nucleotide sequence coding for human DNase I was first isolated from a pancreatic cDNA library in the 1980s. It codes for a 260 amino acid glycoprotein which shows high (77 per cent) homology to bovine DNase I. The cDNA has been expressed in recombinant CHO cell lines, and DNase produced from this source (trade name 'Pulmozyme') has been approved for medical use.

Glucocerebrosidase

Gaucher's disease is a hereditary metabolic disorder caused by a lack of the glycosylated enzyme β-glucocerebrosidase. Like CF, this condition is recessive. The enzyme's native substrates are glucocerebrosides, glycolipids normally found in small quantities in certain cell types. Absence of the enzyme is characterized by the intracellular accumulation of these lipids, particularly in macrophages. Clinical manifestations of the disease include swollen organs (due to accumulation of affected macrophages therein), anaemia, bone pain, and sometimes neuronal damage.

The disease may be treated by enzyme replacement therapy, and glucocerebrosidase purified from human placentae may be used for this

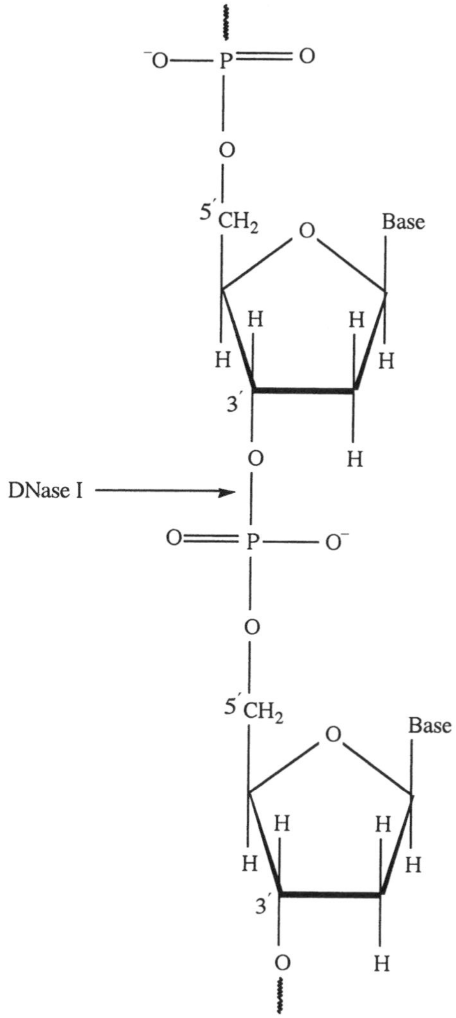

Figure 6.9 Cleavage of DNA by DNase I. Hydrolysis generates two fragments, one with a free 3′ hydroxy terminus, the other with a free 5′ phosphate terminus

purpose. The enzyme is normally administered to sufferers by intravenous injection every 2 weeks. When administered intravenously, a proportion of the enzyme will be taken up by macrophages, as the latter express a range of cell-surface lectins. The lectins bind various glycoproteins avidly, and the resultant complexes are often internalized via an endocytotic mechanism. Concerns relating to cost, source availability and accidental transmission of disease hastened development of a recombinant glucocerebrosidase product. 'Cerezyme' is trade name of such a recombinant product (produced in an engineered CHO cell line) which gained regulatory approval for general medical use in 1994.

FURTHER READING

Books

Aslam, M. (1997) *Bioconjugation*. MacMillan Press, Basingstoke.

Bland, J. (1986) *Digestive Enzymes*. Keats Publishers, New Canaan.

Essand, M. (1995) *Radioimmunotargeting of Cancer Cells*. Acta Universitatis Upsaliensis, Upsaala.

George, A. (2000) *Diagnostic and Therapeutic Antibodies*. Humana Press, New York.

Goding, J. (1996) *Monoclonal Antibodies. Academic Press*, London.

Gosling, G. (1993) *Immunotechnology*. Portland Press, London.

Grossbard, M. (1998) *Monoclonal Antibody-based Therapy of Cancer*. Marcel Dekker, New York.

King, D. (1998) *Applications and Engineering of Monoclonal Antibodies*. Taylor & Francis, London.

Lauwers, A. & Scharpe, S. (1997) *Pharmaceutical Enzymes*. Marcel Dekker, New York.

McCafferty, J. *et al.* (1996) *Antibody Engineering*. IRL Press, Oxford.

Nezlin, R. (1998) The *Immunoglobulins*. Academic Press, London.

Oxender, D. (1999) *Novel Therapeutics from Modern Biotechnology*. Springer-Verlag, Godalming.

Shepard, P. (2000) *Monoclonal Antibodies*. Oxford University Press, Oxford.

Articles

Antibodies

Ansell, P. (2000) Hybridoma technology: a view from the patent arens. *Immunol. Today* **21** (8), 357–358.

Choy, E. *et al.* (1995) Therapeutic monoclonal antibodies. *Br. J. Rheumatol.* **34**, 707–715.

Clark, M. (2000) Antibody humanization: a case of the Emperor's new clothes? *Immunol. Today.* **21** (8), 397–402.

Farah, R. *et al.* (1998) The development of monoclonal antibodies for the therapy of cancer. *Crit. Rev. Eukaryotic Gene Exp.* **8** (3&4), 321–356.

Fischer, R. *et al.* (2000) Antibiotic production by molecular farming in plants. *J. Biol. Regulators Homeos. Agents* **14** (2), 83–92.

Harris, W. (1994) Humanizing monoclonal antibodies for *in vivo* use. *Anim. Cell Biotechnol.* **6**, 259–277.

Hudson, P. (1998) Recombinant antibody fragments. *Curr. Opin. Biotechnol.* **9**, 395–402.

Hudson, P. (1999) Recombinant antibody constructs in cancer therapy. *Curr. Opin. Immunol.* **11** (5), 548–557.

Hudson, P. (2000) Recombinant antibodies: a novel approach to cancer diagnosis and therapy. *Exp. Opin. Invest. Drugs.* **9** (6), 1231–1242.

Little, M. *et al.* (2000) Of mice and men: hybridoma technology and recombinant antibodies. *Immunol. Today* **21** (8), 364–370.

Multani, P. & Grossbard, M. (1998) Monoclonal antibody based therapies for hematologic malignancies. *J. Clin. Oncol.* **16** (11), 3691–3710.

Pluckthun, A. & Pack, P. (1997) New protein engineering approaches to multivalent and bispecific antibody fragments *Immunotechnology* **3**, 83–105.

Russell, C. & Clarke, L. (1999) Recombinant proteins for genetic disease. *Clin. Genet.* **55** (6), 389–384.

Stein, K. (1997) Overcoming obstacles to monoclonal antibody product development and approval. *Trends Biotechnol.* **15**, 88–90.

Verma, R. *et al.* (1998) Antibody engineering: comparison of bacterial, yeast, insect and mamallian expression systems. *J. Immunol. Methods* **216**, 165–181.

Therapeutic enzymes

Estlin, E. *et al.* (2000) The clinical and cellular pharmacology of vincristine, cortosteroids, L-asparaginase, anthracyclines and cyclophosphamide in relation to childhood acute lymphoblastic leukaemia. *Br. J. Haematol.* **110** (4), 780–790.

Koch, C. & Holby, N. (2000) Diagnosis and treatment of cystic fibrosis. *Respiration* **67** (3), 239–247.

Liang, L. & Yang, V. (2000) Biomedical applications of immobilized enzymes. *J. Pharm. Sci.* **89** (8), 979–990.

Mates, J. *et al.* (1999) Antioxidant enzymes and human diseases. *Clin. Biochem.* **32** (8), 595–603.

McIntyre, A. (1999) Dornase alpha and survival of patients with cystic fibrosis. *Hosp. Med.* **60** (10), 736–739.

Pui, C. (2000) Acute lymphoblastic leukaemia in children. *Curr. Opin. Oncol.* **12** (1), 3–12.

7

Hormones and growth factors used therapeutically

INTRODUCTION

A number of hormone preparations have been used clinically for many decades. While some (e.g. adrenaline, peptide hormones and many steroid hormones) may be prepared synthetically, protein hormones used medically were (initially at least) extracted directly from biological source material. Problems associated with source availability and accidental transmission of disease hastened the development of recombinant forms of many such hormones (Table 7.1). In addition, protein engineering has facilitated the development of modified forms of such polypeptide hormones and several such engineered products have recently gained approval for general medical use. The major polypeptide hormones used clinically include insulin (and, to a lesser extent, glucagon), human growth hormone, the gonadotrophins and erythropoietin. These form the basis for most of this chapter.

Table 7.1 Recombinant hormone preparations now approved for general medical use

Product trade name	Description	Major therapeutic use	Year (& region) first approved
Insulins			
Humulin	rh Insulin	Diabetes mellitus	1982 (USA)
Novolin	rh Insulin	Diabetes mellitus	1991 (USA)
Humalog	rh nsulin lispro (modified insulin)	Diabetes mellitus	1996 (USA & EU)
Liprolog	rh Insulin Lyspro (modified insulin)	Diabetes mellitus	1997 (EU)
Insuman	rh Insulin	Diabetes mellitus	1997 (EU)
NovoRapid	rh Insulin aspart (modified insulin)	Diabetes mellitus	1999 (EU)
Glucagon preparations			
Glucagen	rh glucagon	Hypoglycaemia	1998 (USA)
Glucagon for injection	rh glucagon	Hypoglycaemia	1998 (USA)
Growth hormones			
Protropin	rh growth hormone (rhGH)	hGH deficiency in children	1985 (USA)
Humatrope	rhGH	hGH deficiency in children	1987 (USA)

Table 7.1   (*continued*)

Product trade name	Description	Major therapeutic use	Year (& region) first approved
Nutropin	rhGH	hGH deficiency in children	1994 (USA)
Biotropin	rhGH	hGH deficiency in children	1995 (USA)
Genotropin	rhGH	hGH deficiency in children	1995 (USA)
Norditropin	rhGH	hGH deficiency in children	1995 (USA)
Nutropin AQ	rhGH	hGH deficiency in children	1995 (USA)
Saizen	rhGH	hGH deficiency in children	1996 (USA)
Serostim	rhGH	Treatment of AIDS-associated wasting	1996 (USA)
Gonadotrophins Gonal F	Follicle-stimulating hormone (rhFSH)	Anovulation and superovulation	1995 (EU)
Puregon	rhFSH	Anovulation and superovulation	1996 (EU)
Erythropoietins Epogen	Erythropoietin (rhEPO)	Anaemia	1989 (USA)
Procrit	rhEPO	Anaemia	1990 (USA)
Neorecorman	rhEPO	Anaemia	1997 (EU)

rh, recombinant human.

More recently the clinical potential of various growth factors has been realised (Table 7.2). Recombinant platelet-derived growth factor is the first such product to gain regulatory approval (in 1997 for the treatment of diabetic ulcers). Well over a dozen additional growth factor based products remain in clinical trials, and the most prominent of these are discussed in the final section of this chapter.

INSULIN

Insulin is a polypeptide hormone that is produced by the β cells of the pancreatic islets of Langerhans. The hormone was first isolated in 1921, and its amino acid sequence was determined in the late 1950s. Insulin exerts a wide variety of metabolic effects. It plays a central regulatory role in the metabolism of carbohydrates, as well as in the metabolism of protein and lipid. The level of secretion of insulin from the β islet cells

Table 7.2 Some recombinant growth factors (and their intended applications) which are approved or undergoing clinical trials

Growth factor	Intended application(s)	Status
Platelet-derived growth factor	Diabetic/skin ulcers	Approved
Insulin-like growth factors	Amyotrophic lateral sclerosis, burns, hip fractures, type II diabetes	In trials
Transforming growth factors	Diabetic foot ulcers, impaired wound healing	In trials
Epidermal growth factor	Wound healing, skin ulcers, corneal/cataract surgery	In trials
Vascular endothelial growth factor	Cardiovascular disorders, coronary artery disease	In trials
Keratinocyte growth factor	Wound healing, various dermatological conditions	In trials
Neurotrophic factors	Diabetic peripheral neuropathy, amyotrophic lateral sclerosis, various neurological conditions	In trials
Fibroblast growth factor	Chronic soft tissue ulcers, leg and foot ulcers, venous stasis.	In trials

is primarily determined by blood glucose levels. Increases in the concentration of blood glucose induces insulin secretion which promotes the uptake of glucose by a number of tissues, particularly liver, and by muscle. This reduces blood glucose levels to normal values which in turn, decreases the rate of insulin release.

Diabetes mellitus

Failure to produce insulin results in the development of diabetes mellitus (type 1 diabetes), a chronic disease characterised by an elevated level of blood glucose and by the presence of glucose in the urine. Increased rates of glycogenolysis, gluconeogenesis, fatty acid oxidation, ketone body production and urea formation are also observed. Diabetes mellitus also results in a decrease in fatty acid and protein biosynthesis as well as a decreased uptake of glucose by peripheral tissues. Approximately 12 million people world wide suffer from diabetes mellitus, with an annual death rate directly attributable to this condition of some 600 000. Diabetes mellitus is usually caused by irreversible damage to the insulin-producing β pancreatic cells. The underlying molecular events inducing such damage remain to be fully characterized, although autoimmunity, viral infection and genetic pre-disposition are all believed to be contributory factors. Diabetes mellitus may be controlled by the parenteral administration of suitable insulin preparations. Insulin was first administered to diabetics in 1922, just one year subsequent to its initial isolation.

Insulin synthesis *in vivo*

Insulin is initially synthesized in the pancreatic β cells as preproinsulin (Figure 7.1). This molecule contains a 23 amino acid amino-terminal signal sequence, which directs the protein through the rough endoplasmic reticulum (ER) membrane to the lumen of the ER. Here, the leader sequence is removed from the preproinsulin molecule by a specific signal peptidase, yielding proinsulin. The proinsulin molecule remains within the lumen of the ER. Small vesicles containing proinsulin subsequently bud off from the ER and fuse with membranous structures termed Golgi apparatus. Proinsulin-containing vesicles, in turn, pinch off from the golgi apparatus. These vesicles are often termed coated secretory granules, as they exhibit a coat composed of a protein, clathrin, on their outer surface. As they move further away from the Golgi body, these vesicles loose their clathrin coat, thus forming non-coated secretory granules.

The conversion of proinsulin into insulin takes place in the coated secretory vesicles. This process involves the proteolytic cleavage of the proinsulin molecule, yielding mature insulin and the C or connecting peptide. The mature insulin consists of two polypeptide chains, the A and B chains, joined by two disulfide cross-links and has a molecular mass of 5.8 kDa. The insulin A chain normally contains 21 amino acid residues whereas the B chain consists of 30 residues. The A chain contains one intrachain disulfide linkage. Conversion of proinsulin to insulin generates a polypeptide sequence which originally bridged the insulin A and B chains of the proinsulin molecule. Upon the formation of mature insulin, two dipeptides are removed from either end of this bridge peptide yielding a slightly shorter peptide, termed C or connecting peptide.

In addition to mature insulin, secretory granules also contain low levels of proinsulin the C peptide and some proinsulin-derived amino acids. Non-coated secretory granules serve as the storage depot of insulin in the β cells. The insulin is normally stored as a hexamer, consisting of six molecules of insulin stabilized by two atoms of zinc. Indeed the addition of a small quantity of zinc to purified insulin results in the formation of characteristic rhombohedral crystals, with the basic crystal unit consisting of the zinc-containing insulin hexamer. The secretory granules release their contents into the blood by the process of exocytosis. Insulin is released in this manner only upon stimulation by specific secretory signals, the most significant of which is an increase in the blood glucose concentration.

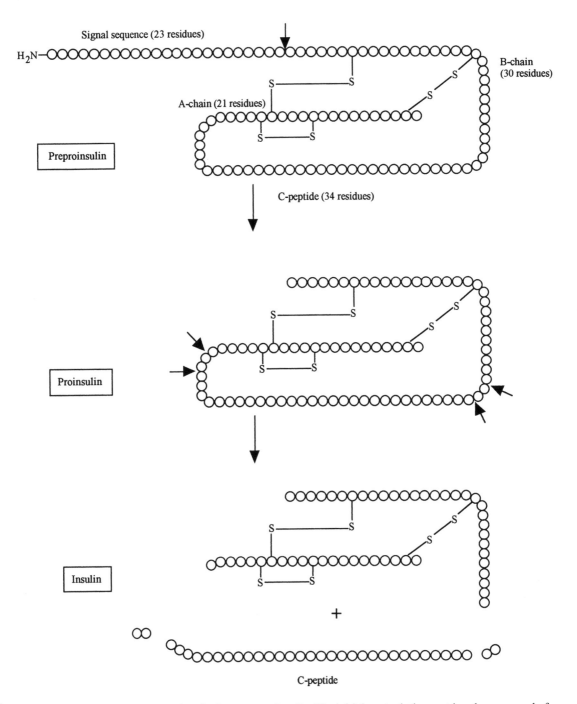

Figure 7.1   Synthesis of human insulin from preproinsulin. The initial proteolytic event involves removal of a 23 amino acid signal sequence from the amino terminal end of preproinsulin, thus yielding proinsulin. Proinsulin is converted into insulin by additional proteolytic events, resulting in the generation not only of insulin but also a 30 amino acid sequence (termed C or connecting peptide) and two dipeptide moieties.

Insulins obtained from various different species are similar, though not identical, in their overall amino acid sequence. Rats and mice synthesize two insulin types which differ slightly from each other in amino acid sequence. Porcine insulin differs in sequence from human insulin by only one amino acid, the C terminal residue of the B chain. Bovine insulin differs from the human hormone by three amino acids while sheep insulin differs by four amino acids. The amino acid sequence of the insulin C peptide, however, differs greatly from species to species.

Conventional insulin preparations

Insulin preparations administered to diabetic patients have traditionally been obtained by direct extraction from the pancreas of healthy slaughterhouse animals, such as pigs or cattle. Porcine insulin is considered by some to have a slight clinical advantage over bovine preparations, as its amino acid sequence is more closely related to the native human molecule. For this reason it would likely be marginally less immunogenic when compared to the bovine product.

Purified insulin was first used therapeutically in the early 1920s. Such preparations were generally extracted from the pancreas of pig or ox by a technique employing acid–alcohol precipitation. The insulin preparations obtained were quite impure. Although insulin was first crystallized in 1926, the crystallization process was poorly understood, and researchers found it impossible to consistently crystalize insulin from crude extracts. The discovery in 1934 that zinc promoted crystallization of insulin allowed commercial producers to grow insulin crystals from crude extracts of pancreatic tissue. The crystals were then isolated and subject to recrystallization in order to further purify the protein. Such insulin preparations are often termed conventional insulins. Although the crystallization process resulted in significant purification of the insulin, such preparations are considered to be relatively crude by modern standards.

In addition to insulin itself, conventional insulin preparations also contain modified insulin molecules such as arginine insulin, desamidoinsulin, insulin esters and insulin dimers. Desamidoinsulins or their ethyl esters are formed during the acid–alcohol extraction of insulin. Under such conditions the amido groups of asparagine residues may be hydrolysed. Conventional insulin preparations also contain appreciable quantities of proinsulin in addition to other polypeptides such as vasoactive intestinal peptide (VIP), somatostatin and glucagon. A high molecular

---

Mature insulin thus consists of two polypeptides, the A and B chains. The B chain contains 30 amino acid residues whereas the A chain contains 21 amino acids. The chains are covalently linked via two interchain disulfide linkages. One intrachain disulfide linkage is also present in the A chain

mass protein fraction is observed on occasions, although the recrystallization process effectively fractionates the insulin from such molecules.

Contaminants present in conventional insulin preparations can adversely affect the insulin content of the final preparation, and may also lead to clinical complications upon its administration. Many of the higher molecular mass contaminants present display protease activity. Preparations containing such unwanted activities are normally maintained in solution at acidic pH values in order to minimize proteolysis of the insulin molecule. Contaminants present are also generally immunogenic. This fact is particularly relevant when one considers that many diabetics require insulin injections several times daily throughout their life. Repeated administrations could promote a strong immunological response, even with weak immunogens. Immunological reactions may also lead to inflammation and destruction of the tissue surrounding the site of injection.

Insulin obtained from cattle, and in particular from pigs elicits only weak immunological responses when administered to humans. Nonetheless, anti-insulin antibodies are often detected in the serum of diabetics. The presence of such neutralizing antibodies may, over time, render necessary the administration of increased doses of insulin in order to achieve the desired biological effect. Furthermore, as antibody-bound insulin is largely resistant to the normal insulin degradative processes, such antibodies may seriously effect the activity-versus-time profile of the administered hormone.

Chromatographically purified insulin

Recrystallized insulin preparations are now usually subjected to additional chromatographic purification steps, in order to further reduce the level of the various contaminants associated with conventional insulin preparations. Gel filtration was the first chromatographic technique used in the further purification of insulin. Such a gel filtration step effectively separates insulin from higher molecular mass contaminants such as proteolytic enzymes, proinsulin and insulin dimers. Contaminants co-eluting with the insulin fraction include desamino insulin and arganine insulin, which tend to be of limited clinical significance. Other contaminants which gel filtration fails to remove include a variety of pancreatic peptides of similar molecular mass to that of the insulin. As insulin is eluted from gel filtration columns in one peak, preparations purified by this method are sometimes referred to as single-peak insulins.

In order to be economically viable, the gel filtration columns used to purify recrystalized insulin must be quite large. Some manufacturers produce a system of stack columns for this purpose (Figure 7.2). The total bed volume of the columns shown in Figure 7.2 is 96 l. A sample volume of 2 l may be applied at any one time, from which 50 g of purified insulin may be obtained. A single column run takes approximately 7 h.

Figure 7.2 An automated system for production scale insulin purification. Capacity is 600 g common recrystallized insulin per week. Equipment (from left to right): 300 l stainless steel buffer tank, 50 l conical stainless steel sample tank, control cabinet, six stack columns, and two collection tanks. The columns can also be arranged in two three-section stacks placed side by side. (Photo courtesy of Pharmacia)

Additional chromatographic steps may be used to further reduce the level of contaminants present in the purified insulin preparations. Process-scale ion exchange chromatographic columns are often utilised for this purpose.

It has been recommended that the proinsulin content of modern insulin preparations should not exceed 10 p.p.m., while the pancreatic polypeptide content should not exceed 1 p.p.m. It is also recommended that the content of high molecular mass protein contaminants should not exceed 10 parts per thousand or 1 per cent.

Human insulin preparations

Over the past three decades methods have been developed which allow the production of relatively large quantities of human insulin. Some methods, such as direct chemical synthesis of the insulin molecule, while possible, are economically unattractive. Other methods however have proven to be both economically viable and technically sound. Such methods include the enzymatic modification of porcine insulin and the expression of the human insulin cDNA in recombinant microorganisms.

Human insulin was first produced by enzymatic modification of porcine insulin in the early 1970s. Human insulin differs from porcine insulin by a single amino acid. Threonine forms the carboxyl terminus (residue number 30) of the human insulin B chain, whereas an alanine residue is found in this position in the porcine molecule. Treatment of intact porcine insulin with trypsin results in the proteolytic cleavage of the B chain between residues number 22 and 23 (and also between residues number 29 and 30), effectively removing the carboxy terminal octapeptide from the B chain. A synthetic octapeptide whose sequence is identical to the analogous human octapeptide may then be coupled to the trypsin treated porcine insulin. This process, outlined in Figure 7.3, effectively converts

(a)

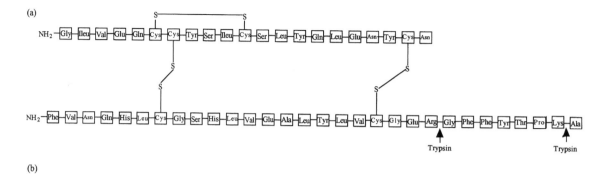

(b)

Figure 7.3   Amino acid sequence of porcine insulin. Trypsin cleavage sites are also indicated. Trypsin therefore effectively removes the insulin carboxy terminus B chain octapeptide. The amino acid sequence of human insulin differs from that of porcine insulin by only one amino acid residue. Porcine insulin contains an alanine residue at position 30 of the B chain whereas human insulin contains a threonine residue at that position. Insulin exhibiting a human amino acid sequence may thus be synthesized from porcine insulin by treating the latter with trypsin, removal of the C terminus fragments generated, and replacing this with the synthetic octapeptide shown in (b)

porcine insulin into human insulin. Human insulin produced by this means has been used clinically for a number of years.

Human insulin may also be produced by the application of recombinant DNA technology. Recombinant human insulin preparations represented the first commercial healthcare product produced by this technology to be approved for widespread clinical use. The successful production of human insulin in microorganisms ensures that future insulin supplies will not depend on the availability of pancreatic tissue from slaughterhouse animals. The quantity of purified insulin obtained from the pancreas of one pig would satisfy the insulin requirement of one diabetic patient for only 3 days. Production in recombinant organisms also eliminates the risk of accidental transmission of viral disease from infected pancreatic tissue to diabetics.

The first insulin preparation obtained by methods of genetic engineering was produced by inserting the cDNAs coding for the insulin A and B chains into individual *E. coli* cells (strain K12). The A and B chains were then purified from the different recombinant systems and subsequently joined by disulfide bond formation. Human insulin may also be produced by expressing the nucleotide sequence coding for human proinsulin in *E. coli*. The proinsulin molecule may subsequently be converted into native insulin by the enzymatic removal of the connecting peptide. This method of production has become somewhat more popular than the method relying on the initial independent production of the two insulin peptide chains in two separate bioreactors. This is largely due to the fact that only a single fermentation and subsequent purification scheme is necessary. Recombinant human insulin is now produced industrially on a large scale, with 50 000 l fermenters being used in routine production runs. Human insulin has also been produced in recombinant yeast systems.

Although the insulin produced by recombinant methods is identical to that of the native hormone, any impurities present in the recombinant preparations will be of microbial rather than animal origin. However, modern protein purification methods minimize the level of any such impurities found in the final product. Insulin extracts obtained from animal sources would of course also be heavily contaminated with non-insulin protein impurities prior to its purification. The insulin-producing β cells of the islet of Langerhans constitute less than 1 per cent of total pancreatic tissue.

A number of additional techniques may be employed to increase still further the purity of insulin preparations. Such techniques include hydrophobic interaction chromatography and reverse-phase HPLC (RP-HPLC). These additional techniques, which serve to complement rather than replace existing purification techniques, have been discussed in Chapter 3. Industrial scale purification of biosynthetic human insulin could thus consist of the series of steps outlined in Figure 7.4.

While the introduction of additional purification steps decreases the overall insulin yield, they serve to increase the purity of the final product. Purity levels approaching or exceeding 98 per cent, with an acceptable overall yield of insulin are now attainable. RP-HPLC represents a particularly powerful purification step. This chromatographic technique is capable of effectively separating the insulin molecule from a wide range of contaminants, many of which are quite difficult to remove by conventional low-pressure chromatographic procedures. Large-scale RP-HPLC columns, with bed volumes approaching 80 l, have been built to facilitate industrial-scale insulin purification. Such columns may be employed to purify several hundred grams of insulin in a single run.

Both HPLC and FPLC are also used extensively on an analytical scale during the production of modern high purity insulin. These techniques play a significant role in in-process quality control, and in the assessment of final product purity. The speed and sensitivity with which these techniques function render them particularly suitable to such tasks. Analytical-scale RP-HPLC, for example, can distinguish between various insulin molecules which differ by a single amino acid residue. It can also detect various modified forms of insulin such as desamido or formyl insulin, or insulin polymers.

Insulin formulations

When injected directly into the bloodstream, insulin has a half life in the order of minutes. Clearly such a short half life would render the clinical management of diabetes extremely difficult. Administration by

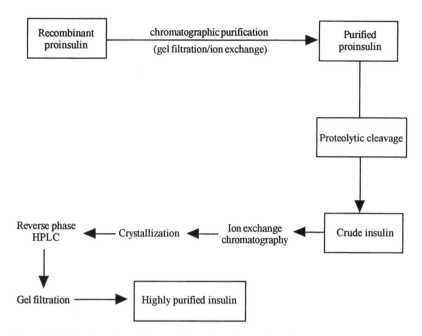

Figure 7.4 Purification of recombinant insulin produced as proinsulin in *E. coli*. Initially the proinsulin is purified chromatographically and subsequently converted into insulin by proteolytic cleavage. The insulin may be further purified by ion exchange chromatography followed by a crystallization step. The redissolved crystals can be applied to a reverse phase HPLC column. A gel filtration step is normally utilized to produce final product

subcutaneous injection facilitates a more prolonged release of hormone from the site of injection into the bloodstream. This has become the preferred method of insulin administration. Insulin preparations may also be formulated in a number of ways in order to increase their duration of action. The hormone may, for example, be complexed with a protein such as protamine which further retards its release into the general circulation. Protamines are a group of basic proteins found in association with nucleic acids in the sperm of certain species of fish. Alternatively, zinc may be added to the final preparation in order to promote the growth of insulin crystals in the resultant zinc–insulin suspension. Soluble insulins are often referred to as short-acting insulins, having a duration of activity of up to 8 h. Peak hormonal activity is observed in such cases after 3–4 h. Long-acting insulin preparations, such as those containing protamine and/or zinc, often exhibit a duration of action of up to 36 h.

Engineered insulins

Native insulin molecules tend to spontaneously interact, forming oligomers (dimers and particularly hexamers). Self association is mainly due to the formation of interchain hydrogen bonds between amino acid residues residing towards the C terminus of the insulin B chains of adjacent insulin molecules. When administered subcutaneously, insulin oligomers must dissociate from each other before individual insulin molecules can leak from the site of injection into the blood. This disassociation appears to be the rate limiting step of insulin absorption. As a consequence diabetics must self-administer insulin preparations at least 30–40 min prior to eating. This is sometimes inconvenient, and can lead to hypoglycaemia if the person subsequently skips the meal for any reason.

Altering the sequence of amino acids found towards the C terminus of the insulin B chain can reduce the propensity of these molecules to self associate. In this way faster acting insulins can be generated. Insulin lispro and insulin aspart represent two such faster-acting insulins now approved for general medical use. These modified insulins are self-administered by diabetics immediately prior to consuming a meal.

Native human insulin contains a proline and a lysine residue respectively at positions B 28 and B 29 (Figure 7.3). Insulin lispro differs in that the positioning of these two amino acids have been reversed. Insulin aspart differs from native insulin in that the proline residue at position B 28 is replaced by an aspartic acid.

GLUCAGON

Glucagon is a polypeptide hormone synthesized in the pancreas by the α cells of the islets of Langerhans. It is also synthesized by related cell types found in the gastrointestinal tract. Glucagon consists of 29 amino acids and has a molecular mass of approximately 3.5 kDa. The amino acid sequence of the hormone isolated from different animal species is almost invariably the same. As in the case of insulin, glucagon seems to be synthesized initially as proglucagon. Proglucagon is converted to glucagon by two separate proteolytic events.

Glucagon is a hyperglycaemic hormone, inducing an increase in the blood glucose concentration of recipients. In this way it opposes one of the major physiological actions of insulin. The increase in blood glucose is promoted by stimulating an increase in the rate of gluconeogenesis and glycogen breakdown in the liver. It also promotes an increase in the rate of lipolysis and a decrease in the rate of glucose utilization by adipose tissue and muscle. Glucagon is occasionally used clinically in order to reverse insulin-induced hypoglycaemia in diabetic patients.

Although glucagon obtained by direct extraction from pancreatic tissue has been available for many years, more recently recombinant glucagon products have come on the market (Table 7.1). In addition to their therapeutic use in counteracting hypoglycaemia these products may also be used as a diagnostic aid in radiological examinations of the stomach. The hormone relaxes smooth muscle in the stomach wall. By reducing stomach contents motility improved images may be obtained during radiological examinations.

GONADOTROPHINS

Gonadotrophic hormones, as the name suggests, exert their primary effect on the male and female gonads. A number of gonadotrophins may be used clinically in the treatment of various human clinical conditions and to induce a superovulatory response in female animals.

The most commonly used gonadotrophins include follicle-stimulating hormone (FSH), and luteinizing hormone (LH), which is also referred to as interstitial cell stimulating hormone (ICSH). Both LH and FSH are synthesized by the pituitary. Another gonadotrophin, human chorionic gonadotrophin (hCG) is produced by the placenta of pregnant women. Yet an additional gonadotrophic hormonal preparation, termed menotrophin, may be isolated from the urine of post-menopausal women. Pregnant mare serum gonadotrophin represents a gonadotrophic hormone which is found exclusively in the serum of pregnant mares.

FSH, LH and hCG

FSH, LH and hCG are glycoproteins of molecular mass 34 , 28 and 36 kDa, respectively. Both LH and FSH consist of approximately 16 per cent carbohydrate while hCG contains over 30 per cent carbohydrate. All are dimers composed of an α and β subunit. In any one animal species, the α subunit of all three is identical, whereas the β subunit differs. It is thus the β subunit which confers on the particular molecule its distinct biological characteristics. The β subunit of hCG exhibits significant amino acid sequence homology with the β subunit of LH. It is therefore not surprising that hCG displays similar biological activities to LH.

In the male, FSH stimulates the production of spermatocytes by the Sertoli cells of the seminiferous tubules. LH stimulates production of testosterone, the principal male sex hormone, by the Leydig cells of the testes. In the female, FSH and LH play critical roles in regulating the reproductive cycle.

The ovary represents the primary female reproductive organ. It houses egg cells (ova) and synthesizes a range of steroid hormones. The ovary contains numerous follicles, each follicle consists of a single egg cell (ovum) surrounded by two layers of cells. The inner layer of cells are termed granulosa cells and these respond primarily to FSH. The granulosa cells produce oestrogens, the major female sex hormones. The outer follicular layer is composed of theca cells. These cells, which fall primarily under the hormonal influence of LH, synthesize a variety of steroids which are subsequently utilized by the granulosa cells during synthesis of oestrogen.

At the commencement of a normal menstrual cycle a group of follicles begin to mature, a process largely stimulated by FSH. Shortly thereafter a single dominant follicle emerges, and the remaining follicles regress. This first half of the menstrual cycle is often referred to as the follicular phase. The growing follicle begins to secrete increasing levels of oestrogens which ultimately trigger a surge in the concentration of LH at mid-cycle, approximately day 14 of the human female cycle. This, in turn, triggers ovulation (i.e. the release of the egg from the mature follicle), with the resultant conversion of the ruptured follicle into a structure known as the corpus luteum. The menstrual cycle now has entered its second phase, also known as the luteal phase. The corpus luteum secretes oestrogens and the steroid hormone progesterone. Progesterone serves to prepare the lining of the uterus, the endometrium, for implantation of the fertilized ovum, and is required to support the growing embryo.

If fertilization does not occur, the maximum life span of the corpus luteum in humans is normally 14 days, during which time it steadily regresses. This in turn results in a decrease in the levels of the corpus luteal hormones, oestrogen and progesterone. Withdrawal of such hormonal support results in the shedding of the endometrial tissue, which is discharged from the body during menstruation.

If follicular rupture is followed by fertilization of the released ovum the corpus luteum does not regress, and continues to produce progesterone. In the pregnant female, the corpus luteum is maintained by human chorionic gonadotropin, hCG, which is secreted by the placenta. hCG is also found in urine, and detection of this hormone forms the basis of pregnancy detection kits (Chapter 9).

FSH and LH essentially regulate the reproductive process in females (Figure 7.5). In addition, FSH and LH also regulate the development and maintenance of male fertility. These hormones may be employed therapeutically to treat some forms of sterility, or other medical conditions which are caused by low circulatory levels of the hormones. Approximately 10 per cent of reproductive age couples in the USA are infertile. In six to seven out of every 10 such cases the underlying problem lies with the female. Causes of female infertility are diverse, but at least 20 per cent of cases are attributable to a dysfunction of the hypothalamic–pituitary–ovarian axis (Figure 7.6). Many of these cases are responsive to administration of exogenous FSH and LH.

Sources of FSH, LH and hCG

FSH may be isolated from the pituitary gland. Supply of human FSH from such sources is obviously very limited and is no longer used due to the potential for accidental transmission of pathogens. FSH is also found in the urine of post-menopausal women. Preparations obtained from this source also contains LH activity, and are termed menotropin. Menotropin is purified from urine by a variety of fractionation procedures, as well

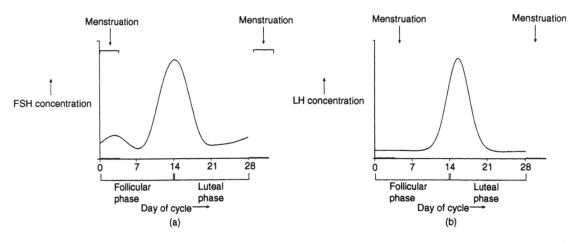

Figure 7.5   Changes in plasma FSH (a) and LH (b) levels during the menstrual cycle of a healthy human female

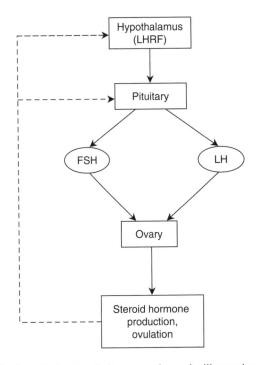

Figure 7.6   The hypothalamic–pituitary–ovarian axis, illustrating the interaction between/effects of various reproductive hormones. Dashed lines represent negative feedback mechanisms

as by chromatographic techniques such as ion exchange chromatography. Menotropin preparations may be subjected to further chromatographic steps in order to remove most or all of the contaminating LH activity.

Menotropin is often used clinically to treat a variety of reproductive complications such as anovulatory infertility, which is caused or exacerbated by low levels of circulating FSH.

Human luteinizing hormone may also be isolated from the pituitary. As in the case of FSH, this does not represent a commercially viable source of this hormone. hCG, as mentioned earlier, exhibits many biological activities which are similar if not identical to that of LH. hCG may be purified relatively easily from the urine of pregnant women. Such hCG preparations have found widespread medical application in the treatment of conditions caused by lack of LH activity and it has been employed clinically to treat infertility and delayed puberty in males. It is also administered to females together with FSH in the treatment of anovulatory infertility. In such cases, the administered FSH stimulates development of the follicle while subsequent administration of LH initiates the mid-cycle LH surge, stimulating final maturation and subsequent rupturing of the follicle.

Recombinant FSH, LH and hCG have all been produced, and two recombinant human FSH (rhFSH) preparations have gained regulatory approval for medical use (Table 7.1). These products are both produced by co-expression of the FSH α and β genes in CHO cell lines. The resulting products exhibit amino acid sequences identical to that of the native human protein, although their glycosylation patterns varies somewhat as compared to urinary derived product.

Pregnant mare serum gonadotrophin

Pregnant mare serum gonadotrophin (PMSG), as the name suggests, is obtained from the serum of pregnant horses. It is present in serum at elevated levels between days 40 and 130 of gestation. PMSG is a gonadotrophic hormone consisting of α and β subunits and containing up to 45 per cent carbohydrate, most of which is associated with the β subunit. The carbohydrate component is particularly high in galactose, glucosamine and sialic acid.

PMSG is a unique gonadotrophin insofar as it exhibits both FSH and LH-like bioactivities. Its biological specificity is conferred upon PMSG by its β subunit. PMSG is secreted by a series of small cup-shaped outgrowths found in the horn of the pregnant uterus. These structures, which are highly enriched with PMSG, are termed endometrial cups and are found only in equines. They first become visible on day 37–40 of gestation, and reach maximum size by approximately day 70, after which they undergo steady regression. The endometrial cups are fetal rather than maternal in origin.

PMSG may be purified from serum by a procedure involving pH fractionation, alcohol precipitation, and several chromatographic steps utilising both gel filtration and ion exchange media. This hormone fails to enjoy widespread clinical use in humans but it is used to induce a superovulatory response in certain animals.

Superovulation in animals

Gonadotrophic hormones are routinely used to induce superovulation in certain animals, most notably cattle. The aim of superovulation is to induce the simultaneous growth of multiple follicles such that several eggs are available for fertilization at the time of mating. After mating the embryos may be collected from the mother by surgical or non-surgical techniques. These harvested embryos may then be implanted in several recipient animals which carry the offspring to term. By employing such technologies the reproductive potential of very valuable animals, or animals with desirable genetic traits is in effect multiplied several-fold. Superovulation and embryo transfer is used by some farmers to genetically improve their herd at an accelerated pace.

Superovulation basically involves administration of exogenous gonadotrophins in order to stimulate follicular growth. In many agriculturally important animal species, only a single follicle at a time develops to maturity. Injection of exogenous FSH increases its circulatory levels well above its normal physiological range. This disturbs the natural hormonal balance which regulates the menstrual cycle, and can result in the development and maturation of several follicles. Gonadotrophins administered in order to induce superovulation include menotrophin, porcine pituitary FSH (P-FSH) or pregnant mare serum gonadotrophin (PMSG).

P-FSH is FSH purified from pituitary glands excised from slaughterhouse pigs. Porcine as opposed to bovine pituitaries are normally employed, as the bulk of the purified product is used to superovulate cattle. The use of FSH preparations obtained from species other than the species of the intended recipient is normally encouraged as it decreases the potential risk of transmission of species-specific diseases associated with the use of biological products obtained from slaughtered animals.

FSH is usually purified from porcine pituitary extracts by a variety of fractionation techniques, including precipitation with salts and/or alcohol. Chromatographic steps are frequently employed. Most FSH preparations available commercially contain several contaminating proteins. The most consistent superovulatory response is obtained when FSH preparations exhibiting a low LH content are used in superovulatory experiments.

Superovulatory regimes employing PMSG normally involve the administration of a single does of this hormone. One dose is sufficient as PMSG is removed from circulation at a very slow rate. In cattle, clearance of PMSG may take up to 120 h. The extended half life of PMSG preparations is due to their high content of *N*-acetyl neuraminic acid. The long half life of PMSG is regarded by some as detrimental to its effectiveness as a superovulatory hormone as it can lead to continued stimulation of follicular growth after ovulation, and lead to a reduced number of recoverable viable embryos. Antibodies raised against PMSG may be administered in order to mop up residual PMSG activity after an appropriate time period. It is now generally accepted that a greater ovulatory response is obtained by administration of FSH rather than PMSG.

Most superovulatory regimes involve administration of FSH preparations twice daily for approximately 4 days. Regular injections of the hormone are required to sustain multiple follicular growth, as FSH has a relatively short half life in serum. An attempt is usually made to prolong the effectiveness of each injection by administering the preparation subcutaneously. This injection programme is subsequently followed by the administration of a single dose of LH, which is considered necessary to promote final maturation of the follicle and its subsequent release from the ovum.

Superovulatory responses may be quite variable even when FSH preparations are employed to induce follicular growth. The number of viable embryos recovered from a superovulated cow can vary between 0 and 50 or over, though five to eight embryos represent the average obtained. The variability associated with superovulatory regimes is caused by numerous factors, only some of which are clearly defined. The general health and condition of the animal influences the process as does the exact protocols used and the experience and technical ability of the personnel involved. The ratio of FSH : LH activity in the preparations administered also seems to effect the final outcome, with preparations of FSH containing reduced LH activity appearing to be most effective. Studies have shown that the FSH : LH ratio in various batches of one brand of (pituitary derived) FSH preparation could vary by up to 20-fold. Purified recombinant FSH preparations may prove useful in this regard.

The cDNAs for the α and β subunits of bovine FSH have been cloned and expressed in a mouse epithelioid cell line. Clones producing biologically active FSH were selected and their fermentation scaled up for routine production purposes. The resultant recombinant FSH may be subsequently subjected to a three-stage purification process, yielding a product which is greater than 95 per cent pure. Such recombinant preparations would of course be devoid of contaminating LH activity.

Inhibin

Inhibin is a glycoprotein hormone produced by the gonads. It is synthesized in the Sertoli cells of the testes and by granulosa cells in the ovary. The hormone is a dimer, consisting of α and β subunits. Two forms of the inhibin molecule have been characterized, inhibin A and inhibin B. The α subunit is identical in both cases while their respective β subunits differ in amino acid sequence.

The synthesis of inhibin is stimulated by increased circulatory concentrations of FSH. Inhibin plays an important role in feedback inhibition of FSH action, as it functions to selectively suppress the secretion of this hormone by the pituitary. Due to its ability to decrease circulatory FSH levels, inhibin is being investigated as a potential contraceptive for both males and females.

LIVERPOOL
JOHN MOORES UNIVERSITY
AVRIL ROBARTS LRC
TITHEBARN STREET
LIVERPOOL L2 2ER
TEL. 0151 231 4022

GROWTH HORMONE

Growth hormone (GH) is a species-specific hormone which is also known as somatotrophin. Somatotrophin is produced by the anterior pituitary and is required for normal body growth and lactation. The hormone promotes general protein synthesis and also plays a role in the regulation of carbohydrate and lipid metabolism, and it promotes bodily retention of various minerals and other elements essential for normal growth.

Secretion of growth hormone from the pituitary is stimulated by a hypothalamic regulatory factor termed growth hormone releasing factor (GHRF, see also Table 7.3). Growth hormone release is inhibited by a second hypothalamic factor termed growth hormone release-inhibiting hormone, or somatostatin.

Growth hormone initiates its anabolic effect by binding to specific cell-surface receptors. The receptor consists of three distinct domains, an extracellular ligand-binding domain, a transmembrane domain and an intracellular effector domain. The exact molecular mechanisms which mediate the effects of growth hormone upon binding to its receptor are poorly understood. A truncated form of the GH receptor, which approximates to the extracellular ligand-binding domain, is found in serum. This serum protein is capable of binding growth hormone and may play an important role in its clearance from the body (Figure 7.7).

Growth hormone receptors have been detected in a number of tissues. The hormone exerts its primary influence on the liver, where it stimulates the synthesis of somatomedin or insulin-like growth factor-1, (IGF-1). IGF-1 directly mediates most of the growth promoting effects of GH, as described later.

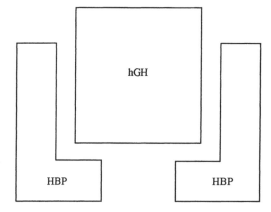

Figure 7.7 Binding of the human growth hormone (hGH) to two molecules of the hormone binding protein (HBP), as occurs naturally in serum. The hormone binding protein equates to the extracellular ligand binding domain of the hGH receptor located in the plasma membrane of responsive cells. Binding of the first molecule of HBP renders more efficient the binding of the second HBP molecule to the growth hormone

Insufficient production of growth hormone in humans leads to dwarfism. Excessive production of the hormone result in gigantism or acromegaly, a condition characterized by an increase in the size of the hands, feet and face. Growth hormone may therefore be employed clinically to treat hypopituitary dwarfism. The species-specific nature of GH makes it essential to use human growth hormone in human medicine.

Human growth hormone (hGH) is a polypeptide consisting of 191 amino acid residues. It contains two intrachain disulfide links and has a molecular mass in excess of 21 kDa. Until the mid 1980s the sole source of hGH was human pituitary glands and for this reason preparations of hGH were not widely available. Furthermore, such preparations were generally not homogeneous, containing contaminants including a number of modified forms of growth hormone in addition to unrelated proteins.

The clinical use of hGH extracted from human pituitaries was banned in 1985. In that year it was shown that several people suffering from Creutzfeldt–Jakob disease (CJD) had contracted the condition via hGH preparations contaminated with the CJD causative agent.

The expression of the cDNA for hGH in *E. coli* facilitated the large-scale production of ample quantities of growth hormone free of such infectious agents. The recombinant product is identical in sequence to the native molecule with the exception of an additional methionine residue, due to the presence of an AUG start codon at the 5' end of the cDNA. This methionyl-hGH exhibits identical biological activity to that of the native molecule and was first approved for clinical use in 1985. Its major application is in the treatment of short stature.

The recombinant growth hormone may be purified to homogeneity by a series of fractionation steps including ammonium sulfate precipitation, ion exchange chromatography on diethylaminoethyl cellulose and gel filtration chromatography on Sephacryl S-200. Immunoaffinity chromatography has also been used to purify both native and recombinant hGH. The ready availability of large quantities of recombinant growth hormone has accelerated investigations assessing its potential applications. Clinical evidence amassed to date suggests that the hormone may be of considerable value in treating burns, peptic ulcers, osteoporosis, hip fractures and some forms of cancer. In addition the general anabolic effects of this hormone may make it useful in inducing muscle growth, and perhaps even in counteracting some of the effects of ageing.

Bovine growth hormone

In addition to hGH, growth hormone obtained from a number of other species has also been produced in recombinant systems. Recombinant bovine growth hormone (rbGH) may be used to increase milk yield from dairy cattle. As was the case with human growth hormone, the advent of recombinant DNA technology allowed the production of industrially significant quantities of the recombinant bovine somatotrophin. Several biotechnology companies have since developed such bovine somato-trophin preparations for use with dairy cattle.

Treatment of dairy cattle with bGH is not considered to pose a threat to the health of persons consuming the milk or meat of such animals. This conclusion is based on a number of important experimental observations. Bovine growth hormone is not biologically active in humans. Furthermore, as is the case for most other proteins ingested, bGH would be degraded by the human gastrointestinal tract. Toxicity studies have shown that recombinant bGH is not orally active in rats, although this species is responsive to bGH when administered paren-terally.

Administration of bGH results in elevated concentrations of IGF-1, see later) in cows' milk. Bovine IGF-1 is identical to human IGF-1. However, toxicity studies have shown that bovine IGF-1 is inactivated when administered orally to rats. Furthermore, the concentration of IGF-1 reported to be present in the milk of cows treated with rbGH is no higher than the levels normally found in human breast milk.

ERYTHROPOIETIN

Erythropoietin (EPO) is a glycoprotein hormone produced by the kidneys. This hormone stimulates the production of red blood cells, the erythrocytes, from their precursor stem cells. Human erythropoietin has a molecular mass in the region of 36 kDa, 60 per cent of which is carbohy-drate. The hormone displays four potential glycosylation sites and the glycocomponent contains a high level of sialic acid and varying amounts of hexosamines and hexoses. Removal of the sugar residues dramatically decreases EPO's *in vivo* biological activity, apparently by promoting its rapid removal from plasma. EPO is found in plasma and low levels are found in urine, from which it may be purified.

Production of EPO *in vivo* is regulated by a number of factors, most notably by tissue oxygen tension. Its plasma concentration can increase by one 100-fold in highly anaemic individuals, due to insufficient oxygen supply to the tissues. EPO production can plummet as a result of certain conditions such as chronic renal failure. Such a decrease in EPO levels can in turn lead to the development of secondary anaemia. The resultant metabolic consequences of inadequate EPO production may be avoided by administration of exogenous EPO preparations.

EPO used clinically is produced by recombinant DNA technology. The cDNA coding for human EPO was first expressed as a heterologous protein product in 1984. Various companies have developed recombinant human EPO preparations which have gained regulatory approval (Table 7.1). The recombinant products prove indistinguishable from native EPO in terms of primary sequence and biological activity. They display clear efficacy in treating their target indications (anaemia and related conditions) and have a combined market value of approximately $2 billion. The recombinant products are produced in eukaryotic cell lines (mainly CHO cell lines) in order to facilitate glycosylation.

EPO is but one member of a large family of regulatory factors which stimulate the growth of blood cells. The various members of this family are collectively referred to as haematopoietic growth factors or haematopoietins. This family of regulatory proteins include not only EPO, but also colony-stimulating factors and various interleukins (see also Chapter 8)

OTHER GROWTH FACTORS

A number of other growth factors are currently generating considerable clinical interest (Table 7.2). One such product (recombinant human platelet derived growth factor, rhPDGF) has already been approved for general medical use. Various preparations of these factors are currently undergoing extensive clinical trials, and many are likely to be approved for general medical use in the near future. All of these growth factors exhibit marked mitogenic effects by inducing cellular proliferation and division. Most clinical trials concentrate on their effectiveness in accelerating the wound healing process and treating ulcers. An ulcer may be described as a break in the skin or a mucous membrane which fails to heal. Natural wound healing depends upon the release of growth factors which initiate the tissue repair process at the site of damage.

Thrombin, factor IIa of the blood coagulation pathway (Chapter 5), stimulates the release of α granules from blood platelets. These granules contain a variety of growth factors including fibroblast growth factor (FGF), PDGF and transforming growth factors (TGF). Such growth factors exert mitogenic effects on certain cell types essential for tissue repair. They also exhibit chemotactic effects, and in this way attract a number of cell types to the site of damage. Released TGF for example attracts macrophages to the area of damage. These macrophages then release a variety of additional factors including PDGF and TGF. EGF is mitogenic for epithelial cells which are essential for rapid wound healing.

Platelet-derived growth factor

PDGF, as the name suggests, was first isolated from platelets. Subsequently however it has been established that this growth factor is synthesized and secreted by a number of other cell types. PDGF represents a powerful mitogen for a variety of cell types, including fibroblasts, glial cells and smooth muscle cells. This growth factor has a molecular mass of 30 kDa, and consists of two polypeptide chains termed A and B, which are covalently linked by an interchain disulfide bond. Three isoforms of the growth factor have been isolated – the homodimers AA and BB in addition to the hetrodimeric form AB. The A chain consists of 124 amino acids whereas the B chain contains 140 amino acids.

Two distinct, although related PDGF receptors have also been identified. These have been termed α and β PDGF receptors, respectively. Both receptor types have tyrosine kinase activity. The α receptor binds all three PDGF isoforms with high affinity. The β receptor on the other hand binds the differing PDGF isoforms with varying degrees of affinity. Binding affinity increases in the order of AA, AB, BB. PDGF seems to play an important role in the process of wound healing. A number of potential therapeutic applications of PDGF are currently being assessed by numerous biopharmaceutical manufacturers. Regranex is the trade name given to the recombinant PDGF product which has recently come on the market. Its approved indication is to help accelerate the wound healing process of diabetic skin ulcers. The product is a homodimer of the PDGF B chain (i.e. rhPDGF-BB). The polypeptide chains are aligned in an antiparallel manner, linked by a single interchain disulphide linkage. It is produced in a recombinant *Saccharomyces cerevisiae* strain containing the PDGF B gene. The final product is formulated in a gel and is administered topically on the wound surface.

Insulin-like growth factors

IGF consist of a family of three related polypeptide regulatory factors; Insulin, IGF-1 and IGF-2. The family members exhibit 50 per cent amino acid homology, and structurally IGFs resemble insulin, but retain the C peptide sequence and display an extended carboxy terminus region. IGFs circulate throughout the body in the blood bound to one of six related IGF binding proteins. The binding proteins modulate IGF biological activity and limit the access of IGFs to specific tissues. Cell surface receptors for both IGF-1 and IGF-2 have been identified. The IGF-1 receptor displays tyrosine kinase activity essential to signal transduction. It can be activated by both IGF-1 and -2. A second receptor which binds only IGF-2 has been identified but it is devoid of any known signal transduction ability and its physiological function is not understood. The IGF-1 receptor is very widely distributed throughout the body. IGFs themselves are synthesized largely in the liver, from where they are released directly into the circulatory system, and hence display true endocrine activity. However, smaller quantities of IGFs can be synthesized locally within many tissue types, where they display paracrine or autocrine regulatory properties.

The biological functions of IGFs are complex and not fully understood. The availability of recombinant human IGF-1 and -2 will help further elucidate their complete biological actions. Both IGF-1 and -2 play an essential role in embryonic development. However, after birth IGF-1 plays a predominant role in regulating growth, whereas the role of IGF-2 is not understood. As already outlined IGF-1 mediates most of the growth promoting effects of growth hormone. It stimulates bone formation, inhibits muscular protein degradation, and also stimulates glucose uptake and protein synthesis in muscle cells. IGF-1 also appears to promote neuronal survival and myelin synthesis. These biological activities suggest possible clinical application of IGF-1 in treating such conditions as short stature, osteoporosis and neurological disorders. Early clinical trials focused upon the potential of IGF-1 in the treatment of Laron Dwarfism. This condition develops in individuals who express a mutated non functional growth hormone cell surface receptor. Treatment of these individuals with IGF-1 generally induces significant growth. Other trials have shown that IGF-1 alters body composition by increasing protein synthesis while reducing fat mass. As such IGF-1 may find application in treating some forms of obesity.

Epidermal growth factor

EGF is perhaps the most extensively characterized of all growth factors. Human EGF was first isolated from urine and was termed urogastrone, due to its ability to inhibit the secretion of gastric acid. EGF from mouse salivary glands has also been isolated and extensively characterized. EGF has proven to exert a powerful mitogenic influence over a wide variety of cell types, including epithelial and endothelial cells as well as fibroblasts. The skin represents EGF's major target tissue, where it plays an important role in growth and development of the epidermal layer. It also serves as a powerful mitogen for a number of cell lines in culture. In addition to its presence in urine, human EGF has also been detected in serum, in the duodenum and in the salivary gland.

EGF isolated from both mouse and man consists of a 53 amino acid polypeptide chain which exhibits 70 per cent homology, and contains three intrachain disulfide linkages. The molecular mass of mouse EGF is 6041 Da while that of the human growth factor is 6201 Da. EGF is initially synthesized as part of a much larger precursor protein of a molecular mass in the region of 128 kDa. Mature EGF is released from this precursor by proteolytic cleavage. Both human and mouse EGF have been produced as heterologous proteins in recombinant bacterial systems. Due to the large size of the mRNA coding for the EGF precursor, scientists have found it easier to synthesize a nucleotide sequence which codes for the 53 amino acids of the mature EGF molecule. The synthetic DNA may then be expressed in *E. coli* or other systems.

The EGF receptor is a transmembrane protein exhibiting tyrosine kinase activity. The receptor has been discovered on the cell surface of many different cell types. Binding of EGF to its receptor induces a mitogenic response in susceptible cells by promoting increased protein and nucleic acid synthesis, by enhancing transport of certain metabolities into the cell and by increasing the rate of various intracellular metabolic activities.

Epidermal growth factor may find an additional application in the defleecing of sheep. Administration of EGF has a transient but marked effect on the sheep's wool follicle bulb cells. This results in the development of a weakened layer in the fleece, which could greatly facilitate harvesting of the wool. Large quantities of EGF would obviously be required should this become a widely adopted agricultural practice. Bulk quantities of the growth factor can be produced by recombinant DNA technology.

Fibroblast growth factor

Fibroblast growth factor was first isolated from bovine brain and pituitary glands. This growth factor was found to exhibit a marked mitogenic effect on fibroblast cell lines in culture. Fibroblasts represent a particular cell type widely distributed in connective tissue. These cells produce the ground substance of such tissues in addition to the precursors of collagen and elastin fibres. It has since been demonstrated that FGF induces growth and division of numerous other cell types including osteoblasts, glial cells, smooth muscle cells and vascular endothelial cells. Their ability to induce blood vessel growth *in vivo* heightens expectations of their potential therapeutic applications.

Two forms of FGF are known to exist. These are termed acidic and basic FGF respectively. Two related forms of acidic FGF have been isolated. One consists of 140 amino acids while the second is a shorter version, lacking the six amino terminal amino acid residues. Acidic FGFs exhibit a pI value in the region of 5–7. Basic FGF is also a single-chain polypeptide, consisting of 146 amino acid residues and having a pI value in the region of 9.5.

Both acidic and basic FGFs are capable of binding to the same receptor, although basic FGF seems to be the more potent of the two. Recombinant DNA technology has facilitated large-scale production of both human and bovine fibroblast growth factors, which has stimulated further research activities seeking to elucidate the mechanism of action of these growth factors, and assessing their potential therapeutic uses. Both acidic and basic FGFs bind heparin. This has facilitated their purification by heparin affinity chromatography. Reverse phase HPLC has also been utilized to further purify acidic FGF.

Transforming growth factors

Various transforming growth factors (TGF) which may be of potential therapeutic significance have also been identified. TGF-α exhibits similar biological activities to EGF. In addition to its presence in a variety of tumour types, TGF-α is synthesized by several normal cell types such as activated macrophages.

A distinct group of transforming growth factors termed TGF-βs have also been characterized. These factors are generally dimeric proteins consisting of two identical polypeptide chains. The two TGF-βs which have been characterized in most detail are termed TGF-β_1 and TGF-β_2. These two factors exhibit a significant degree of amino acid sequence homology.

The TGFs exhibit a variety of biological effects, most of which relate to the modulation of growth of a number of cell types. TGF-β_1 exhibits numerous immunoregulatory effects, including inhibition of B- and T-cell proliferation. TGF-βs are also present in the synovial fluid of arthritic joints. Indeed, recent experiments suggest that TGF-βs may arrest the progression of arthritis. These growth factors are also the subject of intensive academic and industrial research.

Many of the growth factors thus far discovered have proven to be closely related to the polypeptide products of several known oncogenes. Factors quite similar to FGF have been isolated from a number of human tumour types. The transforming protein of the simian sarcoma virus p28[SIS] shares extensive homology with PDGF and EGF shares extensive homology with TGF-α. Moreover, the EGF receptor has been implicated in the development of some tumour types. The product of the oncogene c-erb B exhibits a similar amino acid sequence to that of the EGF receptor.

THYROTROPHIN

Thyrotrophin, also termed thyroid-stimulating hormone (TSH), is a glycoprotein hormone produced by the anterior lobe of the pituitary. TSH is a dimer consisting of α and β subunits. The overall molecular mass of human TSH is in the order of 28 kDa. The α subunit of TSH is identical in sequence to the α subunit of two other glycoproteins produced by the anterior pituitary: FSH and LH. The β subunit thus confers the biological specificity upon TSH. Both the α and β subunits contain several disulfide bridges, making TSH one of the highest sulfur-containing proteins thus far characterized.

Release of TSH from the pituitary gland is controlled by two hypothalamic hormones. Thyrotrophin releasing hormone (TRH), a tripeptide, stimulates the release of thyrotrophin. This was the first of the hypothalamic regulatory factors to be isolated and characterized. TRH may be obtained from hypothalamic extracts or may be synthesized chemically. Several TRH analogues, which exhibit varying degrees of potency, have been synthesized. TRH also stimulates the release of prolactin from the pituitary, as will be discussed subsequently. This releasing factor is used clinically to assess hypothalamic–pituitary disfunction. The release of TSH from the pituitary is inhibited by another hypothalamic factor; growth hormone-releasing inhibiting hormone (somatostatin).

Elevated circulatory levels of thyrotrophin stimulate an increase in iodine uptake by the thyroid gland and an increase in the rate of synthesis and release of the thyroid hormones thyroxin (T_4) and tri-iodothyronine (T_3). Increased circulatory levels of T_3 and T_4 in turn result in a decrease in TRH and TSH secretion. Receptors which specifically bind the thyroid hormones are present in the pituitary gland and possibly in the hypothalamus. Binding of T_3 and T_4 to such receptors mediate this inhibitory effect.

Thyrotrophin has been utilized in the diagnosis of hypothyroidism. However, current diagnosis of this condition relies on direct assessment of circulating levels of thyrotrophin by radioimmunoassay.

CORTICOTROPHIN

Corticotrophin is also referred to as adrenocorticotrophic hormone (ACTH). Corticotrophin, together with several other hormones, is synthesized in the anterior lobe of the pituitary (Figure 7.8). The hormone consists of a single polypeptide, which in humans contains 39 amino acids, and has a molecular mass of 4.5 kDa. ACTH has been purified from a variety of animal species and, in most cases, the first 24 amino acids were found to be identical in sequence in the various species. Moreover, synthetic polypeptides consisting of this 24 amino acid sequence were found to retain the full biological activity associated with intact ACTH. This synthetic peptide may thus be administered to humans as an alternative to the native hormone. Full-length corticotrophin purified from the pituitaries of slaughtered animals have also be used in the past for therapeutic purposes.

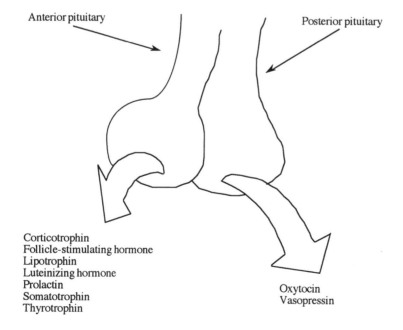

Anterior pituitary

Posterior pituitary

Corticotrophin
Follicle-stimulating hormone
Lipotrophin
Luteinizing hormone
Prolactin
Somatotrophin
Thyrotrophin

Oxytocin
Vasopressin

Figure 7.8 Hormones of the pituitary gland

Injected corticotrophin is degraded relatively quickly in the body. The hormone exhibits a half life in plasma of only 15 min. Its therapeutic effect may be prolonged by intramuscular administration as an ACTH–gelatin or ACTH–zinc formulation. A depot of activity is formed in this way which is released at a slow rate into the circulatory system.

ACTH promotes growth of the adrenal cortex and also stimulates the synthesis and release of the adrenocortical hormones. It promotes an increase in the rate of synthesis of these steroid hormones by increasing the rate of conversion of cholesterol to pregnenolone, which represents the rate-limiting step of adrenocortical hormone synthesis.

ACTH has been used clinically to induce an increase in the circulatory levels of corticosteroids, in particular cortisol and certain mineralocorticoids. Elevated levels of corticosteroids has proven beneficial in the treatment of certain medical disorders. The clinical use of ACTH has decreased in recent years. Most physicians now prefer to administer corticosteroids directly to the patients.

Secretion of corticotrophin from the pituitary is stimulated by the hypothalmic regulatory factor known as corticotrophin releasing factor (CRF). CRF is a peptide containing 41 amino acids. This factor is also known to stimulate the secretion of β endorphin from the pituitary.

PROLACTIN

Prolactin is a single-chain polypeptide hormone synthesized by the anterior pituitary. The hormone has a molecular mass in the range 22–23 kDa and contains 199 amino acids.

Prolactin exhibits a variety of biological effects. In humans this hormone, in conjunction with others such as oestrogens and glucocorticoids, stimulates the growth and development of mammary tissue. It is also responsible for the induction of lactogenesis. Binding of prolactin to its membrane receptor in the mammary tissue of feeding mothers results in an immediate increase in the concentration of cellular mRNAs coding for milk proteins such as a lactalbumin and the caseins. Prolactin also exerts a number of effects on the gonads, where prolactin receptors having been detected in both the ovaries and testes.

Secretion of prolactin from the pituitary is regulated by hypothalamic factors. It is believed that inhibition of secretion is mediated by hypothalamic dopamine. A variety of factors including sleep, stress and pregnancy are known to induce secretion of prolactin. The exact molecular structure of the hypothalamic releasing factor remains to be elucidated.

PEPTIDE REGULATORY FACTORS

This chapter has thus far focused upon polypeptide-based hormones and growth factors. The body also produces a range of peptide regulatory

Table 7.3 Some peptide regulatory factors which have been approved for medical use or are currently undergoing clinical evaluation

Regulatory factor	Medical application
Oxytocin	Induction/augmentation of labour
Vasopressin	Diabetes insipidus
LH-RF	Treatment of some forms of infertility
Somatostatin	Gastrointestinal disorders
GHRF	Diagnosis and treatment of growth hormone deficiency
Thyrotrophin releasing hormone	Diagnosis of disorders of the hypothalamic – pituitary – thyroid axis
Angiotensin II	Increasing blood pressure

LH-RF Luteinizing hormone-releasing factor; GHRF, growth hormone releasing factor.

factors, many of which display actual or potential therapeutic application (Table 7.3). These peptide regulatory factors may be obtained by direct extraction from the native producer source. However, the extremely low levels at which they are produced naturally precludes large scale production by such means. Most such products are thus produced by direct chemical synthesis (see Chapter 2). Chemical synthesis also facilitates the generation of regulatory peptides displaying altered amino acid sequences, and some such analogues have also proven themselves to be medically useful.

Oxytocin and vasopressin

The posterior pituitary serves as a storage and secretory organ for two peptide hormones, namely oxytocin and vasopressin. Both hormones are peptides consisting of nine amino acid residues (Figure 7.9). They are synthesized by specific hypothalamic cells and are subsequently transported in specialized vesicles along axonal pathways directly into the posterior lobe of the pituitary. Both hormones are found in association with specific carrier proteins termed neurophysins.

Vasopressin is also know as antidiuretic hormone (ADH). It is the major antidiuretic hormone found in the body and it exerts a direct effect on the kidney where it promotes an accelerated rate of water reabsorbtion from the distal convoluted renal tubules. Although it may be obtained by

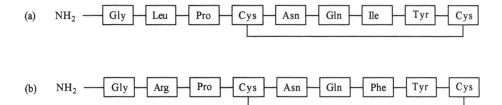

(a)   NH$_2$ — Gly — Leu — Pro — Cys — Asn — Gln — Ile — Tyr — Cys

(b)   NH$_2$ — Gly — Arg — Pro — Cys — Asn — Gln — Phe — Tyr — Cys

Figure 7.9   Amino acid sequence of (a) oxytocin and (b) vasopressin. The similarity in amino acid sequence is striking

purification from pituitary glands it is usually synthesized chemically. Two forms of vasopressin are know to occur naturally. These differ in amino acid sequence by only one amino acid – residue number 8 of the molecule. Arginine vasopressin (argipressin) contains an arginine residue at position number 8 of the peptide and has a molecular mass of 1084 Da. Vasopressin of this amino acid sequence is found in human, ovine, bovine, equine and chicken pituitaries and it is this form of vasopressin that is used medically. Lysine vasopressin, molecular mass, 1056 Da, contains a lysine residue at position 8 of the vasopressin molecule This form is found to be present in porcine pituitaries.

A deficiency in circulatory levels of vasopressin results in the onset of Diabetes insipidus, a relatively rare metabolic disorder in which normal renal water reabsorbtion is impaired. Affected individuals experience a constant thirst and excretes large amounts of dilute urine. The condition may be treated successfully by the administration of vasopressin, either parenterally or as a nasal spray.

Demopressin (1–deamino 8 D-arganine vasopressin) is an analogue of vasopressin that is often used clinically instead of vasopressin, as it exhibits greater antidiuretic activity and boasts a more prolonged period of action.

Oxytocin, like vasopressin is a nonapeptide. It contains one intrachain disulfide linkage and having a molecular mass of 1007 Da. It may be synthesized chemically or purified from the pituitary glands of slaughterhouse animals.

Oxytocin induces contraction of the smooth muscle of the uterus. It is therefore used clinically to induce and maintain uterine contractions associated with labour. It is absorbed from various mucous membranes and may be administered parenterally or as a nasal spray.

Demoxytocin is a synthetic analogue of oxytocin and it is sometimes used as an alternative to the native hormone in the induction of labour, as it exhibits a higher degree of potency and enjoys a longer circulatory half life. Oxytocin also stimulates the contraction of smooth muscle in the lactating breast, thus promoting expulsion of milk from the breast. It is therefore sometimes used to facilitate release of milk during suckling.

Luteinizing hormone-releasing factor

Luteinising hormone-releasing factor (LH-RF) is a decapeptide synthesized by the hypothalamus, and is also known as gonadotrophin releasing hormone, or gonadorelin (Figure 7.10). LH-RF secreted by the hypothalmus is conveyed directly to the pituitary where it stimulates the synthesis and release of LH and FSH.

Hypothalamic regulatory factors are synthesized in minute quantities, thus their preparation from hypothalami in clinically useful quantities is impractical. LH-RF prepared by direct chemical synthesis has been administered (subcutaneously, intravenously or intranasally) to both humans and animals in order to treat certain forms of subfertility or infertility. As administration results in an increase in the secretion of gonadotrophins it improves the conception rate by enhancing normal hypothalamic–pituitary function. It may also be administered to aid in the diagnosis of hypothalamic–pituitary–gonadal dysfunction The releasing factor has also been used clinically in the treatment of certain malignant neoplasms and cryptorchidism – a condition characterized by the failure of the testes to descend into the scrotum during normal development.

In many instances analogues of LH-RF are used clinically. These generally exhibit greater potency or longer duration of action compared to the native releasing factor. Although LH-RF stimulates the synthesis and release of both FSH and LH, changes in the blood concentration of these two pituitary gonadotrophins do not always directly parallel each other. Other agents must exist which can selectively influence the concentration of one or other of these two hormones. One such agent currently receiving much attention is inhibin.

Somatostatin

Somatostatin is a tetradecapeptide of molecular mass 1600 Da. It is produced primarily by the hypothalamus but is also synthesized by the δ cells of the pancreatic islets, and related cells of the gastrointestinal tract. Somatostatin inhibits the release of a variety of pituitary hormones, most notably growth hormone. It is also known to inhibit the release of insulin and glucagon from the pancreas and plays a role in regulating gastric and duodenal secretions. Administration of somatostatin has been found to benefit some patients suffering from gastrointestinal haemorrhage. It has also been used in the management of certain hormone-secreting tumours. Somatostatin may be obtained from hypothalamic extracts but is usually synthesized chemically. The native molecule exhibits a short half life and several analogues with prolonged duration of

Figure 7.10 Amino acid sequence of LHRF

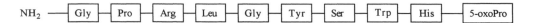

action have been developed. Such analogues include octreotide, an octapeptide which has been used to treat a number of tumour types.

Growth hormone releasing factor, already mentioned in the context of growth hormone, may be used clinically to assist in the diagnosis of growth hormone deficiency, and in some cases may be administered therapeutically to boost natural growth hormone secretion.

FURTHER READING

Books

Anonymous (1997) *Pharmaceutical Biotechnology*. Technomic Publishing.

Bauer, C. *et al.* (1993) *Erythropoietin*. Marcel Dekker, New York.

Bengtsson, B. (1999) *Growth Hormone*. Kluwer, Boston.

Brooks, G. (1998) *Biotechnology in Healthcare*. Pharmaceutical Press, London.

Dawson, J. (1994) *Pharmaceutical Biotechnology*. MDIS Publications.

De Fronzo, R. (1997) *Current Therapy of Diabetes Mellitus*. Mosby, London.

Fauser, B. (1997) *FSH Action and Intraovarian Regulation*. Parthenon Publications, Carnforth.

Juul, A. (2000) *Growth Hormone in Adults*. Cambridge University Press, Cambridge.

Koch, K. *et al.* (1988) *Treatment of Renal Anaemia with Recombinant Human Erythropoietin*. Karger, Basel.

Oppenheim, J. (1994) *Clinical Applications of Cytokines*. Oxford University Press, Oxford.

Oxender, D. (1999) *Novel Therapeutics from Modern Biotechnology*. Springer-Verlag, Godalming.

Pickup, J. (1991) *Biotechnology of Insulin Therapy*. Blackwell Science, Oxford.

Stahnke, N. (1992) *Mammalian Cell Derived Recombinant Human Growth Hormone*. Karger, Basel.

Walsh, G. (1998) *Biopharmaceuticals; Biochemistry and Biotechnology*. John Wiley & Sons, Chichester.

Walsh, G. & Murphy, B. (1999) *Biopharmaceuticals, an Industrial Perspective*. Kluwer, Boston.

Scientific Articles

Insulin

Combettes, S.-M. & Issad, T. (1998) Molecular basis of insulin action. *Diabetes Metab.* (Paris) **24**, 477–489.

Johnson, I.S. (1983) Human insulin from recombinant DNA technology. *Science* **219**, 632–637.

Kroeff, E. *et al.* (1989) Production scale purification of biosynthetic human insulin by reverse-phase high performance liquid chromatography. *J. Chromatogr.* **461**, 45–61.

Nicol, S. & Smith L. (1960) Amino acid sequence of human insulin. *Nature* **187**, 483–485.

Pfeiffer, A. (2000) Current perspectives of biotechnological replacement of insulin secreting cells. *Exp. Clin. Endocrinol. Diabetes* **108** (8), 494–497.

Pharmacia (1983) The large scale purification of insulin by gel filtration chromatography. Pharmacia Fine Chemicals.

Orci, L. *et al.* (1988) The insulin factory. *Sci. Am.* **259**, 50–61.

Ruttenberg, J. (1972) Human insulin: facile synthesis by modification of porcine insulin. *Science* **177**, 623–626.

Schulingkamp, R. *et al.* (2000) Insulin receptors and insulin action in the brain: review and clinical implications. *Neurosci. Biobehav. Rev.* **24** (8), 855–872.

Vajo, Z. & Duckworth, W. (2000) Genetically engineered insulin analogues: diabetes in the new millenium. *Pharm. Rev.* **52** (1), 1–9.

Whitehead, J *et al.* (2000) Signalling through the insulin receptor. *Curr. Opin. Cell Biol.* **12** (2), 222–228.

Gonadotrophins

Boland, M.P. *et al.* (1991) Alternative gonadotrophins for superovulation in cattle. *Theriogenology* **35**, 5–17.

Chappel, S. *et al.* (1988) Bovine FSH produced by recombinant DNA technology. *Theriogenology* **29**, 2135.

Closset, J. & Hennen, G. (1978) Porcine follitropin, isolation and characterization of the native hormone and its a and b subunits. *Eur. J. Biochemi.* **86**, 105–113.

DeJong, F. (1988) Inhibin. *Physiol. Rev.* **68**, 555–595.

Hayden, C. *et al.* (1999) Recombinant gonadotrophins. *Br. J. Obstet. Gynaecol.* **106** (3), 188–196.

Kafi, M. & McGowan, M. (1997) Factors associated with variation in the superovulatory response of cattle. *Anim. Reprod. Sci.* **48**, 137–157.

Lindsell, C. *et al.* (1986) Variability in FSH : LH ratios among batches of commercially available gonadotrophins. *Theriogenology* **25**, 167.

Looney, C. *et al.* (1988) Superovulation of donor cows with bovine follicle-stimulating hormone (bFSH) produced by recombinant DNA technology. *Theriogenology* **29**, 271.

Moore, W. & Ward, D. (1980) Pregnant mare serum gonadotrophin. *J. Biol. Chem.* **255**, 6923–6929.

Prevost, R. & Pharm, D. (1998) Recombinant follicle stimulating hormone: new biotechnology for infertility. *Pharmacotherapy* **18**, (5), 1001–1010.

Growth hormone and growth factors

Furman, T. *et al.* (1987) Recombinant human insulin-like growth factor II expressed in *E. coli. Bio/Technology* **5**, 1047–1056.

Hull, K. & Harvey, S. (2000) Growth hormone – roles in male reproduction. *Endocrine* **13** (3), 243–250.

Hull, K. & Harvey, S. (2001) Growth hormone: roles in female reproduction. *J. Endocrinol.* **168** (1), 1–23.

Juskevich, J. & Guyer, C. (1990) Bovine growth hormone: human food safety evaluation. *Science* **149**, 875–884.

Kallen, K. *et al.* (2000) New perspectives on the design of cytokines and growth factors. *Trends Biotechnol.* **18** (11), 455–461.

Kopchick, J. & Andry, J. (2000) Growth hormone (GH), GH receptor, and signal transduction. *Mol. Genet. Metab.* **71** (1–2), 293–314.

Koveker, G. (2000) Growth factors in clinical practice. *Int. J. of clin. Pract.* **54** (9), 590–593.

Kurokawa, T. *et al.* (1987) Cloning and expression of cDNA encoding human basic fibroblast growth factor. *FEBS Lett.* **213**, 189–194.

LeRoith, D. (1997) Insulin like growth factors. *N Engl. J. Med.* **336** (9), 633–639.

Olson, K. *et al.* (1981) Purified human growth hormone from *E. coli* is biologically active. *Nature* **293**, 408–411.

Robinson, C. (1990) Polypeptide growth factors – a growth area for biotechnology. *Trends Biotechnol.* **8**, 59–60.

tenDijke, P. & Iwata, K. (1989) Growth factors for wound healing. *Bio/Technology* **7**, 793–798.

Thomas, K. & Gimenez-Gallego, G. (1986) Fibroblast growth factors: broad spectrum mitogens with potent angiogenic activity. *Trends Biochem. Sci.* **11**, 81–84.

Vos, P. *et al.* (1998) Insulin like growth factor-1: clinical studies. *Drugs Today* **34** (1), 79–90.

Walsh, G. (1995) Nervous excitement over neurotrophic factors. *Biotechnology* **13**, 1167–1171.

Erythropoietin

Lee-Huang, S. (1984) Cloning and expression of human erythropoietin cDNA in *E. coli*. *Proc. Nat Acad. Sci. USA* **81**, 2708–2712.

Sasaki, R. *et al.* (2000) Erythropoietin: multiple physiological functions and regulation of biosynthesis. *Biosc. Biotechnol. Biochem.* **64** (9), 1775–1793.

Tilbrook, P. & Klinken, S. (1999) Erythropoietin and the erythropoietin receptor. *Growth Factors* **17** (1), 25–35.

8

Interferons, interleukins and additional regulatory factors

REGULATORY FACTORS; CYTOKINES VERSUS HORMONES

For many decades hormones represented the major group of biological regulatory factors know to scientists. As a group, hormones conform to certain defining characteristics. They are biomolecules synthesized in small amounts by ductless (endocrine) glands. They travel via the bloodstream from their site of synthesis to a far distant target tissue. They induce characteristic regulatory effects in target tissues by binding to specific cell surface receptors (or, in the case of steroid hormones, cytoplasmic receptors). In terms of structure, they fall into one of three categories: polypeptide, amide or steroid hormones.

During the second half of the twentieth century researchers began to identify additional regulatory factors that did not fit the classical definition of a hormone. Many of these factors were subsequently labelled 'cytokines', a term originally coined in the 1970s. Initially 'cytokine' referred to regulatory factors promoting the production, activation/regulation of cells which constitute the immune system. Cytokines represented regulatory factors synthesized mainly by leukocytes, which generally targeted other white blood cells. Unlike classic hormones cytokines are exclusively polypeptide-based molecules, and are generally produced by cells which are not organized into discrete anatomical glands. In addition cytokines were seen to primarally act in a paracrine–autocrine (as opposed to endocrine) manner. (Paracrine factors target cells in the immediate vicinity of the cells synthesizing the factor. In the case of autocrine regulatory factors the producer and target cells are one and the same.)

Subsequent research findings continue to blur the definition of a cytokine as well as the distinction between 'cytokines' and 'polypeptide hormones'. Cytokines display a very complex biology. They are usually synthesized by more than one cell type, and also generally target several different cell types. They tend to promote a wide and complex range of responses in their target cells. It is now also clear that many cytokines can act in an endocrine as well as a paracrine–autocrine fashion, and conversely that several hormones can display paracrine activity.

Table 8.1 presents a list of regulatory polypeptides now usually categorized as cytokines. Some (e.g. erythropoietin and polypeptide growth factors) have been considered in Chapter 7. This chapter focuses upon additional cytokines which have found actual therapeutic application, particularly interferons, interleukins, tumour necrosis factor and colony-stimulating factors.

INTERFERONS

The first cytokine to be identified and studied was interferon (IFN), which was discovered initially in 1957. It has since been demonstrated that all vertebrates produce a variety of interferons. These proteins are

Table 8.1 The major polypeptides/polypeptide families which constitute the cytokine group of regulatory molecules

---

Interleukins (IL-1–IL-19)

Interferons (IFN-α, -β, -γ, -τ, -ω)

Colony-stimulating factors (G-CSF, M-CSF, GM-CSF)

Tumor necrosis factors (TNF-α, -β)

Neurotrophins (NGF, BDNF, NT-3, NT-4/5)

Ciliary neurotrophic factor (CNTF)

Glial cell-derived neurotrophic factor (GDNF)

Epidermal growth factor (EGF)

Erythropoietin (EPO)

Fibroblast growth factor (FGF)

Leukaemia inhibitory factor (LIF)

Macrophage inflammatory proteins (MIP-1α, -1β, -2)

Platelet-derived growth factor (PDGF)

Transforming growth factors (TGF-α, -β)

Thrombopoietin (TPO)

G-CSF, Granulocyte colony-stimulating factor; M-CSF, Macrophage colony-stimulating factor; GM-CSF, Granulocyte–macrophage colony-stimulating factor; NGF, Nerve growth factor; BDNF, Brain-derived neurotrophic factor; NT, Neurotrophin.
Reproduced from *Biopharmaceuticals, Biochemistry and Biotechnology*, John Wiley & sons, Chichester (1998), with permission.

generally species specific and most mammals produce at least three types of interferon: interferon alpha (IFN-α), interferon beta (IFN-β) and interferon gamma (IFN-γ). Humans produce at least 16 closely related types of interferon-α but produce only one interferon-β and one interferon-γ.

IFN-α and IFN-β both appear to bind the same cell surface receptor and are both acid stable. These are collectively termed type I interferons. IFN-γ is acid labile and binds to a distinct receptor molecule. This interferon has been termed type II interferon. More recently two new members of the interferon family have been identified: interferon omega (IFN-ω) and interferon tau (IFN-τ). Both are classified as type I IFNs. An overview summary of the biology of interferons is presented in Tables 8.2 and 8.3.

Table 8.2 Traditional human interferons and the cellular sources from which they were initially isolated

Interferon type	Additional name	Cell type from which it was initially isolated
Interferon-α (IFN-αs) (at least 16 distinct but related types exist)	Leukocyte interferon, lymphoblastoid interferon	Leukocytes or lymphoblastoid cells
Interferon-β (IFN-β)	Fibroblast interferon	Fibroblasts.
Interferon-γ (IFN-γ)	Immune interferon	Activated T lymphocytes

Table 8.3 Summary of the biological characteristics of the interferons

Interferon	No. amino acids	Molecular mass (kDa)	Glycosylated?	Major cellular sources	Major biological activities
α (multiple)	Usually 166	16–27	Some are	Lymphocytes Monocytes Macrophages	Confers target cells with viral resistance. Inhibits proliferation of various cell types. Regulates expression of MHC class 1 antigens
β	166	20	No	Fibroblasts, Epithelial cells	Similar to those of IFN-αs
γ	143	40–70	Yes	T-lymphocytes Natural killer (NK) cells	Regulates all phases of the immune and inflammatory response. Displays only weak antiviral and antiproliferative activity
ω	172	20	Yes	Mainly leukocytes	Similar to those of other type I interferons
τ	172	19	No	Trophoblast cells	Sustains the corpus luteum during the early stages of pregnancy. In addition, displays the major biological activities of other type I interferons.

Details provided relate to human interferons, except in the case of interferon tau (IFN-τ), which is only produced by ruminants. MHC, major histocompatibility complex.

Interferon alpha

IFN-α is also known as leukocyte interferon or lymphoblastoid interferon, as it was initially isolated from these cell types. At least 16 different human IFN-αs are known to exist. Although these proteins exhibit significant sequence homologies, they are different in their primary structure and are encoded for by a family of related genes. Most human IFN-αs are single chain polypeptides consisting of 165–166 amino acids and are rich in leucine, glutamic acid and glutamine. They generally exhibit pI values ranging from 5.5 to 6.5. Although many IFN-αs are devoid of carbohydrate side chains, several are glycoproteins, exhibiting varying degrees of glycosylation. The molecular masses of the IFN-αs may range from 16 kDa to in excess of 26 kDa, depending upon the carbohydrate content of the molecule. The enzymatic or chemical removal of the carbohydrate component seems to have little adverse effects on their bioactivity.

Purification of individual native IFN-αs initially proved difficult, due not only to their extremely low expression levels, but also due to the similarity of their physicochemical properties. Several individual IFN-αs were finally purified to homogeneity in the late 1970s and early 1980s, by employing a combination of several high resolution chromatographic techniques, including immunoaffinity chromatography. Isoelectric focusing and sodium dodecyl sulfate-polyacrylamide gel electropharesis (SDS–PAGE) have also been used preparatively in some purification schemes. IFN-αs initiate their antiviral, antiproliferative, anti-inflammatory and immunoregulatory effects by binding to specific cell-surface receptors. Two such receptors have been isolated and cloned. The IFN-αB receptor appears to only bind one specific IFN-α (called IFN-αB or IFN-α8). The other receptor, termed the IFN-α/β receptor, can apparently mediate the biological effects of all type I IFNs. IFN-α receptors appear to be present on most cell types. They are generally present in relatively low numbers, but display high binding affinities for their ligands.

Binding of type I IFNs to the IFN-α/β receptor results in the activation of two intracellular receptor-associated tyrosine kinases: TYK-2 and Janus kinase (JAK)-1. These in turn phosphorylate (and hence activate) three cytoplasmic proteins known as STAT (signal transducers and activators of transcription). The activated STAT interact with an additional cytoplasmic protein (p48), and this entire complex migrates to the nucleus, where it interacts with interferon-stimulated response elements (ISRE). ISRE are upstream regulatory elements of IFN-sensitive genes. The JAK–STAT pathway thus mediates the biological effects of type I IFNs by inducing the synthesis of specific cellular gene products (Figure 8.1). Type I IFNs may induce some of their biological responses by initiating additional signal transduction pathways independent of the JAK–STAT pathway. IFN-αs, for example, are capable of activating phospholipase C, releasing diacylglycerol and arachidonic acid.

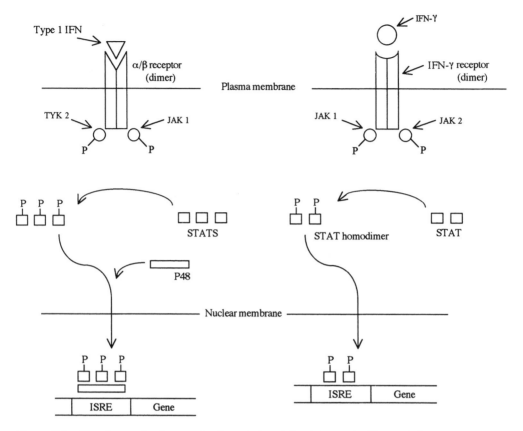

Figure 8.1 Signal transduction mechanisms induced by (a) type I IFN and (b) type II IFN. Binding of the IFN to the appropriate receptor results in receptor dimerization, with subsequent activation of JAK-STAT pathways. Refer to text for specific details

Interferon beta

Human IFN-β, also termed fibroblast interferon, was initially derived from fibroblasts that had been exposed to certain viral particles, or to polynucleotides. It is now produced by recombinant DNA technology. Human IFN-β was the first interferon molecule to be purified to homogeneity. It is a glycoprotein of 166 amino acids and displays a molecular mass in the region of 20 kDa. It exhibits a pI value of approximately 8 and contains one intramolecular disulfide linkage. The human genome has a single IFN-β gene. The gene product displays 30 per cent homology in amino acid sequence to IFN-αs. Native IFN-β has been purified using a combination of various chromatographic techniques. A one-step purification method employing dye affinity chromatography (blue sepharose) has also been developed. IFN-β induces similar biological responses to other members of the type I IFN family, by binding to the type I IFN cell surface receptor (Figure 8.1).

Interferon gamma

IFN-γ, often referred to as immune interferon, is produced largely by activated T lymphocytes and natural killer (NK) cells. Human IFN-γ is a single chain glycoprotein consisting of 143 amino acids. Its molecular mass ranges from 15 to 25 kDa, depending upon the degree of glycosylation (the molecule harbours two potential glycosylation sites). As previously mentioned, IFN-γ differs significantly from type I interferons in terms of its physicochemical properties and biological activities. It regulates almost all aspects of the immune and inflammatory response, including the differentiation and activation of B and T lymphocytes, macrophages and NK cells. Additionally it influences the growth and activation of non-leukocytes such as endothelial cells and fibroblasts. Unlike type I IFNs, IFN-γ displays at best weak antiviral and antiproliferactive activities, but it does potentiate such effect when they are induced by type I IFNs.

Biologically active IFN-γ is a homodimer, formed by non-covalent, anti-parallel association of two IFN-γ monomers. Each subunit displays six major α-helical regions, and is totally devoid of β conformation. Given its characteristic biological activities, it is not surprising that the IFN-γ receptor is present on most white blood cell types, as well as on endothelial and epithelial cells. Erythrocytes are devoid of such receptors. Binding of the dimeric IFN-γ ligand promotes receptor dimerization and subsequent activation of a JAK–STAT pathway, as illustrated in Figure 8.1.

Production and medical application of IFN-α

Up until the 1970s the majority of interferons available for clinical use were sourced from human leukocytes, (white blood cells), obtained directly from transfusion blood supplies. This leukocyte interferon consisted of a variety of related IFN-α molecules. The final IFN preparation obtained was only approximately 1 per cent pure, and vast quantities of transfused blood was required to produce minute amounts of product. Although the potential clinical applications of IFN-α were well recognized at that time, the scarcity of this cytokine rendered its widespread clinical evaluation impossible, never mind its application.

In the late 1970s bulk quantities of IFN-α were first produced by (non-recombinant) mammalian cell culture. A specific strain of a human lymphoblastoid cell line termed the Namalwa cell line was most often employed in this regard. After induction by Sendai virus, this cell line produces appreciable quantities of leukocyte (α) interferon. Industrial-scale production of Namalwa cell leukocyte interferon is undertaken in large culture vessels, many of which exhibit a capacity in excess of 80001. Subsequent downstream processing yields a partially purified product consisting of at least eight distinct molecular species of IFN-α.

Individual IFN-αs are now most conveniently produced in large quantities by recombinant DNA methodologies. Most recombinant preparations were first developed in the 1980s. Several recombinant human IFN-αs have been produced in engineered *E. coli*, yeast and in some eukaryotic cells including cultured monkey cells and CHO cells.

Recombinant human IFN-αs (rhIFN-αs) approved for general medical use include 'Roferon A', 'Alferon A' and 'Infergen' (Table 8.4). Approved indications include the treatment of certain viral-mediated diseases (eg. hepatitis B, C and genital warts) as well as some cancer types (eg. hairy cell leukaemia, renal cell carcinoma, Kaposi's sarcoma and non-Hodgkin's lymphoma). Infragen (approved in the USA and EU in the late 1990s for the treatment of hepatitis C) is unusual in that it is not a naturally occurring molecule. This synthetic IFN-α, produced by recombinant means in *E. coli*, contains the most frequently observed amino acids in each corresponding position of several naturally occurring IFN-α subtypes. On a mass basis this product appears to display higher antiviral, antiproliferative and NK cell activation capability when compared with several of the other approved IFN-αs.

A wide variety of chromatographic techniques have been used in the purification of various IFN-α preparations. Such techniques include affinity chromatography using various immobilized ligands, including

Table 8.4 Interferon preparations currently approved for general medical use

Tradename	Description	Original Indication
Intron A	rhIFN α-2b	Hairy cell leukaemia, genital warts
Roferon A	rhIFN α-2a	Hairy cell leukaemia
Alferon N	rhIFN α-n3	Genital warts
Infergen	rIFN-α	Hepatitis C
Betaferon	rhIFNβ-1b	Multiple sclerosis
Betasteron	rhIFNβ-1b	Multiple sclerosis
Avonex	rhIFN β-1a	Multiple sclerosis
Rebif	rhIFN β-1a	Multiple sclerosis
Actimmune	rhIFN γ-1b	Chronic granulomatous disease

In each case only the original indication for which the product was approved is listed. Additional indications have subsequently been approved for some such products.

lectins, reactive dyes, concanavalin A and phenyl groups, in addition to anti-interferon monoclonal antibodies. Other techniques used have included ion exchange, gel filtration and mineral chelate chromatography. HPLC has proven to be a particularly powerful technique in the preparation of modern, high-purity interferons. The introduction of sensitive and convenient immunoassays also facilitates the rapid purification of IFN preparations. Initial interferon assays relied upon monitoring their antiviral activity in cell culture. Such bioassays were time consuming and complex.

Production and medical applications of IFN-β

RhIFNβ-1a describes a recombinant human IFN-β produced in engineered CHO cell lines. Two such products ('Avonex' and 'Rebif') have thus far been approved for general medical use (Table 8.4). The amino acid sequence of these recombinant products are identical to that of native IFN-β, and like the native product they are glycosylated. RhIFN-β-1b ('Betaferon' and 'Betasteron', Table 8.4), are produced in engineered *E. coli* cells. The resulting lack of glycosylation in itself has no significant effect upon the medical efficacy of these products. In addition to lack of glycosylation, IFN-β-1b differs from native IFN-β by the substitution of the cysteine residue normally found at position 17 with a serine residue. This engineered product, while retaining the characteristic biological activities of native IFN-β, displays enhanced stability.

Both rhIFN-β-1a and rhIFN-β-1b have proven effective in the treatment of multiple sclerosis (MS). MS is a neurological condition characterized mainly by damage to the myelin sheath surrounding neurons. The ultimate cause of the disease remains unidentified, although genetic factors and viral infection are suspected of involvement. Currently there is no cure for the condition. The number of newly diagnosed cases in Europe alone stands in excess of 10 000 per year, with the majority of these individuals being in their early 30s.

MS is now predominantly seen as an immune-mediated disease, with migration of immune system cells to lesion sites in the brain. An inflammatory response follows, which is believed to contribute to neuronal damage. IFN-β likely exerts its positive therapeutic effect at least in part by countering this inflammatory response. This cytokine is known to inhibit release of pro-inflammatory molecules such as IFN-γ and tumour necrosis factor.

Production and medical uses of IFN-γ

Given its central role in promoting immunological and inflammatory responses, it is not surprising that IFN-γ attracts clinical interest. 'Actimmune' is the tradename given to rhIFN-γ produced in engineered

E. coli cells, and this product has been approved for the treatment of chronic granulomatous disease (CGD). CGD is a genetic condition, characterized by the occurrence of recurrent, often life-threatening infections. The genetic defect inhibits an reduced nicotinamide adenine dinucleotide phosphate (NADPH) oxidase system active in phagocytes, thus inhibiting the production of various oxidative substances which phagocytes use to destroy microorganisms.

Clinical trials have established that administration of IFN-γ can reduce the incidence of life-threatening infections experienced by CGD sufferers by over 50 per cent. The cytokine appears capable of maximizing production of oxidative substances in effected phagocytes. Its ability to stimulate a generalized immune response also likely contributes to its therapeutic efficacy.

Interferon omega

IFN-ω is a more recently discovered type I IFN. It displays some 60 per cent amino acid sequence homology with IFN-αs and this 172 amino acid cytokine appears to be produced by most mammals. Like IFN-β, human IFN-ω is a single gene species, and it contains a single site for *N*-glycosylation (asparagine 78). In contrast to humans several other mammals harbour several distinct (although closely related) IFN-ω genes.

Recombinant human IFN-ω (rhIFN-ω) has been produced in engineered CHO (and other) cell lines. Like IFN-αs, the IFN-ω molecules contains two intrachain disulfide bonds which link cysteine residues 1 and 99, as well as 29 and 139. IFN-ω binds the type I IFN-α/β receptor and appears to induce biological responses similar to other type I IFNs. Although it may prove clinically useful, it is unlikely to be commercialized in the near future unless it exhibits an efficacy–safety profile superior to those of IFN-αs.

Interferon tau

IFN-τ is a novel type I IFN most closely related to IFN-ω. It is found only in ruminant animals with sheep and cattle harbouring at least three or four functional IFN-τ genes. Uniquely in the context of IFNs, IFN-τ (formerly known as trophoblastin) acts as a hormone of pregnancy. It is synthesized by the conceptus (the embryo and its associated membranes) during the early stages of pregnancy, and functions to inhibit the luteolytic mechanism (i.e. prevent regression of the corpus luteum, which is required if pregnancy is to be sustained, see Chapter 7). Failure to inhibit luteolysis is the major cause of pregnancy loss in cows, hence IFN-τ continues to be a focus of animal science research.

In addition to its reproductive activity, IFN-τ retains the antiviral, antiproliferative and immunomodulatory activities typical of other type I IFNs. However, it is untypical of type I IFN, in that it is not inducable by viral infection and it does not display significant species specificity. From an applied perspective one of IFN-τ's most interesting characteristics is its remarkable lack of toxicity, even when it is administered at high concentrations. Because of this IFN-τ may yet prove of use in human medicine, although this remains to be established in clinical trials.

INTERLEUKINS

The interleukins represent yet another family of cytokines. At least nineteen different interleukins have thus far been characterized. The biology of the majority of these is summarized in Table 8.5. All are polypeptide regulatory factors which initiate their biological effects by binding to cell surface receptors on sensitive cell types. Some are glycosylated, others are devoid of a carbohydrate moiety. Moreover, removal of the glycocomponent of several glycosylated interleukins (e.g. interleukin (IL)-2 and IL-3) has no significant effect upon their biological activity. Interleukins exhibit many additional properties characteristic of cytokines in general. Most interleukins can be synthesized by a variety of cell types, and all show pleiotropic effects (i.e. promoter multiple, distinct and usually unrelated biological effects). However, they are predominantly produced by cells of the immune system and function mainly to regulate immunity. Interleukins also exhibit redundancy (i.e. several interleukins can independantly promote similar or identical effects). Most function in a paracrine–autocrine manner, although some display endocrine activity.

Some interleukins such as IL-2 and IL-11 are already employed in the treatment of a number of medical conditions while others are currently being assessed in clinical trials (Table 8.6).

Interleukin-2

IL-2 is perhaps the best characterized of all the interleukins. This cytokine, which is also known as T-cell growth factor, is a single chain glycosylated polypeptide of molecular mass 15–20 kDa. The protein consists of 133 amino acid residues arranged in four major and two minor α-helical regions. These are aligned in anti-parallel fashion such that their hydrophilic regions are directed towards the surrounding aqueous environment. Analysis of the crystalline structure of IL-2 preparations also reveal that the molecule is devoid of any β conformational sequences. IL-2 contains one intrachain disulfide bond, linking cysteine residues 58 and 105, and which is essential for biological activity. The molecule exhibits an isoelectric point between 6.5 and 6.8. The human gene coding for IL-2 is comprised of four exons separated by three introns.

Table 8.5 Human interleukins; summary of size, sources and major biological activities

Name	Molecular mass (kDa)	Major cellular producers	Major biological activities
IL-1α	17.5	Monocytes, macrophages, lymphocytes, NK cells, vascular endothelial cells, fibroblasts, smooth muscle cells, keratinocytes	Activation of T cells, endothelial cells & monocytes. Promotes antibody synthesis, induces fibroblast proliferation. Promotes acute phase protein synthesis
IL-1β	17.3		
IL-2	15–20	T lymphocytes	Promotes growth and differentiation of T and B lymphocytes, NK cells, monocytes, macrophages and oligodendrocytes
IL-3	14–30	Activated T cells, mast cells, eosinophils	Haemopoietic growth factor
IL-4	15–19	T cells, mast cells	T and B cell growth and differentiation, promotes synthesis of IgE
IL-5	45 (homodimer)	Mast cells, T cells, eosinophils	Eosinophil differentiation
IL-6	26	T & B cells, macrophages, fibroblasts, keratinocytes astrocytes, endothelial cells	Regulates B and T cell function, haemopoiesis and acute phase reactions
IL-7	20–28	Bone marrow, spleen cells, thymic stromal cells	Stimulates proliferation and differentiation of immature B cells, immature & mature T cells
IL-8	6–8	Multiple cell types including monocytes, lymphocytes, hepatocytes, keratinocytes, fibroblasts, endothelial cells, amongst others	Pro-inflammatory cytokine. Neutrophil chemoattractant and activating factor

IL-9	32–39	Activated lymphocytes, fibroblasts, hepatocytes	Enhances T lymphocyte & mast cell proliferation
IL-10	35–40	Activated T lymphocytes	Stimulates B-cell proliferation, thymocytes & mast cells. Stimulates IgA production by B cells.
IL-11	23	IL-1 stimulated fibroblasts	Haemopoietic growth factor
IL-12	75–80	B cells, monocytes, macrophages	Co-stimulates lymphocyte proliferation, enhances NK cell activity, induces IFN-γ production by T cells & NK cells
IL-13	17	Activated T cells	Promotes B-cell proliferation, inhibits production of pro-inflammatory cytokines
IL-14	60	T lymphocytes	Enhances proliferation of activated B cells, inhibits immunoglobulin synthesis
IL-15	14–15	Epithelial cells, monocytes, skeletal muscle, placenta, lung, liver, heart	Shares many of the biological activities of IL-12

Table 8.6   Some interleukin preparations which are approved for general medical use, or are currently being evaluated in clinical trials

Interleukin	Indication	Status
rhIL-2 (Proleukin)	Renal cell carcinoma, melanoma	Approved (1992)
rhIL-11 (Neumega)	Prevention of chemotherapy-induced thrombocytopenia	Approved (1997)
rhIL-4	Cancer, rheumatoid arthritis	In trials
rhIL-6	Haematological conditions, cancer	In trials
rhIL-12	Cancer, infectious diseases	In trials

IL-2 plays a pivotal role in normal immunological functioning, regulating many aspects of both innate and acquired immunity (Box 8.1). It is synthesized and secreted by T lymphocytes upon their activation, either by an antigen or a mitogen. IL-2 then stimulates the further growth and differentiation of activated T and B lymphocytes. It also potentiates the activity of NK cells, monocytes and macrophages. Together with antigen,

---

Box 8.1   Innate and acquired immunity

Higher animals protect themselves against infection by microorganisms and other foreign substances by a variety of means. In general such protective mechanisms may be divided into two types: innate immunity (also termed natural or native immunity) and acquired immunity. Innate immunity is a constitutive form of immunity, being present from birth. It is non specific (i.e. is not pathogen or antigen specific, but operates against almost any substance which threatens the body). Furthermore it is not enhanced or amplified by exposure to the foreign substance. Elements of innate immunity include:

1. **Physicochemical barriers** such as the skin and mucous membranes, stomach acid, digestive enzymes, bile in the small intestine and the low pH of the vagina. These barriers function by preventing entry of the pathogen into the body or destroying the pathogen before it can colonise within the body.

2. **Phagocytosis**, which refers to the ingestion and destruction of microbes and other foreign particles by phagocytes, a special class of defence cells. Phagocytes include macrophages, monocytes, polymorphonuclear leukocytes (PMN) and natural killer cells (see also Box 8.2).

Acquired immunity is an inducible form of immunity, found only in vertebrates. Acquired immunity is induced by an encounter with a foreign substance or antigen. This type of immune response is antigen-specific and increases in magnitude upon continued/recurrent exposure to the inducing antigen. It is largely mediated by T lymphocytes ('cellular immunity') and antibody-producing B lymphocytes (humoral immunity, see also Box 8.2).

IL-2 is one of the primary molecular effectors capable of inducing an antigen-specific immune response, as proposed initially by Sir MacFarlane-Burnet in his clonal selection theory. MacFarlane-Burnet rightly predicted that an extensive repertoire of lymphocytes was present naturally in the body, one member of which would specifically recognize virtually any antigen that the body might encounter. The presentation of antigen to such a specific lymphocyte would induce the clonal expansion of that particular cell and, in this way, the immune system would mount an immediate and specific response against the offending antigen. When an antigen gains entry into the body some of the antigenic material is invariably ingested by macrophages. These cells then 'present' fragments of the ingested material on their surface, hence the origin of the term 'antigen-presenting cells'. A small proportion of circulatory T lymphocytes recognize and bind presented antigen via a specific cell-surface receptor. These antigen-specific cells then undergo selective clonal expansion while other circulating lymphocytes, which do not recognize the antigen, remain quiescent. This growth and division of such T cells is dependent upon the presence, not only of antigen, but also of IL-1. IL-1 is produced by activated macrophages. Activated T helper cells, in turn, produce IL-2 which stimulates further T-cell differentiation. Along with antigen and IL-6, IL-2 also co-stimulates the proliferation of B lymphocytes (capable of producing antibodies which bind selectively to the foreign moiety). Activated B and T cells also begin to express IL-2 receptors on their surface. Quiescent lymphocytes are normally devoid of such receptors.

In contrast to most lymphocytes, NK cells constitutively express IL-2 receptors on their surface. Binding of IL-2 to such receptors induces the immediate proliferation of NK cells. IL-2 also stimulates increased production and secretion of various NK cell products such as a variety of additional cytokines, including tumour necrosis factor, IFN-γ and colony-stimulating factors. These cytokines serve to further potentiate the overall immunological response. NK cells comprise approximately 10 per cent of circulating lymphocytes. These cells are believed to play a particularly important role in immunological reactions against cancer cells and virally infected cells (Box 8.2).

The IL-2 receptor has also been characterized in detail. The high-affinity receptor consists of three distinct transmembrane polypeptide chains, which interact with each other non-covalently (Figure 8.2). The smaller of the receptor subunits (the α subunit), has a molecular mass of 55 kDa and consists of three segments: a large extracellular domain capable of binding IL-2, a transmembrane domain, and an intracellular domain consisting of 13 amino acids. The larger polypeptide component the of the IL-2 receptor (the β subunit) has a molecular mass of 75 kDa. Again, this polypeptide contains extracellular, transmembrane and intracellular domains. In this case, however, the intracellular domain is relatively large, consisting of 286 amino acids. It is this portion of the molecule which initiates the intracellular response upon binding of

Box 8.2   Some cell types whose biological function is modulated by interleukins or other cytokines

Astrocytes cells found in the central nervous system; likely function to provide nutrients for neurons

Eosinophil a type of polymorphonuclear leukocyte (white blood cell) capable of ingesting particulate matter, and which is also involved in allergic reactions

Epidermal cells cells found in the outer layer of skin

Fibroblasts cells found in conncetive tissue which function to produce the ground substance of connective tissue, as well as the precursors of collagen and elastin fibres

Haemopoietic cells bone marrow stem cells from which all classes of blood cells are derived

Keratinocyte any epidermal cell that produces keratin

Leukocyte any white blood cell, i.e. any blood cell that contains a nucleus

B lymphocyte antibody-producing white blood cells that mediate humoral immunity

T lymphocyte white blood cell responsible for cell-mediated immunity

Macrophages phagocytic cells present mainly in connective tissue and organs, function similar to monocytes

Mast cells cells of connective tissue housing granules containing histamine, heparin and serotonin, released during inflammation/allergic reactions

Monocyte white blood cell type that ingests and destroys foreign particles such as bacteria and tissue debris

Natural killer (NK) cells white blood cells that can kill cancer cells and virally infected cells

Smooth muscle cells muscle cells that produce slow, long- term contractions, of which an individual is unaware. Found mainly in hollow organs (intestine, stomach, bladder and blood vessles)

Vascular endothelial cells single layer of cells that line blood vessels

IL-2. The third element of the intact receptor, the γ-chain, is a 64 kDa transmembrane glycoprotein, which does not interact directly with IL-2 but is necessary for signal transduction. This γ polypeptide also appears to form part of the IL-4, 7, 9, 13 and 15 receptors.

The exact mechanism by which signal transduction is accomplished has not been elucidated, although tyrosine kinase activity appears to play a role in this process. The IL-2 molecule contains two distinct receptor binding sites on its surface. One site recognizes and binds the α receptor subunit while the other binds the β subunit, as illustrated in Figure 8.2. Various studies have revealed substantial differences in the affinity and kinetics of IL-2 binding to the two distinct receptor subunits. IL-2 binds the α (55 kDa) subunit rapidly with association–disassociation rates almost comparable to diffusional values. In contrast, IL-2 binds the β receptor subunit (p75) with greater affinity and with decreased association–disassociation rates. The combination of these two binding characteristics in the intact hetrodimeric species yields receptors of very high affinity.

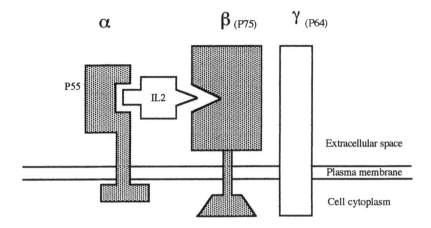

Figure 8.2 The interleukin-2 receptor. Refer to text for further details

IL-2; production and medical applications

IL-2 is secreted in small quantities by activated T helper cells. Several tumour cell lines, such as the Jurkat leukaemia line produce increased quantities of this cytokine. Cell lines such as these provided much of the IL-2 used for its initial characterization. Such a source would, however, be hopelessly inadequate in terms of producing clinically significant quantities of this interleukin. Recombinant DNA technology now facilitates the large-scale production of virtually unlimited amounts of IL-2. Both the native human IL-2 gene and cDNA have been used in recombinant protein production. The resultant protein products exhibit a range of biological activities identical to those of the native IL-2 molecule. Although IL-2 is normally glycosylated, most commercialized recombinant products are produced in *E. coli* and hence are devoid of carbohydrate side chains. Lack of glycosylation seems to have little, if any, adverse effect on the biological activity of the products.

IL-2 has been used for a number of years in the treatment of some forms of cancer (Table 8.6). More recently clinical trials assessing its potential in the treatment of a variety of infectious diseases, including AIDS, have been initiated. The effectiveness of IL-2 in treating such conditions lies in its ability not only to enhance B- and T-cell responses, but also in its ability to mobilize NK cells. Experiments in animal models carried out in the early 1980s illustrated that lung tumours regressed if NK cells were removed from the body, activated by incubation with IL-2 and then reintroduced into the body along with additional IL-2.

As IL-2 plays a pivotal role in the clonal expansion of both activated T and B lymphocytes, this cytokine may also prove to be useful as a vaccine adjuvant. A large proportion of conventional adjuvants are unsuitable for use in humans as they promote a variety of unacceptable side effects. Unfortunately, these are invariably the most potent adjuvants. Many such adjuvants potentiate the immune response by promoting the activation of macrophages. These activated cells release IL-1, which in turn induces the production of a variety of other cytokines, primarily IL-2. IL-2 then orchestrates the subsequent immunological response. Direct administration of IL-2, therefore, seems to be one possible approach to the development of future vaccine adjuvants.

A molecular understanding of IL-2 and its mode of action also allows researchers to logically manipulate IL-2-mediated immune responses. This knowledge has already facilitated the development of a number of ingenious methods which may be used to clinically induce selective immunosuppression. Such treatments could have important consequences for the survival of allografts or in the treatment of autoimmune conditions. Allografts, which is the grafting or transplanting of tissue or organs from a donor to a recipient, result in the immune system of the recipient perceiving the transplant as being non-self, due to the presence of foreign antigens. An immunological response against the transplant is initiated thus leading to transplant rejection. Thus far, the most common means of preventing rejection is to induce broad immunosuppression in the recipient by administering drugs such as glucocorticoids and cyclosporine. These drugs function by inhibiting the production of IL-2.

Selective immunosuppression may be induced by a number of means. IL-2 receptors are expressed solely on activated B and T lymphocytes. Administration of IL-2–toxin conjugates will thus lead to selective destruction of activated lymphocytes. Alternatively, monoclonal antibody preparations which react specifically with IL-2 receptors, but fail to trigger an intracellular response can be administered. Administration of such antibodies would block IL-2 binding and hence prevent IL-2 mediated proliferation of the activated lymphocytes. Soluble IL-2 receptors have also been produced. Parenteral administration of such polypeptides limits or prevents normal IL-2 functioning as they compete with native IL-2 receptors for binding of the cytokine. Other strategies employed in promoting clinical immunosuppression include the development of IL-2 analogues, which bind to the IL-2 receptor, but fail to trigger an intracellular response. Selective immunoenhancement may of course, be induced by the exogenous administration of IL-2. IL-2 analogues may also be developed which exhibit increased immunostimulatory activity.

Interleukin-11

Human IL-11 is a 23 kDa polypeptide predominantly produced by fibroblasts and bone marrow stromal cells. It is a haematopoietic growth factor which stimulates thrombopoiesis (Figure 8.3). cDNA coding for human IL-11 was first isolated in 1991, and from this its amino acid sequence was deduced. The cDNA has been expressed in a number of recombinant expression systems and IL-11's tertiary structure consists of four α-helical bundles with two streaches of β-sheets.

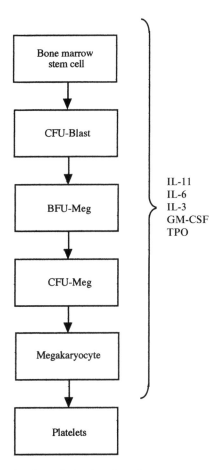

Figure 8.3 The production of platelets from bone marrow stem cells. Stem cells undergo differentiation under the influence of various combinations of cytokines, thereby producing red blood cells (erythrocytes), white blood cells (leukocytes) or platelets. This process is known as haemopoiesis. Thrombopoiesis (i.e. the production of platelets), is described here. Megakaryocytes are the immediate progenator cells from which platelets are derived. Platelet production entails the budding off of small membrane-bound vesicles from the megakaryocyte surface. These vesicles (platelets) then enter circulation. A combination of cytokines serve to promote platelet production. The interleukins and GM-CSF appear to promote differentiation of earlier cells in the sequence, whereas TPO mainly stimulates the final stages of platelet production. Characteristic cytokine redundancy likely results in several of these regulatory factors displaying overlapping activity in the context of platelet production

Note: IL, interleukin; GM-CSF, Granulocyte–macrophage colony-stimulating factor; TPO, thrombopoietin; CFU, colony-forming unit; BFU, burst forming unit.

rhIL-11 produced in an engineered *E. coli* strain has recently been approved for the treatment and prevention of chemotherapy-induced thrombocytopenia, a condition characterized by a low blood platelet

LIVERPOOL JOHN MOORES UNIVERSITY
LEARNING SERVICES

count. Chemotherapy often induces thrombocytopenia as stem cells are particularly sensitive to many chemotherapeutic agents. Prior to the approval of IL-11 for this indication, the only treatment available was direct platelet transfusions. This carried with it the low but real risk of accidental transmission of disease via infected platelet preparations.

Interleukin-1

IL-1 is produced by a number of cell types (Table 8.5), predominantly by phagocytic cells such as macrophages and monocytes. There are actually two distinct IL-1s termed IL-1α and IL-1β. Although these interleukins are the products of two distinct genes and exhibit only limited homology, they bind to the same receptor and induce identical biological responses. Both IL-1α and IL-1β are initially synthesized as precursor molecules with molecular masses of 31 kDa. These precursors are subsequently processed proteolytically, yielding active biological molecules with a molecular masses of approximately 17.5 kDa. Human IL-1α has an isoelectric point in the region of 5.5 while that of IL1-β is close to 7.0.

Two molecular forms of the IL-1 receptor have been identified. These have been termed type I and type II receptors. The type I receptor is found to be present mainly on the surface of T cells and fibroblasts. The type II receptor is present on B lymphocytes. Both IL-1α and IL-1β can bind to both receptors.

IL-1 exhibits a wide range of biological activities (Table 8.7). Such activities include promotion of the proliferation of a wide variety of susceptible cell types, exerting an antiproliferative effect on some cancer cells, induction of fever, stimulation of wound healing and regulation of certain inflammatory events. IL-1 therefore, has also been termed leukocyte-activating factor, lymphocyte-activating factor, T-cell replacing factor and endogenous pyrogen.

IL-1 promotes the release of IL-2 from activated T lymphocytes, and many biological activities attributed to IL-1 may be mediated, at least in part, by IL-2. IL-1 has also been implicated in the progression of a number of disease states. It has been isolated, for example, from the synovial fluid of arthritic patients.

IL-1 preparations were initially obtained from the supernatant fluid of macrophages, activated by incubation with bacterial lipopolysaccharide (LPS). Although small quantities of IL-1 could be purified in this way, large-scale production by such means was obviously impractical.

The cloning and expression of the IL-1 cDNA in *E. coli* has facilitated large-scale production of biologically active IL-1. Large-scale clinical trials designed to assess the effectiveness of IL-1 in treating a variety of cancers in addition to promoting wound healing are currently underway, although initial results have proven somewhat disappointing.

A soluble form of the IL-1 receptor has also been expressed in recombinant bacterial systems. Parenterally administered soluble receptor

Table 8.7 Selected biological activities of interleukin-1

Promotes proliferation of
  thymocytes
  fibroblasts
  haemopoietic cells
  lymphocytes

Promotes wound healing

Displays an anti-proliferative effect on certain cancer cell lines

Promotes inflammation

Induces hepatic acute phase protein synthesis

Stimulates release of prostaglandin and collagenase from
synovial cells

Stimulates release of additional cytokines from selected cells

Induces a fever response

Induces sleep

---

should compete with native receptor for binding to IL-1 and, in this way, lessen its biological effects. This would most likely benefit patients suffering from a number of conditions such as septic shock, where many of the adverse clinical symptoms observed are due to the synthesis and release of inappropriately high concentrations of this cytokine.

Interleukins-4, -6 and -12

Human IL-4 is a 129 amino acid glycoprotein which displays various immunomodulatory activities. It activates B lymphocytes and regulates (in part) immunoglobulin synthesis. It also induces growth and activation of T lymphocytes (particularly cytotoxic T cells) and NK cells. Its effect upon the latter cell types forms the basis for the application of IL-4 in the treatment of cancer (Box 8.2). IL-4 also displays powerful anti-inflammatory properties, which suggests potential application in the treatment of conditions such as rheumatoid arthritis. In addition to the above properties, IL-4 has been implicated in the regulation of haemopoiesis, and its administration to humans can increase haematocrit levels (i.e. the fraction of blood consisting of red blood cells) by up to 20 per cent.

Human IL-6 is a 183 amino acid, 26 kDa glycoprotein produced by a variety of cells (Table 8.5). This cytokine likely plays a central role in combatting infectious diseases and inflammation by co-stimulating

T lymphocytes and inducing (a) antibody secretion by B cells and (b) acute phase proteins (important in destruction of invading pathogens). IL-6 administration has been reported to induce immune-mediated antitumour effects *in vivo* and this has lead to follow-up clinical trials. IL-6 also displays haemopoietic activity. It promotes a significant (up to 20 per cent) increase in blood platelet counts and also increases the production of B and T lymphocytes. As such several clinical trials investigating IL-6's efficacy in the treatment of various haematological conditions have been initiated.

Human IL-12 is a 75–80 kDa heterodimeric protein, consisting of 35 kDa and 40 kDa subunits. Size heterogeneity is due to glycosylation of both subunits. It is a multifunctional cytokine, promoting the cytolytic activity of NK cells and T lymphocytes. It also enhances humoral immunity and promotes the secretion of various pro-inflammatory cytokines such as TNF and IFN-γ. Its broad range of immunoregulatory effects renders it of potential clinical interest, and several trials have been initiated to appraise its effectiveness in treating infectious diseases and cancer.

TUMOUR NECROSIS FACTORS

Tumour necrosis factors (TNFs) are additional members of the cytokine family of regulatory proteins. Two forms of TNF are now recognized: TNF-α and TNF-β. Although both proteins bind the same receptors and elicit broadly similar biological responses, they are distinct molecules and share less than 30 per cent homology. The original protein termed 'tumour necrosis factor', although still referred to as TNF, is more properly termed TNF-α. It is also known as cachectin. TNF-β is also referred to as lymphotoxin. A summary of the physicochemical characteristics of TNF-α and TNF-β is provided in Table 8.8

Table 8.8   Summary of the physicochemical characteristics of human TNF-α and TNF-β

Property	TNF-α	TNF-β
Amino acid content	157	171
Molecular mass (kDa)	52*	25
Glycosylation	No	Yes
Disulfide bonds	1	0
pI	5.6	5.8

*Biologically active human TNF-α is a homotrimer. The monomer, which is biologically inactive, displays a molecular mass of 17.3 kDa.

TNF-α

The eventual discovery of TNF-α stems from observations made at the turn of the nineteenth century by an American surgeon called William Coley. Coley observed that the tumours of some cancer patients regressed or disappeared after they suffered a severe bacterial infection. Coley thus attempted to treat cancer patients by administering live bacteria. This approach however, suffered from several disadvantages, not least of which was the inability to control the ensuing infection in those pre-antibiotic days. In an effort to surmount such difficulties, Coley subsequently developed a vaccine consisting of dead bacterial suspensions, which became known as Coley's toxins. Some clinical successes were recorded when cancer patients were treated with such toxins. Consistent results were never attained and this method of treatment fell out of medical fashion.

The active component of Coley's toxins was later shown to be a complex biomolecule termed LPS. This molecule is found in association with the outer membrane of Gram-negative bacteria. LPS contains both lipid and polysaccharide components, and is also referred to as endotoxin.

LPS itself is devoid of any antitumour activity. The serum of animals injected with LPS was found to contain a factor toxic to such cancerous cells. This factor, which was produced by specific cells in response to LPS, was termed tumour necrosis factor (necrosis refers to cellular death). LPS represents the most potent known stimulant of TNF-α production.

TNF-α is synthesized by a wide variety of cell types, most notably activated macrophages, monocytes, certain T lymphocytes, NK cells, in addition to the brain and liver cells. Although LPS represents the most potent inducer of TNF-α synthesis, various other agents such as some viruses, fungi and parasites also stimulate the synthesis and release of this cytokine. Furthermore, TNF-α may act in an autocrine manner, stimulating its own production.

Native human TNF-α is a homotrimer, consisting of three identical polypeptide subunits tightly associated about a three-fold axis of symmetry. This arrangement resembles the assembly of protein subunits in many viral caspids. The individual polypeptide subunits of human TNF-α are non-glycosylated and consist of 157 amino acids. The molecule has a molecular mass of 17.3 kDa and contains one intrachain disulfide linkage. X-ray crystallographic studies of the TNF-α monomer have shown that the molecule is elongated. Much of its amino acid sequence forms two β pleated sheets, each of which contains five anti-parallel β-strands. Human TNF-α is synthesized initially as a 233 amino acid precursor molecule. Proteolytic cleavage of a 76 amino acid signal sequence releases native TNF-α. TNF-α may also exist in a 26 kDa membrane-bound form.

Biological effects of TNF-α

TNF-α induces its biological effects by binding specific receptors present on the surface of susceptible cells. Two distinct TNF-α receptors have been identified. One receptor (TNF-R55) has a molecular mass of 55 kDa whereas the second receptor (TNF-R75) has a molecular mass in the region of 75 kDa. These two distinct receptor types show no more than 25 per cent sequence homology. TNF-R55 is present on a wide range of cell types whereas the distribution of the type II receptor is more limited. Both are transmembrane glycoproteins with an extracellular binding domain, a hydrophobic transmembrane domain and an intracellular effector domain.

The exact molecular mechanisms by which TNF-α induces its biological effects remain to be determined. Binding of this cytokine to its receptor results in receptor oligomerization. This in turn triggers a variety of events, mediated by G-proteins in addition to the activation of adenylate cyclase, phospholipase A_2 and protein kinases. The exact biological actions induced by TNF-α may vary from cell type to cell type. Additional factors, such as the presence of additional cytokines further modulate the observed molecular effects attributed to TNF-α action on sensitive cells. The major biological activities of TNF-α are summarized in Table 8.9.

Under normal conditions TNF-α is produced in relatively low levels and exerts its influences in a paracrine–autocrine manner. It plays a central role in several physiological processes such as inflammation and the immune response. It upregulates non-specific immunity directly by activating a range of phagocytic cells. It also indirectly influences additional elements of the immune response by promoting the synthesis and release of a range of additional cytokines. It promotes an inflammatory response both directly (by e.g. activating neutrophils and other pro-inflammatory leukocytes) and indirectly (by e.g. promoting synthesis of pro-inflammatory cytokines). Its pro-inflammatory activities, and particularly its stimulation of various elements of innate and acquired immunity also forms the basis of its cytotoxicity towards various transformed cells. However, the ability of TNF-α to induce tumour cell

Table 8.9 Basic biological activities of human tumour necrosis factors

---

Activates a range of phagocytic cells

Promotes an inflammatory response directly and indirectly

Displays cytotoxicity against a range of transformed cells

Induces synthesis and release of a range of additional cytokines, including IL-1, IL-6, IL-8 and colony-stimulating factors

High systemic levels mediates septic shock, cachexia and possibly (in part) various auto-immune conditions

---

necrosis (death) is now known to be a secondary biological activity of the molecule.

Various conditions (e.g. serious microbial infections) can trigger over-production of TNF-α. Under such circumstances this cytokine enters the general circulation and acts in an endocrine like manner. The majority of such systemic (i.e. whole body as opposed to localized) effects have negative health implications

Septic shock is a serious medical condition usually associated with Gram-negative bacterial infection. LPS present in the outer membrane of such Gram-negative bacteria plays a central role in the development of this condition. LPS, as previously mentioned, is the most potent known inducer of TNF synthesis and it has been demonstrated clinically that administration of exogenous TNF-α can induce symptoms identical to those seen in patients suffering from septic shock. Furthermore, it has been shown that pretreatment with anti-TNF-α antibodies exerts a pro-tective effect on animals subsequently challenged with potentially lethal doses of LPS.

Prolonged production of inappropriately elevated levels of TNF-α has also been implicated in the development of cachexia, the wasting syn-drome often associated with chronic parasitic or other infections, and with cancer.

Medical applications of TNF-α

The TNF-α gene has been cloned and expressed in recombinant *E. coli*. The resultant availability of large quantities of purified, biologically active TNF has facilitated clinical evaluation in a number of diseased states, most notably cancer. Many such trials either employing TNF, alone or in combination with IFN-γ, yielded disappointing results.

Such trials usually employed rhTNF-α administered systemically. At therapeutically relevant doses this induced a wide range of unacceptable side effects. High dose TNF-α administration in combination with chemotherapy has, however, proven effective in treatment certain cancers affecting the limbs. In this context the root of the limb is surgically isolated, and the limb is then perfused with TNF-α. This so called isolated limb perfusion technique facilitates administration of high local concen-trations of TNF-α while preventing leakage into the general circulation. This in turn prevents the occurrence of negative systemic effects. 'Bero-mun' is the tradename given to rhTNF-α produced in *E. coli*, which has been approved for this purpose. An overview of its industrial production is presented in Figure 8.4.

TNF-β

TNF-β, previously known as lymphotoxin, was actually discovered prior to TNF-α. At that time however it was not recognized as a TNF. TNF-β

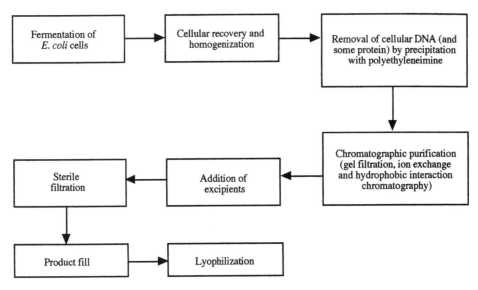

Figure 8.4 Schematic overview of the production of 'Beromun', a rhTNF-α product produced intracellularly in engineered *E. coli* cells

is produced by activated T and B lymphocytes, and is a glycoprotein consisting of 171 amino acids. Although TNF-α and TNF-β exhibit only limited amino acid sequence homology, they both bind the same receptors and exhibit broadly similar biological activities. The TNF-β cDNA has been cloned and expressed in recombinant *E. coli*. Its resultant availability in large quantities allows its biochemical characteristics and clinical potential to be more fully assessed.

COLONY-STIMULATING FACTORS

The serum of healthy individuals normally contains three distinct leukocyte (white blood cell) types: lymphocytes, granulocytes and monocytes. Lymphocytes may be sub-divided into B and T cells, which function to promote antibody- and cell-mediated immunity respectively. Granulocytes may be further sub-categorized as basophils, neutrophils and eosinophils. These cells are capable of detecting and destroying bacteria and other foreign particles (Box 8.1). Some may also play a role in allergic responses. Monocytes are also capable of engulfing and destroying bacteria and other foreign particles, in addition to tissue debris.

As previously discussed, all circulating blood cells are ultimately derived from a single cell type, the bone marrow haematopoietic stem cells. A variety of haemopoietic growth factors stimulate the proliferation and differentiation of such stem cells, ultimately yielding this variety of blood cells (Figure 8.5).

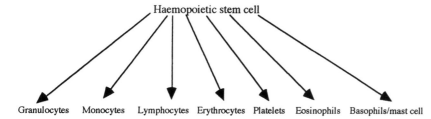

Figure 8.5 Various blood cell types derived from haemopoietic stem cells

Colony stimulating factors (CSF), mainly promote the proliferation and differentiation of certain precursor cells derived from stem cells, ultimately yielding mainly mature granulocytes and macrophages. These two cell types play important roles in defending the body against a wide range of pathogens. Any factor which decreases the circulating level of these cell types renders an individual extremely susceptible to infectious disease.

The term colony-stimulating factor reflects the ability of these substances to promote the *in vitro* growth of various leukocytes in clumps or colonies. A number of different CSF have been identified and characterized (Table 8.10). These include granulocyte–macrophage CSF (GM-CSF), granulocyte CSF (G-CSF), macrophage CSF (M-CSF) and multi-potential CSF (multi-CSF), now known to be IL-3.

The various CSF exhibit little amino acid sequence homology and receptors for all four subtypes have been identified on susceptible cells. All the receptors are transmembrane glycoproteins having an extracellular ligand-binding domain, a transmembrane segment and an intracellular effector domain. CSF signal transduction entails tyrosine phosphorylation, although additional mechanisms may also be involved.

GM-CSF displays haematopoietic activity, specifically promoting the differentiation, growth and activation of granulocytes and monocytes. It also serves as a growth factor for T cells, endothelial cells and megakaryocytes. G-CSF serves as a growth, differentiation and activation

Table 8.10 Range, sources and molecular mass of human colony-stimulating factors

Name	Composition	Molecular mass (kDa)	Sources
GM-CSF	127 amino acid, single chain glycoprotein	22	T cells, macrophages, fibroblasts, endothelial cells
G-CSF	177 amino acid, single chain glycoprotein	21	Macrophages, fibroblasts, endothelial cells
M-CSF	522 amino acid, dimeric glycoprotein	90	Multiple
Multi-CSF (IL-3)	133 amino acid, single chain glycoprotein	14–30	Activated T cells, mast cells, eosinophils

factor for neutrophils and their precursor. These cells display a short circulatory life span (6–8 h) and are normally synthesized in the body at a rate of some 70 000 cells per second. G-CSF also promotes growth of endothelial cells. M-CSF stimulates the growth differentiation and activation of macrophages.

Production and medical applications of CSF

The advent of recombinant DNA technology has facilitated the large-scale production of CSF. In many instances post-translational glycosylation is not required to sustain biological activity. CSF produced in procaryotic expression systems may thus be employed clinically. Recombinant CSF have also been produced in mammalian expression systems which are capable of carrying out post-translational modifications, especially glycosylation.

GM-CSF has been produced in a variety of recombinant expression systems, most notably in CHO cells and in *Saccaromycese cerevisiae*. Leukine (sargramostim) is a recombinant product produced in engineered yeast, and it was approved by the FDA in 1991. It differs from native human GM-CSF with regard to its exact glycosylation pattern and it also contains a leucine at residue 23. It displays identical *in vivo* biological activity to that of the native cytokine. Leukine is administered to patients after autologous bone marrow transplantation in order to accelerate myeloid (i.e. bone marrow derived) cell recovery.

Neupogen is a recombinant form of G-CSF which was first approved by the FDA in 1991, and is used to treat chemotherapy-induced neutropenia. (Neutropenia is a condition characterized by a depressed blood neutrophil count, with a resulting increased suceptability to infection). It is produced in engineered *E. coli*.

CYTOKINE TOXICITY

Many of the adverse clinical symptoms associated with infection or other disease states are often caused by increased cytokine circulatory levels. For example influenza symptoms such as loss of appetite, aches and pains, shivering and fatigue are largely triggered by increased systemic concentrations of IFNs and additional cytokines, released by immune system cells in response to infection. It is therefore not surprising that many of these symptoms are reproduced in clinical studies involving the administration of a variety of cytokines to human patients. Undesirable side effects generally limits the upper cytokine dosage levels that can safely be used in human medicine. This in turn often limits the therapeutic efficacy of these products. A list of side effects sometimes associated with the administration of various cytokines is presented in Table 8.11. The more serious side effects, while fairly infrequently observed at approved

Table 8.11 Some adverse effects sometimes associated with the clinical administration of selected cytokines

Cytokine	Effect
Type I IFNs	Flu-like symptoms (muscle ache, fever, chills, weakness, headache, nausea), vomiting, rash, diarrhoea, vasodilation, psychiatric effects (depression, anxiety), anorexia
IL-2	Flu-like symptoms, diarrhoea, vomiting, rash, respiratory complications, cardiovascular complications (hypertension, vasodilation, tachycardia). Nervous-related conditions (confusion, anxiety)
IL-1	Flu-like symptoms, oedema, headache, nausea, abdominal pain, cardiac and renal complications, confusion
G-CSF	Bone pain, capillary leakage syndrome (rare)
TNF-α	Fever, nausea, vomiting, chills, liver toxicity, cardiac arrhythmia

therapeutic doses, can be potentially fatal. Recipients must be monitored
carefully upon product administration. Treatment can be deferred or
withdrawn should serious side effects begin to develop.

FURTHER READING

Books

Balkwill, F. (2000) *Cytokine Network*. Oxford University Press, Oxford.
Bonivada, B. (1990) *Tumour Necrosis Factor*. Karger, Basel.
Fiers, W. (1993) *Tumour Necrosis Factor*. Karger, Basel.
Garland, J. *et al.* (1997) *Colony-stimulating Factors in Molecular and Cellular Biology*. Marcel Dekker, New York.
Lindenmann, J. (1999) *Interferons*. Springer-Verlag, Godalming.
Maroun, J. (1992) *Colony-stimulating Factors in Clinical Practice*. Royal Society of Medicine, London.
Medcalf, D. (1995) *Haemopoietic Colony-stimulating Factors*. Cambridge University Press, Cambridge.
Mire-Sluis, A. (1998) *Cytokines*. Academic Press, London.
Thomson, A. (1998) *Cytokine Handbook*. Academic Press, London.

Articles

Interferons

Allen, G. (1982) Structure and properties of human interferon-α from Namalwa lymphoblastoid cells. *Biochem. J.* **207**, 397–408.

Allen, G. & Fantez, K. (1980) A family of structural genes for human lympho-blastoid (leukocyte-type) interferon. *Nature* **287**, 408–411.

Aringer, M. *et al.* (1999) Interferon/interleukin signalling, a 1999 perspective. *Immunologist* **7** (5), 139–146.

Borden, E. *et al.* (2000) Second generation interferons for cancer; clinical targets. *Semin. Cancer Biol.* **10** (2), 125–144.

Cosman, D. *et al.* (1990) a new cytokine receptor superfamily. *Trends Biochem. Sci.* **15** (7), 265–270.

Goodbourn, S. *et al.* (2000) Interferons: cell signalling, immune modulation, antiviral responses and virus countermeasures. *J. Gen. Virol.* **81**, 2341–2364.

Gray, P. *et al.* (1982) Expression of human immune interferon cDNA in *E. coli*, and monkey cells. *Nature* **295**, 503–508.

Kerr, I. & Stark, G. (1991) The control of interferon inducible gene expression. *FEBS Lett.* **285** (2), 194–198.

McCarty, M. (2000) Interferons in dermatology. *Curr. Prob. Dermatol. US* **12** (6), 260–264.

Nagata, S. *et al.* (1980) Synthesis in *E. coli* of a polypeptide with human leukocyte interferon activity. *Nature* **284**, 316–320.

Pestka, S. *et al.* (1987) Interferons and their actions. *Ann. Rev. Biochem.* **56**, 727–777.

Platanias, L. & Fish, E. (1999) Signalling pathways activated by interferons. *Exp. Hematol.* **27** (11), 1583–1592.

Taniguchi, T. *et al.* (1980) Expression of the human fibroblast interferon gene in *E. coli. Proc. Natl. Acad. Sci. USA.* **77** (9), 5230–5233.

Interleukins

Auron, P. (1998) The interleukin-1 receptor:ligand interactions and signal trans-duction. *Cytokine Growth Factor Rev.* **9** (3–4), 221–237.

Kaplan, G. *et al.* (1992) Rational immunotherapy with interleukin-2. *Bio/Technology* **10**, 157–162.

Kronheim, S. *et al.* (1986) Purification and characterization of human interleukin-1 expressed in *E. coli. Bio/Technology* **4**, 1078–1082.

Maliszewski, C. & Fanslow, W. (1990) Soluble receptors for IL-1 and IL-4: biological activity and therapeutic potential. *Trends Biotechnol* **8**, 324–329.

Nelson, B. & Willerford, D. (1998) Biology of the interleukin 2 receptor. *Adv. Immunol.* **70**, 1–81.

Scott, P. (1993) IL-12: Initiation cytokine for cell mediated immunity. *Science* **260**, 496–497.

Smith, K. (1988) Interleukin-2, inception, impact and implications. *Science* **240**, 1169–1176.

Smith, K. (1990) Interleukin-2. *Sci. Am.* **262** (3), 26–33.

Trinchieri, G. (1998) Interleukin 12: a cytokine at the interface of inflammation and immunity. *Adv. Immunol.* **70**, 83–243.

Waldmann, T. (1989) The multi-subunit interleukin-2 receptor. *Ann. Rev. Biochem.* **58**, 875–911.

TNF

Aggarwal, B. (2000) Tumor necrosis factor (TNF): A double edged sword. *J. Clin. Ligand Assay* **23** (3), 181–192.

Camussi, G. *et al.* (1991) The molecular action of tumor necrosis factor-α. *Eur. J. Biochem.* **202**, 3–14.

Fiers, W. (1991) Tumor necrosis factor. *FEBS Lett.* **285**, 199–212.

Moreland, L. (1999) Recent advances in anti-tumor necrosis factor (TNF) therapy in rheumatoid arthritis – focus on the soluble TNF receptor p75 fusion protein, etanercept. *Biodrugs* **11** (3), 201–210.

Old, L. (1988) Tumor necrosis factor. *Sci. Am.* **258**, 41–49.

Sprang, S. (1990) The divergent receptors for TNF. *Trends Biochem. Sci.* **15** (10), 366–368.

CSF

Bourette, R. & Rohrschneider, L. (2000) Early events in M-CSF receptor signalling. *Growth Factors* **17** (3), 155–166.

Hartung, T. (1999) Immunomodulation by colony stimulating factors. *Rev. Physiol. Biochem. Pharmacol.* **136**, 1–164.

Plugariu, C. & Williams, W. (2000) Granulocyte–macrophage colony stimulating factor (GM-CSF) antagonists: design and potential application. *Drugs Future* **25** (12), 1295–1305.

Weaver, J. *et al.* (1988) Production of recombinant human CSF-1 in an inducible mammalian expression system. *Bio/Technology* **6**, 287–290.

Wiedermann, F. *et al.* (2001) Recombinant granulocyte colony stimulating factor (G-CSF) in infectious diseases: still a debate. *Wien. Klin. Wochenschr.* **113** (3–4), 90–96.

9

Proteins used for analytical purposes

INTRODUCTION

Enzymes and antibodies have found a range of analytical applications, most notably in the *in vitro* clinical diagnostics sector. An increasing understanding of normal and abnormal metabolic activity allows clinicians to link changes in the concentration of various biomolecules to disease states or to impending events. Sensitive and specific diagnostic assays capable of detecting and quantifying many such marker molecules are now available. These diagnostic systems greatly assist medical practitioners in the accurate diagnosis or prediction of medical abnormalities, thus allowing them to formulate the most appropriate therapeutic responses.

Clinical chemistry is the scientific discipline charged with detecting, monitoring and quantifying a broad variety of marker substances present in biological samples. A wide range of biomolecules are of potentially significant diagnostic value. Such substances include low molecular mass metabolic products such as urea, glucose, cholesterol or steroid hormones. Many substances of higher molecular mass, such as specific proteins which may be released from damaged tissue or whose normal concentration is altered due to a particular metabolic aberration, are also of diagnostic value (Table 9.1). Additional diagnostic tests have been developed which detect supramolecular assemblies including viruses or microorganisms.

It is important to note that most disease conditions are dynamic rather than static in nature. The value of many diagnostic results are often closely correlated not only to the sensitivity and specificity of the test, but also to the speed with which results may be obtained. Indeed, the progression of many disease states is often monitored by performing repeat tests on samples obtained from the patient at appropriate time intervals.

A large number of diagnostic systems have been designed to facilitate their automation. The use of automatic multi-sample analysers in any clinical laboratory increases the speed, efficiency, throughput and

Table 9.1 Some proteins of diagnostic significance and diseases potentially associated with an increase or decrease in the concentrations of such proteins

Protein	Diseases associated with increased concentration of protein	Diseases associated with decreased concentration of protein
Albumin	Blood–brain barrier damage	Malnutrition; liver cirrhosis; severe burns; inflammation; renal disease
α_1-Acid glycoprotein	Inflammatory conditions; malignant neoplasms	Severe hepatic damage; nephrotic syndrome
α_1-Antitrypsin	Acute hepatic diseases; active cirrhosis; inflammatory disease	Genetic deficiency; chronic pulmonary disease
α_2-macroglobulin	Nephrotic syndrome; oral contraceptive use; diabetes	Disseminated intravascular coagulation; stress
Apolipoprotein A1		Atherosclerosis; Tangier disease
Apolipoprotein B	Atherosclerosis; hyperlipidaemias	

Table 9.1 (*continued*)

Protein	Diseases associated with increased concentration of protein	Diseases associated with decreased concentration of protein
C-reactive protein	Inflammation; tumours; tissue destruction; active rheumatoid arthritis	
Ceruloplasmin	Inflammatory conditions; oral contraceptive use	Wilson's disease; malnutrition
Complement C3	Inflammatory disease	Autoimmune disease; lupus; chronic hepatitis; neonatal respiratory tissue injury
Complement C4	Acute inflammatory disease; bacterial infections	Acute glomerular nephritis; chronic hepatitis; autoimmune disease; lupus
Haptoglobin	Inflammation	Haemolytic anaemia; sickle-cell anaemia; hepatic disease
Immunoglobulin A	IgA myeloma; cirrhosis; chronic liver disease; chronic infections	Immune deficiency; non-IgA myeloma
Immunoglobulin G	Chronic infections: IgG myeloma; liver disease; multiple sclerosis; meningitis	Immune deficiency
Immunoglobulin M	IgM myeloma; liver disease; acute hepatitis	Immune deficiency; non-IgM myeloma
Kappa (light chains)	Monoclonal gammopathies; Bence Jones light chain disease	Monoclonal gammopathies
Lambda (light chains)	Monoclonal gammopathies; Bence Jones light chain disease	Monoclonal gammopathies
Microalbumin	Early stage renal disease; diabetes; hypertension	
Prealbumin		Malnutrition; liver disease; acute inflammation
Properdin factor-B		Autoimmune diseases; sickle-cell disease; bacterial diseases
Rheumatoid factor	Rheumatoid arthritis; mixed connective tissue disease; viral and bacterial disorders	
Transferrin	Iron deficiency; pregnancy; acute hepatitis; oral contraceptive use	Inflammation

economy with which diagnostic assays are carried out. Blood and urine normally constitute the most commonly employed biological samples for analysis, although other body products, such as faeces, saliva or sweat, may also be used.

Proteins most often used as analytical tools for diagnostic purposes include a variety of enzymes and antibody preparations. Enzymes are used largely to detect and quantify various medically significant metabolites present in biological samples. Antibodies are normally used to detect and quantify the presence of specific antigens in such samples. In a reversal of this procedure, antigens may be employed to detect the presence of specific antibodies in serum. This latter approach is illustrated by the use of specific antigens associated with the HIV in order to detect the presence of anti-HIV antibodies in blood or serum (described later).

A wide range of alternative techniques also form the basis for numerous diagnostic tests. Such techniques include electrophoresis, conventional chromatography, HPLC, isoelectric focusing and chromatofocusing, all of which are routinely used in modern clinical chemistry laboratories. More recently, a variety of diagnostic tests based upon recombinant DNA technology have been introduced in the clinical arena. Such techniques may be used, for example to detect a variety of genetic abnormalities which may serve as markers, indicating the presence of a number of genetic disorders such as Duchenne muscular dystrophy or cystic fibrosis. DNA tests are characterised by an excellent degree of specificity, sensitivity and rapidity. Such tests are poised to make a significant impact on the future of laboratory medicine.

While the majority of their analytical applications pertain to clinical diagnostics, enzymes and antibodies are also used to detect and quantify analytes important in other sectors (e.g. the food industry and environmental monitoring). Applications to these areas are also briefly considered within this chapter.

ENZYMES AS DIAGNOSTIC/ANALYTICAL REAGENTS

A variety of enzyme preparations have been used as diagnostic reagents for many years. Enzymes may be used to detect and estimate the levels of specific analytes present in biological samples. In certain situations, enzymes may also be used to catalytically remove specific compounds present in such samples, which may interfere with the assay of a particular metabolite. Enzymes are also widely used as labels in enzyme immunoassay systems (EIA). Their high degree of selectivity, coupled with their catalytic efficiency render many enzymatic preparations ideal diagnostic reagents.

Currently in the region of 70 enzymes used for analytical purposes are produced commercially by companies such as Boehringer Mannheim

Diagnostics (now merged with Roche Diagnostics) and Genzyme. Many additional companies purchase enzymes in bulk from these companies for manufacture of finished analytical kits. The total world market for analytical enzymes is well in excess of $100 million, and it is a market which continues to grow steadily. Alkaline phosphatase and peroxidase are the two diagnostic enzymes produced in greatest quantities, with each commanding a market value in excess of $20 million. A number of criteria determine if an enzyme is suited to analytical application. These criteria include:

- *Specificity of reaction catalysed.* The enzyme must display high substrate specificity for the analyte to be measured. It must be devoid of side activities towards additional substances potentially present in the sample to be analysed if such side reactions could in any way interfere with the detection system.

- *Kinetic properties.* The enzyme should display suitable K_m and K_{cat} values, and not be subject to inhibition by any substance likely or potentially present in the samples to be assayed.

- *Source availability and cost.* The enzyme should be produced in modest/high quantities by its (native or recombinant) producer. The cost of upstream and downstream processing should be competitive.

- *pH and temperature versus activity profiles* should be conducive to significant catalytic activity under the assay conditions employed.

- *Enzyme stability.* The enzyme should be stable under the storage conditions specified for several months, and it should be stable under the conditions that the assay is performed.

Enzymes used for analytical purposes are obtained from a variety of plant, animal and microbial sources (Table 9.2). Most are still produced by direct extraction from native producer sources, although some (e.g. alkaline phosphatase and cholesterol oxidase) are now largely produced by recombinant DNA technology. Recombinant systems used for commercial production include *Streptomyces* and *Picia* species, although a whole range of analytical enzymes have been expressed in recombinant *E. coli* as well as in various fungal systems.

Most commercial enzymes used for analytical purposes are required in quantities of 1–10 kg annually. A few (e.g. glucose oxidase) are produced at much higher levels (100 kg or more per year). Most enzyme preparations used for diagnostic purposes have been subjected to some form of purification. Unlike protein preparations destined for parenteral administration, it is not usually necessary to purify diagnostic reagents to homogeneity. However, it is of crucial importance to ensure that the purification procedure employed removes any proteins or other molecules present in the initial preparation which might interfere with the assay or lead to erroneous results.

Table 9.2   Some enzymes used directly or indirectly as diagnostic reagents and their source and likely applications

Enzyme	Source	Application
Acetyl cholinesterase	Bovine erythrocytes	Analysis of organophosphorus compounds such as pesticides
Alcohol dehydrogenase	Yeast	Determination of alcohol levels in biological fluids
Alkaline phosphatase	Calf intestine and kidney, recombinant (*Picca* sp.)	Conjugation to antibodies allows its use as an indicator in ELISA systems
Arginase	Beef liver	Determination of L-arginine levels in plasma and urine.
Ascorbate oxidase	*Cucurbita* species	Determination of ascorbic acid levels; eliminates intereference by ascorbic acid
Cholesterol esterase	Pig/beef pancreas, *Pseudomonas* sp., Recombinant (*Streptomyces* sp.)	Determination of serum cholesterol levels
Creatine kinase	Rabbit muscle, beef heart, pig heart	Diagnosis of cardiac and skeletal malfunction
Glucose-6–phosphate dehydrogenase	Yeast, *Leuconostoc mesenteroides*	Determination of glucose and ATP in conjunction with hexokinase
Glucose oxidase	*Aspergillus niger*	Determination of glucose in biological samples in conjunction with peroxidase; a marker for ELISA systems
Glutamate dehydrogenase	Beef liver	Determination of blood urea nitrogen in conjunction with urease
Glycerol kinase	*Candida mycoderma, Arthrobacter* sp.	Determination of triglyceride levels in blood in conjunction with lipase
Glycerol-3-phosphate dehydrogenase	Rabbit muscle	Determination of serum triglycerides
Hexokinase	Yeast	Determination of glucose in body fluids
Peroxidase	Horseradish	Indicator enzyme for reactions in which peroxide is produced
Phosphoenolpyruvate carboxylase	Maize leaves	Determination of CO_2 in body fluids.
Urease	Jack bean	Determination of blood urea nitrogen; marker enzyme for ELISA systems
Uricase	Porcine liver	Determination of uric acid
Xanthine oxidase	Buttermilk	Determination of xanthine and hypoxanthine in biological fluids

Most enzymatic preparations used in the detection of various molecules of diagnostic significance are free in solution. However, for some applications, the enzyme is used in immobilized format. Examples of the latter format are detailed later.

End point versus kinetic methods

Enzymes used in the detection and quantification of specific analytes normally employ the analyte in question as a substrate. Changes in the concentration of one or other of the co-reactants, cofactors or products of the reaction must be readily monitored. In practice this is most often achieved by spectrophotometry, fluorimetry or (in the case of many biosensors) by electrochemical measurement. As the magnitude of signal generated is proportional to the analyte concentration present, quantitative tests may be developed. For example in the case of spectrophotometric tests the quantity of light absorbed will be related to the substrate concentration by Beer's law ($A = \varepsilon cl$, where A = absorbency measurement obtained, ε = molar absorption coefficient, c = concentration of analyte and l = path length). Enzyme-based analyte quantification is carried out by either end point or kinetic methods.

End point methods may be developed if an enzyme of appropriate characteristics (e.g. substrate specificity and kinetic properties) is identified. Essentially the enzyme must be capable of selectively, speedily and completely converting the target analyte into product in a stoichiometric fashion. In such instances pre-specified quantities of enzyme and test solution are co-incubated under appropriate conditions (of temperature and pH), with subsequent quantitative determination of the signal generated (e.g. change in absorbancy in the case of spectrophotometric tests; Figure 9.1). Some enzymes will achieve incomplete analyte conversion into product if the reaction displays an unfavourable equilibrium constant. End point based analyte quantification is still possible as long as the assay is carried out under conditions which 'shove' or 'pull' the reaction to completion. In practice this can be achieved by increasing the initial concentration of co-substrate (if one exists) or by adding a substance which reacts with one of the reaction products, thereby trapping it (Figure 9.2).

An alternative approach to enzyme-based detection and quantification of a specific analyte is termed the kinetic method. Basic enzyme kinetics illustrate that, for any reaction, if the substrate concentration present is much lower than the enzyme's Michaelis constant (i.e. $[S] \ll K_m$), then the observed reaction rate is linearly proportional to the substrate concentration. The kinetic approach therefore measures the rate at which substrate is being converted (i.e. $-dS/dt$) or the rate at which product (P) is being formed (i.e. $+dP/dt$). This approach required construction of a calibration curve in which varying known substrate concentrations are used (Figure 9.3). Kinetic based assays invariably can be carried out much

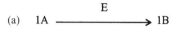

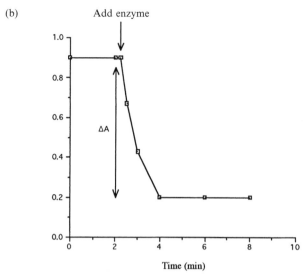

Figure 9.1 Detection and quantification of a specific analyte, A, by an enzyme-based end point method. The enzyme, when incubated under appropriate assay conditions, completely or virtually completely converts A into B in a stoichiometric fashion. In this case one molecule of A is converted into one molecule of B (a). The amount of substrate consumed or product formed is quantified (b). In this case A absorbs (at the wavelength chosen) whereas B does not. The change in absorbance recorded can be directly related to analyte concentration by Beer's law

faster than end point based ones, however an enzyme displaying a high K_m towards the analyte must generally be used.

Some common enzyme-based diagnostic tests

Two of the most commonly employed enzyme types in diagnostic systems are dehydrogenases and oxidases. In the case of dehydrogenases, progression of the reaction may be followed by monitoring the conversion of reduced nicotinamide adenine dinucleotide (NADH) to NAD^+ or vice versa, as illustrated below (S = reaction substrate; P = reaction product).

$$S + NAD^+ \xrightarrow{\text{dehydrogenase}} P + NADH + H^+$$

Oxidation or reduction of the cofactor may be readily monitored spectro-photometrically, as NADH absorbs strongly at 340 nm, whereas NAD^+ does not absorb at this wavelength. Progression of the reaction may also

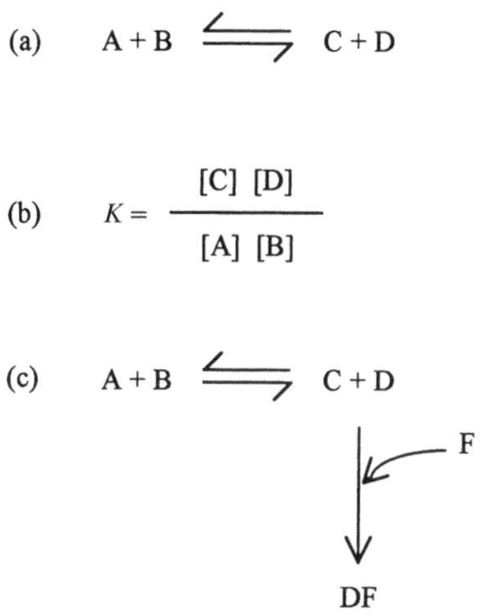

(a) $A + B \rightleftharpoons C + D$

(b) $K = \dfrac{[C]\ [D]}{[A]\ [B]}$

(c) $A + B \rightleftharpoons C + D$

DF

Figure 9.2   Generalized reaction in which an enzyme inter-converts A + B and C + D. The equilibrium constant, *K*, value for the reaction is calculated according to the formula presented in (b). A reaction which, upon reaching equilibrium, has almost completely converted A + B into C + D will display a high *K* value. If the reaction does not go to near completion it can essentially be pulled to completion by addition of a chemical which reacts with one of the products (C or D). This is represented as F in diagram (c). In this example D is removed almost as soon as its formed, so its concentration continually remains low

be followed by fluorescence or luminescence. From the above generalized example, it is clear that one molecule of NADH is formed for each molecule of substrate (*S*) present – the analyte whose concentration is to be determined. Thus, in such a system, the total quantity of NADH formed reflects the quantity of substrate present in the sample analysed.

When considering this example, it becomes evident that all assay reagents should be present in excess, such that it is the concentration of substrate which dictates the final quantity of NADH formed. If insufficient NAD^+ were present for example, the reaction would cease once all the cofactor was converted to NADH, although significant quantities of the substrate, S, might still remain. In such circumstances, the quantity of substrate present in the sample analysed would be underestimated.

Oxidases, which yield hydrogen peroxide as a reaction product, are also commonly employed in diagnostic systems. However in most such instances none of the final products of this primary reaction (e.g. P or H_2O_2 in the example below) are readily quantified. A second enzymatic step is therefore included in the assay system. The second 'linker' enzyme utilizes one of the products of the initial enzymatic step as one of its substrates; H_2O_2 in the generalized example illustrated below. Unlike the primary enzymatic step, one or other of the products of this second

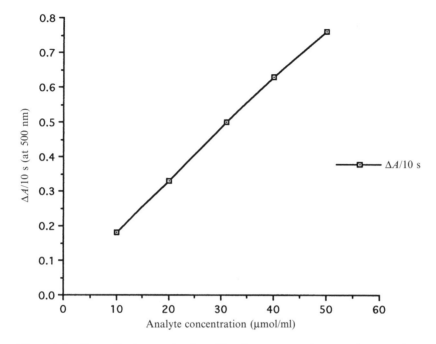

Figure 9.3   Generalized example of a calibration curve used to correlate reaction rate with analyte (i.e. substrate) concentration. In this case the analyte absorbs at 500 nm, whereas the reaction product does not. Assaying a sample containing an unknown concentration of analyte will yield a $\Delta A/10\,\text{s}$ value which can then be correlated to the analyte concentration by reference to the standard curve

reaction is easily measured by use of an appropriate assay. In the example cited, the second enzyme utilizes the H_2O_2 produced in the initial reaction, together with a second colourless substance as substrates, yielding H_2O and a coloured dye as products. The quantity of dye produced can be determined by monitoring the increase in absorbance at the appropriate wavelength. As long as all the reagents present in the assay system are in excess, the quantity of dye produced by such a coupled assay is directly related to the quantity of substrate present in the sample analysed.

The concepts outlined in the generalized examples discussed above are illustrated more clearly by the specific examples as discussed below. Several alternative enzymatic methods have been developed to assay many biological substances of diagnostic interest.

$$S + \frac{1}{2}O_2 + H_2O \xrightarrow{\text{oxidase}} P + H_2O_2$$

$$H_2O_2 + \text{colourless substance} \xrightarrow[\text{e.g. peroxidase}]{\text{second enzyme}} \text{dye(coloured)} + H_2O$$

Assay of blood glucose

The concentration of glucose present in blood is an important diagnostic marker for several disease states, especially diabetes. Blood glucose determinations constitute one of the most common assays carried out in clinical chemistry laboratories. Most such determinations are based upon specific assays, of which there are several available. A system using glucose oxidase was one of the first enzyme based analytical tests developed. The reaction principle is as follows:

$$\text{Glucose} + O_2 + H_2O \xrightarrow{\text{glucose oxidase}} \text{gluconic acid} + H_2O_2$$

$$2H_2O_2 + \text{4-aminophenazone} + \text{phenol} \xrightarrow{\text{peroxidase}} \text{quinoneimine} + 4H_2O$$

In this coupled assay, the level of glucose in the initial sample is directly related to the quantity of quinoneimine formed. Unlike the other reaction products, quinoneimine is a red–violet dye whose concentration can easily be determined by measuring its absorbance at a wavelength of 500 nm.

In the above example the reagents present in the assay system consist of the enzymes glucose oxidase and peroxidase, as well as phenol and 4-aminophenazone. A specified volume of this reagent cocktail would be incubated with a specific volume of the serum sample to be assayed. The reaction would be allowed to proceed for an appropriate time period, typically 5–10 min, at an appropriate temperature, typically 20–37 °C, until all the glucose present in the serum sample has been enzymatically converted to gluconic acid. Several additional assays in which serum samples have been replaced with standard glucose solutions are normally run concurrently. The glucose concentration present in the serum sample can be calculated by comparison of unknown with standard absorbancy values.

A second assay system commonly used to determine blood glucose levels employs the enzymes hexokinase and glucose-6–phosphate dehydrogenase (G6P-DH). The reaction principle is outlined below:

$$\text{Glucose} + \text{ATP} \xrightarrow{\text{hexokinase}} \text{glucose 6-phosphate} + \text{ADP}$$

$$\text{Glucose 6-phosphate} + \text{NAD}^+ \xrightarrow{\text{G6P-DH}}$$
$$\text{gluconate 6-phosphate} + \text{NADH} + H^+$$

The concentration of glucose present in the sample is related to the quantity of NADH formed by the coupled assay. This may be easily monitored by measuring the resultant absorbance at 340 nm.

The above assay systems are capable of rapidly, accurately and conveniently determining the concentration of blood glucose present in any sample tested. Such a result, however, solely reflects the concentration of glucose present at the time the blood sample was taken.

Diagnostic tests that monitor average long-term blood glucose levels are also available. Such tests are invariably based on assessing the level of glycosylated haemoglobin present in a serum sample. Elevated blood glucose concentrations promote the glycosylation of haemoglobin molecules by covalent linkage of a glucose residue to the terminal valine residue of the haemoglobin β chain. This reaction may occur progressively in the erythrocyte throughout its normal 120 day life span. Thus the level of glycosylated haemoglobin present in a blood sample reflects blood glucose levels over a period of several weeks. This assay is invaluable in assessing the long term effectiveness of therapeutic approaches as applied to diabetic patients. Separation of glycosylated haemoglobin from native haemoglobin is normally achieved by electrophoretic or chromatographic procedures.

Determination of the presence of glucose in urine also constitutes an important diagnostic test. Such a system is often employed in the initial detection of diabetes, as will be discussed later.

Assay of blood cholesterol and triglycerides

An association between atherosclerosis and elevated plasma cholesterol levels has been established. This association relates in particular to levels of cholesterol associated with low density lipoproteins (LDL-cholesterol). The condition of atherosclerosis is responsible for the deaths of one in every two individuals in most Western societies.

Due to their largely hydrophobic nature, both cholesterol and triglycerides are transported through the blood stream in the form of soluble complexes termed lipoproteins, (Figure 9.4). The outer layer of lipoprotein particles consists largely of phospholipids and proteins termed

Figure 9.4 Generalized structure of a plasma lipoprotein

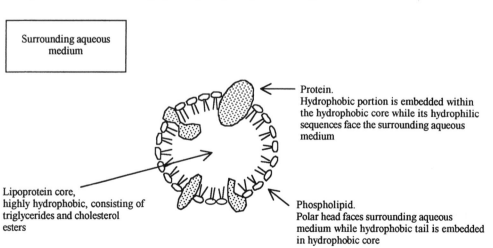

Surrounding aqueous medium

Protein.
Hydrophobic portion is embedded within the hydrophobic core while its hydrophilic sequences face the surrounding aqueous medium

Lipoprotein core, highly hydrophobic, consisting of triglycerides and cholesterol esters

Phospholipid.
Polar head faces surrounding aqueous medium while hydrophobic tail is embedded in hydrophobic core

apoproteins. The hydrophilic portions of these molecules are thus oriented towards the surrounding aqueous environment. The internal portion of such lipoproteins is composed predominantly of triglycerides and cholesterol covalently linked to long chain fatty acids via ester bonds, cholesterol-esters.

Based upon differences in physiochemical properties, lipoproteins may be classified as high density lipoproteins (HDL), low density lipoproteins, LDL or very low density lipoproteins (VLDL) (Table 9.3). Chylomicrons, another group of lipoproteins are found in plasma, particularly after the injestion of a lipid rich meal. Chylomicrons are the largest lipoproteins and consist almost exclusively of triglycerides. They serve to transport dietary fats.

Increases in the circulatory concentration of LDL-cholesterol in particular is linked to the development of artherosclerosis. Elevated levels of HDL-cholesterol on the other hand appear to exert a protective effect against the development of fatty deposits on the inner walls of arteries characteristic of artherosclerosis. This may be due to the fact that HDL functions to transport cholesterol back from peripheral tissues to the liver where it may be converted to bile salts.

Several diagnostic tests capable of measuring serum concentrations of cholesterol are available. Total serum cholesterol levels may be estimated by employing the three-step coupled enzymatic system illustrated below:

$$\text{Cholesterol ester} + H_2O \xrightarrow{\text{cholesterol esterase}} \text{cholesterol} + \text{fatty acids}$$

$$\text{Cholesterol} + O_2 \xrightarrow{\text{cholesterol oxidase}} \text{4-cholestenone} + H_2O_2$$

$$2H_2O_2 + \text{phenol} + \text{4-aminoantipyrine} \xrightarrow{\text{peroxidase}} \text{quinoneimine} + 4H_2O$$

In the initial reaction, cholesterol esterase catalyses the hydrolytic cleavage of cholesterol esters, yielding free cholesterol, which may then be oxidized by cholesterol oxidase. The resultant H_2O_2 can be quantified by the peroxidase system. The end product, quinoneimine, is a red dye which may be easily quantified by measuring its absorbance at 500 nm.

Table 9.3 Approximate content of triglycerides and cholesterol in various lipoproteins

	Triglycerides (%)	Cholesterol (%)
Chylomicrons	85–90	2–4
VLDL	50–60	15
LDL	10	45
HDL	3–4	17–18

Total serum HDL-cholesterol may be specifically quantified by using the above assay method. In this case it is necessary to firstly remove the other lipoproteins from the serum sample. This may be achieved by the addition of phosphotungstic acid and magnesium ions which promote the precipitation of LDL, VLDL and chylomicrons. Following a centrifugal step, the cholesterol content of the HDL-containing supernatant may be assayed by the above procedure.

Elevated levels of serum triglycerides have been linked to atherosclerosis and coronary artery disease. Various enzymatic systems have been developed which facilitate appraisal of triglyceride concentration in serum. One commonly used system which is based upon four sequential enzymatic reactions, is illustrated below:

$$\text{Triglycerides} + H_2O \xrightarrow{\text{lipases}} \text{glycerol} + \text{fatty acids}$$

$$\text{Glycerol} + \text{ATP} \xrightarrow{\text{glycerol kinase}} \text{glycerol 3-phosphate} + \text{ADP}$$

$$\text{Glycerol 3-phosphate} + O_2 \xrightarrow{\text{glycerol 3-phosphate oxidase}} \text{dihydroxacetone phosphate} + H_2O_2$$

$$H_2O_2 + \text{4-aminoantipyrine} + \text{chlorophenol} \xrightarrow{\text{peroxidase}} \text{quinoneimine dye} + \text{HCl} + 2H_2O$$

The concentration of triglycerides in the serum sample provided is obviously proportional to the quantity of dye produced. Quinoneimine levels may be determined by measuring the absorbance at 500 nm.

A variety of additional enzymatic systems, such as the one shown below, have been developed to quantify triglyceride levels in biological samples.

$$\text{Tryglycerides} + H_2O \xrightarrow{\text{lipases}} \text{glycerol} + \text{fatty acids}$$

$$\text{Glycerol} + \text{ATP} \xrightarrow{\text{glycerol kinase}} \text{glycerol 3-phosphate} + \text{ADP}$$

$$\text{ADP} + \text{phosphoenolpyruvate} \xrightarrow{\text{pyruvate kinase}} \text{ATP} + \text{pyruvate}$$

$$\text{Pyruvate} + \text{NADH} + H^+ \xrightarrow{\text{lacate dehydrogenase}} \text{lactate} + \text{NAD}^+$$

Assay of blood urea and uric acid

Determination of urea and uric acid levels are carried out frequently in most clinical laboratories. Elevated levels of these metabolites are often indicative of a variety of metabolic disorders, including diseases of the kidney. Most methods used to estimate urea levels include the enzyme urease, which catalyses the hydrolytic cleavage of urea as shown below.

$$\text{Urea} + H_2O \xrightarrow{\text{urease}} \text{ammonia} + CO_2$$

A variety of methods may then be used to detect and quantify the amount of ammonia formed by this reaction. One such method relies upon the chemical detection of the ammonia formed. Ammonium ions react with phenol and hypochlorite in the presence of sodium nitroprusside, thus forming a blue coloured complex, which absorbs light strongly at 640 nm.

An alternative enzymatic method used to quantify the ammonia produced utilizes the enzyme glutamate dehydrogenase, as illustrated below:

$$\text{Urea} + H_2O \xrightarrow{\text{urease}} 2NH_4^+ + CO_2$$

$$2\alpha\text{-ketoglutarate} + 2NADH + 2NH_4^+ \xrightarrow{\text{glutamate dehydrogenase}}$$
$$2 \text{ glutamate} + 2NAD^+ + 2H_2O$$

The NAD^+ formed by the second reaction may readily be monitored, and it is of course, proportional to the quantity of urea present in the initial sample.

Uric acid may also be quantified by a number of enzymatic systems. One of the more common methods uses a coupled assay system consisting of uricase and peroxidase, as shown below. The end product, the red dye quinoneimine, has an absorbance maximum at 520 nm.

$$\text{Uric acid} + O_2 + 2H_2O \xrightarrow{\text{uricase}} \text{allantoin} + CO_2 + H_2O_2$$

$$2H_2O_2 + 4\text{-aminophenazone}$$
$$+ 3, 5\text{-dichlor-2 hydroxybenzene sulfonic acid} \xrightarrow{\text{peroxidase}}$$
$$\text{quinoneimine} + 4H_2O$$

An alternative method for the quantitation of uric acid is as follows:

$$\text{Uric acid} + O_2 + 2H_2O \xrightarrow{\text{uricase}} \text{allantoin} + H_2O_2 + CO_2$$

$$H_2O_2 + \text{ethanol} \xrightarrow{\text{catalase}} \text{acetaldehyde} + 2H_2O$$

$$\text{Accetaldehyde} + NADP^+ \xrightarrow{\text{aldehyde dehydrogenase}} \text{acetate} + NADPH + H^+$$

The increase in absorbance at 340 nm due to the formation of NADPH is proportional to the quantity of uric acid present in the sample.

Immobilized enzymes as diagnostic reagents

There has been a steady increase in the development and use of diagnostic reagents in immobilized form. Enzymes immobilized on reagent strips are used for selected diagnostic purposes in clinical laboratories, directly by doctors and indeed by patients themselves.

One of the most commonly employed such reagent strips are those designed to detect glucose in body fluids. The wafer-thin plastic strips contain a narrow pad at one end, onto which the sample is applied. The sample material diffuses inwards into the body of the pad, which houses the appropriate diagnostic enzyme, in addition to any ancillary reagents required.

Most such systems are designed such that one of the products produced by the final enzymatic reaction is a chromogen and thus coloured. Thus, most systems are based on visual examination of any colour development in the pad after the sample is applied (Figure 9.5) The intensity of the colour produced reflects the concentration of the metabolite in the sample applied. Such kits may, therefore, be employed in a semi-quantitative fashion.

BIOSENSORS

Biosensors are a class of analytical device which can detect and quantify specific analytes in complex samples. They consist of a biological component, which selectively binds the target analyte in a quantitative fashion, a transducer that quantitatively converts the biological signal into a physical or chemical one, and a detector that gives a quantitative read-out of the transduced signal to the end user (Figure 9.6).

The biological component must obviously exhibit extreme specificity for the target analyte and binding of the analyte must produce a

(a)

(b) $\quad$ Glucose $+ O_2 + H_2O \xrightarrow{\text{Glucose oxidase}}$ gluconic acid $+ H_2O_2$

$\qquad H_2O_2 + \text{dye}_{(red)} \xrightarrow{\text{Peroxidase}} H_2O + \text{dye}_{(ox)}$

Figure 9.5 Detection of glucose in biological (or other) fluids using the dipstick method. The device consists of a thin strip of plastic to which an absorbent pad has been attached at one end (a). Immobilized in the pad are enzyme(s) capable of yielding a coloured product in the presence of glucose. A combination of glucose oxidase, peroxidase and a dye which changes colour or develops colour upon its oxidation are most commonly employed (b)

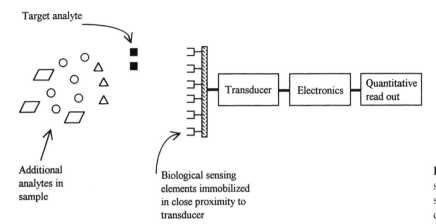

Target analyte

Additional
analytes in
sample

Biological sensing
elements immobilized
in close proximity to
transducer

Figure **9.6** Generalized schematic diagram of a bio-sensor. Refer to text for specific details

detectable response capable of being transduced (converted) into a physical or chemical signal. For the majority of biosensors, the biological component is an enzyme. Most biosensors use electrochemical-based systems as signal transducers, although alternative optical, mass and thermometric systems have been developed. Biosensors display a number of advantages as analyte detection or quantification systems, including:

- reusability, which has favourable cost implications;

- no additional reagents are required to make measurements;

- the assay sample need not be pre-treated to e.g. remove coloured compounds or particulate matter;

- flexibility; portable biosensors can be used in the field. Biosensors can be used to run individual or multiple samples whenever required (it is uneconomic/impractical to run single assay samples on a multianalyte analyser);

- ease of use; biosensors are uncomplicated and can usually be used by personnel with little technical training.

Most biosensors have to date found application in a diagnostics/clinical setting (Table 9.4), although some are used for food analysis (Table 9.5)

Enzyme-based biosensors

A wide variety of immobilized enzymes have been used in the construction of enzyme based biosensors. These biosensors usually employ electrochemical-based detection systems. The principles adopted in the design and construction of an enzyme biosensor are basically straightforward (Figure 9.7). The immobilized enzyme is usually housed within a membranous structure which surrounds the chosen electrode. The pore size of

Table 9.4 Analytes of diagnostic importance for which specific quantitative enzyme electrodes have been developed and the enzymes used in these test systems

Analyte	Enzyme used in its detection
Glucose	Glucose oxidase
Urea	Urease
Uric acid	Uricase
Amino acids	L-Amino acid oxidase
Cholesterol	Cholesterol oxidase
Alcohols	Alcohol oxidase
Penicillin	Penicillinase

Table 9.5 Analytes of importance in the food industry for which biosensors have been developed (although not all have been commercialized)

Analyte	Enzyme
Glucose	Glucose oxidase
Fructose	Fructose dehydrogenase
Sucrose	Invertase and glucose oxidase
Lactose	β-Galactosidase and glucose oxidase
Lactic acid	Lactate oxidase or lactate dehydrogenase
Ascorbic acid	Ascorbate oxidase
Citric acid	Citrate lyase and oxaloacetate decarboxylase
Ethanol	Alcohol oxidase or alcohol dehydrogenase
Asparatame	APBC and aspartate aminotransferase and glutamate oxidase
Sulfite	Sulfite oxidase

In most cases a single enzyme biosensor is sufficient. However in some cases a combination of two or more enzymes are required to generate an easily detectable reaction product. Invertase, for example, hydrolyses the disaccharide sucrose, yielding glucose and fructose. Glucose oxidase then generates a measurable electrochemical signal by oxidizing the glucose produced. In the case of aspartame the enzyme aspartate peptide bond cleaving enzyme (APBC) cleaves the peptide bond of this dipeptide, yielding free aspartate and phenylalanine methyl ester. Aspartate aminotransferase then converts aspartate to glutamate, and glutamate oxidase in turn generates the electrochemical signal.

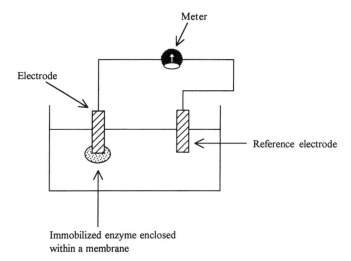

Meter

Electrode

Reference electrode

Immobilized enzyme enclosed
within a membrane

Figure 9.7 Diagrammatic representation of an enzyme electrode

the outer membrane which faces the surrounding aqueous environment of the test sample must be sufficiently large to allow free penetration of all the substrates necessary for the immobilized enzyme reaction. One of these substrates is obviously the target analyte. In order to protect the immobilized enzyme this outer membrane, which is often made from polycarbonate, must exclude entry of proteins, particulate matter and whole cells. The inner membrane, resting between the immobilized enzyme and the electrode surface, must in no way interfere with the diffusion of reaction substrates and products. These molecules must freely come into contact with the electrode surface. This inner membrane is often made from cellulose acetate.

Although most enzyme biosensors employ membrane-based enzyme entrappment systems, additional modalities of enzyme entrapment have also been used (see also the section on enzyme immobilization, Chapter 10). The enzyme can be entrapped within a gel made from polyacrylamide or gelatin, for example. Alternatively the enzyme can be covalently linked to a support material in close proximity to the detecting electrode, or alternatively, directly to the electrode itself. Direct covalent linkage of enzyme generally yields biosensors with longest operational lifetimes.

Several electrode types may be used in constructing the enzyme electrode. Electrodes which specifically detect gases such as O_2, CO_2, NH_3 and various ionic species, are all commercially available. The electrode chosen will depend upon what products are produced by the enzymatic reaction. If for example oxygen is evolved, then the oxygen electrode may be used. The quantity of oxygen molecules impinging on the electrode surface is then converted into some form of detectable signal, such as a flow of electrical current. This signal can then be monitored. Readings obtained when the electrode is immersed in solutions containing known concentrations of the analyte allows a standard curve to be constructed.

In the construction of an enzyme electrode an enzyme is normally chosen which (a) utilizes as substrate the analyte whose concentration is sought, and (b), one or other of whose additional substrates or reaction products may be detected and quantified by the actual electrode.

When the enzyme electrode is immersed in the test sample, the analyte whose concentration is sought diffuses through the outer membrane, comes into contact with the immobilized enzyme and hence is catalytically transformed. The reaction products then pass freely through the inner membrane and impinge on the electrode surface, thus generating a signal. The magnitude of this signal is proportional to the concentration of substrate present in the sample tested.

First-generation enzyme biosensors as described above suffer from a number of drawbacks, including the generation of high background values, due to the presence of interfering ions in assay samples. The incorporation of electron transferring mediators in such biosensor systems largely overcomes such difficulties. Such mediators are low molecular mass substances such as quinones and ferrocenes, which facilitate electron transfer between the electrode and the (redox-based) enzyme employed. The mediator operates at low redox potential allowing operation of the system at low electrode potential. Under such conditions the electrode detects only the transfer from the mediator, which will be directly proportional to the concentration of the target analyte present in the sample.

For most enzyme electrodes the target analyte is a (co)substrate of the enzyme used. A recently developed variation is the inhibited enzyme electrode. In this case the enzyme is chosen on the basis that it is inhibited by the target analyte. The presence of the analyte in a sample therefore results in a reduction of enzyme activity which is quantitatively measured by an electrochemical (or other) detection system. Various pesticides for example can be detected via their inhibition of immobilized acetylcholinesterase.

Non-enzyme-based biosensors

A range of biosensors utilizing a biological component other than an enzyme have also been developed. Alternative biological components include (non-catalytic) proteins, DNA and whole cells. Proteins such as glutathione-S-transferases have been engineered such that they selectively bind specific heavy metals. The engineered proteins have been immobilized directly onto gold electrodes which act as signal transducers. Binding of a heavy metal to the engineered protein results in a change in protein conformation, which in turn alters the system's capacitance in a manner proportional to the concentration of the target metal.

Antibody-based biosensors have also been developed. An obvious strength of this approach is that antibodies can be raised against most antigens of interest. Predictably the major technical hurdle has been the development of suitable transduction systems capable of detecting anti-gen–antibody binding. The tip of an atomic force microscope may be used to monitor biospecific molecular interactions such as antibody–antigen binding. However cost considerations alone renders commercial exploit-ation of this approach unlikely. Optical techniques such as surface plas-mon resonance can also directly measure antibody–antigen interactions occurring at a surface-solution interface (Figure 9.8), and some systems based on this approach have been developed. The general strategy pur-sued in developing DNA based biosensors entails the immobilization of single-stranded DNA sequences. Hybridization with complementary target sensitive sequences could be linked to a system giving suitable optical or electrical signals. Such biosensors display obvious potential in clinical, environmental and forensic science. Biosensors using whole cells have also been developed, and in some instances commercialized. The cells used are invariably microbial, which are more robust under *in vitro* conditions than plant or animal cells. The cells are generally entrapped in a gel or membrane held in close association with an electrode. One of the first such systems to be commercialized was a biosensor used for deter-mining the biological oxygen demand (BOD) of environmental samples. Organic pollutants present in the sample will be metabolized by the

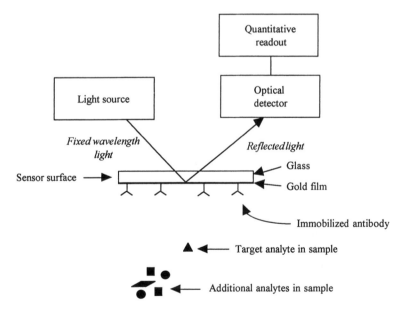

Figure 9.8 Simplified overview of surface plasmon resonance (SPR) mediated detection of antibody–antigen binding. Light of a fixed wavelength is directed at (and reflected from) a thin metal film as shown. Binding of the target analyte to the metal surface (via the immobilized antibody) changes slightly the refract-ive index of the surface layer. This in turn causes a reduction in intensity of the reflected light, which can be measured quantitatively by an appropriate detector

immobilized microbial cells of *Trichosporon cutaneum*. The increase in associated cellular respiration results in increased consumption of dissolved oxygen, which can be measured quantitatively by an oxygen electrode.

ANTIBODIES AS ANALYTICAL REAGENTS

Antibodies are another group of proteins which are frequently employed as diagnostic reagents. Antibodies are favoured as analytical reagents because they exhibit extreme specificity in their recognition of a particular ligand, the antigen which stimulated their production. Antibody preparations are often used in the detection and quantification of a wide variety of specific analytes, many of which are of considerable diagnostic significance. Antibodies thus find widespread analytical use in clinical laboratories, but are also used in other sectors, including in the food industry and for environmental monitoring. It must always be borne in mind that the immunological activity of a substance does not necessarily equate to its biological activity. This is particularly true with regard to protein antigens. Many proteins may retain their immunological identity, even under conditions which promote a reduction of their biological activity. In certain circumstances, antigen molecules may be used as diagnostic reagents in order to detect and quantify the presence of specific antibody species in serum samples.

Assays which employ antibodies to detect and quantify specific substances are generally termed immunoassays. The substance of interest is firstly employed as an antigen and injected into animals in order to elicit the production of antibodies against that particular molecule. If the antigen is too small to elicit an immunological response it may firstly be attached to a large carrier molecule to render it antigenic. Either monoclonal or polyclonal antibody preparations may be used in immunoassay systems.

Antibody molecules themselves have no inherent characteristics which facilitate their direct detection in immunoassay systems. A second important step in developing a successful immunoassay therefore involves the incorporation of a suitable marker. The marker serves to facilitate the rapid detection of antibody-antigen binding. Immunoassay systems which use radioactive labels as marker systems are termed radioimmunoassays (RIA) whereas systems using enzymes are termed enzyme immunoassays (EIA).

Radioimmunoassay

RIA was one of the first immunoassay systems to be developed. Such systems have subsequently been successfully employed in the quantitation of a large variety of biomolecules present in the body at very low concentrations. RIA technology has found widespread application both in basic research and for many applied purposes.

Two basic reagents are required when initially developing an RIA system: (a) antibodies which have been raised against the antigen of interest and (b) highly purified antigen which has been labelled with a radioactive tag such as ^{125}I. In all assay systems the concentration of antibody added must be limiting, while the concentration of radiolabelled antigen must be in excess.

In a typical radioimmunoassay, appropriate volumes of antibody, labelled antigen and sample supplied for assay are incubated together. As the antibody concentration is limiting, the antigen (Ag) present in the sample supplied will compete with the labelled antigen (Ag*) for binding to antibody (Ab):

$$Ag + Ag^* + Ab(limiting) \rightleftharpoons AgAb + Ag^*Ab + Ag + Ag^*$$

In all such systems, the quantity of both antibody and labelled antigen used remain constant. Thus, when equilibrium is reached, the greater the quantity of antigen present in the assay sample, the less labelled antigen that will be bound by antibody.

The next step in the assay progression involves separation of bound from free antigen. Separation is normally achieved by using techniques which precipitate the antibody–antigen complex while leaving unbound antigen, both labelled and unlabelled, in solution. Physical separation may then be achieved by decanting the supernatant from the assay tube, such that only the precipitate remains. In practice, precipitation of antigen–antibody complex from solution is most often achieved by adding a second antibody to the reaction mixture. This thus second antibody has been raised against the first antigen-binding antibody.

Upon separation of free from bound antigen, the level of radioactivity associated with bound antigen is normally assessed. As previously mentioned, the more unlabelled antigen present in the assay sample, the lower will be the proportion of labelled antigen bound and hence the lower the radioactive count obtained.

A standard curve, which relates antigen concentration to levels of radioactivity bound, may be constructed by assaying several samples containing known concentrations of unlabelled antigen. The concentration of antigen present in the samples may thus be calculated by reference to the standard curve.

Enzyme immunoassay

The successful introduction of RIA revolutionized many areas of clinical and other biological sciences over the past 30 years. However, there are a number of disadvantages associated with the use of radioactive elements in such systems. Disadvantages include;

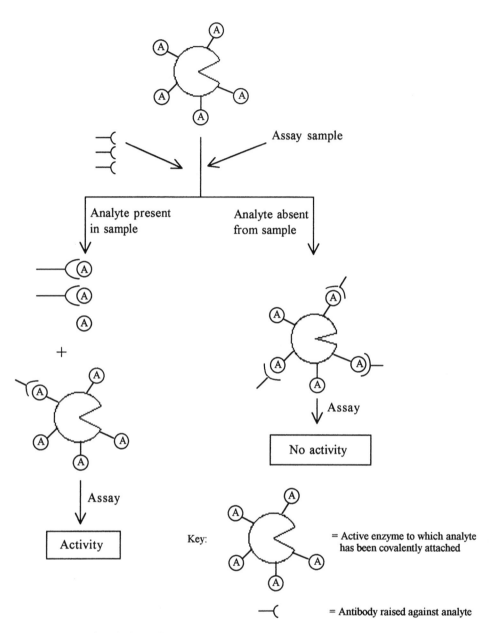

Figure 9.9 Basic principle of the enzyme multiplied immunoassay technique (EMIT). Purified preparations of the analyte to be detected are covalently linked to a chosen enzyme. This
process must not result in enzyme inactivation. Antibody which specifically binds the target
analyte is also required. When running the test the antibody is added in limiting quantities to
the enzyme analyte conjugate, and the sample to be analyzed is also added. If the analyte is
present in the sample it will compete with the conjugated analyte for antibody binding, and
thus the enzyme will remain maximally or near maximally active. Activity is quantified

- the need for radiological protection;

- the generation of radioactive waste;

- the short shelf life of some assays due to the short half-life of some radioactive compounds;

- a requirement for expensive analytical equipment, and often dedicated areas in which to perform the assays.

Such disadvantages have led to the development of a variety of additional immunoassay systems, which use alternative labels. Some such systems employ fluorescent or chemiluminescent tags. However, the development of immunoassay systems employing enzymes as marker substances has proven to be particularly popular. EIA systems take advantage of: (a) the extreme specificity and affinity with which antibodies bind antigens which stimulated their initial production, coupled to (b) the catalytic efficiency of enzymes which facilitates signal amplification as well as straightforward detection and quantification. In general, many EIA exhibit similar sensitivities to RIA but are free from most of the disadvantages associated with RIA.

EIA are sometimes described as heterogeneous or homogeneous systems. Heterogeneous assays involve the separation of bound and free label – as is the case with standard RIA. Certain EIA have been developed in which the activity of the enzyme label is significantly altered by binding of antibody to antigen. In such cases, there is no requirement to separate free from bound. Assay systems of this design are termed homogeneous systems. The enzyme multiplied immunoassay technique (EMIT) is an example of a homogenous immunoassay which has found widespread commercial application (Figure 9.9).

EIA were first introduced in the early 1970s. In most such systems, the antibody was immobilized on a solid surface, such as on the internal walls of the wells in a microtitre plate. (Figure 9.10). As in the case of classic RIA, the immobilized antibody was incubated with a known amount of labelled antigen, in addition to unknown quantities of antigen present in

---

by the use of a chromogenic substrate. However, if no analyte is present in the assay sample, the antibody binds exclusively to the enzyme–analyte conjugate, thereby significantly or totally inactivating the enzyme. A standard curve can be constructed by assaying samples of known analyte concentration; analyte concentration will be proportional to activity levels recorded

Note: Binding of antibody to the analyte–enzyme conjugate likely decreases enzyme activity by a twofold mechanism: (i) It induces a change in enzyme conformation and (ii) because of its high molecular mass, the antibody may sterically hinder substrate access to the active site. For any given EMIT system the level of inhibition achievable will depend upon the exact enzyme and analyte in question, the level of analyte substitution and the mode of attaching the analyte to the enzyme.

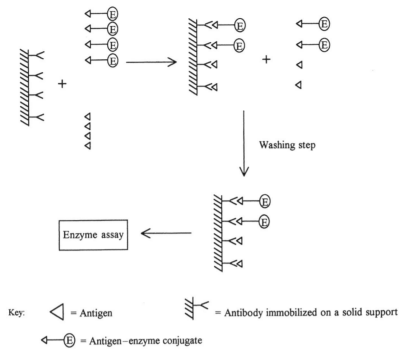

Key: ◁ = Antigen   ⊱⊰ = Antibody immobilized on a solid support

◁—Ⓔ = Antigen–enzyme conjugate

Figure 9.10 Principle of competitive solid phase EIA. The bound enzymatic activity is inversely proportional to the quantity of unlabelled antigen present in the sample assayed

the samples being assayed. As the level of antibody present is limiting, labelled and unlabelled antigen compete with each other for binding. The greater the quantity of unlabelled antigen present in the assay sample, the less labelled antigen will be retained by the immobilized antibody. After allowing antibody–antigen binding to reach equilibrium, unbound antigen is removed by a washing step. The amount of enzyme labelled antigen retained is assayed for enzymatic activity.

An alternative variation of this EIA involves immobilizing the antigen and employing an antibody–enzyme conjugate. The principle involved is outlined in Figure 9.11. In this case, the immobilized antigen and the free antigen present in the assay sample compete for binding to a limited amount of enzyme-labelled antibody. The more free antigen present in the assay sample, the less enzyme–antibody conjugate that will bind the immobilized antigen. A washing step removes all unbound material. A subsequent enzyme assay allows accurate estimation of the quantity of bound enzyme.

A standard curve of antigen concentration versus bound enzymatic activity may be constructed by assaying several samples containing known quantities of antigen. Readings for unknowns may therefore be quantified by reference to this standard curve.

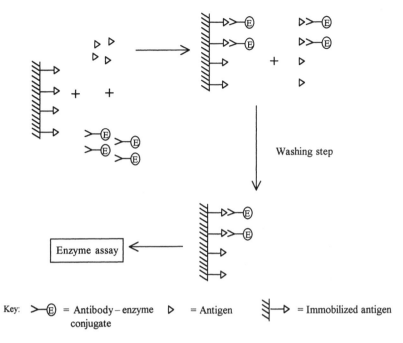

Key: >—Ⓔ = Antibody–enzyme   ▷ = Antigen   ⌇—▷ = Immobilized antigen
conjugate

Figure 9.11 Enzyme immunoassay system using immobilized antigen. The higher the level of antigen present in the sample for assay, the lower the level of enzyme–antibody conjugate retained on the immobilized phase, and hence the lower the level of enzymatic activity recorded

Enzyme-linked immunosorbant assay (ELISA)

Since their initial introduction over 30 years ago many variations on the basic enzyme immunoassay concept have been designed. One of the most popular EIA systems currently in use is that of the ELISA. The basic principle upon which the ELISA system is based is illustrated in Figure 9.12. In this form it is also often referred to as the double antibody sandwich technique.

In the basic ELISA system, antibodies raised against the antigen of interest are adsorbed onto a solid surface – again, usually the internal walls of microtitre plate wells. The sample to be assayed is then incubated in the wells. Antigen present will bind to the immobilized antibodies. After an appropriate time period, which allows antibody–antigen binding to reach equilibrium, the wells are washed.

A preparation containing a second antibody, which also recognizes the antigen, is then added. If monoclonal antibodies are used, this second monoclonal antibody must recognize an epitope on the antigen surface which differs from the epitope recognized by the primary or immobilized monoclonal antibody. The second antibody will also bind to the retained antigen and the enzyme label is conjugated to this second antibody.

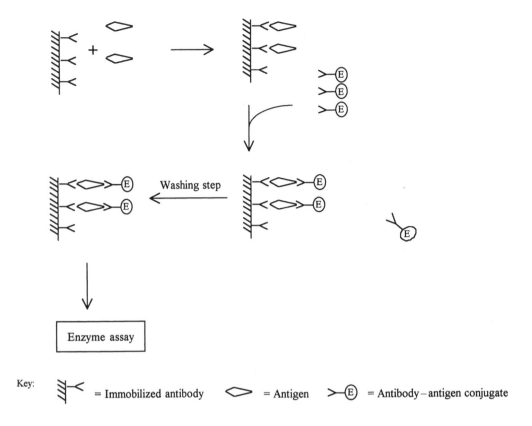

Key: ⥜< = Immobilized antibody   ◇ = Antigen   >—Ⓔ = Antibody–antigen conjugate

Subsequent to a further washing step, to remove any unbound anti-body–enzyme conjugate, the activity of the enzyme retained is assayed. The activity recorded is proportional to the quantity of antigen present in the sample assayed. A series of standard antigen concentrations may be assayed to allow construction of a standard curve. The standard curve facilitates calculation of antigen quantities present in 'unknown' samples.

Figure 9.12 Principle of non-competitive ELISA

Enzymes used in EIA

A wide variety of enzymes have been used as markers in various ELISA systems. Suitable enzymes may be chosen to fit a number of criteria. Apart from its catalytic properties, one obvious criterion is that the activity of the enzyme chosen be easily monitored. Many of the enzymes used produce a coloured product which may be easily monitored by colorimetric methods. Enzymes most often used as labels include alkaline phosphatase and horseradish peroxidase, in addition to β-galactosidase, glucose oxidase and urease. All such enzymes can utilize a suitable chromogenic substrate.

Alkaline phosphatase isolated from calf intestine was one of the first enzymes to be used in ELISA systems. The substrate normally utilized is *para*-nitrophenylphosphate (PNPP). PNPP is enzymatically hydrolysed by alkaline phosphatase, releasing inorganic phosphate and *para*-nitrophenol (PNP), which is yellow and absorbs light maximally at 405 nm. The substrate used with β-galactosidase is normally *O*-nitrophenyl β-D-galactopyranoside, with colour development being measured at 420 nm. Even more sensitive product detection may be obtained for such enzymatic systems if a fluorogenic substrate is used.

The covalent coupling or conjugation of the chosen enzyme to the second antibody used in ELISA systems may be achieved by a number of chemical methods. Perhaps the simplest of such methods involves chemical conjugation with gluteraldehyde (Figure 9.13), which is a homo-bifunctional reagent (its two reactive groups which link two proteins together are identical). Gluteraldehyde reacts irreversibly with the ε-amino group of lysine, forming covalent linkages. The gluteraldehyde method is popular due to its simplicity and is inexpensive. All conjugation methods, however, are likely to result in the inactivation of a proportion of coupled enzyme molecules and indeed antibodies.

Trends in immunoassay development

Several definite trends have been observed in the design of many of the more modern immunoassay systems. An increasing proportion of such immunoassays have incorporated non-isotopic detection systems. The incorporation of enzymatic detection systems in immunoassays has become particularly popular. Horseradish peroxidase is the enzyme most often used in such systems. Its popularity reflects its extremely high catalytic efficiency, its relatively small size and the availability of a

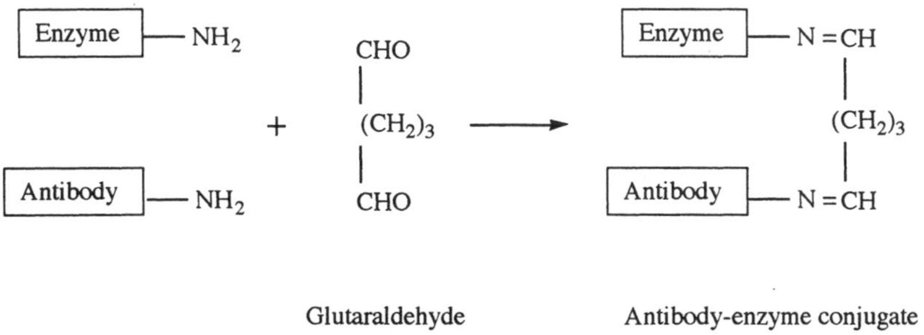

Glutaraldehyde   Antibody-enzyme conjugate

Figure 9.13 Chemical coupling of enzyme to antibody species using gluteraldehyde. Amino groups shown on the free enzyme and free antibody are ε amino groups of lysine residues. The reactants are incubated for a period of one hour or more. Excess free lysine may be added for efficient termination of the coupling procedure. Unreacted gluteraldehyde may be removed by, e.g. dialysis. The procedure will inevitably yield a proportion of antibody–antibody and enzyme–enzyme conjugates in addition to the enzyme–antibody conjugate. Separation of these if required can subsequently be undertaken by, e.g. gel filtration chromatography

variety of specific and sensitive assay systems by which peroxidase activity may be monitored. Alkaline phosphatase is probably the next most popular enzyme employed as a label in EIA.

Increasing numbers of new assays are also based on solid phase systems. This normally renders separation of free from bound antigen quite straightforward. The proportion of new immunoassay systems using monoclonal antibodies has also increased steadily. Monoclonal antibodies exhibit enhanced specificity compared to polyclonal preparations. Monoclonal antibody technology also affords a continuous supply of antibodies of defined specificity.

Recombinant DNA technology also continues to impact upon immunoassay technology. In addition to making various EIA reporter enzymes available by the recombinant route, it allows the generation of engineered antibodies or antibody fragments. An elegant example of the application of recombinant DNA technology in this sector pertains to the development of cloned enzyme immunoassays (CEDIA). CEDIA is a homogeneous competitive immunoassay which utilizes recombinant complementary inactive fragments of β-galactosidase, as described in Figure 9.14.

Modern immunological assays have also become more 'user friendly'. In the clinical laboratory, this is being achieved by the design and installation of automatic and semi-automatic analysers and by the modification of immunoassay methodology to facilitate its applicability to such automated systems. 'User friendly' assay kits have also been developed which personnel with no scientific training may employ outside the laboratory. Typical examples of such immunoassay systems are the pregnancy detection kits which are sold over the counter for home use. Such kits will be discussed at the end of this chapter.

Immunological assays for HIV

World-wide sales of diagnostic kits used to detect infection by HIV currently stand well in excess of US \$100 million. Many early kits were employed largely to screen donated blood or blood products for the presence of HIV. More recently, an increasing proportion of such kits are being used to screen personnel at high risk of acquiring the disease, in addition to persons who must demonstrate they are HIV-negative in order to emigrate to certain countries or be considered for various types of employment.

Assay systems have been developed which are capable of recognizing either HIV-1, the major causative agent of AIDS in the industrialized world, or HIV-2, the major causative factor in Africa. Thus far, however, the majority of diagnostic systems on the market are those which specifically detect HIV-1.

The host's immune function launches an immunological reaction upon infection with HIV. This is characterized, in part, by the presence of anti-HIV antibodies in the serum. The infected individual generally

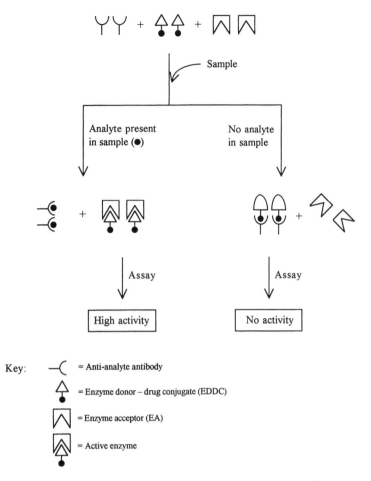

Figure 9.14 Simplified schematic overview of the cloned enzyme immunoassay system, CEDIA. Central to the assay are the complementary fragments of β-galactosidase, produced by recombinant means. One fragment is termed enzyme donator (ED), the other enzyme acceptor (EA). When mixed together they spontaneously form active β-galactosidase. The assay kit contains antibody which specifically binds the target analyte, the enzyme donor conjugated to the target analyte and the enzyme acceptor. If the target analyte is present in the sample to be assayed, it is bound by the antibody. The conjugated ED and the EA are thus free to interact, yielding active enzyme. The enzyme then converts a suitable chromogenic substrate, producing a coloured product which can be quantified spectrophotometrically. In instances where the target analyte is absent form the assay sample the antibody binds the analyte conjugated to the ED. This prevents binding of the EA and hence prevents formation of an active β-galacto-sidase complex. As a result no activity is recorded. Assay of samples containing known increasing concentration of analyte allows the generation of a standard curve

remains asymptomatic and the disease may progress no further for a period of several years before the development of AIDS-related complex and, subsequently, full-blown AIDS. The occurrence of anti-HIV antibodies in the serum of all individuals infected by the virus forms the basis for the vast majority of HIV diagnostic systems. Most such systems are of the enzyme immunoassay type, with the majority of these being ELISA.

In such systems HIV antigens are immobilized on a supporting surface, such as in the wells of microtitre plates. The serum sample to be analysed is then incubated in such wells, thus allowing any anti-HIV antibodies present to bind the immobilized antigen. After washing, a second antibody, which specifically binds human antibody, is added. A suitable enzyme label, such as horseradish peroxidase, is conjugated to this second antibody. Unbound conjugate is subsequently removed by washing and any retained enzymatic activity is detected by a suitable assay system. The level of activity retained is proportional to the quantity of anti-HIV antibody present in the serum sample assayed. This is summarized diagrammatically in Figure 9.15. One disadvantage inherent in this approach is that there exists a window period immediately after infection where no anti-HIV antibodies are detected.

Initial ELISA systems used immobilized antigen which had been partially purified from HIV particles grown in tissue culture. Most modern ELISA systems however, use specific, highly purified HIV antigens which are produced by recombinant DNA technology. Genes coding for HIV proteins such as p24, the core protein, gp41, gp120, and gp160, (envelope proteins), have all been expressed in recombinant systems. Production of antigen by recombinant methods not only eliminates the hazards of working with the actual virus itself, but also ensures the production of a defined viral antigen. This, in turn, has enhanced the sensitivity and specificity of the ELISA and minimized the occurrence of false positive reactions.

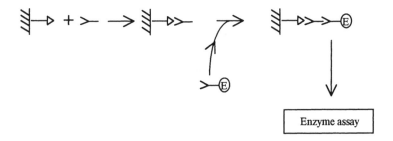

Key: ⊰—▷ = HIV antigen immobilized on a solid support

 ⊱ = Anti-HIV antibody present in human serum sample

 ⊱—Ⓔ = Goat anti-human antibody to which an enzyme has been conjugated

Figure 9.15 ELISA system designed to detect the presence of anti-HIV antibody present in an assay sample

Attempts have also been made to diagnose HIV infection by immuno-assay based detection of anti-HIV antibodies in urine and saliva. Collection of the latter is easier and less hazardous than collecting blood samples. However most immunoassays developed to date are insufficiently sensitive to facilitate this approach. While the mean concentration of IgG found in serum is 15 mg/ml, the level found in saliva and urine average at 35 µg/ml and 4 µg/ml, respectively. Various attempts to develop more sensitive anti-HIV assays are under way. Amongst the most promising is the application of immune complex transfer EIA systems which detect anti-HIV antibody (Figure 9.16).

Additional immunoassay applications

Non-competitive, two-site 'sandwich' immunoassays such as ELISA have gained widespread commercial use. Application of such assays requires the target analyte to be sufficiently large so it can be trapped between the capture and detector antibody (see Figure 9.12). Many low molecular mass substances (e.g. many drugs, metabolites and pollutants) are too small to be assayed via this approach. Competitive heterogeneous assays are normally employed in such instances. The research literature now contains several hundred examples of immunoassays developed to detect and quantify such low molecular mass substances. Commercialization of many such immunoassays has come more slowly for a variety of reasons, including the fact that alternative assay methodologies have already been developed. Traditionally for example, food residue analysis (for example insecticides, herbicides veterinary drugs, plant and microbial toxins, etc.) has relied mainly upon gas or liquid chromatography.

Although satisfactory chromatographic techniques exist for measuring many pollutants, more and more immunoassays are being developed for purposes of food or environmental monitoring (Table 9.6). Increasing environmental awareness, reflected in increasing environmental legislation, now renders commonplace routine environmental testing. This provides a ready and reliable market for appropriate assay techniques.

Immunoassays have also found a place in drug testing. Testing for the presence of drugs (particularly drugs of abuse in e.g. athletes, critically ill patients presenting at hospital emergency departments, etc.) is of obvious importance. Such tests are performed using either serum or (more commonly) urine samples and the most common target analytes include a range of anabolic steroids, amphetamines, opiates, cocaine metabolites, methadone and barbiturates.

Latex-based and other immunoassay systems

A variety of additional immunoassay formats have been developed in addition to the classic RIA and EIA. Latex- and membrane-based immunoassays represent two such alternative systems.

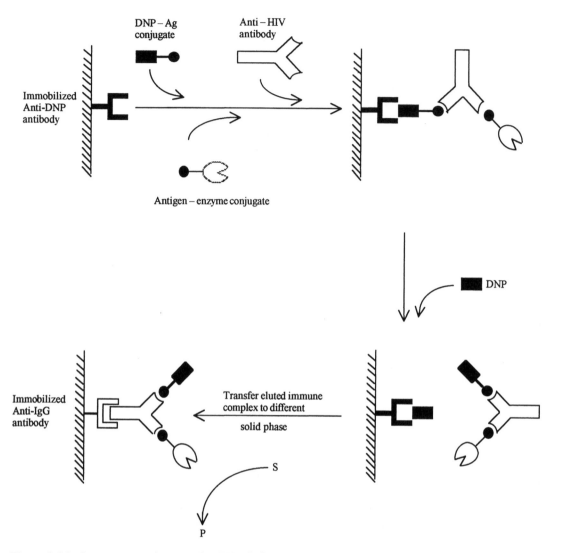

Figure 9.16 Immune complex transfer EIA designed to detect and quantify anti-HIV-1 antibody in biological samples. The biological sample, along with two conjugates (conjugate 1 = HIV antigen conjugated to a 2,4-dinitrophenyl group, conjugate 2 = HIV antigen conjugated to an appropriate enzyme) are incubated in a well containing immobilized anti-dinitrophenyl (DNP) antibody. This traps the three-constituent immune complex as shown in the well. After a washing step the immune complex is eluted from the solid phase by addition of free DNP (or DNP linked to lysine). The eluted complex is then transferred to a new solid phase which contains immobilized anti-human IgG antibody. This results in the immobilization of the immune complex as shown. After another washing step, a suitable substrate is added. Conversion into product, indicating the presence of anti-HIV antibody is normally detected fluorometrically. The transfer step from solid phase to solid phase decreases the non-specific signal to a great extent, helping improve the sensitivity of the assay system

Table 9.6 Some enzyme immunoassay systems developed to detect and quantify selected pesticides in food samples

Pesticide	Food type	Sample pretreatment	Detection limit
Alachlor (Metolachlor)	Grain	Grinding and extraction with methanol-water	20 p.p.b.
Aldrin (Dieldrin)	Eggs	Removal of egg shells, homogenization and dilution	6 ng/ml
Atrazine	Milk, juices	No treatment	0.03 p.p.b.
Benzimidazoles	Bovine liver	Extraction with DMF, water or citric acid	0.3 p.p.b.
Carbaryl	Apple and grape juice	No treatment	2 ng/ml
Difenzoquat	Wheat & barley products	Extraction with HCl	16 ng/g
Levamisole	Meat, milk	Homogenization in buffer (meat)	1 μg/kg
Methoprene	Wheat	Extraction with methanol or acetonitrile	60 p.p.b.
Triadimefon	Fruits	Extraction with methanol	2.4 ng/ml

Some foods (e.g. milk and juice) because of their nature, can be assayed directly. In other cases (e.g. meat) appropriate pesticide extraction procedures must first be undertaken, with subsequent testing of the extract.
Reproduced in modified form from Dankawardt, A. & Hock, B. (1997) *Food Technol. Biotechnol.* **35** (3), 170.

Uniform spherical particles of the polymeric polystyrene-based substance latex have found widespread application in the development of a variety of immunologically based diagnostic systems. Most latexes are manufactured from polystyrene and particle sizes used generally range between 0.1 and 1.1 μm. Individual latex particles therefore are not visible to the unaided human eye. Incubation of latex with a protein solution leads to absorbtion of the protein molecules onto the surface of the latex particle. Alternatively, various reactive groups may be initially introduced into the latex. This permits the covalent linkage of proteins to such particles.

Latex particles coated with antibody can be utilized to detect the presence of a specific antigen in a biological sample. Conversely, latex particles coated with a specific protein antigen, may be used to detect the presence of antibodies that recognize that specific antigen. Latex-based diagnostic systems rely on the inherent specificity of an antibody–antigen reaction, and the bifunctional nature of antibody binding, to promote agglutination of the latex particles. Some diagnostic kits are available in which the latex particles have been replaced by erythrocytes.

As illustrated in Figure 9.17 when latex particles coated with antibody which has been raised against a specific antigen are incubated with a biological sample containing that antigen, the ensuing antigen–antibody

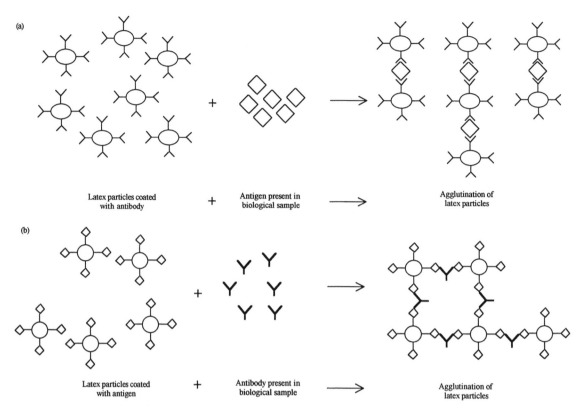

Figure 9.17   Latex agglutination assay system. In (a) latex particles have been coated with antibody. The presence of appropriate antigen in the test sample results in agglutination of the latex particles. The antigen present effectively acts as a bridge between adjacent latex particles. In (b) the latex particles have been coated with antigen. In this case the presence of appropriate antibody in the assay sample results in agglutination of the latex, i.e. the antibody acts as a bridge between adjacent latex particles. Agglutination can easily be detected visibly

reaction results in the formation of large aggregates of latex particles, i.e. the latex agglutinates. The antigen effectively acts as a bridge between adjacent latex particles. This process of agglutination is visibly evident as the newly aggregated particles are clearly seen by the naked eye.

   In practice, such assay systems are initiated by mixing a sample of antibody or antigen coated latex with the biological sample to be analysed. If the sample contains the appropriate antigen or antibody, the latex will agglutinate. The agglutination process is perceived visually as the transformation of a homogeneous milk-like latex mixture into a more granular form. The agglutinated latex particles are similar in appearance to discrete grains of sand.

   Latex-based assays are popular for a number of reasons, most notably the speed with which results are obtained and their requirement for little or no specialized equipment. Latex assays are normally carried out on plastic coated cards. Such cards exhibit a series of slight circular indentations on their surface. The latex, and biological sample to be assayed, are incubated together on one such surface indentation. The card is then gently rocked back and forth to ensure continued mixing of the latex and the sample. Agglutination, should it occur, will be visible within 3–5 min, in contrast to a typical EIA which requires up to 2 h to yield results.

Latex tests are basically qualitative. Semi-quantitative estimations may be obtained by assaying a series of dilutions of the antigen-containing sample until agglutination no longer occurs. The minimum concentration of antigen required to promote agglutination in such a system is then determined by assaying a series of dilutions of an antigen standard. The presence of interfering substances, particularly in undiluted blood or urine samples, may yield false positive results in some cases.

Numerous latex-based assay systems have been developed and many are extensively used. Examples of useful latex-based diagnostic systems include those designed to detect pregnancy, various infectious agents and those which detect various serum factors associated with arthritis.

Shortly after implantation of a fertilized egg, the human placenta begins to synthesize and secrete human chorionic gonadotrophin (hCG, Chapter 7). hCG is not normally synthesized in healthy, non-pregnant females. This hormone, however, is found in both the serum and urine of pregnant women. hCG therefore represents an ideal diagnostic marker for pregnancy. The fact that it may be detected in the urine of pregnant females makes it all the more desirable as urine samples may be conveniently collected.

Many of the initial hCG-based pregnancy detection systems were based on bioassays. This involved injection of urine samples to immature female rabbits or rats. Growth and ripening of the ovarian follicles in such animals indicated the presence of hCG, and hence pregnancy. Modern pregnancy diagnostic systems are virtually all based upon immunological detection of hCG in urine samples. Both RIA and ELISA systems are available, in addition to latex-based particle agglutination assays.

Most latex-based pregnancy detection systems utilize latex coated with anti-hCG antibodies. The presence of hCG in the urine sample thus promotes agglutination of the latex beads. Such systems may be termed direct assay systems. The principle involved is illustrated in Figure 9.18(a).

Alternative indirect latex-based pregnancy detection kits employ latex particles coated with hCG. The three components are incubated together in such indirect assay systems. Initially, anti-hCG antibody and a sample of urine are incubated together. The latex particles are then added and should the sample be obtained from a pregnant female, the hCG present in the urine will bind the anti-hCG antibody. Thus all the free hCG and anti-hCG-antibody is mopped up. In such cases, therefore, no agglutination will be noted upon addition of the latex beads. This is illustrated in Figure 9.18 (a).

If, on the other hand, the sample assayed by this indirect system is from a non-pregnant female, there will be no hCG in the urine sample supplied.

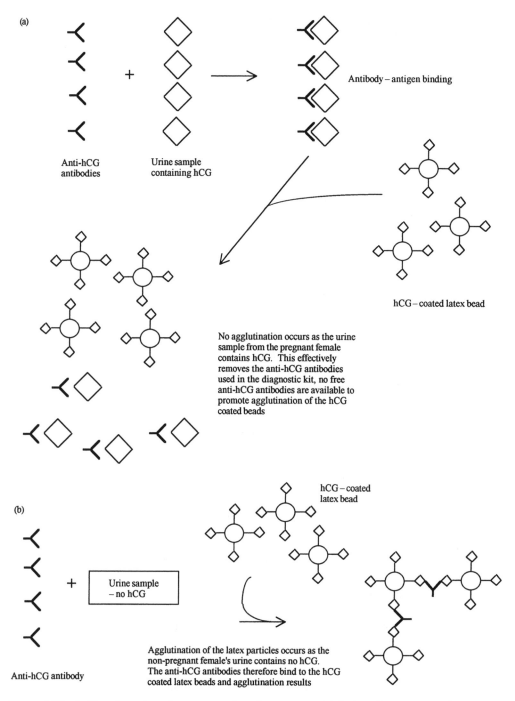

Figure 9.18 Indirect latex-based pregnancy detection systems. (a) Series of events occurring when a urine sample obtained from a pregnant female is tested; (b) series of events occurring when a urine sample from a non-pregnant female is tested

Thus, the binding sites of the anti-hCG antibodies mixed initially with the urine sample will remain empty. In this case, therefore, agglutination will be observed upon the subsequent addition of the hCG-coated latex beads (Figure 9.18 (b)).

The presence of a variety of infectious agents may also be indicated by a number of latex-based systems. Many such systems employ latex particles coated with antibody which recognize antigens associated with the infectious agent. Some systems, however employ latex particles coated directly with antigen. Agglutination of such particles indicate the presence of anti-antigen antibodies in the serum sample analysed. This, of course, indicates that the individual tested has come in contact with the infectious agent in question and has mounted an immunological response to it.

Examples of such latex-based systems which are available commercially include systems that detect hepatitis B, syphilis and HIV. Most hepatitis B latex agglutination tests are based upon incubation of blood samples with latex particles coated with antibody raised against hepatitis B surface antigen. Agglutination indicates the presence of hepatitis B particles in the sample.

Some latex-based syphilis tests utilize latex particles coated with antigens purified from *Treponema pallidum*, the syphilitic causative agent. Agglutination will occur if the blood sample assayed contains antibodies against this bacterium. The presence of such antibodies in the serum sample indicates the sample donor has come into contact with *Treponema pallidum*. Most latex-based HIV tests involve the use of latex particles coated with recombinant HIV-1 antigens.

Several latex-based diagnostic tests also find widespread application in rheumatology. Such systems include those designed to detect the presence of rheumatoid factor (RF) and C-reactive protein (CRP). Rheumatoid factors are auto-antibodies which specifically bind to human IgG. Their presence in serum is usually indicative of rheumatoid arthritis. CRP is a serum protein whose concentration increases several hundred fold subsequent to acute infections.

Membrane-bound diagnostic systems

Antibodies immobilized on membranes such as nitrocellulose may also be used as diagnostic reagents. Thoughtful design of such systems, in particular those designed to detect pregnancy, has contributed greatly to their ease of use. This point is illustrated in the generalized membrane-based hCG detection system outlined below. This, and similiar systems, have become quite popular and are usually sold 'over the counter' by pharmacists or other outlets.

In such systems, two thin lines of antibody are sprayed in the shape of a cross on the surface of the nitrocellulose membrane (Figure 9.19). This process is achieved by specialized industrial spraying equipment. The antibody species applied along one line specifically binds hCG. The

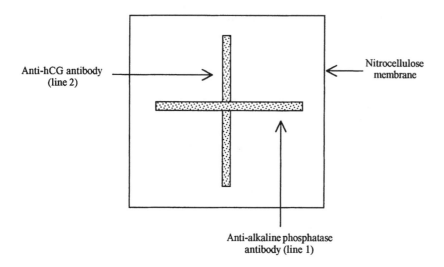

Figure 9.19 Membrane-bound pregnancy (hCG) detection system

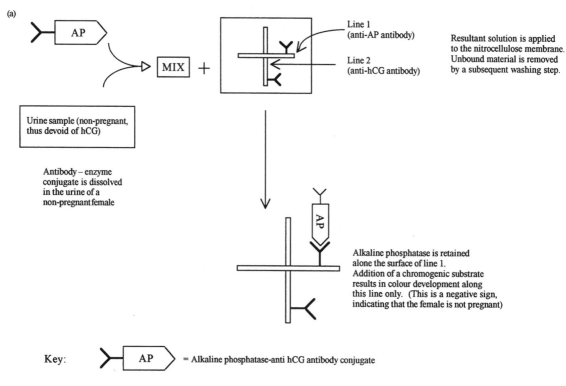

Figure 9.20 Principle of membrane-bound hCG (pregnancy) detection kit. (a) Events occurring if the female is not pregnant; (b) events occuring if the female is pregnant

(b)

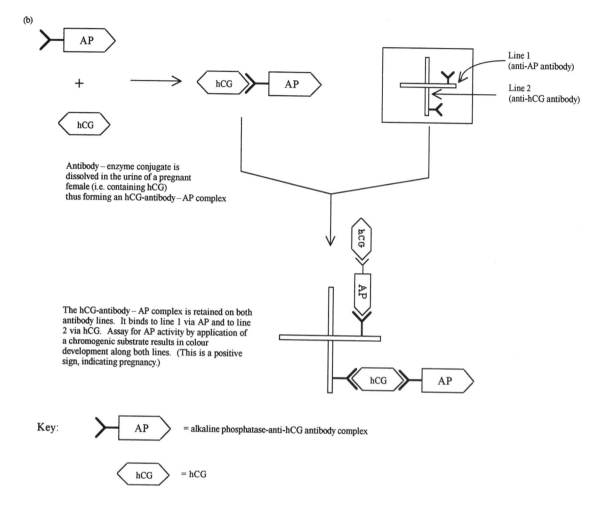

Antibody – enzyme conjugate is dissolved in the urine of a pregnant female (i.e. containing hCG) thus forming an hCG-antibody – AP complex

Line 1 (anti-AP antibody)

Line 2 (anti-hCG antibody)

The hCG-antibody – AP complex is retained on both antibody lines. It binds to line 1 via AP and to line 2 via hCG. Assay for AP activity by application of a chromogenic substrate results in colour development along both lines. (This is a positive sign, indicating pregnancy.)

Key: [AP] = alkaline phosphatase-anti-hCG antibody complex

[hCG] = hCG

antibody applied along the second line binds alkaline phosphatase. At this stage, of course, both such antibody lines are invisible to the human eye. A second component employed in such systems is a freeze-dried preparation of anti-hCG antibody conjugated to alkaline phosphatase (AP). The urine sample obtained for testing is used to reconstitute this antibody–enzyme conjugate. The reconstituted conjugate is then allowed to come into contact with the membrane surface.

A urine sample from a non-pregnant female will contain no hCG. Therefore, the hCG binding sites of the antibody–enzyme conjugate reconstituted by the urine sample remains unoccupied. When this sample comes into contact with the membrane surface, the conjugate is bound via the AP to the anti-AP antibody line (line 1 in Figure 9.19). Nothing binds to the anti-hCG antibody line (line 2, Figure 9.19), as no hCG is present in the urine.

After rinsing the membrane in order to remove unbound material, it is briefly immersed in a solution containing an alkaline phosphatase chromogenic substrate such as PNPP. Colour development is therefore witnessed all along the anti-AP antibody line due to the presence of bound AP (Figure 9.20(a)). No colour development occurs along line 2 as no hCG is present. Thus a minus sign (−) is produced indicating that the female is not pregnant.

On the other hand, the urine of a pregnant female does contain hCG and this hCG is bound by the antibody–enzyme conjugate upon its reconstitution with the urine sample. In this case, the conjugate is bound by both antibody lines upon coming in contact with the membrane surface. The conjugate binds line 1 via the AP moiety. Binding to line 2 (i.e. the anti-hCG antibody line) occurs because the hCG acts as a bridge, as illustrated in Figure 9.19 (b). In this case, immersion of the membrane in a developing solution results in colour development along both lines, i.e. a plus (+) sign is obtained, indicating pregnancy.

FURTHER READING

Books

Edwards, R. (1999) *Immunodiagnostics*. Oxford University Press, Oxford.

Eggins, B. (1998) *Biosensors*. John Wiley & Sons, Chichester.

Gaw, A. *et al.* (1999) *Clinical Biochemistry*. Churchill Livingstone, Edinburgh.

Godfrey, T. (1990) *Industrial Enzymology*. MacMillan Press, Basingstoke.

Gosling, J. (2000) *Immunoassays*. Oxford University Press, Oxford.

Mulchandani, A. (1998) *Enzyme and Microbial Biosensors*. Humana Press, New York.

Ngo, T. (2000) *Biosensors and their Applications*. Plenum Press, New York.

Price, C. (1996) *Principles and Practices of Immunoassay*. MacMillan Press, Basingstoke.

Scott, D. (1998) *Biosensors for Food Analysis*. Royal Society of Chemistry, London.

Sonnleitner, B. (1999) *Bioanalysis and Biosensors for Bioprocess Monitoring*. Springer-Verlag, Godalming.

Wild, D. (2001) *Immunoassay Handbook*. MacMillan Press, Basingstoke.

Articles

Analytical enzymes

Burkhard-Kresse, G. (1995) Analytical uses of enzymes. In *Biotechnology, a Multi-volume Treatise*, 2nd edn, Vol. 9, pp. 138–163. VCH, Weinheim.

Freedman, D. *et al.* (1986) The relation of apolipoproteins A-1 and B in children to parental myocardial infarction. *N. Eng. J. Med.* **315**, 721–726.

Klotzsch, S. & McNamara, J. (1990) Triglyceride measurements: a review of methods and interferences. *Clin. Chem.* **326** (9), 1605–1613.

Kopetzki, E. *et al.* (1994) Enzymes in diagnostics; achievements and possibilities of recombinant DNA technology. *Clin. Chem.* **40** (5), 688–704

MacLachan, J. *et al.* (2000) Cholesterol oxidase: sources, physical properties and analytical applications. *J. Steroid Biochem. Mol. Biol.* **72** (5), 169–195.

Price, C. (1983). Enzymes as reagents in clinical chemistry. *Phil. Trans. R. Soc. London* **B300**, 411–422.

Immunoassay

Beltz, G. *et al.* (1989). Development of assays to detect HIV-1, HIV-2 and HTLV-1 antibodies using recombinant antigents. In: *Molecular Probes: Technology and Medical Applications*, pp. 131–142, Albertini A *et al.* (eds), Raven Press, New York.

Bock, J. (2000) The new era of automated immunoassay. *Am. J. Clin. Pathol.* **113** (5), 628–646.

Borrebaeck, C. (2000) Antibodies in diagnostics – from immunoassays to protein chips. *Immunol. Today* **21** (8), 3789–382

Gosling, J. (1990) A decade of development in immunoassay methodology. *Clin. Chem.* **36** (8), 1408–1427.

Gottfried, T. & Urnovitz, H. (1990) HIV-1 testing: product development strategies. *Trends Biotechnol.* **8**, 35–40.

Hage, D. (1999) Immunoassays. *Anal. Chem.* **71** (12), 294R–304R.

Labeur, C. *et al.* (1990) Immunological assays of apolipoproteins in plasma: methods and instrumentation. *Clin. Chem.* **36** (4), 591–597.

Ronald, A. & Stimson, W. (1998) The evolution of immunoassay technology. *Parasitology* **117**, S13–S27.

Schlageter, M. (1990). Radioimmunoassay of erythropoietin: analytical performance and clinical use in haematology. *Clin. Chem.* **26** (10), 1731–1735.

Biosensors

Ivnitski, D. *et al.* (1999) Biosensors for detection of pathogenic bacteria. *Biosensors Bioelectr.* **14** (7), 599–624.

Rekha, K. *et al.* (2000) Biosensors for the detection of organophosphorus pesticides. *Crit. Rev. Biotechnol.* **20** (3), 213–235.

Rich, R. & Myszka, D. (2000) Survey of 1999 surface plasmon resonance biosensor literature. *J. Mol. Recog.* **13** (6), 388–407.

Scheller, F. *et al.* (2001) Research and development in biosensors. *Curr. Opin. Biotechnol.* **12** (1), 35–40.

Vo-Dinh, T. & Cullum, B. (2000) Biosensors and biochips: advances in biological and medical diagnostics. *Fresenius J. Anal. Chem.* **366** (6–7), 540–551.

10

Industrial enzymes, an introduction

INDUSTRIAL ENZYMES

The preceding chapters focused upon proteins used for therapeutic and diagnostic or analytical purposes. Such proteins are normally produced in small quantities and are highly purified. As with many healthcare products they are often expensive, and economic considerations are not of paramount importance in their production and marketing. This chapter (along with the subsequent two) focuses upon a different group of proteins, that of 'industrial' or 'bulk' enzymes. This group includes amylases, cellulases, lignocellulose degrading enzymes, pectinases, proteases, lipases, phytases, penicillin acylases and cyclodextrin glycosyltransferases. The majority of these enzymes are hydrolytic depolymerases. In contrast to enzymes used for therapeutic or diagnostic purposes, industrial enzymes are produced in large quantities, in the order of thousands to hundreds of thousands of kilogrammes annually, and are normally processed only to a limited degree. Furthermore, in most instances economic considerations such as production costs are of critical importance to their commercial success (Table 10.1).

Bulk enzymes are used in a great many biotechnological processes. Many technologies such as brewing, wine making and cheese manufacture may be traced to the very dawn of history. In such instances individuals unknowingly employed microorganisms as a source of the enzymes required to transform initial substrates into products such as ethanol or cheese. A greater understanding of the molecular basis by which these conversions occur facilitated the subsequent utilization of isolated enzymes for specific industrial purposes.

Enzymatic preparations currently find application in the brewing, breadmaking, starch processing and cheesemaking industries (Table 10.2). The availability of such enzymes has also facilitated the development of numerous additional biotechnological processes which produce a wide range of industrially important commodities. Enzymes are thus also used in the production of sweeteners and in modification of the flavour, texture and appearance of many foodstuffs. They are employed to tenderize meat, clarify beer, wine and fruit juices, and are included in many detergent preparations. Figure 10.1 overviews the major industrial sectors in which industrial enzymes are used.

Table 10.1 Selected attributes of industrial enzymes, as compared to those of proteins used for healthcare purposes

Industrial enzymes	Medical/diagnostic proteins
Produced in large quantities	Produced in small quantities
Partially purified at best	Extensively purified
Economic considerations critical	Economic considerations are of secondary importance to functional excellence of product
Function: catalytic	Function: various (hormones, growth factors, cytokines, other regulatory factors, blood factors, vaccines, antibodies, enzymes)
Source: mainly microbial and recombinant	Source: mainly human or animal and recombinant products
Mainly secreted into the extracellular medium by producer strains	May be intracellular or extracellular

Table 10.2 Some industrially important enzymes, their traditional sources and sample industrial applications

Enzyme	Traditional source	Sample applications
α-Amylase	*Bacillus amyloliquefaciens* *Bacillus licheniformis* *Bacillus subtilis* *Aspergillus oryzae*	Hydrolyzes α 1–4 linkages in starch. Used to liquify starch and reduce its viscosity
β-Amylase	*Bacillus polymyxa* *Bacillus circulans* Barley	Enzymatic degradation of starch yielding the disaccharide maltose
Glucoamylase (Amyloglucosidase)	*Aspergillus niger* *Rhizopus* spp.	Hydrolysis of starch, yielding dextrose
Pullulanase	*Bacillus* spp. *Aerobacter aerogenes* *Klebsiella* spp.	Debranching of starch by hydrolysis of α 1–6 glycosidic linkages
Glucose isomerase	*Bacillus coagulans* *Bacillus stearothermophilus* *Streptomyces* spp. *Arthrobacter*	Production of high-fructose syrup by conversion of glucose to fructose
β-Galactosidase (lactase)	*Bacillus coagulans* *Streptomyces* spp. *Saccharomyces* spp. *Aspergillus* spp.	Hydrolysis of milk lactose yielding glucose and galactose
Invertase (Sucrase)	*Saccharomyces* spp.	Hydrolysis of sucrose, yielding glucose and fructose
Cellulase & hemicellulase	*Trichoderma* spp. *Sporotrichum cellulophilum* *Actinomyces* spp. *Aromonas* spp. *Aspergillus niger*	Enzymatic hydrolysis of cellulose-containing material
Pectinases	*Aspergillus niger* *Fusarium* spp.	Enzymatic hydrolysis of pectin
Proteases	Various bacilli species *Bacillus amyloliquefaciens* *Bacillus subtilis* *Streptomyces* spp. *Aspergillus oryzae* Some animal sources such as calf stomach	Enzymatic hydrolysis of proteins widely used in detergents and in brewing, baking and meat tenderization
Lipases	*Mucor* spp. *Myriococcum* spp. Animal pancreas	Enzymatic hydrolysis of lipids. Used in dairy industry for flavour development in foods and also used in detergents

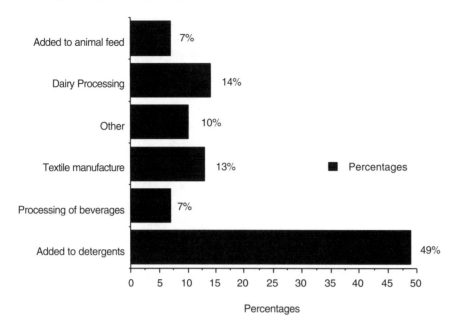

Figure 10.1   The major products and processes in which industrial enzymes are incorporated/used, and their relative share of the total enzyme market. 'Other' includes starch processing (4 per cent), use in bakery/confectionary industry (5 per cent) and leather softening/processing (1 per cent). Reproduced (with modifications) from Rehm, *et al.* (eds) (1999) *Biotechnology a Multi Volume Comprehensive Treatise*, Vol. 5a, p. 191. Wiley-VCH, Weinheim

Sales value of industrial enzymes

By the end of the 1990s the annual worldwide sales value of industrial enzymes stood in the region of US $ 1.5 billion, approximately double the figure recorded a decade earlier. Proteases accounted for almost 50 per cent of the market share, while carbohydrate-degrading enzymes accounted for much of the remainder. (Table 10.3). The strong and continued growth in enzyme sales (Figure 10.2) may be attributable to both economic factors (e.g. a growing world economy) and to technical advances (e.g. the impact of genetic engineering on enzyme production and the development of new enzyme applications). While the combined sales value of 'newer' industrial enzymes such as phytase and cyclodextrin glycosyl transferase ('others', Table 10.3) is modest, their market value is set to grow rapidly over the next few years. The major manufacturers of industrial enzymes are listed in Table 10.4. Novo Nordisk is the world's largest supplier of such enzyme products. They market some 600 different enzyme products, used in the detergent, animal feed, alcohol, wine, brewing, baking, juice, leather, pulp, paper, food and textile industries.

Table 10.3 Estimated annual global sales value of industrial enzymes

Enzyme	Market value (US$, millions)
Proteases	700
Cellulases	180
α-Amylases	135
Lipases	90
Hemicellulases	75
Glucoamylases	60
Pectinases	60
Lactases	15
Glucose isomerase	15
Pullulanases	15
Others	100

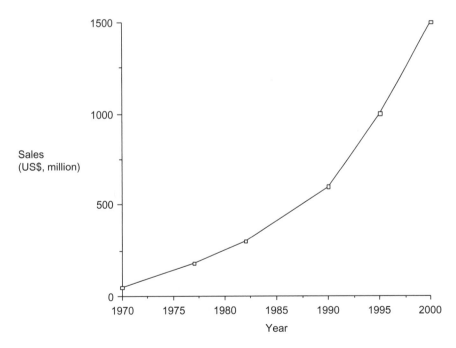

Figure 10.2 Growth of the world industrial enzyme market (in terms of estimated annual global sales value), 1970–2000, compiled from various sources

Table 10.4 Major manufacturers of industrial enzymes and their web addresses

Company	Web address
Novo Nordisk (Denmark)	http://www.novo.dk
Gist-Brocades (Netherlands) (Now part of the DSM group, the Netherlands)	http://www.dsm.nl/
Genencor Inc. (USA)	http://www.genencor.com/gciinternet.nsf
Solvay (Belgium)	Industrial enzyme business sold to Genencor
Danisco Cultor (Finland)	http://www.cultor.fi

Traditional (non-recombinant) sources of industrial enzymes

The majority of industrial enzymes are traditionally obtained from micro-organisms. The producer strains are usually members of a family of microbes classified as GRAS. Such bulk enzymes are produced primarily by bacteria and fungi, most notably by members of the genera *Bacillus* and *Aspergillus* (Table 10.2).

Bacillus Species have traditionally been utilized in the production of industrially important enzymes for a number of reasons:

- With the exception of the *Bacillus cereus* group, they are all GRAS listed and they are widely distributed in nature.

- They are easily cultured in relatively inexpensive media.

- Members of the genera *Bacillus* are capable of producing large quantities of many desirable enzymatic activities, most of which are secreted extracellularly into the fermentation medium. Extracellular production of enzymes obviously simplifies subsequent product recovery and purification (see Chapter 3).

Various species of *Aspergillus* are used as producer organisms as they share many of the positive attributes exhibited by *Bacillus*. A few industrially important enzymes may be obtained from yeast, mainly from species of *Saccharomyces*.

Candidate producer organisms are usually pinpointed by initial screening programmes. Many wild type organisms isolated by such screening systems initially produce relatively small quantities of the enzyme of interest. Traditionally yields were increased by methods of strain improvement such as mutagenesis followed by identification of hyperproducing microbial strains (Chapter 2).

The impact of genetic engineering on enzyme production

Genetic engineering continues to have a major impact upon the industrial enzyme sector. Throughout the 1980s the major enzyme companies initiated research programmes aimed at producing selected industrial enzymes by recombinant means. The first such recombinant products came on stream in the latter half of the 1980s, and today the majority of industrial enzymes are produced by recombinant technology. Recombinant production strategies using both homologous and heterologous expression systems have both been employed (Chapter 2). Selected examples of commercialized recombinant industrial enzymes are provided in Table 10.5

Production of industrial enzymes by recombinant means displays a number of potential advantages, as compared to traditional (non-recombinant) approaches. These advantages include:

Table 10.5 Some industrial enzymes produced by recombinant DNA technology

Product Name	Description	Use	Manufacturer	Original gene source	Gene expressed in
Maxiren	Chymosin	Cheesemaking	Gist-Brocades	Calf Stomach Cells	*Kluveromyces lactis*
Optiren	Protease	Cheesemaking	Gist-Brocades	*Rhizomucor miehei*	*Aspergillus oryzae*
Natuphos	Phytase	Animal feed additive	Gist-Brocades	*Aspergillus niger*	*Aspergillus niger*
Lipolase	Lipase	Detergent additive	Novo Nordisk	*Humicola languinosa*	*Aspergillus oryzae*
Lipomax	Lipase	Detergent additive	Gist-brocades	*Pseudomonas alcaligenes*	*Pseudomonas alcaligenes*
Lumafast	Lipase	Detergent additive	Genencor	*Pseudomonas mendocina*	*Bacillus* sp.
Duramyl	Amylase	Detergent additive	Novo Nordisk	*Bacillus licheniformis*	*Bacillus* sp.
Purafect OxAm	Amylase	Detergent additive	Genencor	*Bacillus licheniformis*	NL
Carezyme	Cellulase	Detergent additive	Novo Nordisk	*Humicola insolens*	*Aspergillus oryzae*

NL, not listed.

- higher expression levels attainable;

- product generally displays a higher degree of relative purity;

- economically attractive;

- heterologous expression facilitates commercialization of enzymes produced naturally by pathogenic/non-GRAS-listed species;

- allows alteration of enzyme's characteristics via protein engineering.

The insertion of multiple copies of the target gene, usually placed under a very powerful promoter, results in high level expression of the recombinant product (Chapter 2). Expression levels generally attained vary between 10–30 per cent of total cellular protein, although levels in excess of 50 per cent have sometimes been recorded. The recombinant products are invariably produced in microbial systems which facilitates their extracellular excretion. High level recombinant production of the target enzyme has a number of obvious technical and economic advantages, including:

- fermentation batch sizes can be substantially decreased, with consequent savings on upstream and downstream processing costs;

- reductions in processing volumes will decrease the quantities of waste or by-products generated, with consequent positive cost and environmental implications;

- high-level expression decreases the ratio of contaminant proteins:desired protein present in the fermentation media, in effect yielding a product of higher initial purity.

The production of α-galactosidase by recombinant versus non-recombinant means, as summarized in Table 10.6, illustrates many of these points.

High-level recombinant expression also renders feasible the commercial production of enzymes expressed naturally at very low levels or produced naturally by non-GRAS listed, pathogenic or novel microorganisms, or produced by microorganisms that are difficult and expensive to cultivate on an industrial scale. For example, a cyclodextrin glycosyl transferase (Chapter 12) produced naturally by a thermophilic anaerobe (*Thermoanaerobacter*) is now produced commercially by Novo Nordisk in *Bacillus macerans* (Table 10.5). Recombinant DNA technology obviously also facilitates the large-scale production of non-microbial enzymes of industrial utility. Recombinant chymosin for example is produced in a variety of microbial systems (Table 10.5).

Engineered enzymes

Recombinant DNA technology allows the incorporation of predefined alterations to the amino acid sequence of enzymes. Protein engineering

Table 10.6 Comparison of the isolation of α-glucosidase when produced by non-recombinant means (in baker's yeast) as compared to its recombinant production

Isolation step	Baker's yeast	Recombinant yeast
Purification steps	10	4
Biomass (tonnes)	236	10
Specific activity (kU/g)	0.4	11
Yeast cell debris (tonnes)	440	12
Ammonium sulfate (tonnes)	1100	25
Potassium phosphate (tonnes)	25	0.5
Alumina absorbent (tonnes)	90	0
Filtration aids (tonnes)	133	5
Water (m^3)	3700	50
Electricity (kWh)	45 000	9000

Reproduced with permission from Kopetzki *et al.* (1994) *Clinical Chemistry* **40**, 688–704.

(Chapter 1) may be used to tailor selected enzymes in order to render them more suited to specific industrial applications. The majority of engineered products thus far commercialized are enzymes made oxidation resistant in order to increase their stability and effectiveness when added to bleach containing detergents (Table 10.7). These are discussed in more detail in

Table 10.7 Selected examples of commercially available detergent enzymes whose amino acid sequence was logically altered by site-directed mutagenesis in order to enhance their industrial utility by rendering them oxidation resistant

Enzyme (Tradename)	Manufacturer	Description
Purafect 0x Am	Genencor Int.	Oxidation resistant α-amylase
Duramyl	Novo Nordisk	Oxidation resistant α-amylase
Lipomax	Gist Brocades	Oxidation resistant lipase
Lipolase ultra	Novo Nordisk	Oxidation resistant lipase
Maxapem	Gist Brocades	Oxidation resistant protease
Everlase	Novo Nordisk	Oxidation resistant protease

subsequent chapters. Additional aims of protein engineering in the context of industrial enzymology includes the development of amylases displaying enhanced thermal stability for application in the starch processing industry (Chapter 11), development of proteases with more relaxed substrate specifically (for detergent use) and development of enzymes displaying enhanced stability in the presence of surfactants (for detergent use).

IMMOBILIZED ENZYMES

The vast majority of enzymes used in industrial processes are not recovered following completion of their catalytic conversion(s). In some instances however enzymes are utilised in an immobilized form such that they may be recovered at the end of the catalytic process. This facilitates reutilization of the enzyme preparation which is of obvious economic benefit. The decision to use immobilized or free enzyme in any given situation depends upon economical, technical and practical considerations. The more expensive the enzyme preparation, the greater the impetus to utilize it in an immobilized form. Immobilization, however, must not adversely effect enzyme stability, kinetic or other relevant properties. In many instances (e.g. enzymes employed therapeutically, or processes employing bulk quantities of crude enzymes), subsequent recovery of an immobilized enzyme is rendered impractical. Some of the more notable industrial processes which do utilise immobilized enzymes are listed in Table 10.8. A variety of methods may be used to immobilize any enzyme and the preparation methodologies generally involve either entrapping the enzyme within a confined matrix or binding the enzyme to an insoluble support matrix (Figure 10.3).

Gel–fibre entrapment

Enzymes may be entrapped within the matrix of a polymeric gel. This is achieved by incubating the enzyme together with the gel monomers and then promoting gel polymerization. Enzymes entrapped within polyacrylamide or polymethacrylate gels serve as two such examples. In order to function successfully, the gel pore size generated must be small enough to retain the entrapped enzyme but must be large enough to allow free passage of enzyme reactants and products. Enzymes may also be entrapped within molecular-sized pockets formed during spinning of industrial fibres such as cellulose acetate.

Encapsulation involves entrapping enzymes within a spherical semipermeable membrane. Smaller molecules, such as enzyme substrates, cofactors and products, must be able to pass freely through such pores, while enzymes or other macromolecules are retarded. Cellulose nitrate and nylon-based membranes have been extensively used in this regard.

Table 10.8 Some industrial enzymes used in an immobilized format, the reactions catalysed, and the industrial processes in which they are used

Enzyme	Process	Reaction catalysed
Glucose isomerase	Production of high fructose corn syrup	Conversion of glucose to fructose
Amino acid acylase	Amino acid production	Deacetylation of L-acetyl amino acids
Penicillin acylase	Production of semi-synthetic penicillins	Removal of side chains from naturally produced penicillin thus yielding 6-amino penicillanic acids
Lactase	Hydrolysis of lactose	Hydrolysis of lactose thus yielding glucose and galactose

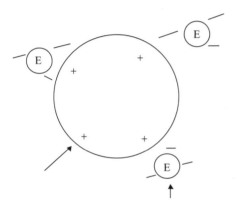

Enzyme adsorption onto solid matrix by e.g. ionic attraction

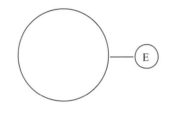

Figure 10.3 Most common means by which enzymes are immobilized

Direct covalent linkage of enzyme to solid matrix

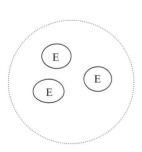

Encapsulation of enzyme in a porous membrane vesicle

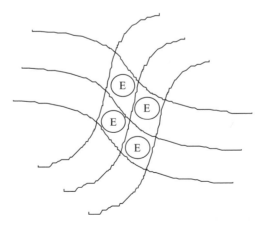

Entrapment of enzyme within a porous gel lattice, or fibre

Immobilization via adsorption

Enzymes may also be immobilized by promoting their binding to an insoluble matrix. A variety of matrices and methods of attachment have been developed. Perhaps the simplest such method involves physical adsorbtion of the enzyme to a suitable carrier substance. This may often be achieved by directly mixing the enzyme and support, under incubation conditions (of ionic strength, pH and sometimes temperature) at which maximal adsorption is observed. These parameters are normally determined empirically. Leakage of enzyme from the support is often a problem, as the forces of attraction retaining the enzyme on its surface are relatively weak. Supports, such as aluminium hydroxide are most often utilized and a variety of enzymes have been immobilized by this procedure.

Enzymes may also be immobilized by promoting ionic interactions with a suitably charged matrix. Ion exchange resins have found particular favour in this regard. Anion exchange resins such as DEAE–cellulose or DEAE–Sephadex have been used in the immobilization of negatively charged enzymes, while cation exchange media such as CM-cellulose may be used to immobilize positively charged enzymes. This method of enzyme immobilization is technically undemanding and economically attractive. Enzyme leakage normally does not present a major problem providing the column is operated under appropriate conditions. Such immobilization systems may also be regenerated by passing through a solution of soluble enzyme.

Perhaps the best known industrial scale enzyme used in immobilized form on DEAE–Sephadex is amino acylase, used in the production of synthetic L-amino acids. Glucose isomerase may also be immobilized by this method. Glucose isomerase, however, is an intracellular enzyme, and immobilized forms generally consist of dead (sometimes lysed) microbial cells which have been crosslinked by gluteraldehyde and subsequently pelleted (Table 10.9). The pelleted cellular preparations may be poured

Table 10.9 Some traditional glucose isomerase preparations immobilized by cross-linking lysed or intact cells (displaying intracellular glucose isomerase activity)

Supplier	Country*	Immobilization method
Novo	DK	*Bacillus coagulans* cells, homogenate, cross-linked with glutyraldehyde (GA), extruded
Gist Brocades	NL	*Actinoplanes missouriensis* cells mixed with gelatine, cross-linked with GA
Miles Labs	USA	*Streptomyces olivaceus* cells, cross-linked with polyamine + GA, extruded
ICI	UK	*Arthrobacter globiformis* cells, coagulated with cationic + anionic polymers
Nagase	J	*Streptomyces phaechromogenes* enzyme, GA cross-linked with intact but dead cells

*Countries: DK, Denmark; NL, the Netherlands; USA, United States; UK, United Kingdom; J, Japan.
Reproduced with permission from Uhlig, H. (1998) *Industrial Enzymes and their Applications.* Wiley Interscience, New York.

into a reactor column, through which the glucose syrup is passed. The main disadvantage associated with this approach is the relatively limited mechanical stability of the immobilized system. In the region of 1500 tons of immobilized glucose isomerase is used to produce in excess of 7 million tons of high fructose corn syrup (Chapter 11) annually.

The most widely used method of enzyme immobilization involves covalent attachment of the enzyme to a suitable insoluble support matrix. Covalent attachment is technically more complex than most other methods of immobilization, and requires a variety of often expensive chemical reagents. Immobilization procedures are also often time consuming. The covalent nature of the binding however, renders the immobilized enzyme preparations very stable and leaching of the enzyme from the column is minimal.

A wide variety of different immobilization procedures have been developed. All promote the formation of a covalent bond between a suitable reactive group present on the surface of the insoluble matrix and a suitable group present on the surface of the protein. The latter functional groups obviously must not play an essential role in the enzyme's catalytic mechanism. Hydroxyl groups present in carbohydrate-based matrices such as cellulose, dextrans or agarose, often participate in covalent bond formation. Amino, carboxyl and sulfhydryl groups present in various amino acids are also generally involved in covalent bond formation (Table 10.10). A selected example of the chemistry of enzyme immobilization (oxirane-based immobilization) is presented in Figure 10.4.

Enzyme safety

Proteolytic enzymes found widespread application in detergent preparations during the 1960s (Chapter 11). This was reflected in an enormous

Table 10.10 Some amino acid residues normally involved in the formation of covalent linkages in enzyme immobilization

Amino acid	Functional group
Lysine	$-NH_2$
Arginine	$-NH_2$
Tyrosine	$-\!\langle\rangle\!-OH$
Aspartate	$-COOH$
Glutamate	$-COOH$
Cysteine	$-SH$
Methionine	$-S-CH_3$

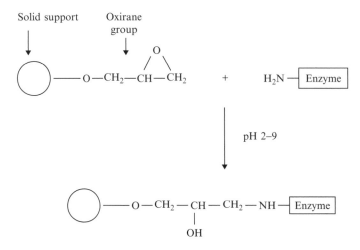

Figure 10.4   Covalent attachment of an enzyme to a solid support containing a reactive oxirane group. The enzyme amino group participating most frequently in the cross-linking reaction is contributed by a surface lysine residue

increase in worldwide protease sales during this period. By the late 1960s, however, the practice of enzyme addition to detergents had become a source of considerable controversy. At that time crude protease products were formulated as a dry, fine powder. Handling the powder (in the enzyme and detergent manufacturing plants, and by the end user) often resulted in significant dust formation. Inhalation of the enzyme-containing dust sometimes resulted in allergic reactions in susceptible individuals, due to the microbial (i.e. foreign) nature of the enzyme.

By the early 1970s, enzymes had been removed from most detergent preparations, and a considerable debate relating to the safety of including enzymes in detergents ensued. This debate widened to include additional industrial enzyme applications, particularly in food processing. However, subsequent scientific investigations concluded that the industrial use of enzymes *per se* was a safe industrial practice. Microbial, plant and animal enzymes have unwittingly been used by man for centuries with no evidence of associated toxicity. Catalytically active enzymes derived from dietary constituents are regularly consumed by humans, with no ill effects, and indeed as proteins they contribute to our nutritional intake. Microbial enzymes used industrially are obtained from GRAS-listed organisms, and hence will contain no contaminants considered dangerous to human health. Furthermore, all ingredients utilized in the formulation of the microbial fermentation media must be non-toxic. The manufacturing process used must ensure that potentially toxic or otherwise harmful substances are not introduced into the product during downstream processing. In the same way enzymes obtained from plants or animals must be obtained from edible, non-toxic species. Enzymes used in food and beverage processing applications are sometimes used in immobilized format (and hence will not be present in the final product). In other instances enzymes introduced in order to mediate a desired effect during food processing are inactivated or destroyed at a subsequent stage of the manufacturing process.

In terms of safe enzyme handling, granulation technologies were developed which minimized dust formation by dry enzyme preparations (Chapter 3). Safe industrial practice however dictates that workers handling enzyme preparations (primarily at pre-granulation stages) should wear protective clothing. Should enzyme-containing aerosols or dust be encountered adequate breathing apparatus should be used, and manufacturing plant ventilation system must be designed to filter out dust particles in order to prevent their recirculation.

The advent of recombinant industrial enzymes has more recently sparked public debate. Thus far the majority of recombinant enzymes developed are added to detergents (Table 10.5) however some (e.g. chymosin) are used in the food processing industry. Once again the recombinant enzymes are produced in GRAS-listed organisms and are considered equally safe to enzymes produced by non-recombinant means.

Industrial enzymes; the future

The range and market value of industrial enzyme products is set to grow very substantially over the coming years. An increased pace in traditional biotechnological research continues to uncover new enzymes from various sources which may prove industrially useful. The recent discovery of microorganisms living in very extreme environmental conditions (extremophiles; see later) provides a particularly rich source of novel, potentially useful enzymes. Rapid screening methods and the advent of genomics and bioinformatics also facilitates the identification of new enzymes of likely industrial significance. The advent of genetic engineering now ensures that any enzyme of industrial relevance can be produced economically and at the scale required. Furthermore an increasing understanding of protein structure–function relationships will continue to allow scientists tailor enzymes to optimize their industrial utility via protein engineering. The commercialization of any genetically engineered enzyme product is subject to tight regulatory control. However, the issue of public perception and acceptance of engineered products is an entirely separate matter. While the market introduction of engineered detergent enzymes has not suffered from adverse public opinion, the engineered enzymes developed for food application is likely to receive closer public scrutiny. Recombinant chymosin has generally obtained considerable market acceptance. However it appears that recombinant products are more quickly accepted in some world regions (e.g. the USA) than in others (e.g. Europe).

Another facet of industrial enzymology currently under investigation is the application of enzymes in processes undertaken in non-aqueous environments. This may allow development of enzyme-mediated biotransformation processes of substrates poorly soluble or insoluble in water, but which are soluble in organic solvents.

EXTREMOPHILES

To date, the majority of microbial enzymes used industrially have been sourced from mesophilic organisms (e.g. *Aspergillus*). With one or two exceptions industrial processes in which they are used operate at temperatures ranging from 35 to 60 °C, and at pH values between 4 and 8. The discovery of extremophiles provided a potential source of industrial enzymes capable of functioning well outside these relatively narrow operational parameters.

Extremophiles represent microorganisms that have successfully colonized ecological niches displaying one or more extreme environmental parameter (temperature, pH, ionic strength, or pressure). An overview of extremophile classifications is presented in Table 10.11. Some extremophiles have colonised niches characterized by two or more extreme environmental parameters. Thermoacidophiles, for example, grow at pH values less than 2 and at temperatures above 75 °C. Some barophiles grow at temperatures below 2 °C, in combination with pressures of up to 1200 bar.

Hyperthermophiles

Thermophiles may be defined as organisms living at high temperatures. Various different life forms thrive or survive over differing temperature ranges, and the maximum temperature at which life is sustainable remains one of the most fascinating (and unanswered) questions in biology. The upper temperature limit for growth (and often survival) of fish and other aquatic species is approximately 38 °C, that of most plants and insect is

Table 10.11   Various categories of extremophiles and the habitats in which they live

Extremophile type	Natural habitat	Relevant growth parameter	Example microbe
Hyperthermophiles	Geothermal habitats (e.g. hot springs)	Opt. temp. for growth at or above 80 °C	*Pyrococcus furiosus*
Halophiles	Hypersaline waters	Grow in upto 5 M NaCl	*Halobacterium halobium*
Psycrophiles	Extremely cold environments (e.g. Antarctic sea water)	Grow at temperatures as low as −2 °C	*Alteromonas* sp.
Alkaliphiles	Alkaline environments	Grow at pH values above 9	*Natronobacterium* sp.
Acidophiles	Acidic environments	Grow at pH values lower than 4	*Thermoplasma acidophilum*

50 °C, while that of eukaryotic microorganisms (certain protozoa, algae and fungi) is in the region of 60 °C. Certain prokaryotic microorganisms (e.g. blue-green algae and photosynthetic bacteria) can grow at temperatures of 70–73 °C. Scientists have long studied thermostable organisms in an attempt to understand how life processes function at elevated temperatures. In addition they have exploited the inherent thermostability of biomolecules produced by such microorganisms by applying them to various industrial processes (e.g. thermostable DNA polymerase as used in recombinant DNA technology, incorporation of thermostable proteases in biologically acting detergents, etc.).

This field of academic and applied endeavour was revolutionized in the mid 1980s with the discovery of microorganisms capable of growth at or above 100 °C. Approximately 20 different such genera are now known (Table 10.12). All the hyperthermophiles thus far discovered have been

Table 10.12 (Hyper)thermophilic archaea and their respective maximum growth temperatures

Order	Genus (max. growth temp.)	
Thermococcales	*Pyrococcus*	(105 °C)
	Thermococcus	(97 °C)
Sulfolobales	*Sulfolobus*	(87 °C)
	Acidianus	(96 °C)
	Desulfurolobus	(87 °C)
	Metallosphaera	(80 °C)
	Styoiolobus	(88 °C)
Thermoproteales	*Pyrodictium*	(110 °C)
	Thermodiscus	(98 °C)
	Desulfurococcus	(90 °C)
	Staphylothermus	(98 °C)
	Thermoproteus	(92 °C)
	Pyrobaculum	(102 °C)
	Thermofilum	(100 °C)
Thermoplasmatales	*Thermoplasma*	(67 °C)
Methanogenic Archaea	*Methanothermus*	(97 °C)
	Methanococcus	(91 °C)
	Methanopyrus	(110 °C)
Sulfate-reducing archaea	*Archaeoglobus*	(95 °C)
Unclassified	*Hyperthermus*	(110 °C)
	ES-1	(91 °C)
	ES-4	(108 °C)
	GE-5	(102 °C)
	GB-D	(103 °C)

isolated from various geothermal habitats. Such habitats are worldwide in distribution, and are associated primarily with tectonically active zones where major crustal movements of the earth occur. In such areas magmatic material is trust close to the Earth's surface, serving as a heat source of sea water or ground water, thus generating deep sea geothermal vents, thermal springs or geysers; all potential sources of hyperthermophiles.

Although a few such hyperthermophiles are aerobic, most are strictly anaerobic heterotrophs, utilizing complex peptide mixtures as a source of energy, carbon and nitrogen. Most are dependent on the reduction of elemental sulfur to H_2S for significant growth. Carbohydrates are utilized by only a few species. Successful culture of these microorganisms generally requires some specialized fermentation equipment and an element of specialized microbiological training.

Considering the extreme environment in which hyperthermophiles grow, it is perhaps not surprising that many aspects of their cellular architecture and metabolism are novel. The plasma membrane of several such archaea has been studied and has proven quite different to mesophilic plasma membranes in terms of composition. The membrane lipids contain almost no fatty acid ester groups for example. Proteins isolated from such hyperthermophiles are inherently thermostable, with some retaining biological activity at temperatures as high as 140 °C.

Enzymes from hyperthermophiles

A range of enzymes displaying remarkable thermal stability have been isolated from various hyperthermophiles (Table 10.13). This has prompted further applied research aimed at identifying high-temperature enzymes suitable for industrial use. Such enzymes would prove desirable in the case of certain industrial processes as high temperature operation would:

- reduce viscosity of processing fluid (e.g. in the case of starch hydrolysis);

- discourage microbial growth (important in food applications);

- render many reaction substrates more soluble/susceptible to hydrolysis.

Furthermore, thermostable enzymes are usually more resistant to inactivation by other denaturing influences such as detergents, organic solvents, chaotropic agents and oxidizing agents. In some cases thermostable enzymes also have proven extremely resistant to proteolysis. However, enzymes from hyperthermophiles should not automatically be assumed to be suited for all industrial applications as:

Table 10.13   Temperature optimum and stability of selected enzymes isolated from various hyperthermophiles

Enzyme	Source	T_{opt} (°C)	Thermostability (T_{50})
Protease	*Thermobacteroides vulgaris*	85 °C	NR
Protease	*Pyrococcus furiosus*	>115 °C	33 h at 98 °C
DNA Polymerase	*Pyrococcus furiosus*	>75 °C	20 h at 95 °C
Amylase	*Pyrococcus woesei*	100 °C	6 h at 100 °C
Amylase	*Pyroccus furiosus*	100 °C	2 h at 120 °C
Amylopullulanase	*Thermoproteus tenax*	118 °C	20 h at 98 °C
Xylanase	*Thermotoga* sp.	105 °C	1.5 h at 95 °C
Cellobiohydrolase	*Thermotoga* sp.	105 °C	1 h at 108 °C
Glucose Isomerase	*Thermotoga neapolitana*	95 °C	NR
α-Glucosidase	*Pyrococcus furiosus*	100 °C	48 h at 98 °C
β-Glucosidase	*Pyrococcus furiosus*	105 °C	85 h at 100 °C or 13 h at 110 °C
Glutamate dehydrogenase	*Pyrococcus furiosus*	95 °C	10 h at 100 °C
Lactate dehydrogenese	*Thermotoga maritima*	>90 °C	1.5 h at 90 °C

Most such enzymes would have potential industrial applications. NR, not recorded. T_{50}, time required to lose 50% of catalytic activity when incubated at the indicated temperature.

- many processes are not amenable to high temperature operation (because of e.g. thermoliability of reactants/products);

- low temperature operation can be more economic due to lower process energy costs incurred;

- in some instances (e.g. detergent applications) commercial trends favour low temperature operation;

- there may be no obvious advantage to increasing the operating temperature;

- sufficiently thermostable enzymes may already be in use;

- for certain food applications (e.g. amylases used in bread making) the added enzyme is heat inactivated after it has achieved its catalytic effect – heat inactivation would prove unrealistic in the case of extremely thermostable enzymes;

- many enzymes derived from (anaerobic) archaeal sources are oxygen sensitive and hence not suited to industrial application.

Most hyperthermophiles are difficult to culture The majority are strict anaerobes, and grow only to a low final cell density. Many produce end products of metabolism that are toxic and/or corrosive (e.g. H_2S production by many anaerobic archaea). As a consequence the industrial production of moderate to large quantities of hyperthermophile-derived enzymes will depend upon recombinant DNA technology. The overall approach to identification and production of potentially useful enzymes from archaea is presented in Figure 10.5 An alternative approach entails interrogation of sequence databases. The entire genome sequence of seven extremophiles had been elucidated or almost elucidated by the turn of the century. Comparative analysis of those sequences (using gene sequences coding for known mesophilic enzymes) can facilitate the quick identification of putitive thermostable enzymes of industrial interest.

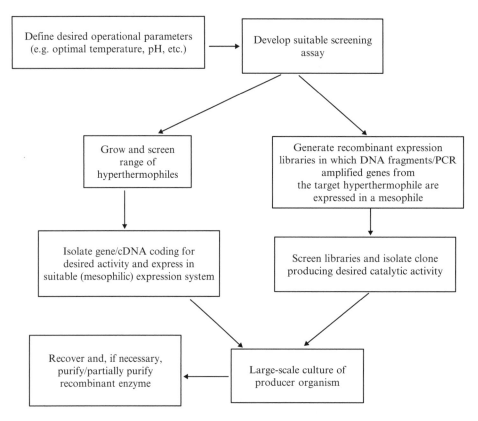

Figure 10.5 Overview of the approach adopted to the identification and production of hyperthermophile derived enzymes of potential industrial utility. Note that partial purification of recombinant heat stable enzymes expressed in engineered mesophiles may sometimes be attained by a simple heat step. Heating (e.g. to 80 °C for several minutes) will inactivate and precipitate many or most mesophile derived proteins, while often having no ill effect on the recombinant protein

Thus far the only such enzymes that have been commercialized are heat stable DNA polymerases (Chapter 12). Continued applied research focuses upon (a) identifying industrial processes suited to the application of extremely thermostable enzymes and matching the process with a suitable enzyme and (b) identifying novel archaeal enzymes that may prove useful in the development of completely new enzyme mediated biotechnological processes.

Enzymes from additional extremophiles

To date, most pure and applied extremophile-based research has focused upon hyperthermophiles. However, the underlining biochemistry and biotechnological potential of additional extremophiles is also now gaining significant attention. Much of the research on psychrophiles thus far has focused upon fish obtained from Antarctic waters which are thus continually exposed to temperatures of about $-1.8\,°C$ (the freezing point of sea water). The freezing of body fluids in these fish is prevented by the presence of 'anti-freeze' glycoproteins, which function to inhibit ice crystal growth. Enzymes isolated from these and other psycrophiles (e.g. Antarctic microorganisms) typically display maximum activity at temperatures close to $0\,°C$. While a proportion of these enzymes display little or no residual catalytic activity at temperatures above $6–10\,°C$, many retain some level of activity up to temperatures approaching $45\,°C$.

The cloning, sequencing and in some cases the three-dimensional structural elucidation of these enzymes continues to assist scientists in pinpointing the molecular adaptations which allow them function under such extreme conditions (Chapter 1). While no cold adapted enzyme has proven itself of actual biotechnological use to date, a number of potential application fields have been preposed, including:

- their use in biotransformation reactions and processes where heat-labile substrate(s) and/or products(s) are consumed or produced;

- use in low temperature food processing, thus discouraging microbial growth;

- use in low-temperature industrial processes, thus reducing processing costs by decreasing energy requirements.

It remains to be seen if any such applications will prove commercially viable.

Various proteins isolated from halophiles have also attracted some interest from the biotechnological community. Both intra- and extracellular enzymes produced by halobacteria (extremely halophilic archaea) are remarkably halophilic or halotolerent. Most such enzymes are quite stable, proving to be more than usually resistant to potential denaturing influences such as elevated temperatures. Stability is a highly desirable characteristic for any industrial enzyme. Perhaps the most likely field of application for halophilic enzymes is in biotransformation reactions carried out in organic solvent-based media. Under such conditions of low water activity some solvents tend to strip the essential hydration layer from the proteins, thus often leading to their inactivation. Halobacterial enzymes naturally function under conditions of very low water activity.

Some halobacteria contain a characteristic 'purple patch' on their plasma membrane, which is largely composed of a bacteriorhodopsin, a transmembrane protein containing retinal as a prosthetic group. Bacteriorhodopsin contains several α-helical stretches separated by short, non-helical sequences. Each α-helical segment, composed of about 20 hydrophobic amino acids, spans the plasma membrane. The protein plays a pseudophotosynthetic function in halobacteria. It (or more specifically its prosthetic group) can absorb photons of light and, as the excited molecules subsequently revert to the ground state, a conformational change in the protein is induced. This results in its pumping protons across the plasma membrane, thus expelling them from the cell. A transmembrane proton gradient is set up, and the energy released as the protons re- enter the cell via a separate transport mechanism is used to drive ATP synthesis from ADP and Pi.

Bacteriorhodopsin, essentially a light driven proton pump, is a very stable protein. It can be embedded in polymers or immobilized on solid surfaces without losing its biological activity. Bacteriorhodopsin molecules placed in synthetic membrane systems remain fully active for periods approaching or surpassing one year. The photochemical reaction it mediates is self-regenerating and the photoelectric signal it generates is extremely reproducable. As such the protein has generated considerable interest as a potential optical switching element in the emerging field of bioelectronics.

Enzymes in organic solvents

Until relatively recently it was assumed that enzymes added to organic solvents (e.g. benzene, acetone, toluene, cyclohexane, etc.) would lose activity and likely be denatured. Proteins have evolved to function in aqueous-based media, and one major force promoting protein folding and subsequent stabilization in its biologically active conformation is that of the 'hydrophobic effect'. This describes the fact that apolar amino acid residues are largely buried in a protein's hydrophobic core, with polar amino acids largely residing on the protein's surface, interacting with the surrounding water molecules (Chapter 1).

Addition of enzymes to non-polar organic solvents should thus prompt the protein to turn inside out. In practice, when many enzymes are added (usually in lyophilised format) to various organic solvents, they retain catalytic activity although there is little or no water in the system. This may be explained by the fact that such solvents have a much lower dielectric constant than that of water. Lowering the dielectric constant strengthens electrostatic forces within the protein, thus stabilising it. An apolar solvent environment also promotes increased intramolecular hydrogen bonding.

The fact that enzymes can retain activity when placed in organic solvents has attracted biotechnological interest. Some potential advantages to industry of using enzymes under such conditions are listed in Table 10.14. One of the main advantages is that lipophilic substrates, sparingly soluble in water, are generally freely soluble in organic solvents. This renders more practical the application of enzymes to the biotransformation of apolar substances. Enzymes also exhibit greater thermostability when present in organic as opposed to aqueous-based media (Table 10.15). Although the molecular basis of increased thermostability

Table 10.14 Some potential advantages of using enzymes in organic solvents for selected biotechnological applications

Increased solubility of non-polar reactants/products

Increased enzyme thermostability

Altered enzyme substrate specificity

Shifting of thermodynamic equilibria to favour synthesis rather than hydrolysis (e.g. formation of peptide bonds by proteases)

Enzyme reusability, due to ease of recovery by filtration or centrifugation

Use of solvents with low boiling points facilitates easier product recovery by solvent evaporation

Most organic solvents discourage/prevent microbial growth

Table 10.15 Comparison of the thermal stability of selected enzymes in aqueous versus non-aqueous media

Enzyme	Solvent	Half life at indicated temperature
Lipase (Porcine)	Aqueous (buffer)	< 1 min at $100\,^\circ$C
	Organic (tributyrin)	12 h at $100\,^\circ$C
Lysozyme	Aqueous (buffer)	8 mins at $100\,^\circ$C
	Organic (cyclohexane)	140 h at $100\,^\circ$C
Chymotrypsin	Aqueous (buffer)	< 1 min at $100\,^\circ$C
	Organic (octane)	270 min at $100\,^\circ$C
Ribonuclease	Aqueous (buffer)	< 10 min at $90\,^\circ$C
	Organic (nonane)	> 6 h at $110\,^\circ$C

has not been fully elucidated, it is probably due to the enhancement of the enzymes intramolecular (stabilizing) electrostatic interactions, as already described. As enzymes are insoluble in most organic solvents they usually form suspensions rather than true solutions when added to such solvents. This allows recovery (and hence re-use) of the enzyme by filtration or centrifugation.

When placed in organic solvents enzymes usually display altered substrate specificity. Specificity can be relaxed and kinetic parameters such as K_{cat} and K_m are generally altered. In some cases reaction equilibria can be reversed (e.g. proteases such as subtilisin can catalyse peptide bond synthesis rather than hydrolysis). Many of these effects are explained, in part at least, by the fact that binding of an enzyme to its substrate(s) is influenced by the solvent. The binding energy of an enzyme to its substrate is determined by the difference between the energy of the enzyme – substrate complex and that of the enzyme and substrate free in solution, each separately interacting with solvent molecules. The natural stereo-specificity exhibited by enzymes is also usually altered by the use of organic solvents. Subtilisin for example can synthesize peptides from D-amino acids as well as L-amino acids in organic media.

One impediment to the more widespread industrial application of enzymes in organic solvents is that most enzymes are less active when placed in such solvents. In some cases the reduction in activity can be relatively modest. However, in many instances a $10^4 - 10^7$ fold reduction in activity is observed. This can render uneconomic the use of an enzyme in some industrial processes. The observed reduction in enzyme activity when placed in organic solvents may be due to a number of factors. The fact that the enzyme is in suspension rather than true solution may impose diffusional limitations on substrate accessibility. The enzyme's conformation may be altered sufficiently in the organic environment to reduce catalytic efficiency. Furthermore, pH has a profound effect on enzyme activity in water, but the concept of pH in the case of organic solvents, has no meaning. Despite these difficulties more and more research and development work is underway aimed at using enzymes in organic solvents for applied purposes, particularly with regard to the use of enzymes in organic synthesis.

FURTHER READING

Books

Adams, M. (1996) *Enzymes and Proteins from Hyperthermophilic Microorganisms.* Advances in Protein Chemistry, Vol. 48. Academic Press, London.

Bickerstaff, G. (1996) *Immobilization of Enzymes and Cells.* Humana Press, New York.

Faber, K. (2000) *Biotransformations in Organic Chemistry.* Springer-Verlag, Godalming.

Gerhartz, W. (1990) *Enzymes in Industry: Production and Applications*. VCH, Weinheim.

Godfrey, T. & West, S. (1996) *Industrial Enzymology*, 2nd edn. MacMillan, Basingstoke.

Herbert, R. (1991) *Molecular Biology and Biotechnology of Extremophiles*. Blackie, Glasgow.

Horikoshi, K. (1998) *Extremophiles*. John Wiley & Sons, Chichester.

Johri, B. *et al.* (1999) *Thermophilic Moulds in Biotechnology*. Kluwer Academic Publishers, Boston.

Kelly, J. (1992) *Applications of Enzyme Biotechnology*. Plenum Publishing Co., New York.

Torchilin, V. (1991) *Immobilized Enzymes in Medicine*. Springer-Verlag, Godalming.

Uhlig, H. (1998) *Industrial Enzymes and their Applications*. Wiley Interscience, New York.

Wiseman, A. (1995) *Handbook of Enzyme Biotechnology*. Ellis Horwood, Chichester.

Articles

Industrial enzymes; general & immobilization

Cosnierm S. (1999) Biomolecule immobilization on electrode surfaces by entrapment or attachment to electrochemically polymerized films. A review. *Biosensors Bioelectr.* **14** (5), 443–456.

Demirjian, D. *et al.* (1999) Screening for novel enzymes. *Biocatalysis – From Discovery to Application* **200**, 1–29.

D'Souza, S. (1999) Immobilized enzymes in bioprocess. *Curr. Sci.* **77** (1), 69–79.

Marrs, B. *et al.* (1999) Novel approaches for discovering industrial enzymes. *Curr. Opin. Microbiol.* **2** (3), 241–245.

Ohdan, K. & Kuriki, T. (2000) An approach for introducing a different function to an industrial enzyme. *Trends Glycosci. Glycotechnol.* **12** (68), 403–410.

Pandey, A. *et al.* (1999) Solid state fermentation for the production of industrial enzymes. *Curr. Sci.* **77** (1), 149–162.

Tischer, W. & Wedekind, F. (1999) Immobilized enzymes; methods and applications. *Biocatalysis – from Discovery to Application* **200**, 95–126.

Enzymes from extremophiles

Hough, D. & Danson, M. (1999) Extremozymes. *Curr. Opin. Chem. Biol.* **3** (1), 39–46.

Huber, H. & Stetter, K. (1998) Hyperthermophiles and their potential in biotechnology. *J. Biotechnol.* **64** (1), 39–52.

Madigan, M. & Oren, A. (1999) Thermophilic and halophilic extremophiles. *Curr. Opin. Microbiol.* **2** (3), 265–269.

Niehaus, F. *et al.* (1999) Extremophiles as a source of novel enzymes for industrial application. *Appl. Microbiol. Biotechnol.* **51** (6), 711–729.

Zeikus, J. *et al.* (1998) Thermozymes: biotechnology and structure–function relationships. *Extremophiles* **2** (3), 179–183.

11

Industrial enzymes; proteases and carbohydrases

PROTEOLYTIC ENZYMES

Functionally and quantitatively proteins represent one of the most abundant classes of biomolecule, and protein-degrading (proteolytic) enzymes constitute the single most important group of industrial enzymes currently in use. Their annual sales value represents over 50 per cent of total sales revenue generated by all industrial enzymes combined. They enjoy a long history of use in the food and detergent industries and more recently have been employed in leather processing and as therapeutic agents. An overview of their industrial uses (excluding medical uses, discussed in Chapter 6) is provided in Table 11.1. Proteolytic enzymes used industrially are obtained from a wide range of source material (Chapter 2), and the most recent innovations in this regard relate to their recombinant production (Chapter 10).

Classification of proteases

Proteases may be classified using various criteria, but are most often classified on the basis of either the positioning of the peptide bond hydrolysed or the molecular mechanism by which such hydrolysis is

Table 11.1 Overview of the various industrial uses of proteolytic enzymes

Industry	Use
Beverage	Solubilization of grain proteins, stabilization of beer
Detergent	To catalytically degrade protein-based stains on clothing
Bread/confectionery	To modify gluten elasticity
Cheese production	To coagulate casein, forming curds. To ripen cheese
Leather processing	To dehair hides; to bate (soften) leather
Meat	To tenderize meat

achieved. Based upon the relative position of the susceptible bond within the protein substrate, proteases may be described as exopeptidases or endopeptidases. Exopeptidases may be further classified into aminopeptidases, dipeptidyl peptidases, tripeptidyl peptidases, carboxypeptidases and peptidyl dipeptidases (Table 11.2). Endopeptidases cleave peptide bonds which are found internally in the protein, usually some distance from the carboxyl or amino termini. Most such enzymes (both endo and exo) also display some level of selectivity towards the peptide bond they hydrolyse, for example chymotrypsin hydrolyses peptide bonds adjacent to aromatic amino acid residues.

On the basis of mechanism of action, proteolytic enzymes may be divided into four categories: serine proteases, cysteine proteases, aspartic proteases and metalloproteases (Table 11.3).

Serine proteases are characterized by the presence of an essential serine residue at their active site (Figure 11.1). They are the most common class of protease, being produced by archaea, bacteria, eukarya and viruses. Although exceptions exist, most serine proteases display a molecular mass

Table 11.2 Peptide bonds hydrolysed by various types of exopeptidases

Exopeptidase	Peptide bond cleaved
Aminopeptidase	$NH_2 - O \overset{\downarrow}{-} O - O - O - (-O-)_n - O - O - O - COOH$
Depeptidyl peptidase	$NH_2 - O - O \overset{\downarrow}{-} O - O - (-O-)_n - O - O - O - COOH$
Tripeptidyl peptidase	$NH_2 - O - O - O \overset{\downarrow}{-} O - (-O-)_n - O - O - O - COOH$
Carboxypeptidase	$NH_2 - O - O - O - O - (-O-)_n - O - O \overset{\downarrow}{-} O - COOH$
Peptidyl dipeptidase	$NH_2 - O - O - O - O - (-O-)_n - O \overset{\downarrow}{-} O - O - COOH$

Table 11.3 The major classes of proteolytic enzymes (as classified on the basis of their mechanism of catalytic conversion) and examples of members of each class

Protease class	Examples
Serine proteases	Trypsin, chymotrypsin, elastase, subtilisins, proteinase K
Aspartic proteases	Pepsin, rennin (chymosin), microbial aspartic proteases
Cysteine proteases	Papain, ficin, bromelain
Metalloproteases	Collagenase, elastase, thermolysin

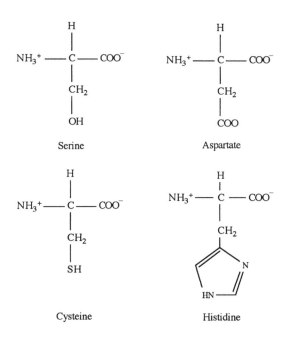

Figure 11.1   Structure of amino acids essential to the catalytic activity of proteases. Serine proteases contain an essential serine (and often an essential histidine and aspartate) residue at their active site. Cysteine proteases contain active site cysteine and histidine residues while aspartic proteases contain an essential aspartate residue at their active site

of between 18 and 35 kDa and are maximally active somewhere between pH 7 and 11. They may be divided into a number of subfamilies, based upon structural similarities. From an applied perspective, the bacterial subtilisins are the subgroup of serine proteases of greatest industrial significance. Subtilisins, particularly those produced by selected *bacilli* have found widespread application as detergent additives.

Aspartic proteases, as the name suggests, are a group of acidic proteases that contain an essential aspartic acid residue at the catalytic site (Figure 11.1). Most aspartic proteases display maximum activity at pH values between 3 and 4, and have isoelectric points between 3 and 4.5. They generally exhibit molecular masses of 30–45 kDa. Pepsin and rennin represent two animal stomach-derived aspartic proteases. Microbial aspartic proteases are produced mainly by *Aspergillius, Penicillium, Rhizopus* and *Mucor* spp., and the best-known application of aspartic proteases is in cheese manufacture.

Cysteine proteases are widely distributed in nature and are characterized by the presence of a cysteine and a histidine residue at the active site, which forms a catalytic dyad essential for biological activity. These proteases, the best-known of which is papain, are generally active under reducing conditions and tend to display optimum activity at neutral pH values. Several members find industrial application, mainly in the food industry.

Metalloproteases, as the name suggests, are a family of proteases characterized by their requirement for (divalent) metal ions to sustain biological activity. Removal of these ions, by for example incubation of the enzyme with chelating agents, abolishes their catalytic activity. Most metalloproteases display maximum activity at neutral to alkaline pH values. The single best-known member of this family is microbial thermolysin, a very heat-stable neutral protease.

Detergent proteases

The bulk of proteolytic enzymes produced commercially are incorporated into detergents. Enzymes were first introduced into detergent preparations at the turn of the nineteenth century. Few such products were successful as the enzymes chosen, generally from animal sources, were invariably inactivated by other detergent components or by the ensuing washing process.

Most clothing becomes soiled by substances such as dyes, biological molecules, soil and miscellaneous particulate matter. Biological 'dirt' includes protein-, lipid- and carbohydrate-based stains. Such dirt components may be derived directly from humans or animals, (e.g. shed skin or blood), or may be derived from other sources such as foodstuffs or grass.

Modern detergent formulations contain a range of ingredients capable of efficiently removing both biological and non-biological dirt. Typical ingredients are listed in Table 11.4. Soaps and surfactants remove the majority of dirt particles from fabrics. Perborates exhibit a bleaching

Table 11.4 Principal ingredients of modern detergent preparations

Detergent component	Function
Soap and surfactants	Removal of dirt particles, especially hydrophobic molecules, from fabrics
Sodium perborate	Removal of dyes and stains from fabrics
Sodium tripolyphosphate	Used to soften water
Enzymes	Removal of biological dirt, which is mainly protein
Sodium carbonate and silicate	Maintenance of an alkaline pH
Polycarboxylates	Help disperse dirt particles in water
Silicones	Control foaming
Perfume	Imparts an appropriate scent to fabrics

action and thus help remove dyes and stains, such as those from tea, coffee and wine. Proteolytic enzymes function to degrade protein dirt such as blood, egg and gravy. Proteins are often denatured and aggregated by the washing process. This renders them even more difficult to remove from clothing fibres. The addition of proteolytic enzymes active under the washing conditions employed greatly facilitates the degradation and subsequent removal of such stubborn stains. Phosphates, such as sodium tripolyphosphate, function to soften the water, thus ensuring maximal detergent efficiency. Hard water contains appreciable quantities of calcium ions. Calcium and other divalent ions such as Mg^{2+} or Fe^{2+} will react with soap, thus forming a precipitate. Precipitate formation in detergent solutions would obviously be undesirable. Calcium-containing water may be softened by a number of means. One popular method entails the addition of sequestering agents such as sodium polyphosphates. The sequestering agent complexes or sequesters the divalent cations in solution and thus prevents precipitate formation. One consequence of employing such sequestering agents is that divalent cation-dependent proteases may not be used. The presence of sodium carbonate in the detergent formulation ensures the maintenance of an alkaline environment, required for maximal cleaning efficiency.

The exact formulation of detergents sold in various world regions can vary somewhat. This largely reflects different washing practices characteristic of these regions. In Europe, for example, automatic washing machines are used with washing cycles of about one hour and at temperature settings of either 40, 60 or 90 °C. In the US, washing cycles are much shorter (10–15 min) and in all world regions, the trend is towards lower washing temperatures.

Proteolytic enzymes incorporated into detergent formulations must exhibit satisfactory catalytic activities in the presence of other detergent components, and under standard washing conditions. Such enzymes must, therefore, be stable at alkaline pH values, at relatively high temperatures and in the presence of sequestering agents, bleach and surfactants. Of the various classes of proteolytic enzymes (Table 11.3), only the serine proteases are potentially suited to application in the detergent industry. Aspartic proteases will be totally inactive at alkaline pH values. Metalloproteases will be inactivated by sequestering agents in the detergent formulation while cysteine proteases will be inactivated by the oxidising environment (due to the bleach) and high pH values.

Screening of serine proteases from various (particularly bacterial) sources quickly identified members of the subtilisin subfamily as being the most appropriate for detergent application. These generally (a) display maximum activity at pH values typical of detergent containing wash water, (b) are modestly or highly stable in the presence of other detergent ingredients and (c) generally display a broad substrate specificity (i.e. cleave peptide bonds linking many different amino acid residues). Currently, the vast majority of detergent proteases are subtilisins isolated from either

Bacillus licheniformis, *B. lentus*, *B. alcalophilus* or *B. amyloliquefaciens* (Table 11.5). These (extracellular) proteases may be produced in large quantities by fermentation technology, using the native producer organism. However, many now are produced by recombinant means using homologous or heterologous systems (Chapter 2) in order to achieve even higher expression levels.

One of the first proteolytic enzymes successfully used in detergent products was that of subtilisin Carlsberg. Subtilisin Carlsberg is a single-chain serine protease produced by *Bacillus licheniformis*. It consists of 274 amino acids and has a molecular mass of 27.5 kDa. It is a single-domain polypeptide whose active site is formed by serine 221, aspartate 32 and histidine 64. The enzyme exhibits typical Michaelis–Menten kinetics and broad substrate specificity. The protein is maximally active at alkaline pH values of 8–10 and is stable at temperatures in excess of 50 °C. The subtilisin gene has been cloned and expressed in various production systems. Its relatively straight forward structure and kinetic mechanism, along with its almost unparalleled industrial importance, has rendered it subject to extensive investigation.

Subtilisin BPN′, produced by *B. amyloliquefaciens* also found early application in the detergent industry. The enzyme consists of 275 amino acids and its three-dimensional structure is quite similar to that of the *subtilisin Carlsberg* enzyme, although their kinetic properties do vary. Both enzymes function effectively when used in liquid detergent formulations (the resulting wash water pH will typically fall between 7.5 and 9.5). Powdered detergent formulations generally produce wash water of even higher pH (usually 9–11). Subtilisin sourced from *B. lentus* is often included in such formulations, as this enzyme exhibits an even more alkaline pH versus activity profile than those of subtilisin BPN′ or Carlsberg. The *B. lentus* subtilisin is slightly smaller (269 amino acids) than

Table 11.5 The major commercialized subtilisin proteases, their sources, properties and manufacturers

Subtilisin source	Act. range (pH)	Act. range (Temp °C)	Trade name (and producer)
Bacillus licheniformis	7–10.5	50–65	Alcalase (Novo Nordisk) Maxatase (Genencor)
Bacillus lentus	9–12	45–70	Savinase (Novo Nordisk) Esperase (Novo Nordisk) Purafect (Genencor)
Bacillus alcalophilus	9–12	45–60	Maxapem (Gist Brocades)
Bacillus amyloliquefaciens	9–12	45–60	Maxacal (Genencor)

Note: the detergent business of Gist brocades was acquired by Genencor in the mid 1990s.

subtilisin Carlsberg or BPN', and it displays in the region of 60 per cent sequence homology with each of the latter proteases.

Although subtilisins display biochemical characteristics making them suitable for detergent application, various modified forms have been generated by protein engineering. Although earlier engineering studies aimed to increase the thermal stability of the enzymes, the general trend towards decreased washing temperatures renders this goal less important. Most commercial attention in this area has focused upon developing detergent proteases resistant to oxidation.

The bleach constituent of all modern detergents invariably promotes the oxidation of sensitive surface amino acid residues such as methionine and cysteine. The subtilisins commonly included in detergent preparations (Table 11.5) all contain a particularly susceptible surface methionine located adjacent to the active site serine 211. Oxidation of the methionine residue (Figure 11.2) reduces the enzyme's catalytic activity, usually quite significantly. Site-directed mutagenesis allows scientists to replace the methionine with a non-oxidizable amino acid residue, and this approach has resulted in the development of second generation oxidation-resistant engineered subtilisins. Such commecialized products include Maxapem (Gist Brocades), Durazyme (Novo Nordisk) and Purafect OXP (Genencor). Although methionine replacement with some amino acids caused an unacceptable decrease in activity, replacement with serine or alanine promoted considerably lower levels of activity loss.

Proteases used in cheese manufacture

Rennin, also termed chymosin, represents another proteolytic enzyme subject to considerable industrial demand. This protease finds application in the cheese manufacturing process. The initial step in cheese making

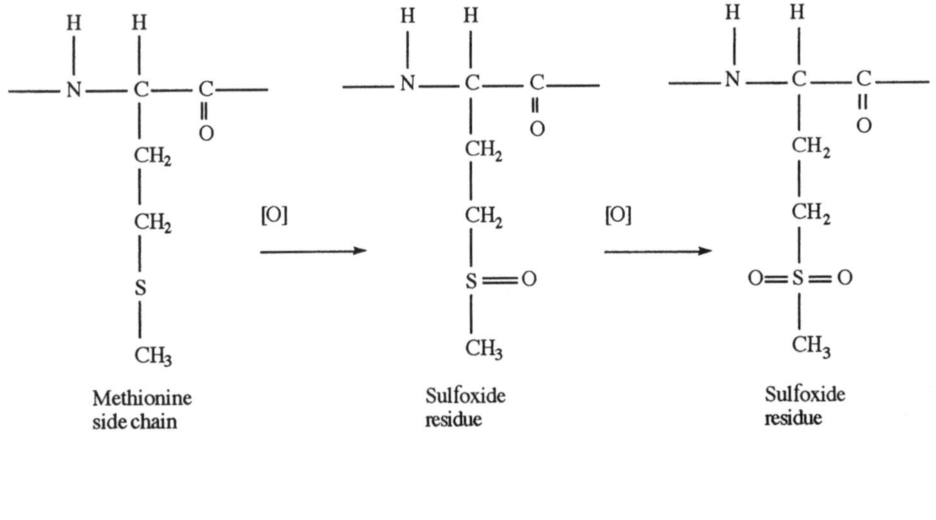

Figure 11.2 Oxidation of a methionine residue, forming a sulfoxide and finally a sulfone

Methionine side chain   Sulfoxide residue   Sulfoxide residue involves the enzymatic coagulation of milk. Rennin catalyzes limited proteolytic cleavage of milk kappa (κ) casein. This destabilizes casein micelles and promotes their precipitation, thus forming curds (three casein subunit types are found in milk, α, β and κ. These aggregate spontaneously, forming spherical particles – micelles). The remaining liquid or whey is removed and the curd is further processed, yielding cheese or other dairy products.

Rennin is obtained from the fourth stomach of suckling calves. Extraction often involves prolonged treatment of dried strips of the calf stomach with a salt solution containing boric acid. The enzymatic extract obtained is termed rennet and it contains a variety of proteolytic and other enzymatic activities. Rennin accounts for no more than 2–3 per cent of such preparations.

Rennin (chymosin), is an aspartic proteinase and is synthesized as preprorennin. The hydrophobic leader sequence that directs the newly synthesized molecule through the endoplasmic reticulum is cleaved yielding prorennin, the inactive rennin zymogen. Prorennin undergoes autocatalytic activation under the acidic conditions associated with the stomach thus yielding the active enzyme. Rennin has a molecular mass of 35 kDa and consists of 323 amino acids. Two isoforms of the enzyme occur naturally, termed chymosin A and chymosin B. Both exhibit acid pH optima in the region of 4.0.

Calf rennin catalyses the proteolytic cleavage of a single peptide bond, the Phe 105-Met 106 bond in the κ casein molecule, thus inducing protein precipitation and curd formation. The extreme specificity of rennin towards its casein substrate renders this enzyme ideally suited to cheese making operations. Further proteolytic cleavage of casein would result in the production of inferior quality unpalatable cheeses.

The rate of slaughter of young calves reflects market demand for veal. Fluctuation in this regard has an obvious effect on the availability and price of rennet. Thus, many alternative sources have been screened in an effort to identify a suitable replacement protease. Pepsin preparations, obtained from a variety of slaughterhouse animals, have had limited success in this regard and are sometimes used in combination with rennet preparations. Of all the microbial enzymes thus far screened, few induce fully satisfactory curd formation. Several members of the *Mucor* species of thermophilic fungi were found to produce acceptable alternative activities to calf rennin. The *Mucor* enzymes may be produced economically and in satisfactory quantities by fermentation technology. Like rennin, these enzymes catalyse limited cleavage of the casein molecule and function adequately under the conditions utilised in the cheese making process. Most such *Mucor* enzymes, however, are thermostable and high temperatures are required to bring about their subsequent inactivation.

An alternative approach involves the expression of the rennin cDNA in recombinant microbial species. The chymosin cDNA has been expressed in a variety of such systems including *E. coli, Aspergillus nidulans,*

Saccharomyces cerevisiae and *Trichoderma reesei*. Recombinant rennin produced as a heterologous protein product in *E. coli* K-12 was the first food ingredient produced by recombinant DNA technology approved for human use (in the US by the FDA in 1990).

The recombinant rennin forms inclusion bodies in *E. coli* and recovery of the inclusion bodies, followed by solubilization and subsequent renaturation, yields active enzyme. The recombinant product displays biological properties identical to those of native rennin. Furthermore, this preparation is in excess of 60 per cent pure. Alternative recombinant systems capable of synthesizing large quantities of rennin produced as an extracellular protein would be an obvious advantage in production.

Gist Brocades have developed a recombinant chymosin (tradename Maxiren) produced in an engineered strain of *Kluyveromyces lactis* (Figure 11.3). *K. lactis* was chosen as the production strain as it has a long and safe history of industrial usage. It is harmless, non-toxic and has GRAS status. It was cultured in the 1960s and 1970s and used directly as a health food and protein supplement. It has also been used for many years in the (non-recombinant) production of lactase.

The chymosin production strain houses the calf preprochymosin gene, spliced to an appropriate yeast leader sequence so that the recombinant product is efficiently excreted from the cell. After fermentation is complete acid is added, promoting autolysis of the inactive chymosin precursor to yield active chymosin. The cell biomass is then removed by centrifugation or filtration, leaving behind the chymosin-rich extracellular

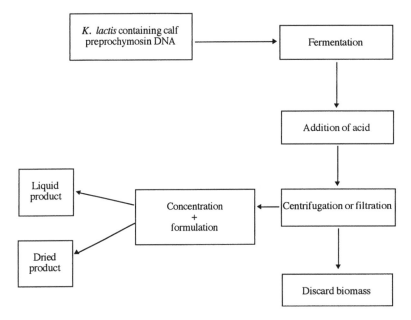

Figure 11.3 Outline of the production of Maxiren (recombinant chymosin) in engineered *Kluyveromyces lactis*. Refer to text for details

media. As *K. lactis* excretes very little protein, the extacellular chymosin is substantially pure. Concentration and formulation is followed by direct marketing of the liquid product, although a substantial proportion of the product is sold in a dried format (as powder, granules or tablets). The recombinant chymosin is biologically and chemically identical to its natural counterpart, and is kosher and halal approved. The Maxiren and other proteases marketed by Gist Brocades for use in cheesemaking is summarized in Table 11.6.

Proteolytic activity is also thought to contribute to flavour development in cheese. A number of peptides are known to impart a characteristic flavour to foodstuffs. So-called bitter peptides have been isolated from a number of cheese types. Most bitter peptides exhibit a highly hydrophobic amino acid content, and their presence in cheese is generally considered undesirable. The existence of peptides which impart a desirable flavour to cheese products still remains to be confirmed.

Proteases and meat tenderization

Tenderness is rated by most consumers as the most important attribute of meat. It is primarily connective tissue collagen that renders meat tough and, hence, necessitates its cooking prior to consumption. Individual collagen molecules in young animals are cross-linked to a low degree and cooking readily promotes their solubilization into gelatin, hence meat from young animals is generally very tender. As animals get older significant additional quantities of collagen is deposited in their connective tissue (to help support their greater bulk), and individual collagen molecules become progressively more cross-linked. The cooking of meat from older animals thus promotes only partial collagen solubilization and hence the cooked meat remains tough.

The tenderness of meat may be maximized by storing fresh carcasses in a cold room for several days (ideally up to 10) post-slaughter. This

Table 11.6 The major proteolytic preparations (recombinant & non-recombinant) produced by Gist Brocades, used to promote curd formation in cheese making

Product brand name	Product description
'Delvoren' or 'Caglio camoscio'	Animal rennet extracted directly from the fourth stomach of young calves. Clotting activity due to a mixture of chymosin and some pepsin
'Fromase'	Microbial product derived from the fermentation of the fungus *Rhizomucor miehei*. The clotting activity is due to a fungal acid protease
'Maxiren'	Recombinant rennin produced in *Kluyveromyces lactis*

process is termed conditioning (also ageing or ripening). During storage, proteolytic and other hydrolytic enzymes are released as the physiological integrity of some muscle cells (i.e. the meat) breaks down.

Artificial tenderization (particularly of meat from older animals) may be achieved using papain, a plant derived cysteine protease. Papain tenderizes the meat by degrading myofibrillar (contractile elements of muscle fibre) and connective tissue proteins. Its tenderization action occurs during the cooking process. Papain displays little activity against native, intact collagen. However, the enzyme has an unusually high optimum temperature (50 °C) and is quite thermostable. At temperatures in excess of 50 °C, the native collagen structure is loosened, facilitating attack by near maximally active papain. Collagen breakdown occurs during cooking when the meat temperature is between 55 °C and 65 °C. The papain is likely fully inactivated when meat temperature reaches 80–90 °C.

Commercially available powdered papain preparations may be rubbed or dusted onto the meat before cooking, although this will mainly promote a surface action. For larger meat cuts, the enzyme must be injected into the joint by commercially available equipment. An alternative approach entails injecting a papain solution into the animal approximately 30 min before slaughter. This facilitates even body distribution of the enzyme. Injection of an active protease into an animal's bloodstream will promote activation of the serum complement system, leading to shock and death. Papain injected into live animals is first chemically oxidized. Oxidation of the essential catalytic site cysteine residue inactivates the enzyme. As long as the animal is alive the papain remains oxidized and hence inactive. After slaughter, cellular glycolysis continues for several hours. This quickly consumes muscle oxygen, resulting in the generation of a reducing environment. This in turn reduces the papain, restoring its catalytic activity. Ficin and bromelain – additional plant proteases – have also been used to tenderize meat. Papain, however, remains the enzyme of choice, mainly on economic grounds.

Proteases and leather production

The production of leather from animal hides is a multistep process (Figure 11.4) with enzymes being utilized in several of the steps. Animal skin (Figure 11.5) is composed largely of the following components: water (64 per cent), protein (mainly collagen, 33 per cent), lipid (2 per cent) and minerals (0.5 per cent). The leather production process essentially involves removing the lipid, water and some of the surface protein (e.g. hair), partial disruption of the collagen, and its subsequent cross-linking during tanning.

A range of proteases are employed during leather manufacture, including animal pancreatic proteases and microbial proteases (acidic, neutral and alkaline bacterial and fungal proteases). Application of plant-derived

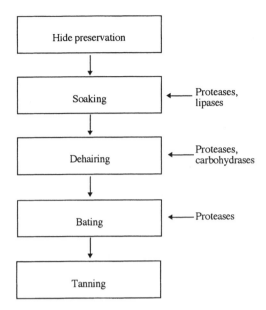

Figure 11.4 An overview of the leather production process, emphasizing steps that are at least partially dependant upon enzymatic activity. Refer to text for specific details

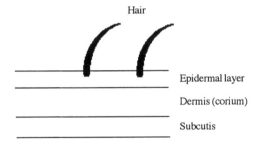

Figure 11.5 The structure of skin. The epidermal layer, which constitutes approximately 1 per cent of the skin houses hair, hair roots, sebaceous (oil/wax-producing) glands and sweat glands. The dermis (which constitutes approximately 84 per cent of skin) largely consists of collagen/connective tissue while the subcutis (constituting approximately 15 per cent of the skin) houses blood vessels, nerves, muscle, connective tissue and fat

proteases (papain and bromelain) has been recorded, as has the use of carbohydrases and lipases, although these latter activities have not gained widespread application in the industry. The enzymes help degrade and hence remove unwanted skin components, but a secondary effect is the consequent reduction in chemical treatments required. This has environmental benefits as less chemical waste is generated.

Raw fresh animal hide is very susceptible to microbial attack. The first step in leather manufacture is thus generally hide preservation (by air drying or dehydrating and curing by packing with dry salt). This allows the hides to be stored or shipped safely. Leather manufacture begins with rehydration of the hides in the so-called soaking process. The hide is immersed in water containing a detergent, preservatives and enzymes for several hours. In addition to rehydrating the hide, this process also helps remove non-fibrous proteins (globulins and albumins) from the hide. Pancreatic proteases have traditionally found most application in this regard, as they efficiently degrade globular proteins, while leaving the collagen intact (finished leather is almost pure collagen). Neutral and alkaline bacterial proteases now also find application in the soaking process. Pancreatin (which contains lipase activity in addition to proteolytic activity) is also sometimes used and research has shown that additional lipase activity can help clean the hide by initiating the degradation of its lipid components.

Dehairing or dewooling was traditionally achieved by rubbing a paste containing a mixture of lime and sodium sulfide onto the outer hide surface. The combination of alkaline conditions and sulfide quickly solubilizes the hair root proteins, discharging the hair from the hide surface. Subsequent alkaline treatment swells the hide. In addition to removing surface hair or wool, these processes also strip away most of the epidermal layer, exposing the corium (Figure 11.5) for further processing. Nowadays, the dehairing process includes the use of proteolytic enzymes (alkaline microbial proteases), which allows significant reductions in the quantities of (polluting) chemicals required. A synergistic effect between proteases and carbohydrases in promoting hair removal has been reported. This may be as a result of carbohydrase-mediated degradation of proteoglycans in the basal membrane area surrounding the hair root.

After dehairing, the hide is usually immersed in acid to reduce its pH to neutrality. The enzyme-mediated bating (softening) process then begins. The purpose of bating is to remove non-collagen proteins from the hide. These proteins include various glycoproteins, proteoglycans, some keratin and elastin. Their removal not only purifies the collagen network remaining, but also loosens it (yielding softer leather). Pancreatic proteases (mainly trypsin) in combination with microbial (*Bacillus-* and *Aspergillus*-derived) proteases find most application in the bating process, as they show least ability to degrade collagen, while degrading the 'contaminant' proteins. Higher concentrations of enzymes are used if highly bated (i.e. very soft, pliable) leather is required. The bated leather is then ready for chemical treatments in the subsequent tanning process, which crosslinks the collagen fibres and yields finished leather.

Synthesis of aspartame

Aspartame is a dipeptide consisting of L-aspartic acid linked via a peptide bond to the methyl ester of phenylalanine (Figure 11.6). It is approximately 200 times as sweet as table sugar (sucrose), and finds extensive use

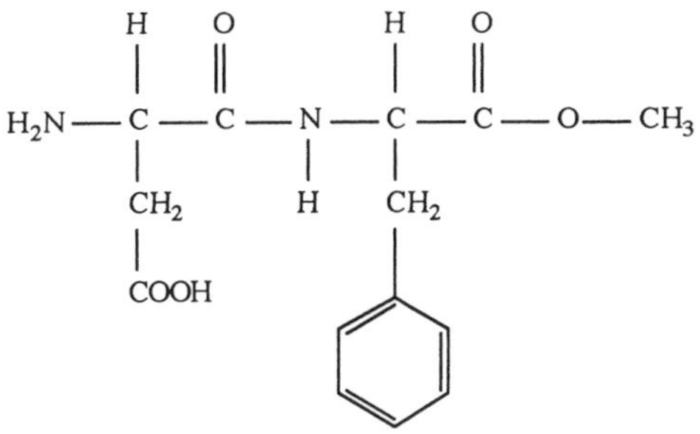

$$H_2N-\underset{\underset{COOH}{|}}{\overset{H}{\underset{|}{C}}}-\overset{O}{\overset{\|}{C}}-\underset{\overset{|}{H}}{N}-\overset{H}{\underset{\underset{CH_2}{|}}{\overset{|}{C}}}-\overset{O}{\overset{\|}{C}}-O-CH_3$$

Figure 11.6 The structure of aspartame

as a low calorie sweetner. Its sweet taste is dependent upon the L-configuration of the two amino acids, and it may be synthesized chemically or enzymatically. Chemical synthesis is costly due to the need to preserve amino acid stereospecificity. Enzymatic synthesis, which automatically preserves stereospecificity has found favour. The neutral metalloprotease thermolysin (sourced from *Bacillus thermoproteolyticus*) is used in immobilized form. Under certain kinetically controlled conditions, the protease synthesizes peptide bonds rather than hydrolysing them.

Protease enzymes used in the brewing and baking industries

The cooling of beer after brewing often promotes haze formation. The haze is composed largely of protein, carbohydrate and polyphenolic compounds. Haze formation can be arrested by addition of proteolytic enzymes to the beer. Although various microbial enzymes have been assessed, plant-derived proteases such as papain and bromelain are most commonly used for such purposes.

Fungal proteases also enjoy limited application in the baking industry. Such enzymes, generally sourced from *Aspergillus species*, are used in order to modify the protein components of flour, and thus alter the texture of the dough. Gluten represents a major protein fraction of flour. It is a complex between two protein types: gliadin and glutenin. When flour is wetted during dough preparation gluten binds a portion of the water and expands to form a lattice-like structure. This promotes a resistance to dough stretching. The addition of low levels of a neutral fungal protease derived from *Aspergillus oryzae* results in a limited degradation of the gluten lattice, thereby reducing the dough's resistance to stretching. This better facilitates retention within the dough of CO_2 produced by yeast fermentation. In turn this influences pore structure of leavened bread and allows the dough to rise uniformly during baking.

Enzymatic conversion of protein waste

The food production and processing industries generate large quantities of waste protein. Such waste includes dead animals and the inedible portions of animals, such as heads, feet, guts and feathers. Such substances are at best of marginal economic value and often are regarded simply as generating a waste disposal problem. Some waste material may be effectively recycled as food by what are termed rendering facilities. These facilities process or render dead animals and animal offal, usually by cooking at high pressure. This yields a valuable source of protein which may subsequently be incorporated into livestock feed. The process also effectively sterilizes the rendered material, thus preventing potential transmission of disease from any infected starting material.

The rendering process requires dedicated, well-equipped facilities. High energy imputs are also required in the cooking process. An alternative method of converting such biological waste into a valuable nutrient source involves the use of degradative enzymes. Such a biological process would require less sophisticated processing facilities and should function with lower energy costs. Enzymatic conversion of waste is also more flexible than traditional rendering processes, and may expand the range of convertible waste products. Enzymatic digestion of poultry feathers represents one such example.

It is estimated that in excess of 400 million chickens are killed each week on a world-wide basis. Typically, each bird has up to 125 g of feathers. The weekly world-wide production of feather waste would thus be 3000 tonnes. Feathers are poorly digested by animals, mainly due to the poor ability of the latter to hydrolyse the highly ordered structure of the α-keratins. An enzymatic process has been developed which facilitates comprehensive digestion of such feather waste. The process is summarized in Figure 11.7. Briefly, it entails blending a slurry of ground feathers with sodium sulfite and an enzyme cocktail consisting predominantly of proteolytic enzymes. The chemical environment generated promotes dissolution of the disulfide bonds. The enzyme activities hydrolyse the ground feather protein (α-keratin), yielding a peptide-rich, more digestible product.

The feather digest may be formulated with another protein source such as soybean meal in order to upgrade the overall nutritional value of the product. The resultant mixture can then be sterilized by heat in order to prevent potential transmission of pathogens. Feather digest products are used in the animal feed industry and are fed to both poultry and pigs. It has also been included in the diets of fish and domestic pets.

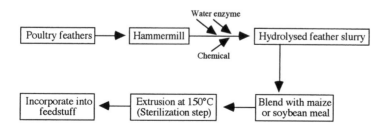

Figure 11.7 Conversion of poultry feathers from waste to a source of amino acids

Animal hair and wool consist almost entirely of α-keratins. Thousands of tonnes of such material are deposited in landfill sites annually. Such waste products also represent potential substrates for enzymatic digestion. A process similar in design to that utilized in feather digestion could be used to convert such substrates into a potential nutrient source.

Large volumes of blood generally accumulate as a by-product in most slaughterhouse facilities. Whole blood may be centrifuged in order to collect red blood cells. This red blood cell fraction contains 70 per cent of total blood protein. Incubation of the collected red cell fraction with alkaline proteases results in their degradation. Enzymatic activity may be arrested simply by reducing the reaction pH to an acidic value. Filtration of the hydrosylate removes the colour. Evaporation is then used to concentrate the protein hydrosylate to 40 per cent solids. This slurry is finally dried by a drum or spray-drying process. The final product is almost 100 per cent digestible and is often incorporated into the rations of young animals.

The fish processing industry also generates large quantities of waste by-products, which are proteinaceous in nature. Such products include not only inedible portions of fish, but also undersized or damaged whole fish. Such waste products may be enzymatically converted into protein hydrosylates of high nutritional value and fish hydrosylates are frequently incorporated into the diets of animals such as mink.

Partially degraded protein preparations also find use as additives in human food. Protein scraps can be recovered from the mechanical fleshing of beef, chicken, turkey or pig bones. Partial proteolytic hydrolysis of such scraps can yield a product containing a high meat extract flavour, which can be added to products such as soups, sauces and prepared meals. Partially hydrolysed vegetable protein (e.g. soybean protein) also find application as a food or flavour additive.

Proteases; additional applications

Proteases are used in smaller quantities for a variety of additional applications. Such uses include the cleaning of contact lenses and the removal of unwanted body hair. During use contact lenses adsorb various solutes present in tear fluid, necessitating their regular cleaning. Tears consist of a salt-containing aqueous fluid secreted by the lacrimal glands, located at the top outer edge of each eye. Blinking spreads the liquid over the eye. Tears largely play a protective role by:

- Keeping the eye surface moist. The moisturizing effect protects the cornea in particular which is easily damaged by dehydration. Tear fluid also contains a lipid component which helps retain moisture by preventing /retarding evaporation from the eye surface.

- Trapping and sweeping away particulate matter or bacteria that enter the eye.

- Tears are rich in antibodies, which helps prevent infections.

- Tears contain lysozyme which promote bacterial lysis by degrading the peptidoglycan-based bacterial cell wall.

The major solutes present in tear fluid which adsorb onto contact lenses include proteins (largely the aforementioned antibodies and lysozyme), mucins (high molecular mass glycoproteins which have a lubricant function), and lipid. Standard lens cleaning agents are at best modestly successful in removing such substances, hence deposit build-up can occur on the lens' surface, reducing its transparency and rendering it uncomfortable to wear. Proteases or a combination of proteases and lipases have proven most effective in removing such deposits and such enzyme-containing cleaning solutions are now commercially available. The cleaning process generally entails immersing the contact lenses in the enzyme containing fluid for several hours (preferably overnight), followed by extensive rinsing of the lens with saline. The rinsing step is essential to physically remove the enzyme breakdown products, as well as any active enzyme present on the lens surface before it is placed back in the eye.

The plant protease papain has also found limited application in slowing or preventing hair regrowth on the body and legs. Removal of unwanted body hair can be undertaken chemically, or by waxing, tweezing or shaving. It has been shown that regular application of a papain-containing solution to the cleared area slows or prevents hair regrowth. The papain enzymatically degrades the growing hair (i.e. α-keratin) and loosens the hair follicle. It also likely helps degrade dead surface skin cells, supposedly leaving the skin softer and smoother. Because of its non-human origin papain can provoke an allergic response in some people, and its initial application to a small area of the skin is encouraged, (to test for sensitivity) before it is applied to large surface areas.

CARBOHYDRASES

Polysaccharide-degrading enzymes represent one of the most significant groups of industrially important bulk enzymes. Such enzymes include amylases, pectinases and cellulases. In addition, several other carbohydrate-transforming enzymes such as glucose isomerase, invertase and lactase also enjoy significant commercial niche markets. A list of some industrially important carbohydrates is presented in Table 11.7.

Amylases

Enzymes that participate in the hydrolytic degradation of starch are referred to as amylolytic enzymes or amylases and such enzymes have

Table 11.7 Some industrially important carbohydrates

Monosaccharides	Disaccharides	Polysaccharides
Glucose	Sucrose	Starch
Fructose	Lactose	Glycogen
Galactose	Maltose	Cellulose
		Pectin

found widespread industrial application (Table 11.8). Specific enzymes classified within this group include α-amylase, β-amylase, glucoamylase (also known as amyloglucosidase) pullulanase and isoamylase. Enzymatic degradation of starch yields glucose, maltose and other low molecular mass sugars. Furthermore, enzymatically mediated isomerization of the glucose yields high fructose syrups.

Abundant supplies of starch may be obtained from seeds and tubers, such as corn, wheat, rice, tapioca and potato. The widespread availability of starch from such inexpensive sources, coupled with large scale production of amylolytic enzymes, facilitates production of syrups containing glucose, fructose or maltose, which are of considerable importance in the food and confectionery industry. Furthermore, they may be produced quite competitively when compared to the production of sucrose, which is obtained directly from traditional sources such as sugar beet or sugar cane.

Table 11.8 The major industrial applications of amylolytic enzymes

Industry/process	Amylolytic enzymes employed
Production of glucose/ maltose syrups	α-Amylases, β-amylase, debranching enzymes
Brewing/alcohol production	α-Amylase, β-amylase, amyloglucosidase
Animal feed additive	α-Amylase
Baking industry	α-Amylase, β-amylase, amyloglucosidase, Debranching enzymes
Laundry detergent additive	α-Amylase
Production of dextrins	α-Amylase
Fruit juice processing	α-Amylase, amyloglucosidase
Textile deisizing	α-Amylase

The starch substrate

Starch represents the most abundant storage form of polysaccharides in plants, and next to cellulose it is the most abundant polysaccharide found on earth. As previously mentioned, it is especially abundant in seeds such as corn and in a variety of tubers. It is stored in granular form in the plant cell.

The starch polymer consists exclusively of glucose units. Two forms exist, namely α-amylose and amylopectin (Figure 11.8). α-Amylose is a long, linear polymer, in which successive D-glucose molecules are linked by an α1 → 4 glycosidic bond. Individual α-amylose chains may vary in length and hence in molecular mass. The larger chains have molecular masses in the region of 500 kDa.

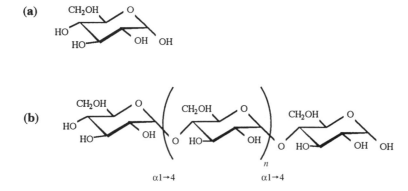

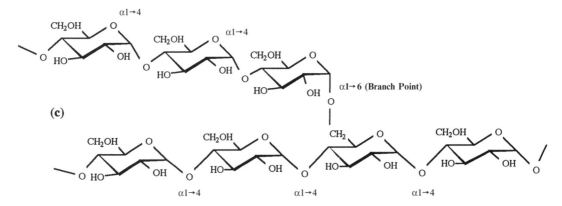

Figure 11.8 Structures of (a) α-D-glucose; (b) segment of amylose chain; (c) section of amylopectin. Glycosidic bonds between successive glucose residues link carbon atom No. 1 of one glucose residue to carbon atom No. 4 of the adjacent glucose residue. The bonds are in the α conformation and, hence, are termed α1→4 glycosidic bonds. At branch points found in amylopectin (c), carbon atom No. 6 of the glucose residue in the main chain is linked to carbon No. 1 of the first glucose residue in the branch. This bond is thus termed an α1→6 glycosidic bond

Amylopectin, on the other hand, is a highly branched molecule. Successive glucose residues are linked via $\alpha 1 \rightarrow 4$ glycosidic linkages along the linear portion of the molecule, with branch points consisting of $\alpha 1 \rightarrow 6$ glycosidic linkages. Such branch points generally occur every 25–30 glucose residues. Starch isolated from most plants consists of 70–80 per cent amylopectin. In some cases, such as waxy rice, the starch granule consists exclusively of amylopectin (Table 11.9). Well over half of all starch utilized by man is used in the manufacture of human food and animal feed. However, modified starches also find application in the textile, pharmaceutical, cosmetic laundry and soap industries.

Starch may be hydrolysed by chemical or enzymatic means. Chemical hydrolysis was used formerly and involves heating in the presence of acid. However, this method has been superseded by the use of specific enzymes. Enzymatic hydrolysis generates fewer by-products and produces higher yields of end product compared to the chemical method.

α-Amylase

The initial step in starch hydrolysis entails disruption of the starch granule. Solubilization of the granules, the process of 'gelatinization', facilitates subsequent catalytic degradation. Gelatinization is normally

Table 11.9 The relative % content of amylose and amylopectin in the starch derived from various plant sources

Source	Amylose content (wt %)	Amylopectin content (wt %)
Corn		
Normal	28	72
Waxy	0	100
High amylose	65–85	15–35
Sorghum	28	72
Tapioca	16	84
Arrowroot	21	79
Sago	26	74
Potato	20	80
Wheat	30	70
Rice		
Normal	20–30	70–80
Waxy	0	100

achieved by heating the starch slurry to temperatures in excess of 100 °C for several minutes. α-Amylase may be added immediately prior to the heating step, in order to render more efficient the process of granule disruption. Once the granules have been disrupted, additional α-amylase is added in order to liqueify the starch. This process reduces the viscosity of the starch solution.

α-Amylase activity is widely distributed in nature. The enzyme may be isolated from microbial sources and from animal and plant tissues. α-Amylase is an *endo*-acting enzyme, catalysing the random hydrolysis of internal α1 → 4 glycosidic linkages present in the starch substrate. These enzymes are incapable of hydrolysing α1 → 6 glycosidic linkages present at branch points of amylopectin chains. One exception to this is the α-amylase produced by *Thermoactinomyces vulgaris*, which can hydrolyze both α1 → 6 and α1 → 4 glycosidic linkages. All α-amylases characterized to date are metalloproteins. These enzymes can also generally catalyse the cleavage of internal α1 → 4 glycosidic bonds in glycogen and a variety of additional oligosaccharides.

Two of the more commonly used α-amylases are those isolated from *Bacillus amyloliquefaciens* and *B. licheniformis*. *Bacillus* amylases exhibit a pH optimum at or around neutrality, and are stabilized by the presence of calcium ions. α-Amylase produced by *B. licheniformis* is particularly suited to industrial applications due to its thermal stability. This enzyme consists of 483 amino acids and has a molecular mass of 55.2 kDa. Its pH optimum is 6.0 and its temperature optimum is 90 °C. Most other α-amylases, including those produced by *B. amyloliquefaciens*, are rapidly inactivated at temperatures in excess of 60 °C.

The advent of recombinant DNA technology has facilitated the cloning and expression of genes coding for various α-amylases in a variety of recombinant organisms. Human, wheat and bacterial α-amylases have, for example, been expressed in *Saccharomyces cerevisiae*. More recently, the gene coding for *B. licheniformis* α-amylase has been expressed in transgenic tobacco plants. This was among the first examples of the production of bulk industrial enzymes in a recombinant plant species. The molecular mass of the recombinant protein was found to be 64 kDa, compared to 55.2 kDa in the native *Bacillus* species. This discrepancy was shown to be due to extensive post-translational glycosylation of the enzyme. Direct application of the α-amylase-containing transgenic seeds in starch liqueficiation studies was found to be highly effective.

α-Amylase activities are also produced by a variety of fungal species. Fungal α-amylases most commonly used industrially are produced by species of *Aspergillus*, most notably *A. oryzae* (Table 11.10).

Glucoamylase

Glucoamylase, also known as amyloglucosidase, is produced as an extra-cellular enzyme by a variety of fungal species, most notably species of

Table 11.10 Sources and characteristics of some industrially significant amylolytic enzymes

Enzyme	*Endo-* or *exo*-acting	Glycosidic bond cleaved	Source	pH opt.	Temp opt. (°C)
α-Amylase	*Endo*	α 1–4	*Bacillus subtilis*	6.0	65
			Bacillus licheniformis	6.0	90
			Bacillus amyloliquefaciens	5.5	60
			Aspergillus oryzae	4.5	60
Amyloglucosidase (Glucoamylase)	*Exo*	α 1–4	*Aspergillus niger*	4.5	60
β-Amylase	*Exo*	α 1–4	*Bacillus* sp.	5.0	60
			Clostridium sp.	5.5	80
Pullulanase	*Endo*	α 1–6	*Klebsiella aerogenes*	5.0	60
Isoamylase	*Endo*	α 1–6	*Pseudomonas* sp.	4.0	55

Aspergillus and *Rhizopus*. The enzyme produced by *Aspergillus niger* is the most widely used on an industrial scale.

Glucoamylases catalyse the sequential hydrolysis of $\alpha 1 \rightarrow 4$ glycosidic bonds from the non-reducing end of the starch molecule. The enzyme also catalyses the hydrolysis of $\alpha 1 \rightarrow 6$ glycosidic bonds present at branch points in amylopectin, although at a much slower rate. Glucoamylase thus catalyses the hydrolysis of starch, yielding glucose. It is normally used industrially after the liquefication of the starch by bacterial α-amylase in order to produce glucose syrup. This process is described as saccharification of starch.

Amyloglucosidase is a relatively thermolabile enzyme and is unstable at temperatures in excess of 60 °C. Thus the temperature of liquefied starch thus must be adjusted downwards before addition of the saccharifying enzyme. pH adjustment to more acidic values is also required to ensure optimal enzyme activity. A glucoamylase displaying a higher optimum activity temperature would be of industrial interest. Glucoamylases are produced by a number of plant and animal species, but the enzyme is most widely produced by microorganisms. Over the past several years a number of glucoamylases have been purified from a variety of aerobic and anaerobic microbial species. Most are maximally active at acidic pH values and few display an optimum activity temperature above 60 °C (Table 11.11). None exhibit physicochemical properties which would make them more suited to industrial application than the current commercial products. Attempts to identify more thermostable glucoamylases entail (a) screening thermophiles and hyperthermophiles and (b) protein engineering of meso-phile-derived glucoamylases in order to improve their thermal stability and optimum activity. The overall scheme entailing production of glucose syrups from starch is summarized in Figure 11.9. The glucose syrup

Table 11.11 Optimum temperature and pH values for glucoamylases isolated from the indicated sources

Source	Optimum temp. (°C)	Optimum pH
Aspergillus awamori	60	4.5
Aspergillus oryzae	60	4.5
Aspergillus terrens	60	4.5
Cephalosporiura charticola	60	5.5
Corticum rolfsii	50	4.5
Clostridium thermosaccharolyticum	70	5.0
Humicola lanuginosa	65	4.9
Mucor rouxianus	55	4.6
Penicillium oxalicum	60	5.0
Rhizopus delemer	40	4.5

produced by this method is often used directly by the food and allied industries, in addition to the production of crystalline glucose. A further application entails conversion of some of the glucose to fructose by the enzyme glucose isomerase, thus producing a high fructose syrup.

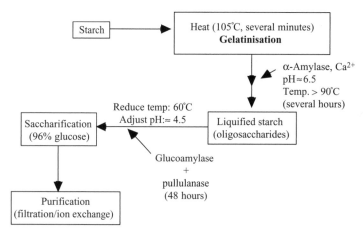

Figure 11.9 Controlled hydrolysis of starch, yielding upto a 96 per cent glucose syrup. The various steps and conditions used are illustrated diagramatically. Pullulanase may be added in addition to glucoamylase during the saccharification process. This enzyme hydrolyzes $\alpha 1 \rightarrow 6$ glycosidic bonds

β-Amylase

In contrast to α-amylase, β-amylases are *exo*-acting enzymes catalysing the sequential hydrolysis of alternate α1→4 glycosidic linkages present in starch, from its non-reducing end. This reaction produces maltose units with inversion to the β form (Figure 11.10). As is the case of α-amylases, β-amylases are incapable of hydrolyzing α1→6 linkages. Conversion of amylopectin to maltose by β-amylase is therefore limited by branch points. The amylose molecule on the other hand, being devoid of such branch points, may be fully degraded by β-amylases. Overall, hydrolysis of most starches by β-amylase yields a product mix of maltose and larger oligosacharides, often termed β-limit dextrans.

β-Amylase is produced by many higher plants. Molecular masses vary from circa 60 kDa (cereal β-amylase) to 150 kDa (sweet potato β-amylase). Plant-derived β-amylases do not require metal ions for activity, are generally maximally active at pH values slightly below neutrality and are inactivated when incubated at temperatures above 60 °C for any length of time. Plant β-amylases used commercially are obtained from barley or, in some world regions, soybeans. Hydrolysis of starch using α-amylase and these β-amylases yields a syrup typically consisting of 2–8 per cent glucose, 40–60 per cent maltose, 10–25 per cent maltotriose and lower levels of β-limit dextrans.

β-Amylase is also produced by a range of microorganisms, most notably bacilli, as well as various species of *Pseudomonas* and *Streptomyces*. Those from bacilli have gained greatest commercial use and again are optimally active in the 55–60 °C range. A more thermostable β-amylase would be of industrial interest and at least one such enzyme has been reported. *Clostridium thermosulfurogenes* β-amylase retains considerable activity up to temperatures of 85 °C.

α1→6 Glucosidases

α1→6 Glucosidases represent a group of amylolytic enzymes which are capable of hydrolysing the α1→6 linkages present at branch points in amylopectin. Amyloglucosidase is one such α1→6 glucosidase. Such debranching enzymes play a central role in the complete degradation of starch as neither α- nor β-amylases possess the catalytic ability to hydrolyse the α1→6 bonds of amylopectin. Several additional α1→6 glycosidases hydrolyse branch-point linkages much more efficiently and rapidly than do amyloglucosidase. The most important such enzymes are pullulanase and isoamylase. These enzymes are often utilized to aid the saccharification process outlined in Figure 11.9.

Although both pullulanase and isoamylase cleave the α1→6 linkages of amylopectin, they may be differentiated by their ability to degrade the polysaccharide pullulan. Pullulanase degrades pullulan, wherease isoamylase does not. Pullulan is a linear polysaccharide consisting of up to 1500

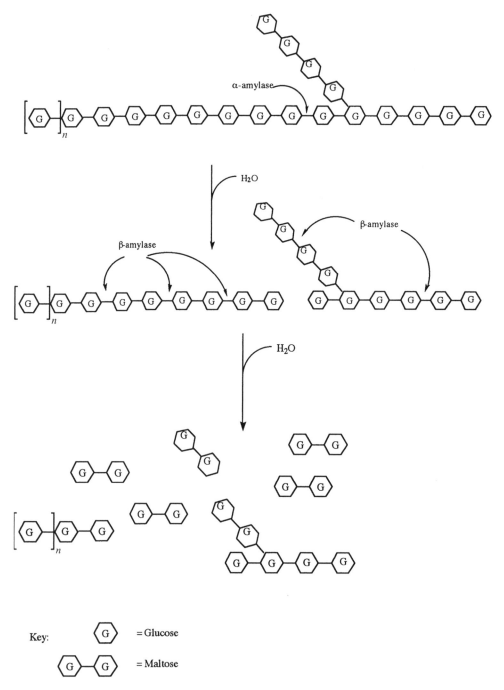

Key:

$\langle G \rangle$ = Glucose

$\langle G \rangle - \langle G \rangle$ = Maltose

Figure 11.10 Hydrolysis of starch, in this example amylopectin, by α-amylase and β-amylase. α-Amylase catalyses the random hydrolysis of internal α1 → 4 glycosidic linkages. It is incapable of cleaving α1 → 6 linkages. β-Amylase catalyses the sequential removal of maltose units from the non-reducing end of the starch molecule. It too fails to hydrolyse α1 → 6 glycosidic linkages found at branch points

glucose molecules. The basic recurring structure consists of three glucose residues linked via two $\alpha 1 \rightarrow 4$ glycosidic linkages. Each such maltotriosyl unit is linked to the next via an $\alpha 1 \rightarrow 6$ bond, as shown in Figure 11.11.

Pullulanase was first discovered in a species of *Aerobacter* in the early 1960s. It is produced by a variety of bacteria including some bacilli and species of streptococci. The pullulanase used commercially is largely sourced from *Klebsiella aerogenes*. This enzyme displays optimum activity at 60 °C and hence is well suited to simultaneous use with amyloglucosidase. Isoamylase, produced by a number of microbial species, was initially isolated from yeast. This enzyme is also synthesized by a variety of bacteria, including some bacilli. Extracellular isoamylase produced in large quantities by a particular mutant strain of *Pseudomonas amyloderamosa* enjoys widespread industrial application.

More recently, a number of pullulanases exhibiting novel activities have been isolated from several thermophilic organisms. Such producer microorganisms include a variety of clostridia and a number of strains of *Thermoanaerobium* and *Thermus*. Some of these enzyme have been produced as heterologous protein products in recombinant systems such as *E. coli* and *Bacillus subtilis*. Most exhibit excellent thermal stability and remain active for prolonged periods at temperatures in excess of 90 °C. Perhaps the single most striking attribute of many such novel pullulanases, sometimes termed amylopullulanases, is their ability to hydrolyse $\alpha 1 \rightarrow 6$ linkages in some carbohydrates such as pullulan and $\alpha 1 \rightarrow 4$ linkages in others such as starch. Conventional pullulanase fails to hydrolyse the $\alpha 1 \rightarrow 4$ glycosidic linkages of either pullulan or amylose. Purified pullulanase from *T. brockii*, for example, hydrolyses only $\alpha 1 \rightarrow 6$ glycosidic bonds in pullulanase but exhibits an almost exclusive preference for $\alpha 1 \rightarrow 4$ bonds in starch.

Glucose isomerase

The enzymatic hydrolysis of large quantities of inexpensive, readily available starch facilitates the economical production of large quantities of glucose syrup. Although glucose syrups may be used directly, many are first converted into high fructose syrups. The conversion of glucose into fructose is desirable in so far as fructose is substantially sweeter than glucose (Table 11.12) and is therefore more attractive industrially

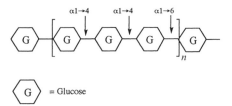

Figure 11.11 Structure of pullulan. Refer to text for specific details

Table 11.12 The relative sweetness of a number of commonly occurring sugars of industrial importance

Sugar	Relative sweetness (%)
Sucrose	100
Glucose	70
Fructose	130
Maltose	40

For comparative purposes, sucrose has been assigned a relative sweetness of 100%.

when utilized as a sweetener in confectionery, ice cream and soft drinks.

Glucose may be converted into fructose by chemical or enzymatic isomerization (Figure 11.12). Chemical conversion relies upon the use of alkali at high temperatures. Low yields and undesirable side reactions (e.g. the production of non-metabolizable sugars such as psicose as well as colour and some off-flavour sugar derivatives) limit the applicability of this particular method. The enzyme glucose isomerase catalyses the required isomerization reaction at ambient temperatures and at near neutral pH values, yielding a syrup containing ca. 42 per cent fructose. Enzymatic isomeration of glucose to fructose was first undertaken on an industrial scale in the late 1960s in the USA. Recently developed process refinements now allow production of syrups of even higher fructose content.

Glucose isomerase obtained from a wide variety of microbial species have found industrial application in the production of high fructose syrups. Bacterial species from which this enzyme may be obtained include *Aerobacter*, *Bacillus* and organisms such as *Streptomyces albus*, *Lactobacillus brevis* and *Actinoplanes missouriensis*. The main sources of commercially available glucose isomerase are *Actinoplanes missousriensis*, *Bacillus coagulans*, and various species of *Streptomyces*.

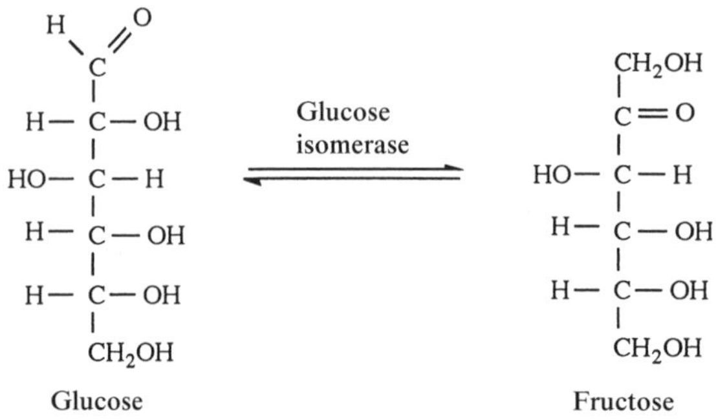

Figure 11.12 The isomerization of D-glucose, an aldohexose, forming D-fructose, a ketohexose

Glucose isomerase from these and many other (mainly microbial) sources have been purified and characterized. From such studies, the following generalizations can be made. Glucose isomerases are usually capable of isomerizing several sugars. The most frequent isomerizations catalysed are the interconversion of D-glucose and D-fructose and the interconversion of D-xylose and D-xylulose. However, the enzyme from various sources also exhibits the ability to isomerize D-ribose, D-allose, L-arabinose and L-rhamnose. Glucose isomerases generally require the presence of a divalent cation (e.g. Mg^{2+}, CO^{2+} or Mn^{2+}) for maximum activity. Molecular masses reported range from 50 kDa to 190 kDa and active glucose isomerases tend to be homodimers or homotetramers, with individual subunits being held together by non-covalent interactions. Optimum temperatures of microbial glucose isomerases tend to range from 60 to 80 °C, while optimum pHs range between 6 and 9.

Glucose isomerase, as produced by most microorganisms, is an intracellular enzyme. Isolation of the enzyme thus requires disruption of the producer cells. Its intracellular location renders the production of purified glucose isomerase more technically and economically demanding than production and isolation of extracellular enzymes.

Many of the initial studies designed to investigate the industrial potential of glucose isomerization used soluble enzymes. However, most industrial-scale isomerization systems now utilise an immobilized form of the enzyme. On an economical level this is quite significant, as it facilitates reuse of the enzyme (Chapter 10).

As with most other enzymes of commercial interest, a range of glucose isomerases have been produced by recombinant means in both homologous and heterologous systems. Protein engineering studies have been initiated, with site-directed mutagenesis studies being aimed at either increasing the enzyme's thermal stability, lowering its pH optimum or altering its substrate specificity. Several studies aimed at increasing the enzyme's thermal stability concentrated on altering surface amino acid residues in order to strengthen non-covalent interactions between individual enzyme subunits (thereby holding them together at higher temperatures). Results obtained were mixed, with the introduction of disulfide linkages and additional salt bridges having no effect, although alterations of surface hydrophobic amino acid residues did enhance thermostability in at least one case.

Industrial importance of starch conversion

Starch may be enzymatically hydrolysed using varying combinations of amylolytic enzymes in order to generate specific end products. The major end products produced commercially are maltodextrins or corn syrup solids, maltose syrup, glucose syrup or high fructose corn syrup (Figure 11.13), as well as cyclodextrins (discussed in the following chapter).

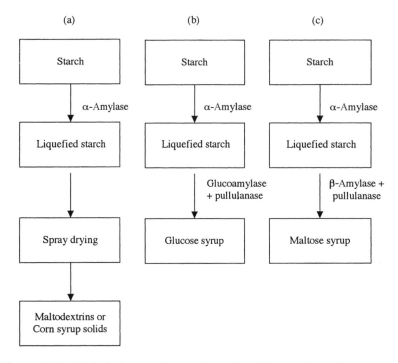

Figure 11.13   Hydrolytic degradation of starch, yielding industrially important end products. In (b) a combination of α-amylase, glucoamylase and pullulanase yields a glucose rich syrup, whereas in (c), a combination of α-amylase, β-amylase and pullulanase yields a maltose syrup. Process parameters such as adjustments of pH and temperature, and addition of stabilizers have been omitted for clarity

Maltodextrins are produced by partial enzymatic hydrolysis of starch, using α-amylase (Figure 11.13). The product is often defined as a non-sweet nutritive saccharide mix, largely containing glucose units linked primarily by α1→4 glycosidic linkages. Although the exact product composition can vary, it generally consists of glucose (approx. 1 per cent), maltose (3–5 per cent), maltotriose (5–10 per cent), maltotetraose (6 per cent), with the remainder (75 per cent or more) being saccharides of higher molecular mass. When the desired degree of starch hydrolysis is attained by α-amylase, the pH value of the slurry is dropped from 6.5 to 3.0 and the slurry is then heated to boiling for about five minutes in order to totally inactivate the α-amylase. Filtration and passage though a carbon column removes any particulates and colour, and the resultant purified maltodextrin mix is evaporated and subsequently spray-dried to yield a powdered product.

Maltodextrins display a number of useful functional properties, including low hygroscopicity, a bland non-sweet flavour and the ability to retard ice crystal growth in ice creams and related frozen products. They are used as an ingredient in soft sweets, where they contribute to viscosity and chewiness. They are also added to many harder sweets, helping to maintain their moisture levels and extending their shelf life. Higher molecular mass maltodextrin preparations (i.e. where 95 per cent of the product consists of penta- and higher saccharides) can be used as a fat replacement in some low fat foods. It appears that such maltodextrins can provide a fat-like texture while containing less than half the calories of fat on a weight basis. In pharmaceutical manufacturing, maltodextrins are used as binding agents in tablet manufacture and as coating agents for some tablets or capsules. Corn syrup solids are similar to maltodextrins, except they are more extensively hydrolysed. In general, they have similar applications to those of maltodextrin products.

Glucose syrups are used to produce crystalline glucose, as a raw material for the production of various organic acids and other chemicals, and as a raw material for the production of high fructose corn syrup. High fructose syrups are as sweet as sucrose (or sweeter depending upon the exact fructose content). On the basis of its sweetening power, high fructose corn syrup is 10–20 per cent cheaper than sucrose. As a result, it has largely replaced sucrose as a food ingredient in many world regions, particularly in the United States. High fructose corn syrups find application as a sweetening ingredient in cakes, confectionery and soft drinks, canned foods, jams, jelly and ketchup. Maltose syrups are characterized by low viscosity and hygroscopicity, good heat stability and mild sweetness. They are used as ingredients in various foods, confectionery, soft drinks and in ice cream where they help control ice crystal formation. Maltose syrups are also sometimes used medically in the intravenous feeding of diabetics.

Starch degrading enzymes are also utilized in the production of alcoholic beverages and in bread making. The production of alcoholic beverages by brewing relies upon the ability of yeast to ferment carbohydrates present in the malted barley and other added sugars. Yeast cells do not possess the enzymatic ability to degrade starch, as they utilize only glucose or other simple monosaccharides and disaccharides as substrates for growth. Germination of the grain is thus promoted prior to the fermentation step. The germinating seeds produce endogenous enzymes capable of hydrolysing not only the stored starch but also cellulose and other structural polysaccharide components of the seed. Germination is subsequently arrested by controlled heat, in order to prevent further seed growth. The seed now contains enzymes such as α- and β-amylases, which are capable of hydrolysing stored starch, thus producing glucose and other sugars which the yeast cells are capable of fermenting. This process is called malting. This traditional process, by which the seeds are induced to produce amylolytic enzymes, may now be supplemented or replaced by the addition of exogenous amylolytic enzymes obtained from microbial sources.

The level of β-amylase present in cereals remains relatively constant but the content of α-amylase is low prior to germination. Milled flour therefore often contains low concentrations of α-amylase. Supplementation of such flour with fungal α-amylase results in a more effective degradation of flour starch, hence rendering the dough easier to work and allowing yeast fermentation to proceed. This, in turn promotes leavening which increases loaf volume, and enhances bread texture.

α-Amylase: detergent applications

Carbohydrate-based food stains constitute one of the most common stain types on soiled clothing. Foodstuffs such as potatoes, oatmeal, gravy, spaghetti, chocolate, puddings and baby food contain a high proportion of starch or modified starch. Satisfactory removal of starch-based stains can be achieved at higher washing temperatures, but the addition of α-amylase to detergent preparations allows starch removal under less aggressive washing conditions.

Amylases were first introduced into detergents in the early 1970s, although this practice only gained widespread favour in the 1980s. By the mid-1990s, well over 90 per cent of laundry detergents and a growing percentage of dishwashing detergents contained α-amylase. Most traditional detergent amylases are produced by fermentation of *Bacillus subtilis*, *B. amyloliquefaciens* or *B. licheniformis*. The α-amylase from *B. licheniformis* has gained most widespread use, because of its thermal stability, resistance to proteolytic degradation and good wash performance. It is available commercially under the tradenames Maxamyl (Genencor Int.) and Thermamyl (Novo). These (and other) traditional detergent amylase preparations were, initially, developed for use in starch processing as opposed to being developed specifically for detergent use. These α-amylases display acceptable storage stability in detergents devoid of bleaching agents. However, most such amylase preparations lose significant activity in bleach-containing detergents (typically greater than a 50 per cent loss of activity is witnessed after storage in such detergents for 6 months).

The three-dimensional structure of several microbial α-amylases have been elucidated and several of these enzymes have been subject to protein engineering studies. As in the case of detergent proteases replacement of oxidation-sensitive amino acid residues in α-amylases should (and does) render them more bleach stable.

In the case of *B. licheniformis* α-amylase, site directed mutagenesis studies indicated that replacement of methionine 197 with a non-oxidizable amino acid results in the production of an oxidation resistant product (Figure 11.14). Two such engineered products – Purafect OxAm (Genencor Int.) and Duramyl (Novo) – were introduced to the market in the mid 1990s.

Additional protein engineering experiments sought to develop a more thermostable α-amylase with improved stability in the presence of low calcium ion concentrations. Ca^{2+} ions serve as cofactors for most bacterial α-amylases, enhancing their thermal stability and increasing resistance to proteolytic degradation. Site-directed mutagenesis was used to introduce

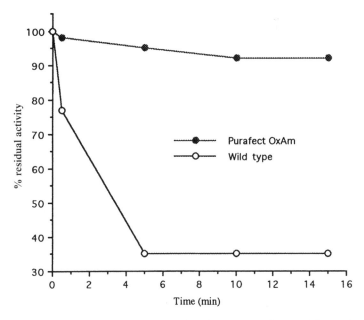

Figure 11.14 Stability of wild type *B. licheniformis* α-amylase, and its engineered form, 'Purafect Ox Am' in the presence of peracetic acid, the active component of a common detergent bleach

additional negatively charged amino acid residues adjacent to the enzyme's calcium binding site. The rationale was to generate an enzyme with increased affinity for Ca^{2+} ions. Several such modified enzymes did show increased thermostability, even at low calcium concentrations. These may prove useful in the starch processing industry and as detergent amylases. Detergent sequestering agents ensures that free Ca^{2+} concentrations in the washing liquor are very low and these agents could even potentially strip weakly bound Ca^{2+} ions from the enzyme surface, thereby destabilizing it.

α-Amylase: applications in textile desizing

Modern methods of textile weaving places considerable mechanical stress on the fabric threads. To prevent breakage, the fibre strands are nearly always coated with a substance known as a 'size'. The size serves to strengthen the fibre prior to weaving. Essential attributes of sizing materials include good adhesion to the textile threads, ease of removal after weaving (it is necessary to remove the size after weaving as it would subsequently prevent proper dyeing or bleaching of the finished product) and inexpensiveness. A range of natural and synthetic substances have been used in textile sizing (Table 11.13). Starch remains a popular sizing

LIVERPOOL JOHN MOORES UNIVERSITY
LEARNING SERVICES

Table 11.13 Various substances used in textile sizing

Natural	Synthetic
Starch	Methylcellulose
Gelatin	Carboxymethyl cellulose
Guar gum	Polyvinyl alcohol
Carbo bean meal	Methacrylate

agent due to its low cost and ready availability. In Europe potato starch is normally used, whereas cornstarch and rice starch find wider application in the USA and far East, respectively.

Subsequent removal of the starch (i.e. 'desizing') may be achieved by steam heating in the presence of NaOH, or by oxidants. However, such treatments can damage the textile and will generate a process effluent which must be treated before disposal. Desizing using α-amylase has thus become popular. Generally, thermostable bacterial α-amylases (e.g. *B. licheniformis* α-amylase) are mainly used. Depending upon the enzyme concentration and environmental parameters chosen, the desizing process may last from several minutes to several hours.

Like textiles, paper is also often sized using starch. Sizing of paper protects it from mechanical damage during manufacture, and also enhances the stiffness and feel of the finished product (in this case desizing is not subsequently carried out). Natural unprocessed starch slurries are too viscous to be used in paper sizing. α-Amylase is used to partially degrade the starch in order to yield a product of appropriate viscosity for the task at hand.

Lactase and sucrase

A number of other sugar degrading enzymes have also found important, if limited, industrial application. Lactase (β-galactosidase, Figure 11.15) for example, catalyses the hydrolytic cleavage of lactose, the major sugar of milk, yielding glucose and galactose. Bovine milk typically contains 4.5–5 per cent lactose. Hydrolysis of lactose is of interest for a number of reasons. Lactose present in ingested milk must be hydrolysed to its constituent monosaccharides prior to being absorbed across the small intestine. Although high levels of intestinal lactase activity is present in infants, some adult human populations are virtually devoid of this enzyme. Adults of northern European origin, however, do produce significant levels of intestinal lactase activity throughout their lives. Those adults exhibiting little or no lactase activity are lactose intolerant, and cannot digest dietary lactose. In such instances the ingested lactose causes

(a)

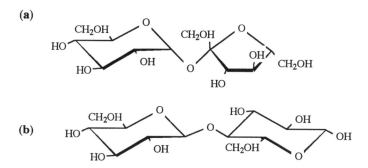

(b)

Figure 11.15 Structure of the disaccharides, sucrose (a) and lactose (b)

stomach upsets and promotes diarrhoea. This phenomenon severely restricts the use of milk as a nutrient source by many adult populations. Prior enzymatic hydrolysis of the milk lactose overcomes this difficulty, and immobilized preparations of lactase have been successfully used in this regard. Furthermore, lactase-containing tablets may be taken orally by lactose-intolerant individuals prior to ingestion of milk or milk products.

The cheesemaking industry produces large quantities of a by-product termed whey, which is difficult to dispose of. Whey, not surprisingly, contains a relatively high content of lactose. Hydrolysis of the lactose could potentially render this by-product a useful food or feed supplement. Hydrolysis of whey lactose renders the resultant syrup considerably sweeter (but it still displays only 70 per cent of the sweetness of sucrose). The monosaccharide mix is also considerably more soluble than intact lactose, and the increased solubility renders the product more suitable for use in food products. Lactose hydrosylate is sometimes used as an ingredient in the manufacture of ice cream, desserts, baked goods, and soft drinks. Lactase is produced by a wide range of plants (e.g. peach and apricot plants), by mammals and by microoganisms. Lactase preparations used commercially have been sourced from *Kluyveromyces lactis*, *K. fragilis* or *Aspergillus niger* (Table 11.14). *Aspergillus* lactase is used for the hydrolysis of acid whey, whereas *Kluyveromyces* lactases are used to hydrolyse whole milk (pH 6.6) or sweet whey (pH 6.1). Hydrolysis may be achieved in batch systems using free enzymes or by using immobilized lactase.

The lactases used to date display poor thermal stability at temperatures in excess of 60 °C, and hence enzymatic hydrolysis is normally undertaken at temperatures as low as 40 °C. At such temperatures, microbial contamination becomes a real problem, particularly when milk is used as a substrate. In fact, milk is normally sterilized prior to the hydrolysis step. Process hygiene and effective cleaning and sanitation of columns and process equipment between production runs is, therefore, particularly critical in the context of lactose hydrolysis. More recently, several thermostable lactases (e.g. *Thermus acquaticus* lactase, Table 11.14) have been identified and expressed in various recombinant expression systems. Such thermostable lactases may prove more attractive for industrial applications than those currently in use. In addition to technical

Table 11.14 Microbial sources and selected properties of some lactases

Source	pH optimum	Temp. Optimum (°C)	Molecular mass (kDa)
Kluyveromyces lactis	7.0	35	135
Kluyveromyces fragilis	6.6	37	201
Aspergillus niger	3–4	55–60	124
A. oryzae	5.0	50–55	90
E. coli	7.2	40	540
Bacillus stearothermophilus	6–6.4	65	215
Thermus aquaticus	4.5–5.5	80	570

considerations, process economics are also central to the large-scale industrial application of this enzyme.

Sucrase (invertase) is also used industrially to promote hydrolysis of sucrose (Figure 11.15), forming glucose and fructose. This enzyme is often utilized in the confectionery industry to form the semisolid filling present in the centre of some soft-centred sweets. It has also been used in the manufacture of ham, artificial honey and marzipan. Invertase used on an industrial scale is usually obtained from selected strains of *Saccharomyces cerevisiae* or *Aspergillius*.

LIGNOCELLULOSE DEGRADING ENZYMES

The most abundant carbohydrate reserves on the planet are present in plant biomass. Biomass has been defined as everything that grows. Plant biomass consists largely of three polymeric substances: cellulose (40 per cent), hemicellulose (33 per cent) and lignin (23 per cent). Perpetual renewal of plant biomass via the process of photosynthesis ensures an inexhaustible supply of such material. It has been estimated that approximately 4×10^{10} tonnes of cellulose is synthesized annually by higher plants. In practice, this means that in excess of 70 kg of cellulose is synthesized per person per day. Enzymes capable of degrading cellulose therefore attract obvious industrial interest.

The substrate cellulose

Cellulose is a linear, unbranched homopolysaccharide consisting of glucose subunits linked via $\beta1 \rightarrow 4$ glycosidic linkages. Individual cellulose molecules vary widely with regard to polymer length, with some molecules containing as few as 3000–4000 glucose residues, whereas others may contain as many as 20 000 units. The majority of cellulose molecules consist of between 8000–12 000 glucose molecules. Each glucose molecule present in cellulose is rotated at an angle of 180° with respect to its immediately adjacent residue. The actual repeating structural subunit, therefore, is cellobiose (Figure 11.16).

The underlying molecular structure confers upon cellulose properties which are very different to those of amylose which also consists exclusively of glucose residues linked by $\alpha 1 \rightarrow 4$ glycosidic linkages. Because of the β form of the linkage and the spatial arrangement of alternate glucose molecules, cellulose adopts an extended conformational structure. Furthermore, individual cellulose molecules are usually arranged in bundles or fibrils consisting of several parallel cellulose molecules which are held in place by an extensive network of intermolecular hydrogen bonds. Glucose residues within the cellulose molecule also seem to engage in intramolecular hydrogen bonding which contributes to the overall rigidity of the molecule. Within cellulose fibrils, there are extended areas exhibiting a completely ordered structure (crystalline areas) which are water insoluble and very resistant to chemical or enzymatic attack, and smaller areas which are less well ordered (amorphous regions).

Vertebrates are devoid of endogenous enzymatic activities capable of hydrolysing the cellulolytic $\beta 1 \rightarrow 4$ linkages, and therefore are incapable of digesting cellulose and utilizing it as a nutrient source. Some microorganisms, in particular various fungi, do synthesize cellulase enzymes. Ruminant animals, by virtue of the fact that their digestive system contains cellulolytic microorganisms, can indirectly gain nutritional benefit from ingested cellulose.

Hemicellulose and lignin

While cellulose is the principal constituent of the plant cell wall, it is rarely found in pure form as it is in intimate association with other polymeric substances termed hemicelluloses and lignin.

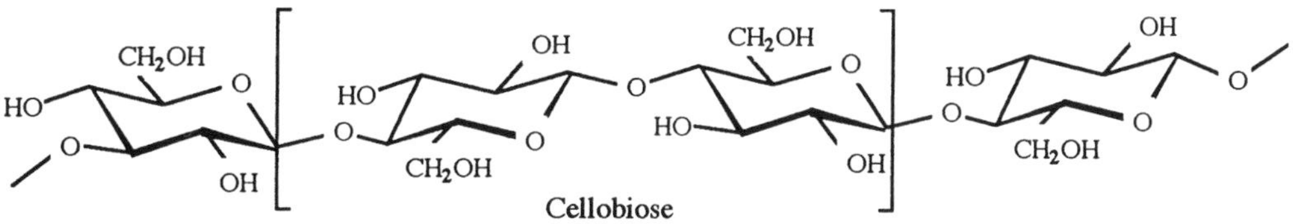

Figure 11.16 Structure of a portion of the cellulose backbone. Individual glucose molecules are linked via $\beta 1 \rightarrow 4$ glycosidic bonds. Successive glucose residues are rotated at an angle of 180 °C relative to the preceding glucose residue

Hemicelluloses are generally lower molecular mass polysaccharides. They consisting predominantly of D-xylose, D-mannose, D-glucose, D-galactose, L-arabinose and 4-*O*-methyl-D-glucuronic acid. The most abundant hemicellulose types present in the cell walls include glucans, mannans and xylans. Xylan, the single most abundant hemicellulose, is a homopolymer, consisting of β1→4 linked D-xylosyl residues.

Lignin is a complex aromatic polymer found in higher plants, predominantly located within the plant cell wall, interspersed with hemicellulose. This mixture forms a cement-like matrix in which the ordered cellulose fibrils are embedded. The presence of hemicellulose and lignin serves to increase the overall structural strength of the cell wall. The woody portion of tree trunks consists of over 20 per cent lignin. In such woody tissues, lignin is also present in the intercellular spaces, the middle lamella, where it serves as an adhesive, holding adjacent cells together. Unlike cellulose or hemicellulose, lignin is not a polysaccharide. It is a polymeric molecule composed mainly of three alcohol subunits: coumaryl alcohol, coniferyl alcohol and sinapyl alcohol (Figure 11.17).

Seemingly random cross-linking of such alcohols yields highly dispersed lignin molecules. Angiosperm lignins generally contain equal quantities of coniferyl alcohol and sinapyl alcohol monomers. Gymnosperm lignins, on the other hand, consist largely of coniferyl units alone whereas grass lignins contain all three alcohols.

Cellulases

Enzymes capable of hydrolysing cellulose are termed cellulases. Serious research on cellulases dates back to the Second World War, when the US

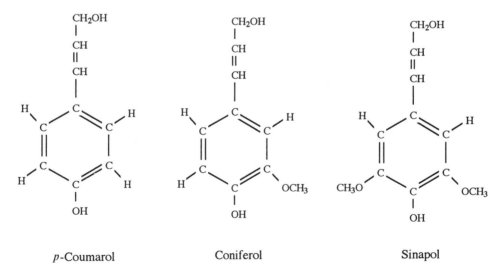

Figure 11.17 Structure of the alcohol molecules found in lignin

Army noted a rapid deterioration of cellulose-based materials (e.g. clothes, tents, etc.) while fighting in the South Pacific. Investigations revealed cellulase-producing fungi to be the major culprit.

Agricultural and household waste contain appreciable quantities of cellulose. Wood and wood pulp consists is over 40 per cent cellulose, straw and bagasse contains 30–50 per cent cellulose, while newspapers can contain up to 80 per cent cellulose. Cotton is almost pure cellulose. The complete hydrolysis of cellulose yields glucose. Any process which could efficiently and economically achieve conversion of cellulolytic material to glucose would be of immense industrial significance.

Cellulose is not degraded by a single enzyme but by a combination of enzymes which function in a concerted manner. Degradation of cellulolytic material occurs slowly in nature for a number of reasons:

- Very few microbial species actually produce complete cellulase systems capable of total and systematic degradation of cellulose to glucose molecules.

- The ordered crystalline structure of individual cellulose molecules present in cellulose fibres renders enzymatic attack very difficult. Amorphous, non-structured areas of cellulose, on the other hand, are degraded more rapidly.

- The close natural association of cellulose with hemicellulose, lignin and sometimes pectin, further reduces the accessibility of cellulases to their substrates.

Cellulolytic materials may be subject to various pretreatments in order to render the cellulose molecules more accessible to enzymatic attack. Adoption of such treatments on an industrial scale would render glucose production from cellulose uneconomical. Cellulolytic enzymes are synthesized by a number of microorganisms, most notably fungi (Table 11.15). Some bacterial species also exhibit cellulose degrading ability.

Fungal cellulases

Fungal cellulases have received most attention. Some fungal species, most notably *Trichoderma* species such as *T. viride*, *T. reesei* and *T. koningii*, as well as *Penicillium funiculosum*, produce cellulases capable of degrading at least in part, crystalline regions of native cellulose.

Most cellulolytic enzymes produced by fungi may be classified as one of three major types: (a) endocellulases (*endo*-β1 → 4-D-glucanases, often simply called cellulases); (b) cellobiohydrolases; and (c) β-glucosidases. Any one fungal species capable of degrading cellulose may produce multiple forms of each of these three enzymatic activities. Some such multiple forms are genetically distinct whereas others may result from partial proteolysis or differential glycosylation of a single protein. All of

Table 11.15 Major microbial sources of cellulolytic enzymes

Fungal sources	Bacterial sources
Trichoderma viride	*Bacillus* spp.
T. reesei	*Cellulomonas* spp.
T. koningii	*Celluibro* spp.
Penicillium pinophilum	*Thermomonospora* spp.
P. funiculosum	*Clostridium thermocellum*
Sporotrichum pulverulentum	*Acetivibrio cellulolyticus*
Talaromyces emersonii	*Bacteroides cellulosolvens*
Sporotrichum thermophile	*Bacteroides succinogenes*
Humicola insolens	*Ruminococcus albus*
Chaetomium thermophile	*R. flavefaciens*

these differing cellulolytic activities act synergistically to solubilize the cellulose substrate.

The endocellulases (endoglucanases) catalyse the random internal hydrolytic cleavage of the cellulose molecule. As many as six endocellulase activities may be associated with some fungi. Endocellulases appear to hydrolyse cellulose chains primarily within amorphous regions and display low hydrolytic activity towards crystalline cellulose.

Cellobiohydrolases generally catalyse the sequential removal of cellobiose units from the non-reducing end of the cellulose molecule. The β-glucosidases, on the other hand, hydrolyse both short-chain oligosaccharides derived from cellulose in addition to cellobiose, yielding glucose monomers. Many cellobiohydrolases exhibit product inhibition, as their activity is decreased in the presence of increasing concentrations of cellobiose. β-Glucosidase activity thus prevents such product inhibition.

Bacterial cellulases

A wide range of both aerobic and anaerobic bacteria produce cellulase enzymes. Cellulolytic bacteria mainly produce endocellulases, with only a few species being capable of producing cellobiohydrolase. β-Glucosidase is produced by some, but it is often cell bound. A subset of bacteria (most notably *Clostridium* species capable of degrading highly crystalline cellulose) produces a high molecular mass enzyme complex termed a cellulosome. The cellulosome normally consists of several endoglucanases as well as some exoglucanases, which are all bound to a large non-catalytic protein (the 'scaffolding subunit') via short sequences ('docking domains'). This highly ordered structure probably allows more efficient degradation of the cellulose substrate. Cellulosomes may be anchored on the surface of the producer bacteria and these bacteria may, therefore, be required to dock on the cellulose surface in order to commence substrate hydrolysis (Figure 11.18).

Cellulase structure

Limited proteolysis of many endocellulases and some exocellulases produce two protein fragments, one capable of binding to cellulose (the cellulose binding domain) and a catalytic domain (Figure 11.19). Not all cellulases contain cellulose binding domains, and the size and three-dimensional shape of those that do can vary significantly. By targeting the cellulase to the substrate, the cellulose binding domains help achieve more efficient cellulose hydrolysis. Cellulose binding domains are linked to the enzymes catalytic domain by a short linker sequence. Depending upon the cellulase studied, this linker sequence can vary in length from 6–59 amino acid residues. Linker sequences are frequently rich in proline and hydroxylated amino acids, and are often highly glycosylated. Catalytic domains, based on comparative amino acid sequences, display considerable variability from cellulase to cellulase. They are invariably much larger than the cellulose binding domains, typically accounting for 70 per cent or more of the cellulase mass.

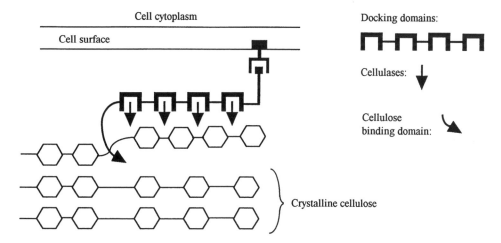

Figure 11.18 Generalized proposed structure of a bacterial cellulosome. A cellulose binding domain (see later) attached to the docking platform binds a cellulose strand in the cellulose microfibril, thereby disturbing the latters crystalline structure. This in turn makes this cellulose stretch accessible to the various cellulolytic activities of the cellulosome. Refer to the text for further details

Catalytic domain

Cellulose binding domain

Figure 11.19   Schematic representation of the structure of many cellulases. Refer to the text for further details

This summary outlining cellulose hydrolysis is somewhat simplistic. The enzymatic degradation of this polymer is complex and controversy still exists regarding the number of enzymes required in this process and the exact role that each enzyme plays in the overall degradative pathway. Cellulase systems produced by bacteria are less well understood than their fungal counterparts.

Industrial application of cellulose hydrolysis

The potential industrial applications of the cellulases and related enzymes are enormous. Glucose produced from the cellulose substrate could be used directly in animal/human food. Alternatively, the glucose product could be used as a substrate for subsequent fermentations or other processes which could yield valuable end products such as ethanol, methanol, butanol, methane, amino acids, organic acids, single cell protein and other bioreactors. Cellulolytic enzymes could also be used directly to increase the digestibility of food having a high-fibre content, and to enhance food flavour, texture or other qualities.

Despite the abundance of such potential applications, cellulases have as yet only been used in a few industrial processes:

- detergent additive;

- stonewashing of denim;

- maximizing extraction of juice from fruits (in conjunction with pectinases);

- clarification of fruit juices (in conjunction with pectinases);

- removal of external coat from soybean during production of fermented soybean foods;

- to improve rehydration of dried soups/vegetables;

- as an additive in animal feed (Chapter 12);

- as an additive in silage making;

- as an additive in the gelatinization of seaweed.

This is due to both technical difficulties and economic factors. As previously outlined, cellulose degrading organisms synthesize a very complex complement of cellulolytic activities. In most cases, the exact mechanism by which the molecule is sequentially degraded remains poorly understood. The crystalline arrangement of cellulose molecules into fibrils, and the association of other polymeric substances such as lignin, pectin and hemicellulose with such fibrils, greatly retards the process of its enzymatic degradation. Efficient hydrolysis necessitates expensive pretreatments which are economically unattractive and the current high costs associated with production of cellulolytic enzymes further decreases its economic feasibility. Cellulases are produced in relatively low quantities by most wild type producer strains and the enzymes generally exhibit disappointing specific activities. While mutational approaches have yielded hyperproducing strains, their specific activity still remains low. Many cellulases are strongly inhibited by the products they form. Such product inhibition can further retard the degradation rate.

Commercial cellulase preparations have been sourced from fungi such as *Trichoderma longibrachiatum* and *Humicola insolens*. Commercial bacterial preparations are generally sourced from bacilli. Specific recombinant endocellulases have also found application, chiefly in the detergent and textile industries, as described below.

Detergent applications

Cellulases were first added to commercial detergent preparations in Japan in the late 1980s. This practice was subsequently adopted in Europe and the USA in the 1990s. Not all cellulases display physicochemical properties conducive to detergent use, although several have been identified which are active at alkaline pH values and are stable in the presence of typical detergent components. Prominent cellulase preparations used for detergent application are presented in Table 11.16.

A screening programme undertaken by the KAO Company in Japan identified a detergent-compatible cellulase from an alkalophilic *Bacillus*

Table 11.16 Sources, characteristics and industrial producers of major detergent cellulases

Name	Company	Source	pH range	Temp (°C)
KAC	KAO Corp. Japan	Alkalophilic *Bacillus* sp.	7–10	40 (optimum)
Celluzyme	Novo Nordisk	*Humicola insolens*	7–9	50 (Max)
Carezyme	Novo Nordisk	Recombinant *H. insolens* endoglucanase	10	50 (Max)

(*Bacillus* KSM-635). Chromatographic purification using a combination of anion exchange and gel filtration steps yielded two peaks of cellulase (endoglucanase) activity. The endoglucanases displayed molecular masses of 130 kDa and 103 kDa, were maximally active at 40 °C and at pH 9.5 and were stable when incubated at pH values ranging from 6 to 11. These activities were very resistant to incubation with various surfactants and chelating agents and were found to be inhibited by Hg^{2+} and Cu^{2+}. They were activated by CO^{2+} and required either CO^{2+}, Ca^{2+}, Mg^{2+} or Mn^{2+} for thermal stability. Novo Nordisk's cellulase preparation 'Celluzyme' (Table 11.16) is sourced from the thermophilic fungal strain *Humicola insolens* DSM 1800. The preparation contains a mix of endoglucanases, cellobiohydrolase and β-glucosidase activities. Novo Nordisk's 'Carezyme' product (Table 11.16) is a recombinant endoglucnase isolated from *H. insolens*, but whose gene is expressed in an engineered strain of *Aspergillus oryzae*.

Cellulases added to detergents promote a number of effects, including stain removal, colour revival, depilling and fabric softening. They achieve these effects by promoting a very limited hydrolysis of cellulose fibres in cellulose-based fabrics (cotton/cotton blends, rayon and flax). It appears that a proportion of 'dirt' molecules in a stain on cotton fabrics may be trapped within the amorphous regions of cellulose microfibrils. This makes their effective removal difficult. Cellulase activity in detergent preparations is believed to partially cleave cellulose molecules in the amorphous regions of fibres, thus allowing the dirt to be removed more easily while not promoting extensive degradation of the cotton fibre.

The effects of colour revival, depilling and fabric softening are based on the same mechanism of action. Repeated wearing and washing of garments results in mechanical damage to the cotton fibres. As a result, microfibrils protrude from the surface of the originally smooth cotton fibre. This results in a reduction of colour brightness which is independent of bleaching of the colour. This appears to be due to the now uneven fibre surface dispersing incident light, thereby promoting a dulling effect. Treatment with cellulases enzymatically removes the damaged microfibrils, thus restoring a smooth surface to the cellulose fibre (it appears that mechanical damage sustained at the anchor point of the microfibril renders the area more susceptible to cellulase activity).

On-going mechanical damage can result in the microfibrils gathering into little balls called pills. Again cellulases can promote a depilling effect by enzymatic hydrolysis of the pills' anchor points on the fabric surface. Removal of microfibrils also appears to improve the textue and grip of the fabric, by preventing interlocking of individual fibres. This effect is perceived as fabric softening.

Stonewashing and biopolishing

Cellulases also find application in the textile industry, promoting stonewashing of denim and the biopolishing of cotton fabrics. Stonewashing is the process whereby denim is treated to give it a worn and rugged look. Traditionally, this was undertaken by washing the denim in the presence of pumice stones, which achieved their effect through mechanical abrasion. Cellulases can achieve a similar effect enzymatically and have largely replaced the use of pumice stones. This prevents mechanical damage to the washing machines and the absence of the stones allows more denim to be washed in each cycle. When denim is dyed (particularly with indigo), most of the dye molecules are retained on the surface of the denim yarn. Treatment with cellulases promotes a very partial hydrolysis of the cellulose fibres, thus releasing some of the dye, giving the fabric a worn or faded look.

The biopolishing process involves treating new cotton fabrics with cellulases. The aim is to remove any loose or protruding microfibrils which may be present on the cotton fibres' surface. By ensuring that the fibres are as smooth as possible, the process of pill formation and colour dulling is delayed. As a result, the clothing looks newer for longer.

Enzymatic deinking

A growing concern relating to the environmental impact of human activity has given impetus to recycle as much waste as possible. Wastepaper recovery and reuse has become a prominent example of such recycling. It is estimated that approximately 50 per cent of wastepaper is now recovered in the USA. Successful reuse of paper requires effective removal of inks from the original paper stock. Deinking can be achieved with relative ease in the case of some paper grades (e.g. newspapers printed using oil-based inks). However, in most cases deinking can be more problematic. Deinking involves dislodging the ink particles from the paper fibre surface, followed by separation of the dislodged ink from the fibre by washing. This can be achieved chemically but such processes generate waste effluents which must be disposed of carefully. Enzyme-mediated deinking is more environmentally friendly and appears to be economically viable. Lipases may be used to degrade and hence dislodge vegetable-oil-based inks. Other synthetic inks, however, are less susceptible to enzymatic degradation. In such cases cellulases (or sometimes a mixture of cellulases and hemicellulases) are used. These promote deinking by achieving a very limited hydrolysis of the cellulose and hemicellulose fibres, thereby releasing ink molecules trapped in the fibre surface.

PECTIN AND PECTIC ENZYMES

Pectin is yet another structural carbohydrate found in higher plants. It is located primarily in the cell wall and in the middle lamella, where it serves to bind adjacent cells. Various enzymatic activities capable of degrading pectin may be isolated from both plant and microbial sources. Degradation of pectin plays an important role in the growth of plant cells and the ripening of fruit. Microbial pectic enzymes, mainly produced by fungi, are utilized in many large-scale industrial processes. Advances in recombinant DNA technology has facilitated the detection, cloning and sequencing of genes coding for many such enzymes, and enabled their expression in heterologous systems.

The pectic substrate

Pectic substances are a relatively diverse group of polysaccharides, which vary in both their composition and molecular mass. Galacturonic acid is the major molecule present, constituting up to 60–80 per cent of some pectic preparations. Other sugars often present in pectic preparations include rhamnose, arabinose, galactose and xylose, (Figure 11.20).

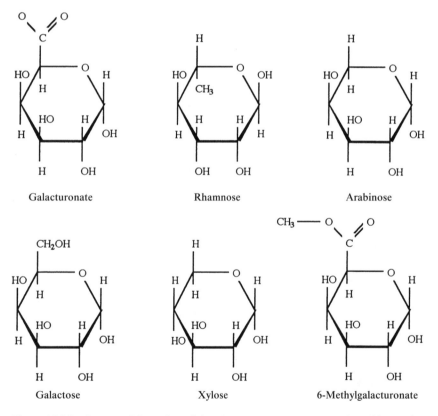

Figure 11.20   Structural formulae of the more common monomers found in pectin. In native pectin, 75 per cent or more of the galacturonic acid units are esterified with methanol, thus forming methyl galacturonides

The pectin molecule, present in intact immature plant tissue, is often referred to as protopectin. Protopectin is insoluble, which seems to be due to its polymer size and its association with calcium and other divalent cations. All other pectic substances which are soluble are derived from protopectin by hydrolysis. Pectins may thus be described as polysaccharides composed mainly of galacturonic acid, at least 75 per cent of which is esterified with methanol. Enzymatic de-esterification of such galacturonide yields a polymeric substance often termed pectic acid. Pectic substances are often classified as galacturonans, rhamno-galacturonans, arabinans, galactans and arabinogalactans.

Rhamnogalacturonans represent the major constituent of pectic substances. The polysaccharide backbone of rhamnogalacturonans consists mainly of α-D-galacturonate units linked via α1 → 4 bonds. Molecules of L-rhamnose are interspersed in the backbone, occuring on average every 25–30 galacturonate units. The rhamnose units are linked via β1 → 2 and β1 → 4 bonds to the D-galacturonate residues. Side chains of variable length, often consisting of galacturonans, galactans, arabinans or arabinogalactans, branch off from the main chain (Figure 11.21).

Pectic enzymes

There are two broad groups of pectic enzymes: the first group are termed pectin esterases, also known as pectin methyl esterases; the second group are depolymerases. Pectin methyl esterases remove methoxy groups from methylated galacturonides. Pectin esterase activity is present in all higher plants and is particularly abundant in citrus fruits and vegetables. Esterase activities are also found with a variety of microorganisms, most notably fungi.

The depolymerases catalyse the cleavage of glycosidic bonds via hydrolysis (hydrolases), or via β-elimination (lyases). In many instances, pectin esterase must firstly remove methoxyl groups from the galacturonide before depolymerase activity can commence.

Polygalacturonases catalyse the hydrolysis of α1 → 4 linkages between β-galacturonic acid residues. A number of distinct polygalacturonase activities have been recognized, both in higher plants and in microbes. Endopolygalacturonases catalyse the hydrolysis of internal α1 → 4 glycosidic linkages in stretches of polygalacturonic acid. As cleavage sites are chosen more or less at random, a number of lower molecular mass oligosaccharides are produced. Cleavage by endopolygalacturonases requires the prior removal of methoxy groups by pectin methylesterase. Exopolygalacturonases catalyse the sequential removal of galacturonic acid residues from the non-reducing end of polygalacturonate.

Lyases are a group of pectin-degrading depolymerases produced almost exclusively by microorganisms. These enzymes catalyse the cleavage of α1 → 4 glycosidic bonds which link galacturonic acid residues by

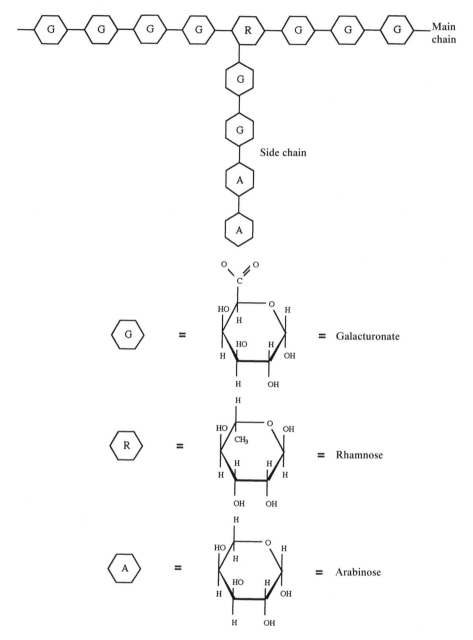

Figure 11.21   Structure of a segment of rhamnogalacturonans. Regions containing a high density of side chain are termed 'hairy regions'. Such hairy regions are normally separated by extended sequences devoid of side chains, the smooth regions

β-elimination. Endopectate lyases catalyse the cleavage of internal glycosidic bonds in regions of polygalacturonic acids devoid of methoxy groups. Such enzymes have been isolated from a number of microbial plant pathogens. They have a high pH optimum and require the presence of calcium ions to maintain activity.

Exopectate lyases are found mainly in bacteria. Most such enzymes cleave the penultimate glycosidic bond of galacturonans, thus releasing dimeric molecules composed of galacturonic acid. Some such lyases, however, split the terminal galacturonin glycosidic bond, yielding single galacturonic acid moieties. As in the case of endopectate lyases, most exopectate lyases preferentially attack polygalacturonic acid which is essentially free of methyl ester groups. Pectin lyases display a preference for highly esterified polygalacturonic acid sequences, (galacturonins) as substrates. These enzymes are produced mostly by fungi and usually catalyse the random internal cleavage of glycosidic bonds thus producing esterified oligogalacturonates.

Industrial significance of pectin and pectin-degrading enzymes

Pectin enjoys widespread industrial application, as solutions of pectin are viscous, and readily gel when heated in the presence of sugar under acidic conditions. This renders pectin useful as a gelling agent, emulsifier and thickener in the production of a number of foods. Pectin used for such purposes has been classified as high-methoxy or low-methoxy pectin. As the name suggests, the majority of galacturonic acid residues present are methoxylated in high-methoxy pectin. In low-methoxy pectin, many such methoxy groups have been removed. While gelation of high-methoxy pectin is dependent upon the presence of significant quantities of sugar, low-methoxy pectin may be induced to gel by the addition of certain metal ions, even in the absence of sugar. Use of the latter pectin substrate thus makes possible the manufacture of jams and jellies of low sugar content. Commercial pectin is normally produced by extraction from the rind of citrus fruits or from sugar beet pulp, both of which serve as particularly rich sources of this substance.

Pectic enzymes are also used in a number of industrial processes. Such enzymes have found particular favour within the fruit juice extraction and clarification industries. Fruit juices are normally manufactured by the mechanical pressing of the relevant fruits. In many instances, the physical characteristics of the fruit hinder maximal juice extraction. The addition of commercial preparations of pectic enzymes generally facilitates a greatly enhanced juice yield.

Enzymatic degradation of pectin by fungal pectinase preparations is routinely used to maximize juice yields from grapes and apples. Juice extraction from most soft fruits, such as raspberries, strawberries and blackberries, is also increased by supplementation with exogenous pecti-nolytic enzymes.

Commecial pectinase preparations are routinely employed to clarify 'sparkling' fruit juice preparations, such as apple juice and pear juice. When freshly pressed, most such juices contain relatively high levels of soluble pectins which contribute greatly to the characteristic viscosity and haziness associated with such products. Haze formation often reflects a decreased solubility of one or more of the components present in the juice – with which pectins associate. Partial degradation of the pectins is achieved by addition of commercial pectinase preparations, which results in a significant drop in product viscosity. Furthermore, partial degradation of the pectin destabilizes the haze particles, resulting in their coagulation and precipitation from solution. Subsequent removal of the precipitate is easily achieved by centrifugation or filtration, yielding the sparkling, clear juice. The decrease in solution viscosity observed upon treatment with pectic enzymes permits the production of concentrated juice extracts.

Pectic enzyme preparations are also used in the maceration of fruits and vegetables. Maceration usually entails the conversion of fruit or vegetable tissue into a suspension of individual intact cells. This may be achieved by selectively hydrolysing the pectin present in the middle lamella, which binds plant cells. Maceration is normally achieved by treatment with pectic enzyme preparations exhibiting high levels of polygalacturonase activity. This enzyme has also been directly associated with the process of fruit softening during ripening. Enzymatic maceration may be used in the production of fruit nectars, 'pulpy' drinks, and in the preparation of some baby foods. Pectic enzymes are also used to aid extraction of various citrus oils and pigments from orange and lemon peel.

Most commonly available pectinase preparations are obtained from fungal sources such as various species of *Aspergillus* or *Penicillium*. A source of pectin (such as apple pomace, citrus peel or dried sugar beet pulp), is normally included in the fermentation medium used to culture pectinase-producing fungi. This enhances not only pectinase production, but also promotes increased secretion of these enzymes from the mycelium. The enzymes produced are usually concentrated and partially purified by techniques such as precipitation. Stabilizers, preservatives and other additives are then incorporated into the final enzymatic preparation, which may be marketed in liquid or powdered form. Commercial pectinase preparations usually contain a variety of pectinolytic activities. Most also contain appreciable quantities of additional enzymes such as cellulases and hemicellulases.

FURTHER READING

Books

Amylase Research Society of Japan (1988) *Handbook of Amylases and Related Enzymes*. Pergamon Press, Oxford.

Barrett, A. *et al.* (1998) *Handbook of Proteolytic Enzymes*. Academic Press, London.

Beynon, R. (2000) *Proteolytic Enzymes*. Oxford University Press, Oxford.

Bott, R. (1995) *Subtilisin Enzymes*. Plenum Publishing Co., New York.

Claeyssens, M. (1998) *Carbohydrases from* Trichoderma reesei *and other Micro-organisms*. Royal Society of Chemistry, London.

Fogarty, W. (1990) *Microbial Enzymes and Biotechnology*. Elsevier, Amsterdam.

Gerhartz, W. (1990) *Enzymes in industry: production and applications*. VCH, Weinheim.

Godfrey, T. & West, S. (1996) *Industrial Enzymology*, 2nd edn. MacMillan, Basingstoke.

Himmel, M. (2001) *Glycosyl Hydrolysases in Biomass Conversion*. American Chemical Society.

Tsao, G. (1999) *Recent Progress in Bioconversion of Lignocellulosics*. Springer-Verlag, Godalming.

Uhlig, H. (1998) *Industrial Enzymes and their Applications*. Wiley Interscience, New York.

Wiseman, A. (1995) *Handbook of Enzyme Biotechnology*. Ellis Horwood, Chichester.

Yamamoto, T. (1995) *Enzyme Chemistry and Molecular Biology of Amylases and Related Enzymes*. CRC Press, Boca Raton.

Articles

Proteases

Bryan, P. (2000) Protein engineering of subtilisin. *Biochem. Biophys. Acta – Prot. Struct. and Mol. Enzymol.* **1543** (2), 203–222.

Daneil, R. & Toogwood, H. (1996) Thermostable proteases. *Biotechnol. Genet. Eng. Rev.* **13**, 51–100.

Rao, M. *et al.* (1998) Molecular and biotechnological aspects of microbial proteases. *Microbiol. Mol. Biol. Rev.* **62** (3), 597–635.

Siezen, R. & Leunissen, J. (1997) Subtilases: The superfamily of subtilisin-like proteases. *Protein Sci.* **6**, 501–523.

Siezen, R. *et al.* (1991) Homology modelling and protein engineering strategy of subtilases, the family of subtilisin like proteinases. *Protein Eng.* **4** (7), 719–737.

Carbohydrases

Bertoldo, C. & Antranikian, G. (2001) Amylolytic enzymes from hyperthermophiles. *Hyperthermoph. Enzymes, A* **330**, 269–289.

Bhosale, S. *et al.* (1996) Molecular and industrial aspects of glucose isomerase. *Microbiol. Rev.* **60** (2), 280–300.

Gekas, V. & Lopez-Leiva, M. (1985) Hydrolysis of lactose: a literature review. *Proc. Biochem.* **Feburary**, 2–12.

Guzman-Maldonado, H. & Paredes-Lopez, O. (1995) Amylolytic enzymes and products derived from starch: a review. *Crit. Rev. Food Sci. Nutr.* **35** (5), 373–403.

Hashida, M. & Bisgaard-Frantzen, H. (2000) Protein engineering of new industrial amylases. *Trends Glycosci. Glycotechnol.* **12** (68), 389–401.

Himmel, M. *et al.* (1999) Cellulase for commodity products from cellulosic biomass. *Curr. Opin. Biotechnol.* **10** (4), 358–364.

James, J. & Lee, B. (1997) Glucoamylases: microbial sources, industrial applications and molecular biology – a review. *J. Food Biochem.* **21**, 1–52.

Kashyap, D. *et al.* (2001) Applications of pectinases in the commercial sector; a review. *Bioresource Technol.* **77** (3), 215–227.

Lang, C. & Dornenburg, H. (2000) Perspectives in the biological function and the technological applications of polygalacturonases. *Appl. Microbiol. Biotechnol.* **53** (4), 366–375.

Nielsen, J. & Borchert, T. (2000) *Biochemi. biophysi. Acta – Protein Struct. Mol. Enzymol.* **1543** (2), 235–274.

Ohmiya, K. *et al.* (1997) Structure of cellulases and their applications. *Biotechnol. Genet. Eng. Rev.* **14**, 365–407.

Prade, R. (1996) Xylanases: from biology to biotechnology. *Biotechnol. Genet. Eng. Rev.* **13**, 101–132.

Prade, R. *et al.* (1999) Pectins, pectinases and plant-microbe interactions. *Biotechnol. Genet. Eng. Rev.* **16**, 361–391.

Savchenko, A. *et al.* (2001) Alpha-amylases and amylopullanase from pyrococcus furiosus. *Hyperthermoph. Enzymes A* **330**, 354–363.

Schulein, M. (2000) Protein engineering of cellulases. *Biochem. Biophys. Acta – Protein Struct. Mol. Enzymol.* **1543** (2), 239–252.

Singh, A. & Hayashi, K. (1995). Microbial cellulases, structure, molecular properties and biosynthesis. *Adv. Appl. Microbiol.* **40**, 1–39.

12

Additional industrial enzymes

LIPASES

Lipases are enzymes that catalyse the hydrolysis of lipids, yielding glycerol and fatty acids. The reaction is generally reversible (Figure 12.1). They exhibit little or no activity against soluble substrates in aqueous solutions. Instead they become activated only at the water–substrate (i.e. lipid) interface, a process termed interfacial activation. Lipases have been purified from a wide variety of mammalian, plant, fungal, yeast and bacterial sources. The three-dimensional structure of several has been elucidated. These enzymes generally exhibit molecular masses ranging

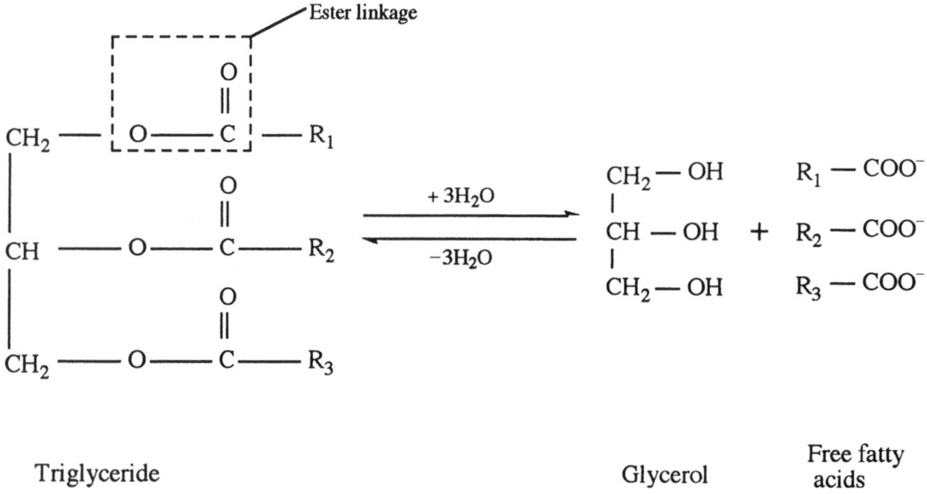

Triglyceride   Glycerol   Free fatty acids

Figure 12.1 Reaction catalysed by lipases

from 20 to 60 kDa, and despite relatively low sequence homology, they all share similar aspects of three-dimensional architecture. For example they display a characteristic α/β hydrolase fold, with the enzyme's core composed of a β-sheet consisting of up to eight stretches of β-conformation, connected by α-helical segments.

Lipases are often termed serine hydrolases as they possess a catalytic triad somewhat similar to serine proteases (Chapter 11). The lipase triad is generally composed of a serine and histidine residue along with an aspartic or glutamic acid residue. Active site topography is somewhat unusual in that it is surrounded by a large hydrophobic region. This entire area is in turn shielded or covered by a lid-like polypeptide loop composed of polar (charged, i.e. hydrophilic) amino acid residues. The presence of such a 'lid' likely explains why the enzyme displays no activity against soluble substrates in water. When the enzyme makes contact with the lipid interface a conformational change is induced, making the active site freely accessible to substrate molecules (Figure 12.2), hence explaining the aforementioned phenomenon of interfacial activation.

Industrial demand for lipases has grown steadily over the last two decades, and from a modest base lipases now account for some 7 per cent of industrial enzyme sales. The single most extensive application of lipases is their inclusion in detergent preparations. Lipases also find application in the food industry, in organic synthesis and in the paper and pulp processing industry. The most common sources of commercial lipase preparations are various species of *Candida*, *Pseudomonas* and *Rhizopus*, although some commercial preparations are sourced from *Bacillus subtilis*, *Aspergillus niger*, *Mucor*, *Humicola* and *Penicillium* species, as well as porcine pancreas. Some commercial lipases are produced by direct extraction of the enzyme from the native producer. Greater emphasis however is now placed upon recombinant production, in order to overcome barriers such as low natural enzyme expression levels.

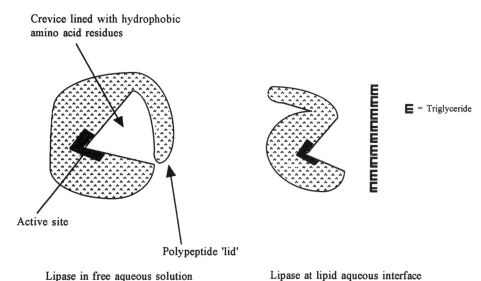

Crevice lined with hydrophobic
amino acid residues

Active site

Polypeptide 'lid'

E = Triglyceride

Lipase in free aqueous solution

Lipase at lipid aqueous interface

Figure 12.2 Binding of lipase to the lipid–aqueous interface promotes a conformational change in the enzyme which makes the active site available to the substrate. Refer to text for details

Detergent applications

The first recorded application of a lipase in the laundry industry dates back to 1913, when Rhom added pancreatic extracts to detergent preparations. Surfactants and additional detergent components however inactivated the pancreatic enzymes and it was not until the 1970s that researchers began to identify lipases suited to detergent application. The main detergent lipases are now produced by recombinant means (Table 12.1).

Lipid-based stains (Table 12.2) have long been recognized as the most difficult stain type to remove from soiled clothing. In the past these stains

Table 12.1 Major detergent preparations now commercially available, their sources and manufacturers

Enzyme brand name	Manufacturer	Lipase gene sourced from	Lipase gene expressed in
Lipolase	Novo Nordisk	*Humicola lanuginosa*	*Aspergillus oryzae*
Lipomax	Gist Brocades	*Pseudomonas alcaligenes*	*Pseudomonas alcaligenes*
Lumafast	Genencor	*Pseudomonas mendocina*	*Bacillus* sp.
Lipolase Ultra	Novo Nordisk	Protein engineered variant of lipolase	
Lipo Prime	Novo Nordisk	Protein engineered variant of lipolase	

Table 12.2 Lipid-based stains commonly found on soiled clothing/laundry

Food derived	Body derived
Butter and fat	Sebum
Edible oils	Sweat
Chocolate	**Cosmetics derived**
Salad dressing	Lipstick
Mayonnaise	Mascara
Spaghetti sauce	

were best removed by washing at high temperatures. The trend towards milder washing conditions (20–40 °C) thus renders effective lipid removal even more difficult. Detergent lipases catalytically degrade water-insoluble lipid-based stains, yielding more water-soluble products such as mono- and diglycerides, fatty acids and glycerol. These breakdown products are then more easily removed by detergent action into the washing liquor.

The inclusion of a low amount of a detergent-compatible lipase (usually added up to levels of 0.1 per cent of total detergent) has proven effective in removing such lipid-based stains. The cleaning action however does not become evident at the end of the first wash, with the lipid-based stain only being completely removed after one (or sometimes more) subsequent wash cycles. This phenomenon is known as the multi-cycle effect. It occurs because lipases are generally only poorly active on fully wetted textile. Maximum lipolytic activity (i.e. stain degradation) appears to occur during the spinning and drying of the fabric, when the textile water content has decreased to between 10 and 40 per cent (w/w) (Figure 12.3). As a result the bulk of lipid breakdown products are still present on the fabric after the first wash, and will only be physically removed during a subsequent washing cycle.

Like other detergent enzymes various lipases have been subject to protein engineering in an attempt to develop more effective products. Targeted end points of such experiments include increasing the enzyme's stability in the presence of detergent constituents, increasing its specific activity or relaxing its substrate specificity (rendering it capable of degrading a greater variety of lipid based stains). In some instances a non-specific approach was undertaken, for example the substitution of various amino acids at or close to the enzyme's active site, with screening of the variants produced to determine the effect on lypolytic activity. One commercial product (Lipomax, Table 12.1) was identified in this way. In this instance it was discovered that replacement of the methionine residue

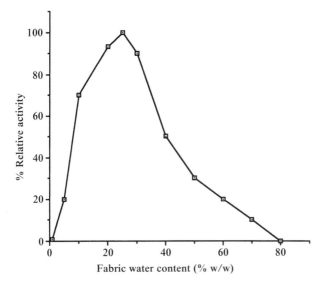

Figure 12.3 The effect of water content of fabric on the activity of the detergent lipase 'Lipolase'

at position 21 of the lipase with leucine significantly enhanced wash performance. *In vitro* studies revealed that this substitution increased the enzyme's specific activity by upto 50 per cent.

More specific approaches to protein engineering of lipases have also yielded commercial products. Lipid-based stains are presumed to have a nett negative charge (contributed by fatty acids and by anionic surfactants bound to the stain surface). It was thus hypothesized that the replacement of negatively charged amino acid residues on the lipase surface with neutral (or positively charged) amino acids could enhance binding of the enzyme to the stain by minimizing charge repulsion. Novo researchers found that replacement of Asp 96 of *H. lanuginosa* lipase (i.e. lipolase) with Leu resulted in an engineered lipase with enhanced cleaning ability. This product was subsequently marketed under the tradename Lipolase Ultra (Table 12.1).

Additional applications

The content and structure of constituent triglycerides has a major impact on the nutrition, flavour and sensory value of foodstuffs. Lipases have found application in the alteration/generation of flavour in certain foods. Enzyme modified cheeses, for example, are produced by controlled proteolysis or lipolysis of traditional cheeses. The resultant production of flavour-rich peptides and free fatty acids generates a more intensely flavoured product. Milk-derived cream can be converted into a product known as lipolysed milk fat by treatment with lipases. The free fatty acid mix generated yields a more intense cream–butter-like flavoured product, which in turn can be added to various products to give them such flavours.

Lipases also find increasing use in the pulp and paper industries, to facilitate pitch removal. 'Pitch' is a term used to describe hydrophobic constituents of wood. It is most prevalent in pine and other softwoods, contributing 0.5–3.0 per cent of the wood by weight. Pitch is largely composed of triglycerides and waxes. These substances lend a sticky property to wood which can cause serious processing problems during paper manufacture. Some of the pitch is removed by pulping and chemical bleaching. An alternative approach entails its enzymatic degradation using lipases. Lipase sourced from *Candida cylindracea* has found most application in this regard, and the enzyme is now employed commercially by some paper manufacturers.

Lipases also show potential in organic synthesis. The stereospecificity of the enzyme is a major advantage in this regard, particularly in the production of some pharmaceutical products and intermediates. Many drugs used in medicine are produced by direct chemical synthesis. If the component synthesized contains a chiral center, a racemic mixture (equal quantities of D and L isomers) is generally produced. Usually only one isomer will promote the desired biological effect, with the other isomer being biologically inactive. In some cases, however, this 'other' isomer can exert a negative biological activity and must be removed from the final product. Enzymes can potentially be used to separate isomers or to selectively synthesize the required isomer (see also the section on aminoacylase). Lipases, for example, show potential in this context for the production of captopril, an angiotensin-converting enzyme inhibitor. While lipases exhibit potential in the production or resolution of chiral compounds, few such processes have thus far been commercialized.

PENICILLIN ACYLASE

The discovery of penicillin heralded a revolutionary advance in the medical control of bacterial disease. Subsequently, a variety of other compounds exhibiting antimicrobial activity have been introduced into the clinical arena. The penicillins and the structurally related cephalosporins remain the most popular antibiotic prescribed by the medical community. The passage of time has witnessed an increasing number of bacterial populations which have become resistant to the antimicrobial action of penicillin. Resistance has been counteracted in part by the development of semisynthetic penicillins, to which many such resistant strains remain sensitive. The enzyme penicillin acylase plays an essential role in the production of such semisynthetic penicillins, and hence is the subject of significant industrial demand.

The chemical structure of some naturally occurring and semisynthetic penicillins are illustrated in Figure 12.4. All contain an identical core ring structure termed 6-aminopenicillanic acid. Different penicillin types differ in their attached side chains.

(a)

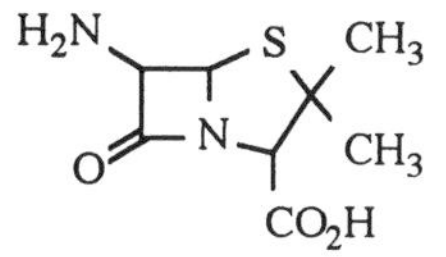

(b)

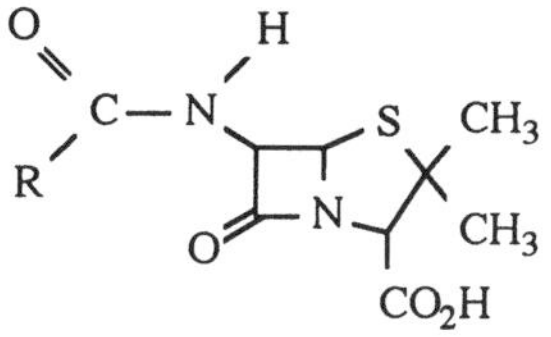

(c)

Name	Substituent (R)
Penicillin G (NATURAL)	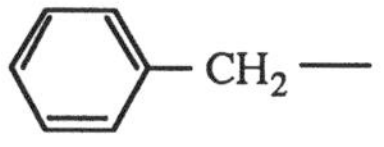
Penicillin V (NATURAL)	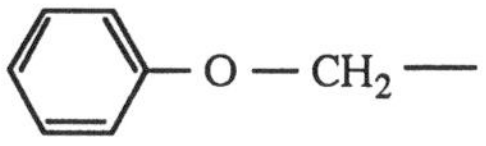
Methicillin (Semisynthetic)	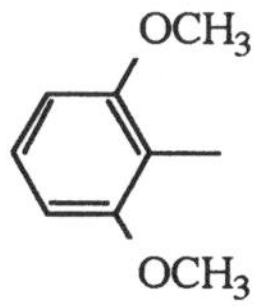
Ampicillin (Semisynthetic)	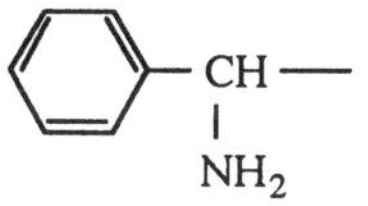

Figure 12.4 Structure of 6–aminopenicillanic acid (a); generalized penicillin structure (b); and the side groups present in two natural penicillins and two semisynthetic penicillins

Semisynthetic penicillins may be produced by the enzymatic removal of the side chain of native penicillins, with subsequent attachment of a novel side chain to the resultant 6-aminopenicillanic acid core. The removal of the side chain from penicillin G is illustrated diagrammatically in Figure 12.5. The reaction catalysed by penicillin acylase (penicillin amidase or penicillin amidohydrolase) is generally reversible. The hydrolytic reaction is catalysed under alkaline conditions, while, the acylation (biosynthetic) reaction is favoured at neutral–acidic (pH 4–7) values. In this way penicillin acylase may be used convert native penicillins to semisynthetic products. Annual global production levels of the 6-aminopenicillanic acid intermediate is estimated to be in the region of 7000 tonnes.

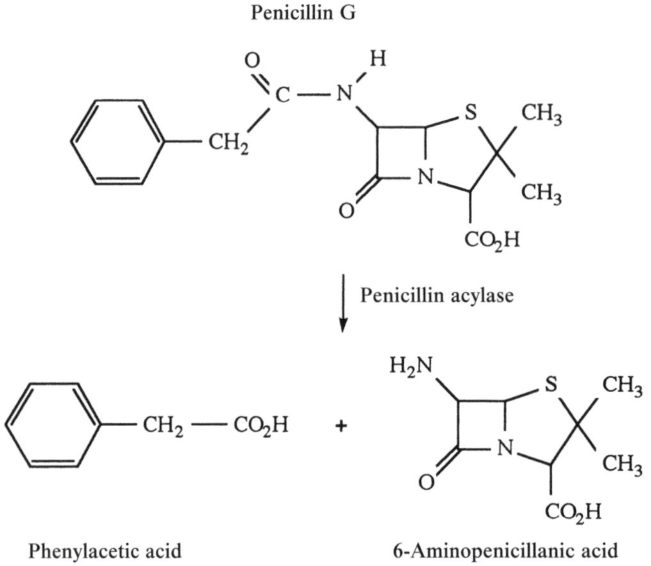

Penicillin G

Penicillin acylase

Phenylacetic acid   6-Aminopenicillanic acid

Figure 12.5 Action of penicillin acylase on penicillin G, a natural penicillin produced in large quantities by fermentation. The reaction proceeds in the direction indicated under alkaline conditions. As one of the reaction products is an acid, the pH must be continually adjusted to maintain alkaline values. Upon completion of the conversion, the pH value may be adjusted downwards to a value of 4.2–4.3. At this pH, 6-aminopenicillanic acid precipitates from solution and thus may be harvested. Some penicillin acylase preparations are also capable of catalysing the reverse reaction under mildly acidic conditions. Such enzyme preparations may thus be employed in the synthesis of semisynthetic penicillin from 6-aminopenicillanic acid and the relevant side chain group

Quite a number of microorganisms produce penicillin acylase (Table 12.3). Among these, the enzyme produced by *E. coli* has received most attention. The enzyme is a heterodimer, composed of a 20–23 kDa subunit which houses the (penicillin) side chain binding site, and a 60–69 kDa subunit housing the catalytic site, which contains a serine reside essential to activity. Although it is possible to utilize the free enzyme, an immobilized form is generally used in the production of semisynthetic penicillins as this is more attractive economically, facilitating the reuse of the enzyme over many production runs.

The increasing incidence of bacterial resistance to natural penicillins renders crucial the continued production of semisynthetic penicillins. In addition to overcoming the problem of resistance, several semisynthetic penicillins exhibit improved clinical properties. Many inhibit a greater variety of bacterial pathogens than, for example, penicillin G. Others are more acid stable and hence are particularly suited for oral administration.

The advent of recombinant DNA technology has not left the discipline of antibiotic production untouched. This technology has facilitated the

Table 12.3 Various microorganisms that produce penicillin acylase

---

Escherichia coli

Pseudomonas spp.

Proteus rettgeri, P. morganii

Brevibacterium spp.

Bacillus spp.

Flavobacterium spp.

Streptomyces spp.

Aerobacter spp.

Kluyvera spp.

---

detailed molecular elucidation of biosynthetic pathways for β-lactam antibiotics, such as penicillin and cephalosporins. A number of genes coding for specific enzymes involved in β-lactam biosynthesis have now been cloned and characterized. Introduction of additional copies of such genes into producing microorganisms facilitates logical strain improvement. Such genetic manipulations have already lead to enhanced production of certain industrial strains of antibiotic producing fungi. An increased understanding of β-lactam biosynthetic pathways could also make possible alteration of such pathways at the molecular level in order to produce novel antibiotic molecules.

AMINO ACYLASE AND AMINO ACID PRODUCTION

Unlike many microorganisms most animals cannot synthesize all the naturally occurring amino acids required for protein synthesis. Those essential amino acids which animals are incapable of producing must be procured from dietary sources. Nutritionists have long recognized that protein dietary constituents must contain a well-balanced amino acid composition, in terms of essential amino acids, in order to ensure optimal animal growth.

Both lysine and threonine constitute essential amino acids in the mammalian diet. They also generally represent the first and second limiting amino acid for pigs and poultry fed a cereal-based diet. Maximal utilization of cereal-derived protein requires addition of almost 4 kg of crystalline lysine and almost 2 kg of threonine per tonne of feed. Other essential amino acids which may become limiting, in particular if animals are fed diets high in soybean, include the sulfur-containing amino acids cysteine and methionine, and also tryptophan (Figure 12.6).

The problem of attaining a well-balanced dietary amino acid complement could be overcome by generating transgenic animals capable of synthesizing their own essential amino acids. An alternative strategy entails the generation of transgenic plants capable of producing proteins of altered amino acid composition. The major proteins present in soybean are deficient in methionine and cysteine. A number of alternative seed proteins rich in methionine have been identified. Production of these proteins in transgenic soybeans could negate the necessity to supplement soybean-based rations with methionine, and as cysteine can be synthesized from methionine it also would not be required. Such projects however are long-term in nature. In the meantime nutritionists must ensure that animals are fed rations containing an optimal amino acid balance. This often necessitates supplementation of the diet with specific amino acids.

In addition to the amino acids required in bulk quantities by the animal feed industry a number of other amino acids are required in bulk for a variety of additional industrial processes. Glutamic acid, for example, is widely used in the food industry. Inclusion of its sodium salt form (monosodium glutamate), enhances the natural flavour associated with many food types. Hundreds of thousands of tonnes of this amino acid are now produced annually. The development of the sweetener aspartame (Chapter 11) has substantially increased the industrial demand for its two amino acid constituents, phenylalanine and aspartic acid. Other amino acids are also required in moderate quantities for medical, research and other specialist purposes.

Amino acids may be produced either chemically or enzymatically. Chemical production has traditionally been the method of choice, mainly on economic grounds. Chemical production, however, yields a racemic

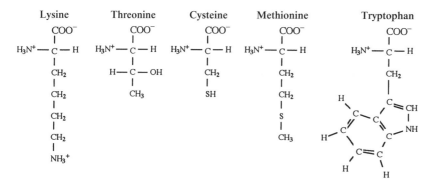

Figure 12.6 Structure of amino acids most likely to become limiting when monogastric animals are fed cereal-based diets. In contrast to such animals, ruminants procure a significant level of all naturally occurring amino acids from ruminal microorganisms as these organisms can synthesize such essential amino acids *de novo*

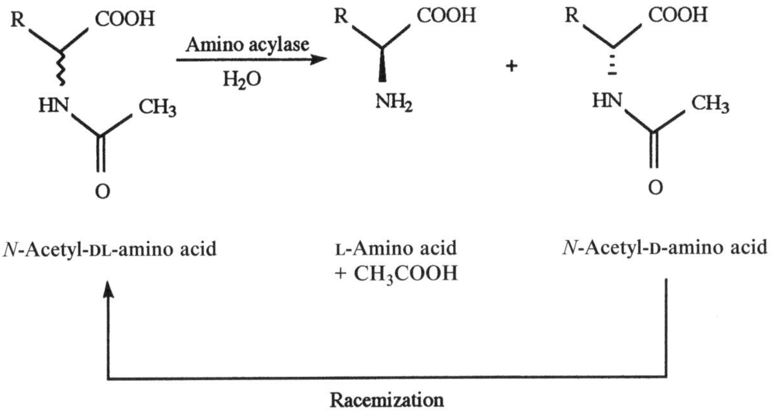

Figure 12.7 Reaction catalysed by aminoacylase

mixture consisting of the D and L isoforms of the amino acids. As only the L form is biologically utilizable by most higher species, separation of L from D is pursued subsequent to chemical synthesis. This is most readily achievable by utilizing the enzyme L-amino acylase.

The racemic amino acid mixture is firstly chemically acetylated, usually by reaction with acetyl chloride or acetic anhydride. The mixture is then passed over a bed of immobilized L-amino acylase. This enzyme will deacylate only the L-form of the acylated amino acids, yielding a free L-amino acid and the intact *N*-acetyl D-amino acid (Figure 12.7). These can be conveniently separated via ion exchange chromatography or crystallization. The *N*-acetyl D-amino acid can then be re-racemized either chemically (by addition of acetic anhydride under alkaline conditions) or enzymatically (using a racemase). The racemate can then again be passed over the immobilized amino acylase. This process, now used to produce several 100 tonnes of enantiomerically pure L-amino acids each year, was first commercialized in 1969. The amino acylase used industrially is sourced from *A. oryzae*. It is a 73 kDa, dimeric zinc-containing metalloprotein, which is activated by cobalt ions. It displays a pH optimum of 8.5 and is stable up to temperatures approaching 60 °C. A more thermostable amino acylase has been isolated from *Bacillus stearothermophilus*. It exhibits an activity optimum at 70 °C and has been expressed in various recombinant systems. D-Amino acylases may be used to produce enantiomerically pure amino acids, although the demand for the D isomer is very significantly lower that for L-amino acids. D-amino acylases have been purified from various species of *Pseudomonas*, *Streptomonas*, and *Streptomyces*.

CYCLODEXTRINS AND CYCLODEXTRIN GLYCOSYLTRANSFERASE

Cyclodextrins are cyclic oligosaccharides enzymatically derived from starch. Three main cyclodextrin types have been identified (Figure 12.8).

(a)

(b)

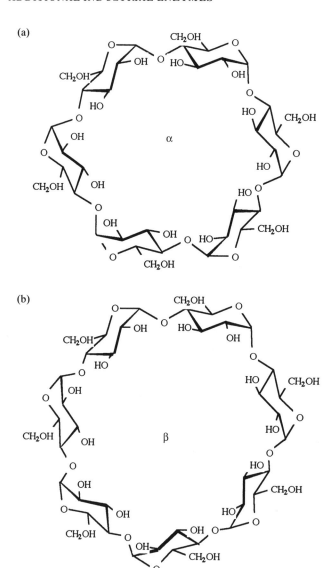

Figure 12.8 Structures of (a) α-, (b) β- and (c) γ-cyclo-dextrins

α-Cyclodextrins are composed of six glucose molecules, β-cyclodextrins are composed of seven glucose molecules, while γ-cyclodextrins consist of eight glucose molecules. In each case, individual glucopyranose units are linked to each other via α1 → 4 glycosidic linkages characteristic of linear starch molecules.

Cyclodextrins were first studied over 100 years ago, when their presence was noted in bacterial digests of starch. It is only within the recent past, however, that these substances have begun to enjoy widespread industrial demand. They are utilized in the pharmaceutical industry as well as in the food, cosmetic and allied industries.

(c)

Cyclodextrins are doughnut-shaped molecules, the outer surface of which is hydrophilic in nature while their internal cavity is apolar. When dissolved in aqueous media, the internal cavity is occupied by water molecules. The polar nature of water renders this energetically unfavourable. Added substances which are less polar than water, and which are of appropriate molecular dimensions, will replace the water as 'guest molecules' within the cyclodextrin cavity. Such guest molecules are retained within the cavity solely by non-covalent interactions. The overall guest molecule–cyclodextrin structure is termed an inclusion complex.

Only molecules of an appropriate size will form stable inclusion complexes. The internal cavity of α-cyclodextrins is obviously the smallest. This cavity will only accommodate molecules/elements of low molecular mass – such as chlorine, bromine or iodine. Much larger molecules such as steroids or antibiotics may be accommodated in the cavity of γ-cyclodextrin molecules. Specific side groups present in macromolecules may also interact with the cyclodextrin cavity and, in this way, form a complex.

Of the three basic cyclodextrins, the β form is by far the most commonly employed on an industrial scale. The internal dimensions of the β-cyclodextrin cavity makes it ideally suited for a variety of applications. β-Cyclodextrins are also economically attractive as they are the least expensive to produce. Native β-cyclodextrins however exhibit poor solubility characteristics. A 14 per cent solution of α-cyclodextrins and a 23 per cent solution of γ-cyclodextrins are readily achievable in aqueous media. However, the maximum solubility of β-cyclodextrin in water is of the order of 1.8 g/100 ml, i.e. 1.8 per cent. Such poor solubility characteristics limit the industrial potential of the native β-cyclodextrin molecule.

This limitation has been overcome by the introduction of modified β-cyclodextrins. Substitution of the cyclodextrin hydroxyl groups with a variety of alkyl, ester or other residues dramatically improves solubility while having no effect on its complex forming ability. Species such as hydroxypropyl β-cyclodextrin and hydroxyethylated β-cyclodextrin are now subject to increasing industrial demand.

Interaction of suitable substances with cyclodextrins effectively results in their molecular encapsulation. Encapsulation may be advantageous for a number of reasons. The guest molecule is protected from a wide variety of chemical and other reactants which might otherwise lead to its destruction. Such protected molecules are less likely to undergo polymerization or autocatalytic reactions. Complex formation may also mask undesirable tastes or odours normally associated with the guest molecule. Volatile compounds may be effectively stabilized by complexation with appropriate cyclodextrins. Fox example, volatile aroma concentrates sensitive to oxygen, light or heat are often stabilized in this way.

Crystallization of cyclodextrin complexes can effectively convert into powder guest molecules originally present in liquid form. This is often of particular significance in the pharmaceutical industry. Another significant consequence of complex formation is the solubilization of hydrophobic substances in aqueous solution. A variety of medically important drugs are hydrophobic in nature and are known to form stable inclusion complexes. Such hydrophobic drugs can be safely transported through the bloodstream and gastrointestinal tract in such a format. The hydrophobic drug is then released at the cell surface due to the hydrophobic nature of the plasma membrane's lipid bilayer. Cyclodextrins are therefore used medically to improve the bioavailability of a range of poorly soluble medicinal substances. This can often allow administration of lower dosage levels with consequent economic, therapeutic and other benefits.

Cyclodextrins may also be used to stabilize certain proteins. The macromolecular structure of the smallest polypeptide precludes complete inclusion-complex formation. However, virtually all proteins contain amino acid subunits containing non-polar or hydrophobic side chains. Cyclodextrins may freely interact with such side chains and in this way become intimately associated with the protein molecule. In some instances, and at high concentrations, cyclodextrins may actually destabilize or denature proteins. Under most circumstances, however, protein–cyclodextrin interactions serve to enhance protein stability. Cyclodextrins have been shown to reduce loss of enzymatic activity due to chemical or physical influences such as heating, freeze-drying, storage or the presence of oxidizing agents.

Upon standing in aqueous solution, many proteins undergo limited aggregation. Powdered protein preparations, such as freeze-dried products, also sometimes form aggregates upon reconstitution in an aqueous media. Aggregate formation prevents intravenous administration of any preparation. Formulation of biopharmaceutical products in order to minimize or eliminate protein–protein interactions does not always yield satisfactory results. Many important biopharmaceutical products, such as growth hormone, urokinase and interleukin-2, still exhibit a marked tendency towards aggregate formation upon reconstitution. In many instances, inclusion of a suitable cyclodextrin preparation will eliminate such undesirable intermolecular interactions.

The ever-increasing industrial demand for cyclodextrins is reflected in an increased demand for cyclodextrin glycosyltransferase preparations. Cyclodextrin glycosyltransferase (CGTase) is the enzyme used in the synthesis of cyclodextrins from substrate starch molecules. Cyclodextrins may also be synthesized chemically but their enzymatic production is technically and economically more attractive.

Three major types of cyclodextrin glycosyltransferases have been identified: α, β and γ. As the name suggests, α-CGTase predominantly yields α-cyclodextrins, whereas β-and γ-CGTases yield β and γ-cyclodextrins, respectively. In all cases, however, prolonged reaction times result in the formation of a mixture of all three cyclodextrin types, with β-cyclodextrin representing the predominant reaction product.

Most cyclodextrin glycosyltransferases identified to date are derived from various bacilli:

- *Bacillus stearothermophilus*;

- *B. megaterium*;

- *B. circulans*;

- *B. subtilis*;

- *E. coli* (recombinant).

The genes coding for many of these enzymes have been identified, sequenced and expressed in a variety of recombinant systems. Elevated levels of heterologous CGTase has been produced in host species such as *E. coli* and *Bacillus subtilis*. Increased production capacity of CGTases should help reduce the overall cost of cyclodextrin preparations which in turn is likely to promote increased utilization of cyclodextrins in a variety of industrial applications.

Current annual cyclodextrin production levels are in excess of 100 tonnes. Production of modified β-cyclodextrin is also increasing rapidly. Production levels secure future large-scale demand for cyclodextrin glycosyltransferases, thus these enzymes may now be classified as 'bulk' industrial enzymes.

ENZYMES AND ANIMAL NUTRITION

Higher organisms have developed sophisticated digestive systems by which they procure nutrients from ingested matter. Degradation of polymeric nutrients, such as proteins, carbohydrates and lipids, is usually a prerequisite to efficient nutrient assimilation. Most such degradative events are mediated by specific digestive enzymes. Evolutionary pressure has ensured the development of an efficient digestive system and the advent of modern intensive livestock production practices places added pressure on this digestive process. Biotechnological intervention can serve to redress any digestive imbalance caused by such modern production methods.

Weaning of young animals such as piglets serves as a good example. Modern production practices transform weaning from a gradual process to an abrupt event. The sudden alteration in dietary composition from a milk-based feed to one of a more complex nutritional composition frequently causes digestive upsets in young animals. Young piglets often display a physiological deficiency in gastric acid production. Stomach pH may therefore be above values required for optimal digestive function.

In addition the overall digestive capability of such animals may not have fully developed at the time of weaning. Poorly digested food results in suboptimal nutrient assimilation by the weaned animal, and incompletely digested matter also promotes vigorous growth of many microorganisms in the large intestine. Such factors contribute to digestive upsets and an increased incidence of post-weaning scours.

Digestive difficulties associated with weaning may be averted by feeding specially formulated, readily digestible rations, and inclusion of acidifiers such as citric acid in the diet to generate a low stomach pH value. The addition of exogenous enzymes, capable of hydrolysing more complex components of the weaned animal's rations, also promote increased feed digestibility. Proteolytic enzymes, in addition to cellulases and hemicellulases, have all been used as digestive aids for weanlings. Not surprisingly, the beneficial effects of exogenous enzyme addition is most noticeable when animals are fed rations containing more complex dietary components.

Removal of antinutritional factors

Various enzymatic preparations have also been used to bring about the removal of specific antinutritional factors from animal feeds. The addition of β-glucanase to barley-containing poultry feed is perhaps the best-known example. The inclusion of pentosanase (xylanase) in wheat-based poultry diets serves as an additional example. Incorporation of phytase in cereal-based animal feedstuffs represents a particularly exciting concept as it not only removes a major antinutritional factor but may also considerably reduce the pollutive effect of animal wastes (as discussed later).

β-Glucan is a non-starch polysaccharide associated in particular with barley. Structurally it consists of glucose units linked via β1 → 3 and β1 → 4 linkages. The level of β-glucans present in barley can vary considerably and is influenced by factors such as soil type, growing conditions and time of harvesting. Although it may be present at levels below 2–3 per cent, values in excess of 10 per cent of grain content have also been recorded. When ingested, the β-glucans present in barley become solubilized in the gut. Animals are devoid of endogenously produced digestive enzymes capable of hydrolyzing the β-glucan molecule.

Soluble β-glucans form highly viscous solutions (Figure 12.9), thus their presence in the diet promotes formation of highly viscous digesta. This has a particularly negative effect on the digestive function of poultry. Maximum nutrient utilization is impeded and the animals develop difficulty in passing faeces. Due to its viscous nature, excreted faeces adheres to the birds' feathers, to their bedding and to the eggs of laying hens. The multiple negative effects associated with ingested β-glucans has traditionally limited the level of incorporation of barley in poultry feed, although on a food cost basis it may appear attractive to do so.

Some microbial populations produce enzymes termed β-glucanases. These enzymes are capable of hydrolysing β-glucans and incorporation of β-glucanase preparations in poultry diets now facilitates the use of barley in such diets. Upon ingestion, the microbially-derived β-glucanase enzyme degrades β-glucans present thus destroying their antinutritive properties and providing sugars for energy. This can significantly increase animal growth performance (Table 12.4).

Wheat contains an antinutritive factor called pentosan. Pentosan is a non-starch polysaccharide consisting predominantly of sugars such as xylose and raffinose. The digestive complement of animals is incapable of hydrolysing this substance and as in the case of β-glucans from barley, solubilization of pentosans in the gastrointestinal tract results in increased viscosity of digesta. Addition of microbial pentosanase (xylanase)

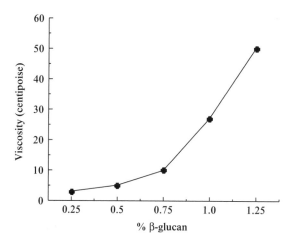

Figure 12.9 The effect of increasing β-glucan concentrations on the viscosity of an aqueous solution

Table 12.4 Effect of supplementing barley-based diets with β-glucanase on performance of poultry

Dietary treatment	Feed consumption per bird (g/day)	Weight gain per bird (g/day)	Feed:gain ratio
Wheat-based diet	102.3	49.9	2.05
Barley-based diet	91.5	41.5	2.2
Barley and β-glucanase	96.3	50.2	1.92

The wheat-based diets serves as a control. Bird performance may be assessed by the feed to weight gain ratio. The lower this value, the more efficiently the bird has utilized the feed provided.

preparations to wheat-based diets will destroy these antinutritional factors. The presence of pentosans does not, however, present such acute dietary difficulties as does the presence of β-glucans.

Phytase and phytic acid

Over 75 per cent of the total phosphorus present in most cereals is in the form of phytic acid, also known as phytin or *myo*-inositol hexaphosphate (Figure 12.10). The phosphate groups present in phytic acid are biologically unavailable to non-ruminant animals. Such animals lack the enzymatic activity required to release the phosphate groups from the core ring structure. In ruminant species, phytic acid is degraded by microbial populations within the rumen. Lack of phosphorus availability in monogastric animals, such as pigs and poultry, renders likely the possibility of dietary phosphorus deficiency. This scenario is generally avoided by the addition of inorganic phosphorus in the form of dicalcium phosphate to such animal feeds. Phytic acid is also an antinutritional factor as it binds a range of essential minerals such as calcium, zinc, iron, magnesium and manganese within the digestive tract and, hence, renders these unavailable for absorption.

As mentioned before, the digestive complement of animals is devoid of an enzymatic activity capable of degrading phytic acid. Several such phytase enzymes (*myo*-inositol hexaphosphate hydrolases) are expressed in plants and in a variety of biological systems. Phytase activity has been

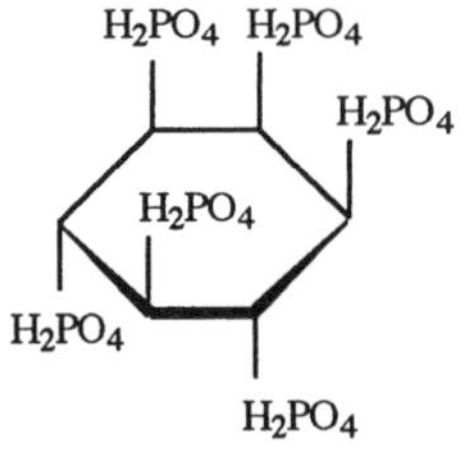

Figure 12.10 Structure of *myo*-inositolhexaphosphate (phytic acid or phytin)

detected in a wide range of plants, particularly in germinating seeds. These sources include cereals such as wheat, corn, barley and triticale as well as beans (e.g. navy, mug and dwarf beans). Wheat, rye and triticale are amongst the richest known sources of plant phytase. Phytase is also produced by various bacteria, including *E. coli*, bacilli, *Klebsiella* and *Pseudomonas*. Bacterial phytases are generally produced at low levels, are intracellular and display activity optima at neutral to alkaline pH values. The only bacteria known to produce extracellular phytase is *Bacillus subtilis*. The extracellular phytase of *B. subtilis* (strain natto N-77) has been purified to homogeneity using a combination of gel filtration and ion exchange (DEAE) chromatography. The active enzyme is a monomer, displaying a molecular mass of 36 kDa, a pI of 6.25 and an optimum activity at 60 °C.

Unlike bacteria, fungi produce a range of extracellular (as well as intracellular) phytases. *Aspergillus* species are the most prolific producers, particularly *A. niger*, *A. flavus* and *A. candidus*. Various species of *Rhizopus* and *Penicillium* also have phytase activity. The extracellular fungal phytases tend to exhibit molecular masses ranging from 35 to 100 kDa, pI's of 4–6.5, optimum temperatures of 35–65 °C and pH optima ranging from 2.5 to 7. Various yeasts such as *Saccharomyces cerevisiae* also produce phytase. Synthesis of most microbial phytases is repressed by high levels of inorganic phosphate in the microbial growth media.

Addition of microbial phytase to dietary rations has been shown to promote degradation of phytic acid within the digestive tract. Such enzyme-mediated degradation has several associated beneficial effects. Degradation of phytate destroys its anti nutritive effect, thus promoting increased mineral absorption. The phosphate groups released from the *myo*-inositol ring are rendered biologically available to the animal (Table 12.5). This facilitates a significant reduction in the quantities of supplemental inorganic phosphorus which must be added to the diet.

Phosphate pollution derived from the animal production sector has long been a source of concern. It is estimated that over 100 million tonnes

Table 12.5 Effect of phytase on phosphorus balance in pigs

Parameter	Phytase diet	Control diet
Phosphorus intake (g/day)	8.3	7.3
Phosphorus absorbed (g/day)	5.4	3.5
% absorption	65	48
% retained	60	47

'Phytase diet' was similar to 'control diet' but was supplemented with phytase. Both phosphorus intake and absorption was monitored. Inclusion of phytase resulted in increases in absorbed and retained phosphorus. All differences recorded were found to be statistically significant (Pointillard *et al.*, (1987) *J. Nutr.* **117**, 907).

of animal manure is generated in the USA each year. Such quantities would contain in excess of one million tonnes of phosphorus. Although animals fail to digest phytic acid, many microorganisms present in faeces display appreciable levels of phytase activity. This form of phosphorus may therefore greatly contribute to the pollutive effect of animal slurry.

Early phytase preparations used in animal trials were sourced from various strains of phytase-producing fungi. Initial screening studies identified *A. niger* NRRL 3135 (formally known as *A. ficuum*) as the microbial strain producing phytase in greatest quantities, and hence phytase was mainly obtained from this source. *A. niger* 3135 produces two phytases, A and B. Phytase A (phy A) is an 85 kDa, 448 amino acid glycoprotein containing a total of nine glycosylation sites. The enzyme displays activity over a pH range of approx. 2–7, with two distinct activity peaks at pH 2.5 and 5.5. Phytase B (phy B) on the other hand displays activities over a narrower pH range (pH 2–3). *A. niger* NRRL 3135 also produces a third extracellular phosphatase, but this is devoid of phytase activity (i.e. it dephosphorylates phosphorylated substances other than phytic acid).

Phytase production often entails fermentation of *A. niger*, either by surface culture (using e.g. wheat-bran based media) or sometimes by submerged fermentation (using e.g. corn starch media). Low natural production levels coupled with phosphate repression of enzyme synthesis renders this product expensive. Although traditional mutational and selection techniques has generated mutants producing increased quantities of enzyme, the vast bulk of phytase now added to animal feed is produced by recombinant means.

'Natuphos' is the tradename given to recombinant phytase A (originally) derived from *A. niger* NRRL 3135. Developed by Gist Brocades, the coding sequence for the enzyme was placed under a powerful promoter (the glucoamylase promoter sourced from an industrial strain of *A. niger* used by Gist Brocades to produce glucoamylase at levels of several grams per litre). Multiple copies of the construct were inserted into *A. niger* (self cloning). High level expression (and removal of phosphate repression) renders possible large scale economic production of the product (Fig 12.11). Phytase A was chosen as it appears to best fulfil various criteria characterizing an enzyme suitable for in-feed application (as discussed later).

A phytase gene obtained from *A. niger* has more recently been expressed in transgenic tobacco seeds. The recombinant enzyme was expressed as 1 per cent of the soluble protein in the mature seed. Animal trials confirmed that direct application of these seeds to animal feed promoted enhanced phytic acid phosphorus utilization by recipient animals.

Factors affecting feed enzyme efficacy and stability

The addition of specific enzymes to animal feedstuffs in order to achieve a particular effect is one of the most innovative advances of modern

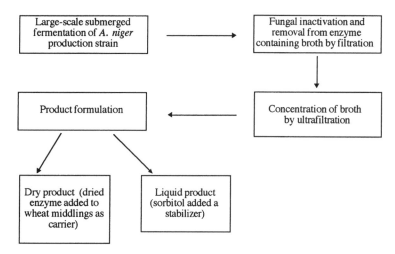

Figure 12.11   Overview of the process used by Gist Brocades to produce their recombinant phytase product, Natuphos

agricultural biotechnology. The physicochemical properties which renders an enzyme maximally suitable for inclusion in animal feed have yet to be fully elucidated. However, the following characteristics are likely to be the most important.

- *Enzyme thermostability.* Virtually all animal feed is heat treated prior to its sale. Heat treatment can form part of the pelleting process, where moist heat is applied followed by mechanical pressing. Heat treatment is essential for both pelleted and non-pelleted feed in order to diminish or eliminate the possibility of accidental transmission of pathogens via infected feed. The process typically entails heating the feed to temperatures of 75–90 °C for a period of up to 2–3 min. Enzymes added to animal feed must therefore be thermostable. (It is possible to apply an enzyme to feed post-heat treatment by spraying. However, the spraying equipment is expensive and uneven enzyme application can sometimes occur).

- *Activity versus temperature profile.* The enzyme must display significant activity at 37 °C, the temperature at which it is destined to function when ingested by the animal.

- *pH versus activity profile.* The enzyme must be active at pH values typical of one or more regions of the digestive tract. Most fungal-derived feed enzymes display maximal activity at acidic pH values and hence are likely to be most active in the stomach and upper portion of the small intestine (the duodenum). Bacterial enzymes (e.g. bacterial amylases or proteases) on the other hand exhibit appreciable activity at neutral or near neutral pH values, rendering the small intestine the most likely site of activity for these.

- *Stability in the presence of gastrointestinal tract influences.* Enzymes added to animal feed must remain stable in regions of the digestive tract up to (and including) the area in which they promote their intended effect. In all cases this means that the enzyme must be stable under the acidic conditions encountered in the stomach (stomach pH can drop as low as 1.5), and must be resistant to digestive proteases such as pepsin.

The majority of enzymes added to animal feed display optimum activity at temperatures of 50–60 °C. When heated in buffer to temperatures in excess of 70 °C such enzymes rapidly lose activity. However, when present in feed they appear significantly more thermostable, probably due to a protective effect exerted by the bulk feed constituents, and the lower water content of the system. Bulk feed constituents may well also exert a protective influence on enzymes in the stomach environment.

Detection of enzymes after their addition to feed

Regulatory requirements as well as concerns relating to the effect of, for example pelleting on enzyme activity makes it necessary to develop suitable assays to allow detection and quantification of enzyme activity after addition to feed. Assay procedures entail initial extraction of the supplemental enzyme from the feed by agitating it in the presence of buffer, with subsequent assay of the buffer contents. A number of technical hurdles make development of such feed assays technically challenging. Enzymes are generally added to feed in low amounts (often as little as 50 g of enzyme per 1000 kg feed). The low initial additional levels and subsequent extraction procedures produce very dilute enzyme solutions, requiring sensitive assays and prolonged incubation times.

Enzymes are often most conveniently assayed by incubation in the presence of their substrate, with subsequent quantification of the amount of product formed over a given time (Table 12.6). All feedstuffs naturally contain high levels of the reaction products that feed enzymes normally produce. These substances co-extract with the enzyme into the extraction buffer. In effect this generates enzyme blank values so high as to render this assay approach impractical.

Various alternative assay strategies have been adopted to overcome this difficulty, the most successful of which entail the use of chromgenic substrates or assay by methods of radial enzyme diffusion. Chromogenic substrates are synthetic or semisynthetic substances containing a coloured group, which is released by action of the appropriate enzyme. The colour released can be quantified spectrophotometrically subsequent to termination of the assay. A number of chromogenic substrates suitable for assaying enzymes extracted from feed are available (Table 12.7).

Assay of enzyme activity by radial enzyme diffusion entails incorporation of the enzyme substrate into an agar gel, with subsequent

Table 12.6 Enzymes most often added to animal feed, and the natural substrates which are most often used in their assay

Enzyme	Substrate	Reaction product monitored
Phytase	Phytic acid	Inorganic phosphate
β-Glucanase	β-Glucan or lichenan*	Reducing sugars
Xylanase	Xylan	Reducing sugars
Cellulase	Cellulose/modified cellulose (e.g. carboxymethyl cellulose)	Reducing sugars
Protease	Haemoglobin, casein, albumin	Amino acids/peptides
Amylase	Starch	Monitoring of starch disappearance

Note: lichenan* is an inexpensive glucan obtained from *Cetraria islandica* which, like β-glucan, consists of glucose molecules linked via $\beta 1 \rightarrow 3$ and $\beta 1 \rightarrow 4$ glycosidic linkages.

Table 12.7 Synthetic chromogenic substrates that may be used to assay the enzymes indicated

Enzyme	Chromogenic substrate
β-Glucanase	Azo β-glucan
Xylanase	Xylan coupled to remazol brilliant blue (RBB)
Cellulase	RBB–cellulose
Protease	Azoalbumin
Amylase	Starch coupled to ostazin Brilliant red

introduction of the enzyme-containing feed extract into wells which have been generated in the gel. The gel (usually housed in a Petri dish) is then incubated for an appropriate time period and at an appropriate assay temperature. During this time the enzyme diffuses outward through the gel in a radial manner, degrading substrate molecules it comes into contact with. The diameter of this zone of substrate hydrolysis is proportional to the enzyme activity levels present (Figure 12.12). This assay approach thus monitors substrate degradation rather than product formation. If a chromogenic substrate is used, the zone of hydrolysis is immediately apparent. If native substrate is used, a stain must be subsequently added to the gel which will bind (only) to undegraded substrate. Stains such as congo red for example will bind undegraded β-glucan or cellulose.

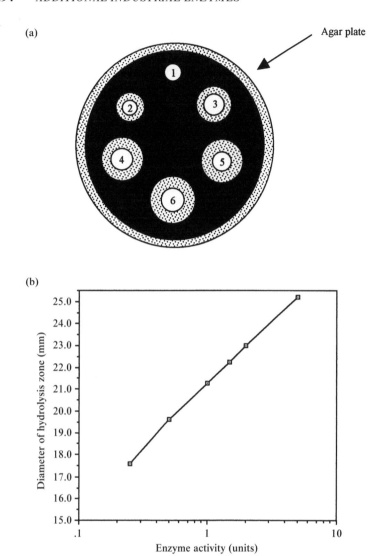

Figure 12.12   Schematic representation of the radial enzyme diffusion approach to the detection and quantification of enzyme activity. (a) A series of wells are punched in a substrate-containing agar gel, and enzyme is placed in the wells. Well 1 contains no enzyme, wells 2–6 contain increasing levels of enzyme activity. The diameters of the zones of substrate hydrolysed are measured. When plotted against the log of enzyme activity a linear relationship is observed (b)

Enzymes and animal nutrition; future trends

The animal feed industry is one of the fastest growing market outlets for industrial enzymes. Current research in this area focuses upon a number of different objectives. Screening of microorganisms continues in an effort to identify enzymes exhibiting physicochemical characteristics rendering them more suited to animal feed application than the enzymes currently used. Enzymes displaying enhanced thermostability while retaining significant activity at 37 °C would be particularly desirable. Production of enzymes by recombinant means is also likely to come to the fore, as enzyme production costs could be reduced, which would have beneficial implications for the price-sensitive animal feed industry. Genetic engineering also facilitates engineering of feed enzymes in order to confer on them some novel or improved characteristics (such as increased thermostability). The industry is also evaluating the potential of additional enzymatic activities not currently used in feed applications. Such enzymes would aim to achieve specific objectives, such as the destruction of specific anti-nutritional factors (Table 12.8). An additional research approach entails incorporation of genes coding for desired enzyme activities into transgenic plants which could subsequently be fed to animals (e.g. the previously discussed expression of phytase activity in tobacco plants). An alternative approach entails the generation of transgenic animals capable of synthesizing novel digestive enzymes. For example, a cellulase gene obtained from *Clostridium thermocellum* has been successfully expressed in the pancreas of transgenic mice. The enzyme was secreted into the small intestine where it was found to be resistant to proteolytic inactivation by endogenous proteases. Such research demonstrates the feasibility of generating monogastric animals with the endogenous capacity to digest plant structural polysaccharides.

OXIDOREDUCTASES

Oxidoreductases represent a large class of enzymes, a few members of which have generated industrial interest. These enzymes catalyse oxidation–reduction (redox) reactions and contain a cofactor (e.g NAD, Flavine adenine dinucleotide (FAD) or a metal ion) at the active site, which acts as the acceptor/donor of electron equivalents (eg. NAD^+ accepts a hydride ion, H^-, which represents two electron equivalents). Some cofactors are covalently attached to the redox enzyme, while other interact with it non-covalently and can diffuse from the enzyme's active site. The requirements of a cofactor usually complicates the industrial use of oxidoreductases. Permanent disassociation of the cofactor from the active site (especially if the enzyme is used in immobilized format) results in loss of activity. Furthermore a suitable means by which the cofactor can be regenerated must also be available (Figure 12.13).

A number of oxidoreductases display potential or actual industrial utility. Lipoxygenases (LOX), in conjunction with additional enzymes such as hydroperoxide lyase, display potential application in flavour biotechnology. Lipoxygenases are non-haem iron-containing enzymes that catalyse the oxidation of unsaturated fatty acids containing a 1-*cis*, 4-*cis*-pentadiene double bond system (eg. linoleic acid, Figure 12.14).

Table 12.8 Various antinutritional factors found in association with animal feed, and the negative effects they promote upon ingestion by farm animals

Antinutritional factor	Major *in vivo* effect
Proteins	
Trypsin inhibitors	Reduction of activity of (chymo) trypsin Pancreas hypertrophy, increased secretion of pancreatic enzymes
Lectins	Gut wall damage, immune response, increased loss of endogenous protein
Amylase inhibitors	Interference with starch digestion
Antigenic proteins	Interference with the gut wall integrity, immune response
Polyphenols	
Tannins	Formation of protein–carbohydrate complexes, interference with protein and carbohydrate digestibility
Glycosides	
Vicine/convicine	Haemolytic anaemia, interference with the fertility and hatchability of eggs
Saponins	Haemolysis, effects on intestinal permeability
Glucosides	
Glucosinolates	Impaired iodine utilization, affected thyroid and liver, reduced palatability and growth
Alkaloids	
Quinolizidine (lupin alkaloids) Scopolamine and hyoscyamine (Alkaloids from Datura)	Neural disturbances, reduced palatability
Other antinutritional factors	
Phytate	Forms complexes with minerals and protein, depresses absorption of minerals
Gossypol	Anaemia due to Fe-complexation, reduced egg weight
Sinapins	Fishy odour in eggs (taint)
Flatulence factors	Gastrointestinal discomfort

Reproduced with permission from *Recent Advances in Animal Nutrition*, 1992, p7. Butterworth Heinemann, Oxford.

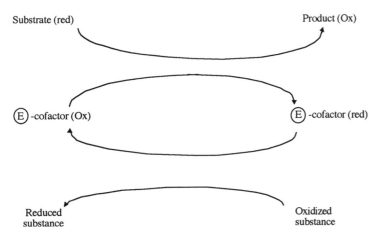

Figure 12.13 REDOX reactions and the regeneration of the enzyme's cofactor. In the generalized example provided a substrate (reduced) docks at the active site of the REDOX enzyme. The enzyme's cofactor (oxidized form) accepts reducing equilivants from the substrate, thereby oxidizing it (i.e. forming product), itself being reduced in the process. In order for the enzyme to promote the catalytic conversion of another substrate molecule its cofactor must first be re-oxidized. Thus a terminal electron acceptor (described above as the 'oxidized substance') must thus also be present

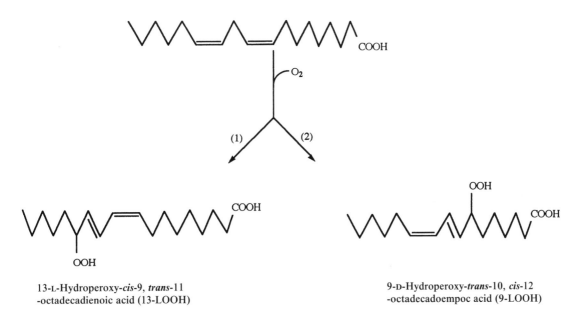

13-L-Hydroperoxy-*cis*-9, *trans*-11
-octadecadienoic acid (13-LOOH)

9-D-Hydroperoxy-*trans*-10, *cis*-12
-octadecadoempoc acid (9-LOOH)

Figure 12.14 The oxidation of linoleic acid as catalysed by (1) Soya type 1 lipoxygenase and (2) tomato lipoxygenase. Reproduced with permission from Godfrey &West (eds) (1986) *Industrial Enzymology*, 2nd edn. MacMillan Press Ltd., Basingstoke

Fungal, and particularly plant-derived, lipoxygenases have thus far gained most attention, and the three-dimensional structure of LOX (iso-enzyme L-1) from soybean has been elucidated. The presence of LOX in plants can result in the production of either desirable or undesirable flavours and aromas. For example, it leads to the generation of off flavours in peas, but plays a central role in the production of the compounds responsible for the characteristic aroma of mushrooms and cucumber in these latter food sources. The exact profile of flavours and aromas produced in any food source will depend upon the substrate (i.e. fatty acid) range present, the exact LOX specificity and the presence of additional enzymes which participate in multi-enzyme 'flavour producing' pathways (Figure 12.15).

Glucose oxidase and catalase represent two oxidoreductases which have found actual industrial application. They are used as components of various diagnostic kits (Chapter 9). In addition these enzymes are used

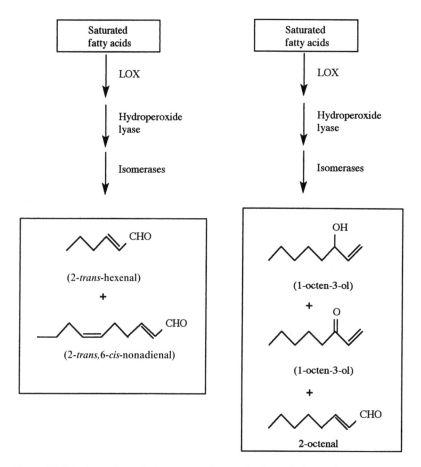

Figure 12.15 Overview of the enzymatic production of the main aroma compounds from (a) cucumber and (b) mushroom, from saturated fatty acid precursors via a lipoxygenase mediated mechanism

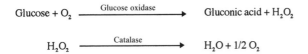

Glucose + O_2 $\xrightarrow{\text{Glucose oxidase}}$ Gluconic acid + H_2O_2

H_2O_2 $\xrightarrow{\text{Catalase}}$ $H_2O + 1/2\,O_2$

Figure 12.16 Enzyme-mediated removal of residual oxygen from (glucose-containing) foodstuffs, as achieved by using a combination of glucose oxidase and catalase. Refer to text for further details

in conjunction with each other in the food industry to remove residual oxygen from food products present in sealed containers. The presence of oxygen in stored foodstuffs can result in spoilage, either via promoting aerobic microbial growth or by promoting fatty acid oxidation, thereby causing rancidity. The mechanism by which oxygen is consumed by this approach is illustrated in Figure 12.16.

ENZYMES IN MOLECULAR BIOLOGY

The term molecular biology has come to describe the study and manipulation of nucleic acids, be it for pure or applied purposes. Many of the basic techniques underpinning molecular biology rely upon the use of specific enzymes. The isolation of specific sequences of chromosomal DNA and the amplification of these sequences (DNA cloning) requires the use of the two enzyme types, restriction endonucleases and DNA ligases. The enzyme reverse transcriptase plays a central role in the generation of complementary DNA (cDNA) from a mRNA template, while DNA polymerase allows the catalytic amplification of specific DNA segments via the polymerase chain reaction (PCR).

Restriction endonucleases and DNA ligase

Most microorganisms have developed a variety of mechanisms by which they protect themselves from invading viral and other pathogens. One such method relies upon specific microbial enzymes termed restriction endonucleases. These enzymes are capable of cleaving foreign double stranded DNA and, hence, preventing its replication. Host microbial DNA may be modified, usually by methylation of specific DNA sequences, and in this way is protected from destruction by endogenous restriction endonucleases.

Some 800 restriction endonucleases have been identified thus far. They belong to one of three types. Types I and III restriction enzymes are complex, consisting of a number of subunit types and requiring a variety of cofactors to maintain activity. These enzymes cleave DNA at sites removed from the DNA sequence which the enzyme recognizes. Type II restriction endonucleases are less complex, generally consisting of a single subunit, require only Mg^{2+} for activity and cleave DNA at the sequence which the enzyme recognizes and binds. Type II restriction endonucleases have thus found wide application in both pure and applied molecular biology.

Restriction endonucleases recognize, bind and cut DNA sequences which exhibit a defined base sequence (Table 12.9). These sequences normally exhibit a 2-fold symmetry around a specific point and are usually 4, 6 or 8-base pairs in length. Such areas are often termed palindromes. In general, the larger the recognition sequence the fewer such sequences present in a given DNA molecule and, hence, the smaller number of DNA fragments that will be generated. Depending upon the specific restriction endonuclease utilized, DNA cleavage may yield blunt ends or staggered ends – the latter are often referred to as sticky ends.

Table 12.9 Some commercially available restriction endonucleases, their sources, DNA recognition sites and cleavage points

Restriction enzyme	Source	DNA recognition sequence and cleavage site
Bcl I	*Bacillus caldolyticus*	$5' - T \downarrow GATCA - 3'$ $3' - ACTAG \uparrow T - 5'$
Bgl II	Recombinant *E. coli* carrying *Bgl*II gene from *Bacillus globigii*	$5' - A \downarrow GATCT - 3'$ $3' - TCTAG \uparrow A - 5'$
Bsa AI	Recombinant *E. coli* carrying *Bsa*AI gene from *B. stearothermophilus* A	$5' - PyAC \downarrow GTPu - 3'$ $3' - PuTG \uparrow CAPy - 5'$
Bsa JI	*B. stearothermophilus* J	$5' - C \downarrow CNNGG - 3'$ $3' - GGNNC \uparrow C - 5'$
Bsi EI	*B. stearothermophilus*	$5' - CGPuPy \downarrow CG - 3'$ $3' - GC \uparrow PyPuGC - 5'$
Eco RV	Recombinant *E. coli* carrying *Eco*RV gene from the plasmid J62 plg 74	$5' - GAT \downarrow ATC - 3'$ $3' - CTA \uparrow TAG - 5'$
Mwo I	Recombinant *E. coli* carrying cloned *Mwo*I gene from *Methanobacterium wolfeii*	$5' - GCNNNNN \downarrow NNGC - 3'$ $3' - CGNN \uparrow NNNNNCG - 5'$
TSP 509 I	*Thermus* sp.	$5' - \downarrow AATT - 3'$ $3' - TTAA \uparrow -5'$
Xba I	Recombinant *E. coli* carrying *Xba*I gene from *Xanthomonas badvii*	$5' - T \downarrow CTAGA - 3'$ $3' - AGATC \uparrow T - 5'$
Xho I	Recombinant *E. coli* carrying *Xho*I gene from *X. holcicola*	$5' - C \downarrow TCGAG - 3'$ $3' - GAGCT \uparrow C - 5'$

G, Guanine; C, Cytosine; A, Adenine; T, Thymine; Pu, any purine; Py, any pyrimidine; N, either a purine or pyrimidine. Arrow indicates site of cleavage.

Each microorganism expresses one or more restriction endonuclease of defined recognition sequence. Restriction endonucleases used either alone or in combination find a ready market in research and industrial applications relying on DNA manipulation. Many are biologically labile and must be transported and stored at temperatures below −20 °C. Several such enzymes commercially available are listed in Table 12.9.

Restriction endonucleases are also used to cleave chromosomal DNA in the first step of restriction fragment length polymorphism (RFLP) analysis. An alternative technique to RFLP for the detection and analysis of mutations or modifications in nucleotide sequences in cleavase fragment length polymorphisms. This entails the additional use of an endonuclease which specifically cleaves single stranded DNA. In this case the cleavage site is not dictated by base sequence, but by single-stranded DNA secondary structure (Box 12.1).

DNA ligases catalyse the formation of phosphodiester bonds in DNA. As such they are mainly used in recombinant DNA technology to seal or 'ligate' a target DNA fragment into a chosen vector as part of a DNA cloning procedure. Ligases available commercially include T4 ligase, a 55 kDa, 487 amino acid enzyme sourced from the phage T4, and *E. coli* ligase, a 73 kDa, 671 amino acid enzyme. More recently a thermostable DNA ligase (Taq DNA ligase) has become available commercially. This ligase is sourced from a recombinant strain of *E. coli* containing a ligase gene isolated from the thermophile *Thermus aquaticus*.

DNA polymerase

DNA polymerases catalyse the synthesis of new DNA strands and thus function naturally to promote DNA replication and (in some instances) DNA repair. The enzymes find application in molecular biology for purposes such as the amplification of specific fragments of DNA via PCR. DNA polymerases most suited to this application should be heat stable as the PCR process entails repeated alternating cycles of (DNA polymerase mediated) polynucleotide synthesis and heat denaturation (95 °C, used to separate DNA double strands). Taq DNA polymerase was the first such thermostable DNA polymerase made commercially available. The enzyme, sourced from *Thermus aquaticus*, displays optimum activity at 75 °C, but is quite stable at much higher temperatures. It displays a half life of 90 min at 95 °C. Subsequently a number of alternative thermostable DNA polymerases have come on the market (Table 12.10). These are sourced from various hyperthermophiles (Chapter 10) but are invariably produced by recombinant DNA technology, usually via heterologous expression in *E. coli*. These enzymes display half lives of the order of hours at 95 °C (e.g. 'Deep Vent' DNA polymerase, Table 12.10, exhibits a half life of 23 h at 95 °C and 8 h at 100 °C.

Box 12.1   The basis of CFLP (cleavase fragment length polymorphism) analysis

Streaches of single stranded DNA refold on themselves, producing hairpin-like secondary structures. Cleavase is the trade name given to a commercialized thermostable endonuclease which recognizes such secondary structures and cleaves the (single-stranded) polynucleotide sequence immediately adjacent to such structures. Mutations, insertions and deletions can affect secondary structure and hence can alter the endonuclease-mediated single-stranded DNA digestion profile. This can be detected by separating the DNA fragments generated on a gel, with subsequent visualization of the DNA banding pattern by appropriate means. In the example provided, a mutation to the wild-type DNA results in the formation of only one hairpin loop (as opposed to two in the native wild-type molecule). The resulting bonding pattern produced thus differs.

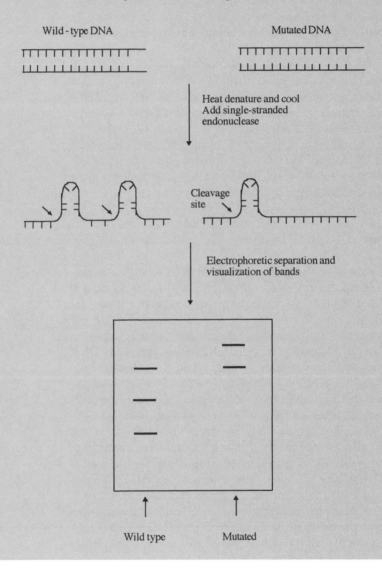

Table 12.10   Some DNA polymerases available commercially

Tradename	Source
Taq DNA polymerase	*Thermus aquaticus*
Tli DNA polymerase	*Thermococcus littoralis*
Pfu DNA polymerase	*Pyrococcus furiosus*
TfL DNA polymerase	*Thermus flavus*
Tth DNA polymerase	*Thermus thermophilus*
Pwo DNA polymerase	*Pyrococcus woesei*
Vent DNA polymerase	*Thermococcus littoralis*
Deep Vent DNA polymerase	*Pyrococcus* sp.
9 °Nm DNA polymerase	*Thermococcus* sp.

Note that the sources listed are the microorganisms which naturally produce the indicated enzymes. The commercial products however are usually produced in *E. coli* via recombinant DNA technology. Most of the listed enzymes display molecular masses in the region of 90–95 kDa.

FURTHER READING

Books

Malcata, F. (1996) *Engineering of/with Lipases*. Kluwer, Dordrecht.
Marquardt, R. (1997) *Enzymes in Poultry and Swine Nutrition*. International Development Research Centre, Canada.
Sivak, M. (1997) *Advances in Food and Nutrition Research*. Academic Press, London.
Woolley, P. (1994) *Lipases*. Cambridge University Press, Cambridge.

Articles

Lipase
Muralidhar, R. *et al.* (2001) Lipases in raceimic resolutions. *J. Chem. Technol. Biotechnol.* **76**, (1), 3–8.
Pandey, A. *et al.* (1999) The realm of microbial lipases in biotechnology. *Biotechnol. Appl. Biochem.* **29**, 119–131.
Planas, N. (2000) Bacterial 1,3-1,4-beta glucanases: structure, function and protein engineering. *Biochem. Biophys. Acta – Protein Struct. Mol. Enzymol.* **1543**, (2), 361–382.
Saxena, R. *et al.* (1999) Microbial lipases; potential biocatalysts for the future industry. *Curr. Sci.* **77** (1), 101–115.

Svendsen, A. (2000) Lipase protein engineering. *Biochem. Biophys. Acta – Protein Struct. Mol. Enzymol.* **1543** (2), 223–238.

Xu, B. (2000) Production of specific-structured triacylglycerols by lipase-catalyzed reactions; a review. *Eur. J. Lipid Sci. Technol.* **102** (4), 287–303.

Enzymes in animal nutrition

Bedford, M. (2000) Exogenous enzymes in animal nutrition-their current value and future benefits. *Anim. Feed Sci. Technol.* **86** (1–2), 1–13.

Bedford, M. & Schulze, H. (1998) Exogenous enzymes for pigs and poultry. *Nutr. Res. Rev.* **11**, (1), 91–114.

Liu, B. *et al.* (1998) The induction and characterization of phytase and behyond. Enzyme and microbial technology. **22** (5), 415–424.

Mullaney, E. *et al.* (2000) Advances in phytase research. *Adv. Appl. Microbiol.* **47** 157–199.

Sebastian, S. *et al.* (1998) Implications of phytic acid and supplemental microbial phytase in poultry nutrition; a review. *Worlds Poultry Sci. J.* **54** (1), 27–47.

Walsh, G. *et al.* (1993) Enzymes in the animal feed industry. *Trends Biotechnol.* **11** (10), 424–430.

Wodzinski, R. & Ullah, A. (1996) Phytase. *Adv. Appl. Microbiol.* **42** 263–302.

Additional enzymes

Bruggink, A. *et al.* (1998) Penicillin acylase in the industrial production of beta lactam antibiotics. *Org. Proc. Res. Devel.* **2** (2), 128–133.

Hingorani, M. & O'Donnell, M. (2000) DNA polymerase structure and mechanisms of action. *Curr. Org. Chem.* **4** (9), 887–913.

Koob, M. (1992) Conferring new cleavage specifities on restriction endonucleases. *Methods Enzymol.* **216** 321–329.

Kovall, R. & Matthews, B. (1999) Type II restriction endonucleases; structural, functional and evolutionary relationships. *Curr. Opin. Chem. Biol.* **3** (5), 578–583.

May, S. (1999) Applications of oxidoreductases. *Curr. Opin. Biotechnol.* **10** (4), 370–375.

Shewale, J. *et al.* (1990) Penicillin acylases – applications and potentials. *Proc. Biochem.* **25** (3), 97–103.

Steitz, T. (1999) DNA polymerases; structural diversity and common mechanisms. *J. Biol. Chem.* **274** (25), 17395–17398.

13

Non-catalytic industrial proteins

INTRODUCTION

The preceding three chapters have focused upon industrial-grade enzymes. A variety of non-catalytic bulk proteins are also produced in very significant quantities by the biotech sector (Table 13.1). These proteins, which are largely derived from food sources and/or find application in the food sector, for the basis of this chapter. Such proteins are added to food either for obvious nutritional reasons or more commonly because they exhibit specific desirable functional properties.

Table 13.1   Principal non-catalytic proteins which find bulk industrial application

Animal derived	Casein
	Whey proteins
	Gelatin
	Egg proteins
Plant derived	Soy protein
	Wheat protein
Microbial	Single cell protein

Although most are used as/added to human food or animal feed, some also have non-food applications in e.g. the pharmaceutical and cosmetics sectors.

FUNCTIONAL PROPERTIES OF PROTEINS

Protein functionality may be broadly defined as the non-nutritive properties of a protein which effects its utilization as a food ingredient. The properties influenced are often responsible for or affect the appearance, structure and texture, mouthfeel and flavour retention of food. The major functional properties of proteins which render them of interest for food application are summarized in Table 13.2. Most such attributes are either hydration-related properties (e.g. solubility, dispersibility, viscosity and gelation) or surface-related properties (e.g. foaming and emulsification).

The physicochemical properties of proteins which affects their functional properties are diverse and include amino acid composition and sequence, protein size and conformation, as well as pI and molecular flexibility. Although the physicochemical characteristics of the major

Table 13.2   Functional properties exhibited by some proteins which renders attractive their addition to selected foods

Functional property	Protein (example)	Food (example)
Viscosity	Gelatin	Soups, gravies, desserts
Gelation	Milk and egg proteins	Confectionery, some meat products
Cohesion	Whey and egg proteins	Bakery products, pasta, sausages
Elasticity	Cereal products	Bakery products, meats
Fat/flavour binding	Milk, egg, cereal proteins	Confectionery, bakery products
Emulsification	Milk and egg proteins	Sausages, soups, cakes
Foaming	Milk and egg proteins	Cakes, ice creams, whipped toppings

proteins added to food are characterized in detail, it is still not possible to automatically predict the exact effects of their addition to a given foodstuff. This is likely because:

- food processing can lead to protein denaturation, hence altering their physicochemical characteristics;

- varying methods of production and extraction of food proteins can result in variation of product composition, particularly with regard to minor 'contaminants' (e.g. the presence of lipid in milk protein products);

- the added protein will interact with other food ingredients thereby influencing the functional effects observed.

Viscosity and thickening

Various foodstuffs (e.g. soups and gravies) usually contain added viscosity or thickening agents. Soluble polymers of high molecular mass (e.g. polysaccharides and proteins) generate viscous solutions when dissolved in water, even at low concentrations. The level of viscosity depends upon properties such as polymer size and shape as well as level of hydration and range and extent of intermolecular interactions formed. Viscosity generally increases with increasing molecular mass, and randomly coiled or extended polymeric structures exhibit greater viscosity than tightly folded polymers of the same molecular mass. As such various carbohydrates, gums and proteins of extended (as opposed to highly globular) structure (e.g. gelatin or partially denatured globular polypeptides) can be used as thickening agents in the food sector.

Gelation; cohesion and elasticity

Most gels are composed of polymers which are cross-linked, thereby forming an interconnected three-dimensional network structure, immersed in a liquid medium. Food-based gels almost always consist of cross-linked carbohydrates or proteins (or sometimes a mixture of both) immersed in water. Partial polymer unfolding or denaturation is a prerequisite to gel formation for most food proteins. This is most often achieved by heating but can also be attained by mechanical agitation or by altering the protein solution's pH or ionic strength. The unfolded or extended protein structures then interact with each other, forming the extended three-dimensional network. The range and relative importance of intermolecular forces which stabilizes the gel structure depends upon the protein type used. Both covalent (disulfide and γ-glutamyl) bonds and non-covalent forces (hydrogen bonding, electrostatic and hydrophobic interactions) all contribute to gel stability (Figure 13.1).

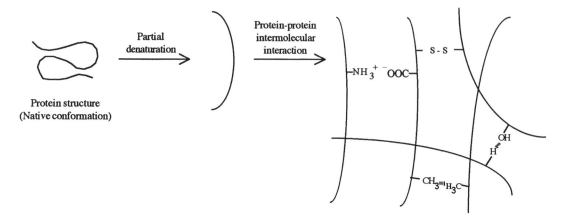

Figure 13.1   Overview of the process of protein-based gel formation. Although not shown, the gel structure will also be influenced or stabilized by hydrogen bonding between the protein strands and the solvent water molecules

Cohesion and elasticity describe functional properties, which also depend upon intermolecular interactions between individual (sometimes partially unfolded) protein molecules. Cohesive forces are by and large non-covalent (i.e. hydrogen bonds, ionic and hydrophobic interactions), which hold a loose network of protein molecules together. The non-covalent nature of these attractions allow the cohesive forces to be broken relatively easily. The functional property of elasticity depends upon the formation of both covalent and non-covalent intermolecular linkages between entangled protein monomers. This property will be subsequently discussed more fully in the context of wheat proteins.

Fat and flavour binding, emulsification and foaming

Many flavour/aroma compounds present naturally in (or added to) food are hydrophobic molecules of relatively low molecular mass. These can be retained or stabilized in food by binding to hydrophobic patches on the surface of proteins, as well as hydrophobic stretches exposed on partially denatured proteins. Food and aroma molecules displaying a polar character may also bind food proteins via hydrogen bonds or electrostatic attractions.

The amphipathic nature of most proteins (i.e. they contain both hydrophobic and hydrophilic regions) facilitates their use as emulsifying agents or foam stabilizing agents in the food industry. Most foods contain an emulsified hydrophobic and hydrophilic ingredient mix. Examples include milk, butter, ice creams and sausages. Many other foods are in fact foams or started out as foams (systems in which tiny air bubbles are dispersed in an aqueous phase; examples include whipped cream, soufflés, mousses, marshmallows and cakes). Both product types are characterized

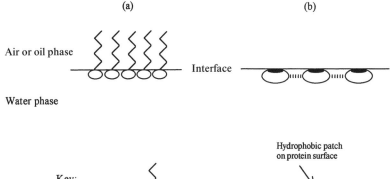

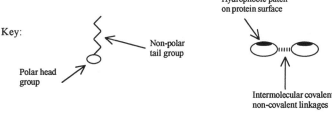

Figure 13.2 The interaction of (a) low molecular mass and (b) protein-based surfactants with an air– or oil – water interface. Refer to text for details

by an apolar phase (i.e. lipid or air) being dispersed in an aqueous-based phase.

Such oil – water or air – water interfaces display high interfacial tension. They collapse as soon as they are formed unless a stabilizing agent (an emulsifier) is added to the system. The emulsifier (surfactant or foaming agent), having hydrophobic and hydrophilic regions, will align along the interface, such that its hydrophobic groups are in contact with the apolar phase and its hydrophilic groups are in contact with the polar phase. This reduces the interfacial tension and stabilizes the emulsion.

Low molecular mass surfactants (e.g. mono- and diglycerides, sorbitan monostearate and phospholipids) and high molecular mass surfactants (proteins and some gums) find widespread application in the food industry as emulsifiers or foaming agents. Lower molecular mass surfactants tend to be more effective at lowering interfacial tension. They pack more tightly along the interface than can proteins (the larger size of proteins causes stearic hindrance (Figure 13.2). However, protein surfactants generally generate more stable foams and emulsions. In addition to lowering interfacial tension, the continuous film they form around the oil – air droplet is further stabilised by intermolecular disulfide and non-covalent linkages. Congregation at an aqueous – apolar interface also promotes partial denaturation of most proteins.

MILK AND MILK PROTEINS

The principal constituents of milk are presented in Table 13.3. Milk is secreted by all female mammals, of which there are in excess of 4000

Table 13.3 Average composition (% w/v) of the milk of humans and various agriculturally important species

Species	Lactose	Fat	Protein	Total solids
Human	7.0	3.8	1.0	12.0
Cow	5.0	3.7	3.3	12.5
Sheep	4.8	7.5	4.5	19.0
Goat	4.0	4.5	3.0	12.3

species. In addition to the major constituents, milk contains several hundred minor constituents, including various vitamins and minerals, flavour components, etc. Milk is a very variable substance in terms of exact composition. Interspecies differences are clearly evident. However, the breed, health and nutritional status, etc, of animals within a single species can also affect the composition of the milk.

The disaccharide lactose represents the major sugar present in milk. Traces of other sugars (e.g. glucose and fructose) are also often present. The milk of most species contains 2.5–5.0 per cent (w/v) lactose, although no lactose is present in the milk of some seals and sea lions. The milk of humans and monkeys displays higher than average lactose levels (7–10 per cent). Lactose is the major constituent of dried milk and whey powder.

The lipid content of milk can also vary very significantly; that of most agriculturally important species displays a fat content of 3–7 per cent, although the milk of dolphins can contain up to 33 per cent fat while that of the harp seal contains over 50 per cent fat. As the major biological function of milk fat is to provide a source of energy for the neonate, it is not surprising that mammals inhabiting cold or marine environments secrete more fat in their milk. The major biological function of most milk proteins (discussed in detail in later sections) is to provide a dietary source of amino acids for neonate protein synthesis, although some have alternative functions (e.g. the protective function of milk immunoglobulins).

From a physicochemical standpoint, milk is a complex fluid, containing three major phases. Lactose and most of the minor constituents are in true aqueous solution. Proteins are dispersed in this solution, some almost completely (e.g. whey proteins) while others (caseins) form large colloidal aggregates, often with diameters of up to 0.5 μm. The lipid fraction exists in an emulsified state, generally as globules with diameters of up to 20 μm. Milk and milk proteins have an extremely wide range of food uses (Table 13.4), however, we now focus specifically upon the biochemistry, industrial production and applications of the major milk proteins, which are:

Table 13.4 The major dairy based products, how they are produced and their uses

Process	Primary product	Further products
Centrifugal separation	Cream	Butter, butter oil, ghee Creams: various fat content (HTST pasteurized or UHT sterilized, coffee creams, whipping creams, dessert creams Cream cheeses
	Skim milk	Powders, casein, cheese, protein concentrates
Concentration thermal evaporation *or* ultrafiltration		In-container or UHT-sterilized concentrated milks; sweetened condensed milk
Concentration *and* drying		Whole milk powders; infant formulae; dietary products
Enzymatic coagulation	Cheese	1000 varieties; further products, e.g. processed cheese, cheese sauces, cheese dips.
	Rennet casein Whey	Cheese analogues Whey powders, demineralized whey powders, whey protein concentrates, whey protein isolates, individual whey proteins, whey protein hydrolysates, neutraceuticals. Lactose and lactose derivatives.
Acid coagulation	Cheese Acid casein	Fresh cheeses and cheese-based products. Functional applications, e.g. coffee creamers, meat extenders; nutritional applications
	Whey	Whey powders, demineralized whey powders, whey protein concentrates, whey protein isolates, individual whey proteins, whey protein hydrolysates, neutraceuticals.
Fermentation		Various fermented milk products, e.g. yoghurt, buttermilk, acidophilus milk, bioyoghurt.
Freezing		Ice-cream (numerous types and fomulations)
Miscellaneous		Chocolate products

HTST, high temperature, short time; UHT, ultrahigh temperature.
Reproduced from Fox, P. & McSweeney, P. (1997) *Dairy Chemistry and Biochemistry*. Blackie Academic and Professional, London, p17.

- Caseins;

- α-Lactalbumin;

- β-Lactoglobulin;

- Serum albumin;

- Immunoglobulin;

- Lactophorin;

- Lactoferrin;

- Various enzymes (e.g. peroxidase, Lysozyme).

Caseins; biochemistry

Bovine milk contains four different casein types (four related but distinct gene products). These are termed α_{s1}, α_{s2}, β and κ-casein. The total casein concentration in such milk is approximately 25 g/l and the four caseins occur in the approximate ratio of 4:1:4:1. The caseins interact to form casein micelles, particles of diameters of 50–250 nm, which also contain appreciable levels of bound calcium phosphate. Biochemically caseins:

- Display molecular masses of approx. 20–25 kDa (α_{s1} is a 23.5 kDa, 199 amino acid protein, α_{s2} is a 25.2 kDa, 207 amino acid protein, β-casein is a 24 kDa, 209 amino acid protein, while κ-casein is a 19 kDa, 169 amino acid protein).

- Are phosphoproteins (the R groups of several of their serine residues are substituted with phosphate groups. These groups bind Ca^{2+} ions strongly).

- Are predominantly hydrophobic proteins (hydrophobic residues in their amino acid backbones generally occur in clusters). This is particularly true for the hydrophobic β-casein (where the majority of hydrophobic residues occur from residues 40–209) and κ-casein (residues 1–105 are predominantly hydrophobic, while the remainder of the molecule is highly hydrophilic).

- All contain unusually high amounts of proline residues.

- κ-Casein molecules display variable levels of glycosylation (up to half of the κ-casein molecules remain unglycosylated). The other casein subtypes are devoid of a glycocomponent.

It appears that individual casein molecules do not possess a rigid three-dimensional conformation (a state conducive with a high proline content), but interact with each other to form a roughly spherical micellar structure. The exact arrangement of casein molecules in the micelle remains the subject of debate. Individual caseins likely interact via hydrophobic interactions, as well as via hydrogen bonding and electrostatic interactions. The calcium phosphate present also likely provides stabilising interactions with polar or charged species present in the caseins. Unlike α_{s1} and β-caseins, α_{s2} and κ-caseins each contain (two) cysteine residues and are therefore capable of forming intermolecular disulfide linkages, which also likely stabilizes overall micellar structure.

It has been established that the vast majority of κ-casein molecules are present on the micelle surface (much lower quantities of the s-caseins are also found on the surface, but the highly hydrophobic β-caseins are buried in the micelle's centre. The κ-casein molecules are mainly oriented such that their hydrophilic, glycosylated C-terminal ends are protruding outwards from the micelle, while their hydrophobic N-terminal ends protrude inwards, interacting with internal caseins (presumably mainly via hydrophobic interactions). The amphipathic nature of κ-casein in particular therefore plays the major role in stabilizing micellar structure in its normal aqueous environment (Figure 13.3).

The structure of the casein micelle also explains why the addition of chymosin to milk destabilizes micellar structure, causing aggregation and thus curd formation (refer to chymosin and cheese-making, Chapter 11). Chymosin selectively cleaves the Phe^{105}—Met^{106} peptide bond of κ-casein. This releases the hydrophilic component of the κ-casein, responsible for micellar stabilization, resulting in aggregation of the modified micelles (Figure 13.4). Biologically, caseins appear to simply play a nutritive function, providing the recipient neonates with a source of amino acids. In addition, casein micelles also serve as a source of calcium, phosphate and small amounts of sugars (derived from the glycocomponent of κ-casein molecules).

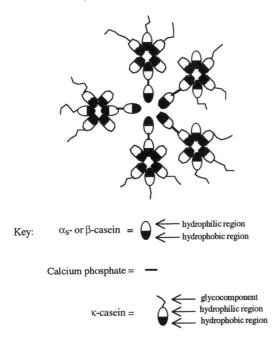

Key: α$_S$- or β-casein = ← hydrophilic region
                                ← hydrophobic region

        Calcium phosphate = —

        κ-casein =   ← glycocomponent
                     ← hydrophilic region
                     ← hydrophobic region

Figure 13.3 Schematic diagram of a likely model of the structure of casein micelles in milk. Variations from this model have also been proposed. The arrangement of sub-micelles and the elongated glycocomponent of the κ-casein molecules would explain the 'hairy raspberry' appearance of many micelles as visualized by electron microscopy

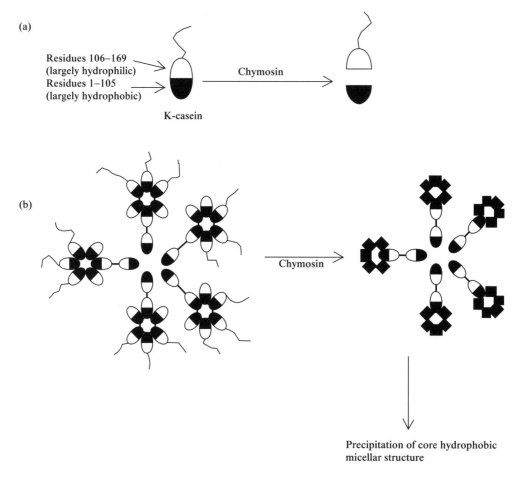

Figure 13.4   Effect of chymosin (rennin) on individual κ-casein molecules (a), and the resulting destabilizing effect on the casein micelles, leading to their aggregation (b)

Casein; industrial production and uses

Caseins can be prepared on a laboratory scale from whole or skimmed milk by a number of means (Table 13.5). Technical and economic considerations have, however, largely limited its industrial scale production to isolation, precipitation or enzymatic (chymosin-based) coagulation (Figure 13.5). Casein has been produced commercially for almost 80 years. Its current annual production level is in the region of 250 000 tonnes. Production begins by removal of the fat from whole milk by centrifugation, yielding skim milk. Prior fat removal is undertaken in order to avoid the subsequent development of off flavours in dried casein preparations, due to oxidation of contaminant lipid.

Table 13.5 Methods by which micellar caseins may be extracted from milk on a laboratory scale

Method	Details
Centrifugation	Micelles can be sedimented by centrifugation at 100 000 g for 1 h. Prior addition of $CaCl_2$ (0.2 M) causes more extensive aggregation of micelles, allowing their recovery at lower centrifugal speeds
Isoelectric precipitation	Acidification of milk to pH 4.6
Salting out	Addition of ammonium sulphate (to a concentration of 260 g/l) precipitates casein (and also some whey proteins)
Ethanol precipitation	Addition of ethanol to a final concentration of 40% selectively precipitates casein from solution.
Cryoprecipitation	Cooling of milk to $-10\,°C$ precipitates casein
Ultrafiltration	Ultrafiltration of whole milk will concentrate/retain both casein and whey proteins
Treatment with chymosin	Aggregated casein micelles precipitate and are easily recovered from milk. κ-Casein component is obviously modified
Chromatographic purification	Methods such as ion exchange and gel filtration chromatography may be used to partially/fully separate caseins from other milk poteins

Product recovery by isoelectric precipitation entails acidification of the milk, producing 'acid caseins'. Acidification (to pH 4.6) can be achieved either by direct addition of mineral acids (usually HCl) or, more slowly by fermentation using lactic acid bacteria. The latter method usually entails addition of a *Lactococcus* starter culture with subsequent incubation at 22–25 °C for 14–16 h. (The bacteria metabolize some of the milk lactose, producing lactic acid.)

When the required pH is reached, the milk is heated to 50 °C by steam injection. This encourages maximal casein precipitation (i.e. curd formation). The curd is then separated from the residual liquid (the whey), either by filtration or low speed centrifugation. Before drying, the casein product is normally washed with hot water in order to remove residual whey constituents. The drying process begins with mechanical pressing (or sometimes centrifugation). The product is then fully dried (to a moisture content of below 12 per cent). This is usually achieved by placing the casein on vibrating perforated stainless steel trays or conveyors, and passing warm air up through the perforations. The product is then milled (ground) before final packaging in order to reduce and standardize the casein particle size.

Enzyme-based casein production differs from the above process only by the means that casein is precipitated from the skim milk. This process entails addition of chymosin to the milk without prior adjustment of the

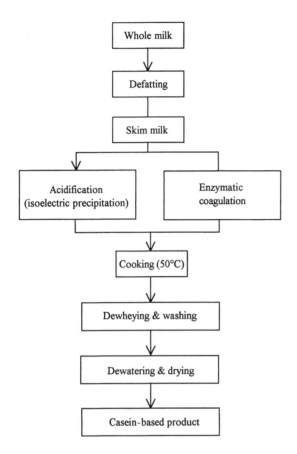

Figure 13.5 Overview of the major methods by which casein is produced on an industrial scale

pH (which is about 6.7). The enzyme containing milk is then held at a temperature of approx. 30 °C for a period usually in excess of 1 h, during which time coagulation occurs (see also Figure 13.4). This so called rennet casein product is then further processed by means similar to that of acid casein (Figure 13.5).

Acid (and rennin) caseins are largely insoluble in water. Soluble casein products (caseinates) are produced industrially by solubilizing acid casein in alkali, with subsequent spray-drying of the product. Solubilization with NaOH is usually undertaken, yielding sodium caseinate. However, KOH, $Ca(OH)_2$ or NH_3 are sometimes used, yielding potassium, calcium or ammonium caseinate.

The acid-or enzyme-based methods of casein production described yield a relatively pure casein product. Another milk-protein based product manufactured commercially is termed 'co-precipitate'. Co-precipitate production is outlined in Figure 13.6. Initial heating of skimmed milk to 90 °C results in the denaturation of most non-casein milk proteins. The denatured whey proteins form a complex with casein, and the entire complex is then precipitated from solution by acidification to pH 4.6.

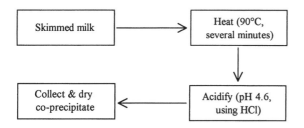

Figure 13.6 Overview of industrial scale co-precipitate production. Refer to text for details

Such casein–whey protein co-precipitate contains in excess of 90 per cent of total milk protein.

Caseins produced commercially obviously consist of a mixture of the four casein subtypes. Although they may be fractionated chromatographically, this is not required in the context of their industrial use. Casein and caseinates have found widespread application in the food industry (Table 13.6). Their emulsifying properties are the functional characteristics that render them most useful industrially. Emulsions based on caseinates or individual casein molecules are very stable (emulsions based on intact casein micelles are less stable). Their strongly amphipathic nature coupled to their conformational flexibility allow them to congregate in an extended manner along an oil–water interface. Moreover their charged nature (mainly due to the fact they are phosphoproteins) ensures that individual emulsified droplets are of like charge, and thus repel each other. Caseins are also unusual in that they retain a stable emulsion in the presence of a high concentration of ethanol. Because of this they also find application as emulsifying agents for cream liquors.

In contrast to their emulsification properties, caseins are less effective as foaming agents. Although casein solutions will foam effectively, the gas–liquid interface of the air cells quickly collapse. This again is a reflection of casein biochemistry. Although they absorb strongly to the air–liquid interface, the hydrophilic portion of individual casein molecules protruding inward to the aqueous phase do not interact with each other. This lack of a stabilizing intermolecular interaction results in a mechanically weak interface (Figure 13.7). For this reason, caseins are not widely used on their own as foaming agents.

Whey proteins; biochemistry

The solution which remains after casein is precipitated from milk is termed whey. Acid whey is obtained after isoelectric precipitation of casein while 'rennet' (i.e. 'sweet') whey is produced by rennet-mediated coagulation of casein. Whey is therefore a major by-product of cheese manufacture. The typical composition of rennet derived whey is provided in Table 13.7. The composition of acid whey would be similar although the exact concentration of individual constituents could differ slightly.

Table 13.6   Major applications of milk derived proteins in the food industry

Bakery products

Caseins/caseinates/co-precipitates

Used in: Bread, pastries/cookies, breakfast cereals, cake mixes, pastries, frozen cakes and pastries, pastry glaze

Effect: Nutritional, sensory, emulsifier, dough consistency, texture, volume/yield

Whey proteins

Used in: Bread, cakes, muffins, croissants

Effect: Nutritional, emulsifier, egg replacer

Dairy products

Caseins/caseinates/co-precipitates

Used in: Imitation cheeses (vegetable oil, caseins/caseinates, salts and water)

Effect: Fat and water binding, texture enhancing, melting properties, stringiness and shredding properties

Used in: Coffee creamers (vegetable fat, carbohydrate, sodium caseinate, stabilizers and emulsifiers)

Effect: Emulsifier, whitener, gives body and texture, promotes resistance to feathering, sensory properties

Used in: Cultured milk products, e.g. yoghurt

Effect: Increase gel firmness, reduces syneresis

Used in: Milk beverages, imitation milk, liquid milk fortification, milk shakes

Effect: Nutitional, emulsifier, foaming properties

Used in: High-fat powders, shortening, whipped toppings and butter-like spreads

Effect: Emulsifier, texture enhancing, sensory properties

Whey proteins

Used in: Yoghurt, Quarg, Ricotta cheese

Effect: Yield, nutritional, consistency, curd cohesiveness

Used in: Cream cheeses, cream cheese spreads, sliceable/squeezable cheeses, cheese fillings and dips

Effect: Emulsifier, gelling, sensory properties

Beverages

Caseins/caseinates/co-precipitates

Used in: Drinking chocolate, fizzy drinks and fruit beverages

Effect: Stabilizer, whipping and foaming properties

Used in: Cream liquers, wine aperitifs

Effect: Emulsifier

Used in: Wine and beer industry

Effect: Fines removal, clarification, reduce colour and astringency

Whey proteins

Used in: Soft drinks, fruit juices, powdered or frozen orange juices

Effect: Nutritional

Used in: Milk-based flavoured beverages

Effect: Viscosity, colloidal stability

Table 13.6   (*continued*)

Dessert products
Caseins/caseinates/co-precipitates
Used in: Ice-cream, frozen desserts
Effect: Whipping properties, body and texture

Used in: Mousses, instant puddings, whipped toppings
Effect: Whipping properties, film former, emulsifier, imparts body and flavour

Whey proteins
Used in: Ice-cream, frozen juice bars, frozen dessert coatings
Effect: Skim-milk solids replacement, whipping properties, emulsifying, body/texture

Confectionery
Caseins/caseinates/co-precipitates
Used in: Toffee, caramel, fudges
Effect: Confers firm resilient, chewy texture; water binding, emulsifier

Used in: Marshmallow and nougat
Effect: Foaming, high temperature stability, improve flavour and brown colour

Whey proteins
Used in: Aerated candy mixes, meringues, sponge cakes
Effect: Whipping properties, emulsifier

Pasta products
Used in: Macaroni, pasta and imitation pasta
Effect: Nutritional, texture, freeze–thaw stability, microwaveable

Meat products
Caseins/caseinates/co-precipitates
Used in: Comminuted meat products
Effect: Emulsifier, water binding, improves consistency, releases meat proteins for gel formation and
                water binding

Whey proteins
Used in: Frankfurters, luncheon meats
Effect: Pre-emulsion, gelation

Used in: Injection brine for fortification of whole meat products
Effect: Gelation, yield

Convenience foods
Used in: Gravy mixes, soup mixes, sauces, canned cream soups and sauces, dehydrated cream soups and
                sauces, salad dressings, microwaveable foods, low lipid convenience foods
Effect: Whitening agents, dairy flavour, flavour enhancer, emulsifier, stabilizer, viscosity controller,
                freeze–thaw stability, egg yolk replacement, lipid replacement

Textured products
Used in: Puffed snack foods, protein-enriched snack-type products, meat extenders
Effect: Structuring, texturing, nutritional

continues overleaf

Table 13.6   (*continued*)

Pharmaceutical and medical products
Special dietary preparations for
  Ill or convalescent patients
  Dieting patients/people
  Athletes
  Astronauts

Infant foods
  Nutritional fortification
  'Humanized' infant formulae
  Low-lactose infant formulae
  Specific mineral balance infant foods
  Casein hydrolastes: used for infants suffering from diarrhoea, gastroenteritis, galactosaemia, malabsorption, phenylketonuria
  Whey protein hydrolysates used in hypoallergic formulae preparations
  Nutritional fortification

Intraveneous feeds
  Patients suffering from metabolic disorders, intestinal disorders for postoperative patients

Special food preparations
  Patients suffering from cancer, pancreatic disorders or anaemia

Specific drug preparations
  β-caseinomorphins used in sleep or hunger regulation or insulin secretion
  Sulfonated glycopeptides used in treatment of gastric ulcers

Miscellaneous products
  Toothpastes
  Cosmetics
  Wound treatment preparations

Reproduced from Fox, P. & McSweeney, P. (1997) *Dairy Chemistry and Biochemistry*. Blackie Academic and Professional, London.

Approximately 20 per cent of total (whole) milk proteins are retained in whey. Collectively, these are termed 'whey proteins', 'serum proteins' or 'non-casein nitrogen'. The major whey proteins include β-lactoglobulin, α-lactalbumin, serum albumin and immunoglobulins (mainly IgG), as summarized in Table 13.8.

β-Lactoglobulin constitutes approx. 50 per cent of total whey protein (about 12 per cent of total milk protein). Four genetic variants of this 18 kDa protein have been characterized. Each differs from the others by one or two amino acid substitutions. β-Lactoglobulin is particularly rich in cysteine residues and X-ray diffraction studies reveal a compact globular structure, consisting of 10–15 per cent α-helix and 45 per cent β-conformation. The protein is relatively heat labile, being denatured at temperatures above 65 °C. This exposes highly reactive sulfhydryl groups, which is important in terms of its food functionality properties. The

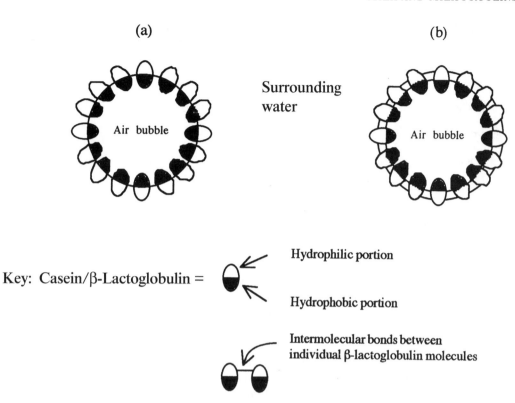

(a)   **(b)**

Surrounding water

Air bubble

Air bubble

Key: Casein/β-Lactoglobulin =

Hydrophilic portion

Hydrophobic portion

Intermolecular bonds between individual β-lactoglobulin molecules

Figure 13.7 Comparison of (a) casein and (b) β-lactoglobulin based foams. β-lactoglobulin-based foam structures are stabilized by the formation of intermolecular non-covalent bonds and disulfide linkages

Table 13.7 Composition of rennin-derived whey

Substance	%
Lactose	4.77
Protein	0.82
Fat	0.07
Lactic acid	0.15
Calcium	0.05
Potassium	0.13
Sodium	0.07
Phosphorus	0.06
Water	93.0

Table 13.8   Some physicochemical characteristics of the major (bovine) whey proteins

	β-Lactoglobulin	α-Lactalbumin	Serum albumin	Immunoglobulin-G
Concentration (g/l)	2–4	0.5–1.5	0.4	0.5–1.0
Molecular mass (kDa)	18	14	66	~150
No. amino acid residues	162	123	582	~1360
No. cysteine residues	5	8	35	16
Isoelectric point (pI)	5.2	4.2–4.8	4.7–4.9	5.5–8.3
Glycosylated?	No	No	No	Yes

quaternary structure of β-lactoglobulin is pH dependent. At pH values below 3.5 and above 7.5, it exists in monomeric form. At pH values 3.5–5.5, it forms an octameric structure, while at pH 5.5–7.5, it exists as a dimer.

Although present in the milk of most agriculturally significant species, β-lactoglobulin is absent from the milk of some mammals, including humans. There is still considerable debate with regard to its natural biological function, although it is known to bind vitamin A tightly in a hydrophobic crevice close to the protein's surface.

α-Lactalbumin is the second most abundant protein in whey derived from the milk of cows and many other mammals. It has been purified from a range of mammalian milks and sequence studies reveal a highly conserved primary structure. The milk of most species of cattle contains a single α-lactalbumin variant (lactalbumin α − 1$_a$), although some species produce two variants (α − 1$_a$ and α − 1$_b$). These differ from each other by a single amino acid. α − 1$_a$-Lactalbumin contains a glutamate residue at position 10, whereas α − 1$_b$-lactalbumin contains an arginine in this position.

α-Lactalbumin contains higher than average levels of tryptophan and sulfur-containing amino acids (cysteine and methionine). Its isoelectric point is in the region of 4.8, at which pH it is least soluble. As in the case of β-lactoglobulin, α-lactalbumin displays a compact globular structure, consisting of 25 per cent α-helix and 14 per cent β-conformation. Its tertiary structure is very similar to that of lysozyme, which is a reflection of the considerable sequence homology that exists between these two proteins.

α-Lactalbumin appears more heat stable than β-lactoglobulin. It is denatured at temperatures much in excess of 60 °C; however it renatures upon cooling. The protein binds calcium and other divalent cations tightly at a surface cleft which houses three asparagine residues, and removal of bound calcium renders it more thermolabile.

Biologically, α-lactalbumin plays a direct and novel role in the synthesis of lactose. The final step in lactose synthesis, entailing the condensation of uridine diphosphate (UDP)–galactose and D-glucose, is catalysed by lactose synthetase (Figure 13.8). This enzyme is a heterodimer, consisting of A and B subunits. The A subunit catalyses the transfer of galactose from UDP galactose to the acceptor molecule. In the absence of subunit B, a wide range of sugars can act as the acceptor. The presence of the B protein, however, confers specificity to the reaction, rendering glucose as the only acceptor molecule used, and thus ensuring only the synthesis of lactose. The concentration of lactose in milk is directly proportional to the concentration of α-lactalbumin present. The milk of animals devoid of this protein is also devoid of lactose (e.g. many marine mammals).

Cows' milk is also found to contain bovine serum albumin (BSA), at concentrations usually varying between 0.1 and 0.4 g/l. The protein is identical to BSA found in blood serum, and probably enters the milk by leakage from the vascular system. The (66 kDa) molecule is significantly larger than β-lactoglobulin or α-lactalbumin. Its amino acid sequence has been determined and it exhibits a globular structure featuring three defined domains. It displays 17 intramolecular disulfide linkages as well as one cysteine with a free thiol (SH) group. Incubation under reducing conditions renders the protein less conformationally stable by breaking disulfide linkages. Heating a BSA solution to temperatures in excess of 40 °C results in its precipitation, which is most likely caused by protein unfolding with subsequent formation of intermolecular hydrophobic interactions. BSA also undergoes acid denaturation at lower pH values (~4). This is likely caused by mutual repulsion of positively charged amino acid residues at such pH values (of its 582 amino acids, BSA contains 59 lysine residues, 23 arganines and 17 histidines). Although BSA plays a well known transportational role in blood, its functional significance in milk is unclear.

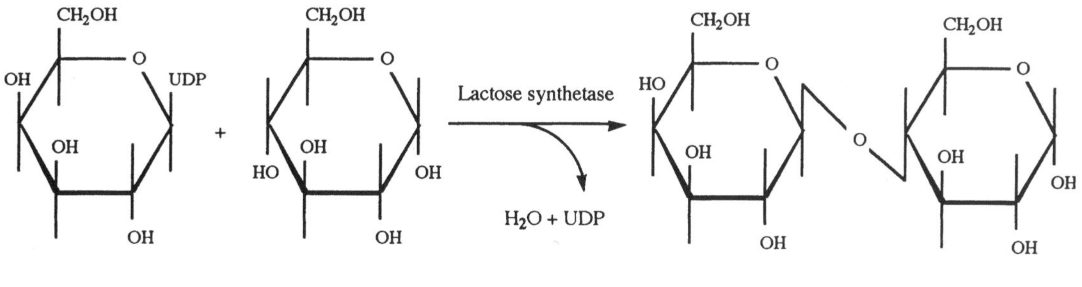

β-D-UDP-Galactose   α-D-gulcose

Lactose
(β-D-galactopyranosayl-(1→4)-α-D-glucopyranose)

Figure 13.8   The final step in the biosynthesis of lactose, as catalysed by lactose synthetase. Overall milk lactose is synthesized from blood-derived glucose via a four-step enzymatic pathway. One of the enzymes is an epimerase which converts UDP-glucose into UDP-galactose

The fourth major protein type found in milk are the immunoglobulins (Ig). IgG is the predominant immunoglobulin subtype found in bovine milk whereas IgA predominates in human milk. The biological function of these proteins is obviously an immunological one, serving to provide neonates with early and immediate immunological protection. In isolated form, Ig display a high unfolding temperature. However, they are far more thermolabile in the presence of other whey proteins. This may be due to the formation of intermolecular disulfide linkages with β-lactoglobulin and/or BSA. Additional minor whey proteins include several which display antimicrobial actions. These include the enzymes lysozyme and lactoperoxidase as well as lactoferrin.

Whey proteins; industrial production and uses

The worldwide manufacture of cheese results in the annual production of in excess of 90 million tonnes of whey by-product, which contains in the region of 630 000 tonnes of whey protein. The low protein concentration present initially rendered whey protein production relatively expensive, but the development of ultrafiltration, in particular, has helped reduce processing costs. Whole whey protein-based products are manufactured essentially by concentration of the whey, often with subsequent drying. Concentration can be achieved by evaporation but ultrafiltration is now the method of choice (Figure 13.9). The application of ultrafiltration in diafiltration mode (Chapter 3) also allows separation of low molecular mass constituents (e.g. lactose and minerals) from the whey proteins.

The initial step in whey processing invariably entails the removal of residual lipid (in order to prevent fouling of ultrafilter membranes and the development of lipid-based off flavours in the final product). This is generally achieved by the addition of $CaCl_2$ to whey, with pH adjustment to 7.3 and heating to 50 °C. Under these conditions, calcium phospholipoprotein complexes are formed and precipitate out of solution. They (along with, for example, casein fines) can then be removed by microfiltration or low-speed centrifugation. A combination of ultrafiltration and diafiltration, using membranes with molecular mass cut-off values of 10–50 kDa results in the production of whey protein concentrate. Concentrates containing 30–75 per cent (w/v) protein are most often produced commercially. If required, the concentrates can be dried fully, most often by spray-drying.

Whey protein isolates usually contain greater than 90 per cent protein. The manufacturing process entails the use of ion exchange chromatography. The whey pH is initially adjusted to a value lower than 4.6 (if cation exchange resins are used) or greater than 4.6 (if anion exchange resins are used). Application of the whey results in binding of the protein fraction. A subsequent washing step removes unbound material (e.g. lactose and minerals). Appropriate subsequent readjustment of the

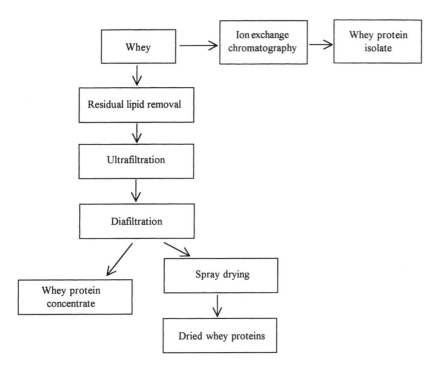

Figure 13.9 Overview of an ultrafiltration process used to produce a whey-protein based product. Refer to text for further details

pH then results in elution of the whey proteins. This whey protein isolate may also be spray-dried if required.

Whey protein products are finding increasing use as animal feed additives (mostly whey protein concentrates of less than 55 per cent protein content), for human nutrition, or are used in the food industry for their functional properties (Table 13.6). From a functional point of view, β-lactoglobulin and α-lactalbumin are the most significant whey protein constituents, being responsible for surface active functional properties (emulsifying and foaming) as well as hydration and gelling properties.

ANIMAL AND MICROBIAL PROTEINS

Gelatin

Gelatin is partially degraded or unstructured collagen. Collagen represents the principal constituent of connective tissue, being found in greatest quantities in tendons, cartilage, bones, blood vessels and skin. The basic structural unit of collagen, tropocollagen, consists of a triple helix of individual collagen polypeptides, each containing about

1000 amino acids. Collagen has an unusual amino acid composition, containing high levels of glycine (23 per cent). It also contains high levels of proline or modified proline (3-hydroxyproline and 4-hydroxyproline) as well as another modified amino acid: 5-hydroxylysine. The collagen triple helix is stabilized by intermolecular covalent and non-covalent bonds. The covalent linkages are mainly formed between the modified amino acids mentioned above. Individual tropocollagen molecules are held together by yet more intermolecular linkages in the overall collagen fibril-based structure.

The industrial-scale production of gelatin is relatively straightforward. The raw material (usually skin and bone) is first stripped of non-collagenous compounds. The material is then heated in water, which promotes extraction of the collagen, and its conversion to gelatin. Conversion entails disassembly of the highly ordered fibril structure and a significant collapse of the tropocollagen triple helix. Limited cleavage of the backbone of some individual collagen polypeptides may also occur. The gelatin extracted is then normally dried and milled to give a flaky or ground granular product.

Gelatin finds extensive application in the food industry. Due to its unbalanced amino acid composition (Table 13.9), it is a poor nutritional source. Its application in foods stems instead from its functional properties. The major notable functional property of gelatin is its ability to form a gel structure. Gelatin readily dissolves in hot water yielding random collagen-based coils. As the temperature is allowed to decrease to 35–40 °C, some of the individual collagen polypeptides begin to interact with one another, forming short segments of triple helix. These junction zones effectively cross-link gelatin polypeptides in a three-dimensional lattice structure, thereby forming the gel. Reheating to temperatures above 40 °C reverses these interactions, to yield a molten product once again. A gelatin solution can go through many such cycles of melt and solidification without damage. Gelatin finds most extensive application in the food sector as a gelling agent (e.g. gelatin-based deserts). The protein also displays weak amphipathic properties and, as such, can help stabilize emulsions and foams. It is on this basis that gelatin is often added to ice creams and marshmallow products.

Gelatin also finds application in non-food areas, particularly in the pharmaceutical and medical sectors. Pharmaceutically, it is used to produce gelatin-based capsules as well as a thickening agent in some paste-based medicinal products. It is also used as a gelling agent for the manufacture of suppositories and pessaries. Its ability to absorb five to ten times its own weight of water forms the basis for its use in some surgical procedures, as gelatin sponges can mop up significant quantities of blood. Gelatin which is more extensively degraded (by, for example, prolonged heating) displays adhesive properties. Such degradation products form the basis of many animal-based glues and binders. A summary of the major industrial applications of gelatin is presented in Table 13.10.

Table 13.9 Amino acid composition (%) of gelatin

Amino acid	%
Glycine	23.0
Alanine	7.8
Valine	2.25
Leucine	2.9
Isoleucine	1.26
Cysteine	0.09
Methionine	0.9
Phenylalanine	2.0
Proline	16.2
Hydroxyproline	12.7
Serine	0.36
Threonine	1.7
Tyrosine	0.45
Aspartate	6.0
Glutamate	10.2
Arginine	7.3
Lysine	3.7
Histidine	0.72

Egg proteins

Eggs form part of the staple diet in many world regions and represent a good source of dietary protein in particular. Layer hens typically produce in the region of 250 eggs per annum. Exact egg weight will vary, depending upon breed, age and nutritional status of the bird, but will usually be in the region of 55–60 g. In terms of composition, whole eggs consist of approximately 10 per cent shell, 60 per cent egg white and 30 per cent yolk. Many processed foods contain egg white and yolk as ingredients. Although contributing to its nutritive value, they are most often added to food for their functional characteristics.

Table 13.10   Major food and non-food applications of collagen-derived gelatin

Food applications	Gelling agent for suppositories
Gelling agent	Gelatin based surgical sponges
Stabilizer	**Other uses**
Thickener	Glue manufacture
Emulsifier	Binder in match heads
Foaming agent	Sizing agent for paper/textile manufacture
Pharmaceutical/medical	Photographic films
Gelatin-based capsules	Manufacture of rubber substitutes
Thickening agents for pastes	

The initial step of egg processing for food application entails mechanical breaking followed by sieving (to remove e.g. egg shell particles). This is carried out at refrigerated temperatures in order to discourage microbial growth. If whole egg is required, homogenization follows. Alternatively, mechanical separation of egg white and yolk can be undertaken, followed by independent homogenization/blending. Pasteurisation is next undertaken. For whole egg or egg yolk, this is usually carried out by holding at 64.4 °C for 150 s. The lipid content of the yolk exerts a protective effect on the thermolabile proteins (mainly egg white proteins) present. Isolated egg white is pasteurised at lower temperatures (ca. 57 °C), although the natural presence of lysozyme helps prevent microbial spoilage in this product. After pasteurization is completed, the processing temperature is reduced and the product is either frozen, spray-dried, salted or acidified in order to prevent microbial spoilage. Further processing to yield purified egg protein, for example, is not undertaken.

Egg white proteins contribute significantly to the overall functional characteristics of egg-based products. Egg white consists of approximately 86 per cent water and 11 per cent protein with lower levels of carbohydrates, fat and minerals being present. The major proteins found in egg white are presented in Table 13.11. Ovalbumin is found at highest concentration in egg white. It is a 385 amino acid, 45 kDa phosphoglycoprotein. The carbohydrate component is attached to the polypeptide backbone by an *N*-glycosyl linkage involving asparagine 292. Three variant forms of the ovalbumin can be separated by, for example, isoelectric focusing. These represent unphosphoylated, monophosphorylated and diphosphorylated forms of the protein. The phosphate groups reside on serine 68 and/or serine 344. Ovalbumin is also acetylated at its N-terminus and contains four cysteine residues.

Table 13.11   Major protein types found in egg white

Protein	% of egg white	Molecular mass (kDa)	pI
Ovalbumin (Egg albumen)	52	45	4.6
Ovotransferrin (Conalbumin)	13	78	6.6
Ovomucoid	11	28	4.0–4.3
Lysozyme	3.5	14	10.7
Ovomucin (α and β)	1.5	18 (α) 400 (β)	
Immunoglobulins (various)	<1.0	Vary	Vary

Immunoglobulins are found in egg white and yolk. See text for details.

Ovalbumin, along with some other egg white proteins undergoes denaturation and coagulation when heated. A gel-like structure results which is stabilized by intermolecular hydrophobic interactions. Gel formation impacts upon food texture and water holding capacity. Ovalbumin also contributes to the foaming properties of egg white, although when compared to proteins such as gelatin and casein, it displays only modest foamability. Whipping of ovalbumin solutions induces congregation of the protein molecules at the air–water interface. This induces their partial denaturation, exposing sulfhydryl groups normally buried internally in the protein. These are readily oxidized, forming intermolecular disulfide linkages, which helps stabilize the foam. Non-covalent interactions also play an important role in ovalbumin-based foam stability.

Ovomucoid and ovomucin represent additional proteins found in egg white. Ovomucoid is a 28 kDa trypsin inhibitor. It is a relatively thermostable glycoprotein whose conformation displays three separate domains linked by intradomain disulfide linkages. Ovomucin is a sulfated glycoprotein that displays viscous or gel forming properties and plays a central role in the foaming ability of egg white. It consists of two subunits, α and β, of molecular mass 18 and 400 kDa, respectively. Both subunits are glycosylated and approximately 50 per cent of the β-subunit's molecular mass is contributed by the carbohydrate component.

Soy proteins also constitute an important group of functional proteins used in the food industry. Seed proteins derived from soybean and other legumes are classified as globulins or albumins. Globulins can account for up to 90 per cent of the protein fraction and serve a storage function. They may be subdivided on the basis of their sedimentation coefficients into 75 and 115 globulins (Table 13.12). Glycinin and β-conglycinin are the soy proteins that demonstrate gel forming functional properties. Soy proteins also display emulsification properties and are used to help stabilize oil in water emulsions of soup, sausages and coffee whitener. As in the

Table 13.12   The major globulin-based seed proteins found in soybean

7S Globulins
β-Conglycinin, a 150–200 kDa glycoprotein
γ-Conglycinin, a 170 kDa glycoprotein
Basic 7S globulin, a 168 kDa glycoprotein
11 S Globulins
Glycinin, a 300–380 kDa protein

case of egg proteins, soy proteins used to alter food functionality are not purified prior to their use.

Single-cell protein

Microorganisms can constitute a valuable source of dietary protein. Indeed the production of such single-cell protein (SCP) has received attention since the turn of the century. Production of microbial biomass as a source of protein is attractive for a number of reasons. Most microbial populations have short generation times compared to traditional, animal or plant protein sources. Fermentation technology required to produce SCP is well established. This process may be carried out independantly of seasonal or climatic conditions, and may utilize raw materials which are readily available and are inexpensive. In addition the protein content of most microbial speces is quite high, up to 50 per cent in some cases, and genetic intervention may be employed to enhance the quality and quantity of the protein produced.

Any microorganism used in the production of SCP must be GRAS-listed (Chapter 2). While such products may be consumed directly by humans they may also be utilized in the animal feed industry. Thus far only yeasts are used as sources of SCP but bacteria, algae and fungi are all likely additional future sources. A steadily increasing world population is likely to render the future microbial production of protein and other nutrients more significant.

SWEET AND TASTE MODIFYING PROTEINS

The production of glucose and fructose syrups and their use as sweeteners has been reviewed in Chapter 11 Sweetness, however, is not a sensation upon which sugar-based substances have a monopoly. Several proteins (and peptides) have been identified which, if tasted, are perceived as being intensely sweet. The sweetest natural substances thus far discovered are in fact two plant proteins, thaumatin and monellin (Table 13.13). These, and the majority of other sweet proteins thus far characterized, are derived from the fruit or berries of selected African plants. These fruits and berries have long since been consumed or used as a food sweetener by indigenous African populations. Their intense sweetness render them attractive as, for example, low-calorie sweeteners. Unlike sugars, they do not promote tooth decay and can be safely used as sweeteners for diabetics.

Thaumatin is a protein found in the berries of the West African plant, *Thaumotococcus danielli*. Although its existence was documented in the 1850s, it was first isolated, characterised and used commercially in the 1970s. Two variants of thaumatin are known to exist, thaumatin I and II. The former exhibits a molecular mass of 22 209 Da while the latter exhibits a mass of 22 293 Da. Both are 207 amino acid polypeptides differing in sequence by only five residues. Their three-dimensional structure is globular and they display several significant stretches of β-conformation. They are extremely thermostable, retaining their sweet taste even after boiling for 1 h. The presence of eight disulfide linkages likely exerts a major stabilizing influence, but the proteins may also undergo partial unfolding during boiling, only to refold upon subsequent cooling. Thaumatins trigger their sweet sensation by binding specific taste bud receptors on the tongue. They are 3000 times sweeter than sucrose on a weight basis, which equates to 100 000 times sweeter on a molar basis.

Thaumatins were first commercialized in the 1970s by Tate and Lyle under the tradename 'Talin'. The natural producer grows poorly or produces little berries outside its native Africa, and Talin production entails local harvest of the berries with shipping to the final production site after freezing. Up to 50 per cent of the berry mass may be accounted for by the protein, which is extracted in water after crushing.

Table 13.13 Some sweet or taste modifying proteins and their sources

Protein	Source	Molecular mass (kDa)	Sweetness (relative to sucrose on a weight basis)
Thaumatin	Fruit of *Thaumotococcus danielli*	22.2	3000
Monellin	Fruit of *Dioscoreophyllum cumminsii*	11.1	3000
Mabinlin	Seeds of *Capparis masaikai*	14.0	400
Pentadin	Berries of *Pentadiplandra brazzenna*	12.0	500
Miraculin	Berries of *Richadella dulcifica*	28.0	Tasteless
Curculin	Fruits of *Curculigo latifolia*	27.8	550

The soluble extract is then concentrated by ultrafiltration and spray-dried–freeze-dried yielding the final 'purified' product.

The thaumatin genes have been isolated and expressed in various recombinant microbial systems, including *E. coli, Bacillus subtilis, Streptomyces levidans, Saccharomyces cerevisiae* and *Kluyveromyces lactis*. The thaumatin genes have also been expressed in transgenic plants and extracts from such plants have exhibited a sweet taste. Production by recombinant means exhibits obvious advantages in overcoming problems of source availability. It also allows the production of variant forms exhibiting altered amino acid sequences. This may allow identification of even sweeter thaumatins or ones displaying altered taste properties. The main technical obstacle to recombinant thaumatin production on a commercial basis has been the low expression levels thus far achieved.

Thaumatin is GRAS-listed and, at concentrations below its sweetness threshold, it acts as a flavour enhancer. It has found commercial application as a sweetener and flavour enhancer in animal feed, human food and beverages, chewing gum and breath fresheners. At high levels, it also imparts a strong licorice-like taste to foods and this property has somewhat limited its widespread food application.

Monellin was first purified in 1972 from the fruit of the West African plant, *Dioscoreophyllum communisii*. It is a heterodimer consisting of a 44 amino acid residue A chain and a 50 amino acid residue B chain, held together by non-covalent interactions. The overall molecular mass is 11 kDa and the protein displays a pI value of 9.0–9.4. Monellin displays little sequence homology to thaumatin; however polyclonal antibodies raised against one protein cross-react with the other. Monellin displays a similar level of sweetness to thaumatin. It is less stable than thaumatin and the sweet taste is not perceived for several seconds after consumption. The sweetness sensation then increases slowly followed by a slow decline, taking up to an hour. For such reasons, it is unlikely that monellin will gain the commercial acceptability that thaumatin has.

Pentadin is yet another sweet tasting protein. It was first extracted in small quantities from the berries of the African plant, *Pentadiplandra brazzenna* in 1989. The protein displays a taste profile somewhat similar to monellin.

Curculin is a homodimeric protein displaying a molecular mass of about 28 kDa. It is produced in the fruit of the herb, *Curculigo latifolia*, which grows in western Malaysia. The protein is over 500 times sweeter than sucrose on a weight basis but also has taste modifying properties.

Finally, miraculin is a 191 amino acid 28 kDa glycoprotein originally isolated from the berries of the West African shrub, *Richadella dulcufica*. This protein, which by itself is tasteless, has the ability to convert sour-tasting substances into sweet-tasting ones. Lemons, for example, when consumed after chewing miraculin-containing berries, taste like sweet oranges. Again, a disadvantage of this protein from an applied perspective is that it remains bound to taste receptors for a prolonged period. The sweet taste perceived, therefore, can last for up to 2 h.

FURTHER READING

Books

Anonymous (1985) *Whey*. National Dairy Council.

Damodaran, S. (1997) *Food Proteins and Lipids*. Plenum Publishing, New York.

Fox, P. & McSweeney, P. (1997) *Dairy Chemistry and Biochemistry*. Blackie Academic and Professional, Glasgow.

Gillies, M. (1974) *Whey Processing and Utilization*. Noyes Data Corporation, USA.

Hittiarachchy, N. (1994) *Protein Functionality in Food Systems*. Marcel Dekker, New York.

Pinder, A. (1980) *Milk and Whey Powders*. Society of Dairy Technology.

Whitaker, J. (1999) *Functional Properties of Proteins and Lipids*. American Chemical Society.

Zayas, J. (1997) *Functionality of Proteins in Food*. Springer-Verlag, Godalming.

Articles

Proteins; functional properties

Dickinson, E. (1997) Properties of emulsions stabilized with milk proteins; overview of some recent developments. *J. Dairy Sci* **80** (10), 2607–2619.

Dickinson, E. (2001) Milk protein interfacial layers and the relationship to emulsion stability and rheology. *Colloids Surf. B* – Biointerf. **20** (3), 197–210.

Fligner, K. & Mangino, M. (1991) Relationship of composition to protein functionality. ACS Symposium Series, **454**, 1–2.

Gosal, W. & Ross-Murphy, S. (2000) Globular protein gelation. *Curr. Opin. Colloid Interf. Sci.* **5** (3–4), 188–194.

Hansen, A. & Booker, D. (1996) Flavor interaction with casein and whey proteins. *ACS Symposium Series* **633**, 75–89.

Lucey, J. & Singh, H. (1997) Formation and physical properties of acid milk gels; a review. *Food Res. Int.* **30** (7), 529–542.

Morr, C. & Ha, E. (1993) Whey protein concentrates and isolates-processing and functional properties. *Crit. Rev. Food Sci Nutr.* **33** (6), 431–476.

Nishinari, K. *et al.* (2000) Hydrocolloid gels of polysaccharides and proteins. *Curr. Opin. Colloid Interf. Sci.* **5** (3–4), 195–201.

Taylor, S. & Lehrer, S. (1996) Principles and characteristics of food allergens. *Crit. Rev. Food Sci. Nutr.* **36**, S91–S118.

Wilde, P. (2000) Interfaces; their role in foam and emulsion behaviour. *Curr. Opin. Colloid Interf. Sci.* **5** (3–4), 176–181.

Milk proteins

Bounous, G. *et al.* (1991) Whey proteins in cancer prevention. *Cancer Lett.* **57** (2), 91–94.

Clare, D. & Swaisgood, H. (2000) Bioactive milk peptides; a prospectus. *J. Dairy Sci.* **83** (6), 1187–1195.

Creamer, L. & MacGibbon, A. (1996) Some recent advances in the basic chemistry of milk proteins and lipids. *Int. Dairy J.* **6** (6), 539–568.

Ginger, M. & Grigor, M. (1999) Comparative aspects of milk caseins. *Comp. Biochem. Physiol. B – Biochem. Mol. Biol.* **124** (2), 133–145.

Imafidon, G. *et al.* (1997) Isolation, purification and alteration of some functional groups of major milk proteins; a review. *Crit. Rev. Food Sci Nutr.* **37** (7), 663–689.

Jayaprakasha, H. & Brueckner, H. (1999) Whey protein concentrate; a potential functional ingredient for the food industry *J Food Sci. Technol.* **36** (3), 189–204.

Rosen J. *et al.* (1999) Regulation of milk protein gene expression. *Annu. Rev. Nutr.* **19**, 407–436.

Wong, D. *et al.* (1996) Structures and functionalities of milk proteins. *Crit. Rev. Food Sci. Nutr.* **36** (8), 807–844.

Sweet and other proteins

Anonymous (1997) Gelatin. *Adv. Polym. Sci.* **130**, 160–193.

Anupama, R. (2000) Value added food; single cell protein. *Biotechnol. Adv.* **18** (6), 459–479.

Faus, I. (2000) Recent developments in the characterization and biotechnological production of sweet tasting proteins. *Appl. Microbiol. Biotechnol.* **53** (2), 145–151.

Gibbs, B. *et al.* (1996) Sweet and taste modifying proteins; a review. *Nutr. Res.* **16** (9), 1619–1630.

Izmailova, V. *et al.* (2000) Properties of interfacial layers in multicomponent systems containing gelatin. *Colloid J.* **62** (6), 653–675.

Kuhad, R. *et al.* (1997) Mocroorganisms as an alternative source of protein. *Nutr. Rev.* **55** (3), 65–75.

Kurihara, Y. (1992) Characteristics of antisweet substances, sweet proteins and sweetness inducing proteins. *Crit. Rev. Food Sci. Nutr.* **32** (3), 231–252.

Zemanek, E. *et al.* (1995) Issues and advances in the use of transgenic organisms for the production of thaumatin, the intensely sweet protein from thaumatococcus danielli. *Crit. Rev. Food Sci. Nutr.* **35** (5), 455–466.

Index

Index compiled by Christine Boylan

LIBRARY
TECHNISCHE UNIVERSITEIT
DELFT